Engineering Materials

For further volumes:
http://www.springer.com/series/4288

Holm Altenbach · Andreas Öchsner
Editors

Plasticity of Pressure-Sensitive Materials

Springer

Editors
Holm Altenbach
Institut für Mechanik
Otto-von-Guericke-Universität Magdeburg
Magdeburg, Sachsen-Anhalt
Germany

Andreas Öchsner
Griffith School of Engineering
Griffith University
Southport
Australia

and

Faculty of Engineering
and Built Environment
The University of Newcastle
Callaghan
Australia

ISSN 1612-1317 ISSN 1868-1212 (electronic)
ISBN 978-3-642-40944-8 ISBN 978-3-642-40945-5 (eBook)
DOI 10.1007/978-3-642-40945-5
Springer Heidelberg New York Dordrecht London

Library of Congress Control Number: 2013948747

Printed on acid-free paper

Springer is part of Springer Science+Business Media (www.springer.com)

Preface

Pressure-Sensitive Materials are widely used in modern engineering applications. As usual they are lightweight structures and have analogs in nature (for example bones). But there are a lot of other application fields. This monograph is devoted to the modeling and simulation of pressure-sensitive materials since the standard methods and equations are partly not applicable. For problems of manufacturing of such materials we refer to the special literature.

The monograph summarizes new trends and established methods in the field of pressure-sensitive materials. It contains six chapters prepared by different research groups. "Basic Equations of Continuum Mechanics" gives an overview on the general continuum mechanics in which the modeling of pressure-sensitive materials is embedded. In addition, some examples of special constitutive equations for incompressible and compressible materials are presented. These examples are mostly related to rubber-like materials. "Phenomenological Yield and Failure Criteria" presents classical and improved criteria of limit states of materials. The classical criteria are applicable only in the cases of ideal ductile or absolutely brittle materials. Various possibilities to develop improved criteria are described. Finally, three examples (gray cast iron, polyoxymethylene (POM), and polyvinyl chloride (PVC) hard foam) demonstrate the application of different approaches in modeling certain limit behavior. "Plasticity of Cellular Metals (Foams)" presents cellular metals, e.g., made by solidification of molten metal foam, which have interesting mechanical properties, among them high specific strength and stiffness coupled with inflammability and good damping properties. The analysis of such materials is not a trivial problem, especially beyond the elastic range, since the micro-mechanical behavior has a great influence on the macroscopic properties. "Transmission Conditions for Thin Elasto-Plastic Pressure-Dependent Interphases" is devoted to the behavior of thin soft elasto-plastic interphases. The case of pressure-independent (von Mises) as well as pressure-dependent yield condition is theoretically treated and finite element calculations are presented. "Effect of Pressure-Dependency of the Yield Criterion on the Strain Rate Intensity Factor" presents several rigid plastic models, for which the equivalent strain rate (quadratic invariant of the strain rate tensor) approaches infinity in the vicinity of maximum friction surfaces. In this case special methods in the analysis are necessary. "Mechanical Response of Porous Materials: the Gurson Model" is devoted to a special model applicable for pressure-sensitive materials under the influence of damage effects: the Gurson model. Some theoretical and numerical aspects are discussed.

Magdeburg, Skudai
August 2013

Holm Altenbach
Andreas Öchsner

Contents

Contributors

Sergei Alexandrov A.Yu. Ishlinsky Institute for Problems in Mechanics, Russian Academy of Sciences, Moscow, Russia; Department of Mechanical Engineering, Advanced Institute of Manufacturing with High-tech Innovations, National Chung Cheng University, Chia-Yi, Taiwan, e-mail: sergei_alexandrov@spartak.ru

Holm Altenbach Lehrstuhl für Technische Mechanik, Fakultät für Mechanik, Institut für Mechanik, Otto-von-Guericke University Magdeburg, Universität splatz 2, 39106 Magdeburg, Germany, e-mail: holm.altenbach@ovgu.de

Alexandre Bolchoun Abteilung Werkstoffe und Bauteile, Fraunhofer Institut für Betriebsfestigkeit und Systemzuverlässigkeit LBF, Bartningstr 47, 64289 Darmstadt, Germany, e-mail: alexandre.bolchoun@lbf.fraunhofer.de

Guillermo J. Creus Instituto Mercosul de Estudos Avançados, Universidade Federal da Integração Latino-Americana, Foz do Iguaçu, Paraná, Brazil, e-mail: creus@ufrgs.br

Luiz A. B. da Cunda Escola de Engenharia, Universidade Federal do Rio Grande, Av. Itália km 8, Campus Carreiros, Rio Grande, Rio Grande do Sul 96203-900, Brazil, e-mail: luizcunda@furg.br

Thomas Daxner CAE Simulation & Solutions, Pitkagasse 2/1/16, Vienna 1210, Austria, e-mail: daxner@cae-sim-sol.at

Victor A. Eremeyev Lehrstuhl für Technische Mechanik, Fakultät für Mechanik, Otto-von-Guericke University Magdeburg, Universität splatz 2, 39106 Magdeburg, Germany; South Scientific Center of RASci and South Federal University, Milchakova Street 8a, Rostov on Don, Russia, 344090, e-mail: victor.eremeyev@ovgu.de

Yeau-Ren Jeng Department of Mechanical Engineering, Advanced Institute of Manufacturing with High-tech Innovations, National Chung Cheng University, Chia-Yi, Taiwan, e-mail: imeyrj@ccu.edu.tw

Vladimir A. Kolupaev Deutsches Kunststoff-Institut, Schloßgartenstr. 6, 64289 Darmstadt, Germany, e-mail: VKolupaev@dki.tu-darmstadt.de

Elena Lyamina A.Yu. Ishlinsky Institute for Problems in Mechanics, Russian Academy of Sciences, Moscow, Russia, e-mail: lyamina@inbox.ru

Gennady Mishuris Department of Mathematics and Physics, IMPACS, Penglais, Aberystwyth University, Aberystwyth, Ceredigion SY23 3BZ, UK, e-mail: ggm@aber.ac.uk

Wiktoria Miszuris Department of Mathematics and Physics, IMPACS, Penglais, Aberystwyth University, Aberystwyth, Ceredigion SY23 3BZ, UK, e-mail: wim@aber.ac.uk

Andreas Öchsner Griffith School of Engineering, Griffith University, Southport 4222, Australia; The University of Newcastle, Callaghan, NSW 2308, Australia, e-mail: andreas.oechsner@gmail.com

Andrea Piccolroaz University of Trento, via Mesiano 77, Trento 38123, Italy; Department of Mathematics and Physics, IMPACS, Penglais, Aberystwyth University, Aberystwyth, Ceredigion SY23 3BZ, UK, e-mail: andrea.piccolroaz@ing.unitn.it

Basic Equations of Continuum Mechanics

Holm Altenbach and Victor A. Eremeyev

Abstract The modeling of the behavior of pressure-sensitive materials is embedded in the general continuum mechanics. The basic equations of continuum mechanics can be split into the material-independent and the material-dependent equations. The starting point is the introduction of the kinematics based on pure mathematical considerations. In addition, the velocities and the accelerations of the relevant kinematical variables are presented. The next section is devoted to the introduction of the action on the continuum and the inner reaction. Starting with such properties like forces and stresses finally the static equilibrium is stated. The last part of the material-independent equations is the introduction of the balances. Limiting our discussions by thermo-mechanical actions only, the balance of mass, momentum, moment of momentum, energy and entropy are deduced. The specific properties and features of the pressure-sensitive materials are presented in the next sections. Within this chapter the general ideas of material modeling (deductive approach) are given. Finally, some examples of special constitutive equations for incompressible and compressible materials are presented. These examples are mostly related to rubber-like materials.

H. Altenbach (✉) · V. A. Eremeyev
Lehrstuhl für Technische Mechanik, Institut für Mechanik, Fakultät für Maschinenbau, Otto-von-Guericke-Universtät Magdeburg, Universitätsplatz 2, 39106 Magdeburg, Germany
e-mail: holm.altenbach@ovgu.de

V. A. Eremeyev
South Scientific Center of RASci and South Federal University, Milchakova St. 8a, Rostov on Don 344090, Russia
e-mail: victor.eremeyev@ovgu.de

H. Altenbach and A. Öchsner (eds.), *Plasticity of Pressure-Sensitive Materials*,
Engineering Materials, DOI: 10.1007/978-3-642-40945-5_1,

1 Kinematics and Deformations

1.1 Lagrangian and Eulerian Description

A body $\mathscr{B}$ is an assemblage of material points, which is bounded by boundary points that means the surface of $\mathscr{B}$. Material bodies are introduced in continuum mechanics with the help of the method of sections. By this method the body $\mathscr{B}$ can be separated from the surrounding. The introduction of the surface and the body is arbitrary, which is helpful for the formulation of the balance equations (see Sect. 3).

The movement of material bodies can be presented by the motion of their material points which should be identified. If the material points are related to points in the Euclidean space $\mathbb{E}^3$ and if one point 0 is fixed in this space, then the position of the material points is determined by the position vector $\boldsymbol{x}(t)$ at arbitrary time t. To distinguish the material points of the body $\mathscr{B}$ each of them gets a label: at the time $t = t_0$ the material point is characterized by the position vector $\boldsymbol{x}(t_0) \equiv \boldsymbol{X}$. t_0 is the natural initial state, which changes should be computed. In many cases it holds $t_0 = 0$.

In the Cartesian coordinate system with the origin 0 and the basis vectors $\boldsymbol{e}_i$ $(i = 1, 2, 3)$ the movement of the material point $\boldsymbol{X}$ can be presented as follows

$$\begin{aligned} &\boldsymbol{x} = x_i \boldsymbol{e}_i, \qquad \boldsymbol{X} = X_i \boldsymbol{e}_i, \qquad \boldsymbol{x}(\boldsymbol{X}, t_0) = \boldsymbol{x}_0 \equiv \boldsymbol{X}, \\ &\boldsymbol{x} = \boldsymbol{x}(\boldsymbol{X}, t) - \text{trajectory of } \boldsymbol{X}, \\ &\boldsymbol{X} = \boldsymbol{X}(\boldsymbol{x}, t) - \text{point } \boldsymbol{X}, \text{ which is in the moment } t \text{ located at } \boldsymbol{x} \end{aligned} \tag{1}$$

In Fig. 1 the trajectory of $\boldsymbol{X}$ is shown. If the Jacobi determinant is not equal to 0

$$\det\left(\frac{\partial x_i}{\partial X_j}\right) \equiv \left|\frac{\partial x_i}{\partial X_j}\right| \neq 0, \tag{2}$$

the following unique and invertible relation exists

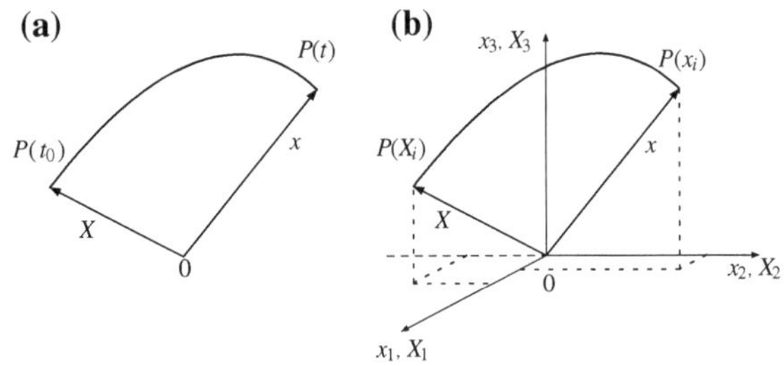

Fig. 1 Trajectory of a *material point*: **a** position vectors, **b** Cartesian coordinates

$$x(X, t) \Longleftrightarrow X(x, t) \tag{3}$$

Now we can introduce the configuration (sometimes called placement).

Definition 1.1 (Configuration) The configuration of a body is defined if we have for any time t an invertible unique mapping for the material points $\boldsymbol{X}$ with $\boldsymbol{x}$. For the initial $t = t_0$ one has the reference configuration, for the actual time t one has the actual configuration.

The placement of the body is defined by its configuration. The motion is presented as follows.

Definition 1.2 (Motion) The motion is given by the changes of the configurations $\boldsymbol{x} = \boldsymbol{x}(\boldsymbol{X}, t)$ (t is a parameter).

For further discussions let us introduce two approaches for the description of continuum mechanics problems.

Definition 1.3 (Lagrangian or material description) The changes of properties prescribed to the material point will be given with respect to $\boldsymbol{X}$. In this case the properties are functions of $\boldsymbol{X}$ and t.

Definition 1.4 (Eulerian or spatial description) The changes of properties prescribed to the material point properties will be given with respect to $\boldsymbol{x}$. In this case the properties are functions of $\boldsymbol{x}$ and t.

1.2 Time Derivatives and Nabla Operator

The properties prescribed to an arbitrary material point can be given in the material or spatial description. For the scalar function φ presenting a property one gets with respect to Eq. (3)

$$\begin{array}{ll} \varphi = \varphi(\boldsymbol{X}, t) = \varphi(X_1, X_2, X_3, t) & \text{material description,} \\ \varphi = \varphi[\boldsymbol{X}(\boldsymbol{x}, t), t] = \varphi(\boldsymbol{x}, t) = \varphi(x_1, x_2, x_3, t) & \text{spatial description.} \end{array}$$

Similar formulations can be introduced for tensorial functions of arbitrary order. In dependence of the assumed presentation of the function φ two time derivatives can be defined: the spatial one and the material one. The spatial derivative is given as

$$\frac{\partial \varphi(\boldsymbol{x}, t)}{\partial t} = \left.\frac{\partial \varphi(\boldsymbol{x}, t)}{\partial t}\right|_{\boldsymbol{x} \text{ fixed}}$$

The material derivative is

$$\frac{\partial \varphi(\boldsymbol{X}, t)}{\partial t} = \left.\frac{\partial \varphi(\boldsymbol{X}, t)}{\partial t}\right|_{\boldsymbol{X} \text{ fixed}}$$

Below the material derivative is denoted by

$$\frac{\mathrm{d}\varphi}{\mathrm{d}t}$$

For the material velocity and acceleration the following definitions hold true.

Definition 1.5 (Material derivatives) The material derivative of the position vector $\boldsymbol{x}(\boldsymbol{X}, t)$ results in the velocity vector $\boldsymbol{v}(\boldsymbol{X}, t)$, the derivative of $\boldsymbol{v}(\boldsymbol{X}, t)$ in the acceleration vector $\boldsymbol{b}(\boldsymbol{X}, t)$

$$\boldsymbol{v}(\boldsymbol{X}, t) = \frac{\mathrm{d}}{\mathrm{d}t}\boldsymbol{x}(\boldsymbol{X}, t) = \dot{\boldsymbol{x}}(\boldsymbol{X}, t), \qquad \boldsymbol{b}(\boldsymbol{X}, t) = \dot{\boldsymbol{v}}(\boldsymbol{X}, t) = \ddot{\boldsymbol{x}}(\boldsymbol{X}, t) \tag{4}$$

The spatial descriptions of $\boldsymbol{v}$ and $\boldsymbol{b}$ one gets, if $\boldsymbol{X}$ will be substituted by $\boldsymbol{x}$

$$\boldsymbol{v} = \boldsymbol{v}[\boldsymbol{X}(\boldsymbol{x}, t), t] = \boldsymbol{v}(\boldsymbol{x}, t), \qquad \boldsymbol{b} = \boldsymbol{b}[\boldsymbol{X}(\boldsymbol{x}, t), t] = \boldsymbol{b}(\boldsymbol{x}, t) \tag{5}$$

Using the nabla operator the material derivative can be introduced

$$\frac{\mathrm{d}\varphi}{\mathrm{d}t} = \left.\frac{\partial \varphi}{\partial t}\right|_{x \text{ fixed}} + \boldsymbol{v} \cdot \boldsymbol{\nabla}\varphi|_{x \text{ fixed}} = \left.\frac{\partial \varphi}{\partial t}\right|_{x \text{ fixed}} + \boldsymbol{v} \cdot \nabla_{x}\varphi \tag{6}$$

1.3 Strains and Deformation Gradient

Let us discuss the transform of line, surface and volume elements from the reference to the actual configuration. The introduction of the deformation gradient (gradient of the position vector) $\boldsymbol{F}$ can be helpful in this case.

Definition 1.6 (Deformation gradient) If the deformation of a body can be with the help of the equation of motion

$$\boldsymbol{x} = \boldsymbol{x}(\boldsymbol{X}, t)$$

transferred from the reference to the actual configuration, the following equation presents the material deformation gradient

$$\boldsymbol{F} = [\nabla_{X}\boldsymbol{x}(\boldsymbol{X}, t)]^{\mathrm{T}} \tag{7}$$

$\boldsymbol{F}$ transform a material line element $\mathrm{d}\boldsymbol{X}$ of the reference configuration into a material line element $\mathrm{d}\boldsymbol{x}$ of the actual configuration, i.e.

$$\boldsymbol{F} \cdot \mathrm{d}\boldsymbol{X} = \mathrm{d}\boldsymbol{x}$$

With the help of the deformation gradient one can express the relations for the transform of a surface or volume element from the reference to the actual configuration. The surface element $\mathrm{d}A_0$ in the reference configuration has the size $\mathrm{d}X_1\mathrm{d}X_2$. Considering the orientation of the surface element one can write

$$\mathrm{d}\boldsymbol{A}_0 = \mathrm{d}\boldsymbol{X}_1 \times \mathrm{d}\boldsymbol{X}_2$$

The transform into the actual configuration of $\mathrm{d}\boldsymbol{A}_0$ results in the element $\mathrm{d}\boldsymbol{A}$

$$\mathrm{d}\boldsymbol{A} = \mathrm{d}\boldsymbol{x}_1 \times \mathrm{d}\boldsymbol{x}_2 = (\boldsymbol{F} \cdot \mathrm{d}\boldsymbol{x}_1) \times (\boldsymbol{F} \cdot \mathrm{d}\boldsymbol{x}_2)$$

With respect to

$$(\boldsymbol{F} \cdot \mathrm{d}\boldsymbol{X}_1) \times (\boldsymbol{F} \cdot \mathrm{d}\boldsymbol{X}_2) = (\det \boldsymbol{F})(\boldsymbol{F}^{\mathrm{T}})^{-1} \cdot (\mathrm{d}\boldsymbol{X}_1 \times \mathrm{d}\boldsymbol{X}_2)$$

finally one obtains

$$\mathrm{d}\boldsymbol{A} = (\det \boldsymbol{F})(\boldsymbol{F}^{-1})^{\mathrm{T}} \cdot \mathrm{d}\boldsymbol{A}_0 \qquad (8)$$

The volume element in the reference configuration is defined as

$$\mathrm{d}V_0 = |(\mathrm{d}\boldsymbol{X}_1 \times \mathrm{d}\boldsymbol{X}_2) \cdot \mathrm{d}\boldsymbol{X}_3|,$$

and in the actual configuration as

$$\mathrm{d}V = |[(\boldsymbol{F} \cdot \mathrm{d}\boldsymbol{X}_1) \times (\boldsymbol{F} \cdot \mathrm{d}\boldsymbol{X}_2)] \cdot (\boldsymbol{F} \cdot \mathrm{d}\boldsymbol{X}_3)|$$

After some manipulations one obtains

$$\mathrm{d}V = |\det \boldsymbol{F}|\mathrm{d}V_0 \qquad (9)$$

It holds true $\det \boldsymbol{F} \neq 0$ for all t if the continuity w.r.t. t can be assumed. In addition, for $t = t_0$ we obtain $\det \boldsymbol{F} = 1$.

1.4 Velocities, Velocity Gradients

The velocity $\boldsymbol{v}(\boldsymbol{X}, t)$ of a material point $\boldsymbol{X}$ is defined by

$$\boldsymbol{v}(\boldsymbol{X}, t) = \frac{\mathrm{d}}{\mathrm{d}t}\boldsymbol{x}(\boldsymbol{X}, t) \equiv \dot{\boldsymbol{x}}(\boldsymbol{X}, t) = \frac{\partial}{\partial t}\boldsymbol{x}(\boldsymbol{X}, t)$$

Since

$$\boldsymbol{x}(\boldsymbol{X}, t) \Longleftrightarrow \boldsymbol{X}(\boldsymbol{x}, t)$$

in the spatial description of $\boldsymbol{v}$ can be presented as

$$\boldsymbol{v}[\boldsymbol{X}(\boldsymbol{x}, t), t] = \boldsymbol{v}(\boldsymbol{x}, t)$$

For the acceleration we get

$$\boldsymbol{b} = \dot{\boldsymbol{v}} = \frac{\partial^2}{\partial t^2}\boldsymbol{x}(\boldsymbol{X}, t) = \ddot{\boldsymbol{x}}(\boldsymbol{X}, t)$$

For the velocity in the Eulerian description we have

$$\boldsymbol{v}(\boldsymbol{x}, t) = \boldsymbol{v}(\boldsymbol{x}, t) = \frac{\mathrm{d}\boldsymbol{x}}{\mathrm{d}t}$$

and for the acceleration

$$\boldsymbol{b} = \boldsymbol{b}(\boldsymbol{x}, t) = \frac{\mathrm{d}\boldsymbol{v}}{\mathrm{d}t}$$

Let us introduce the gradient of the velocities $\boldsymbol{L}$.

Definition 1.7 (Velocity gradient tensor) The spatial velocity gradient tensor $\boldsymbol{L}$ of a given velocity field $\boldsymbol{v} = \boldsymbol{v}(\boldsymbol{x}, t)$ is defined by

$$\boldsymbol{L}(\boldsymbol{x}, t) = [\nabla_{\boldsymbol{x}}\boldsymbol{v}(\boldsymbol{x}, t)]^{\mathrm{T}} \tag{10}$$

With the help of $\boldsymbol{L}$ the time derivatives of the material line, surface and volume elements can be computed in the actual configuration

$$(\mathrm{d}\boldsymbol{x})^{\cdot} = \boldsymbol{L} \cdot \mathrm{d}\boldsymbol{x}, \tag{11}$$

$$(\mathrm{d}\boldsymbol{A})^{\cdot} = [(\nabla_{\boldsymbol{x}} \cdot \boldsymbol{v})\boldsymbol{I} - \boldsymbol{L}^{\mathrm{T}}] \cdot \mathrm{d}\boldsymbol{A}, \tag{12}$$

$$(\mathrm{d}V)^{\cdot} = (\nabla_{\boldsymbol{x}} \cdot \boldsymbol{v})\mathrm{d}V \tag{13}$$

$\boldsymbol{I}$ is the second-order unit tensor.

1.5 *Strains and Strain Measures*

The deformation gradient $\boldsymbol{F}$ is related to the whole motion that means it contains the rigid body motions. The strains can be obtained if we can find expression which are free from the rigid body motions. In this situation the theorem on the polar decomposition can be helpful.

Theorem 1.1 (Polar decomposition) *Any non-singular second-order tensor* $\boldsymbol{T}$ *with* $\det \boldsymbol{T} \neq 0$ *can be represented in a unique manner by a decomposition into the positive definite symmetric tensors* $\boldsymbol{U}$ *or* $\boldsymbol{V}$ *and orthogonal tensor* $\boldsymbol{R}$

$$T = R \cdot U = V \cdot R$$

$R \cdot U$ is named the right $V \cdot R$ the left polar decomposition.

The deformation gradient tensor F is always non-singular ($\det F \neq 0$). So we get

$$F = R \cdot U = V \cdot R \tag{14}$$

The polar decomposition of F yields the following deformation tensors

$$\begin{aligned} U &= \left(F^{\mathrm{T}} \cdot F\right)^{1/2} && \text{right stretch tensor,} \\ V &= \left(F \cdot F^{\mathrm{T}}\right)^{1/2} && \text{left stretch tensor,} \\ C &= U^2 = \left(F^{\mathrm{T}} \cdot F\right) && \text{right Cauchy-Green tensor,} \\ B &= V^2 = \left(F \cdot F^{\mathrm{T}}\right) && \text{left Cauchy-Green tensor} \end{aligned} \tag{15}$$

For application purposes, for example the formulation of constitutive equations, it is better to introduce a strain measure resulting in the reference configuration for pure rigid body motions the value zero instead of one. One possibility is the Green-Lagrange strain tensor

$$G(X,t) = \frac{1}{2}[C(X,t) - I] = \frac{1}{2}(F^{\mathrm{T}} \cdot F - I) = \frac{1}{2}\left(U^2 - I\right) \tag{16}$$

In many continuum mechanics problems one needs the volume strain as a characteristic property.

Definition 1.8 (Volume strain) If we divide the difference of the material volume elements $\mathrm{d}V$ and $\mathrm{d}V_0$ in the actual and the reference configurations by $\mathrm{d}V_0$ we get the volume strain ε_{V}

$$\varepsilon_{\mathrm{V}} = \frac{\mathrm{d}V - \mathrm{d}V_0}{\mathrm{d}V_0} \tag{17}$$

With $\mathrm{d}V = (\det F)\mathrm{d}V_0$ yields

$$\frac{\mathrm{d}V - \mathrm{d}V_0}{\mathrm{d}V_0} = \frac{(\det F - 1)\mathrm{d}V_0}{\mathrm{d}V_0} = \det F - 1$$

and

$$\varepsilon_{\mathrm{V}} = \det F - 1 = \sqrt{\det(F^{\mathrm{T}} \cdot F)} - 1 = \sqrt{\det C} - 1 = \sqrt{\det(2G + I)} - 1 \tag{18}$$

Definition 1.9 (Volume conservation) Volume conservation (isochoric motion) is related to the constraint

$$\varepsilon_{\mathrm{V}} \equiv 0$$

With Eq. (18) it follows $\det F = 1$.

The starting point for the analysis of the strain rates is the spatial rate tensor $\boldsymbol{L}(\boldsymbol{x}, t)$. The following equation is valid

$$\boldsymbol{L}(\boldsymbol{x}, t) = [\nabla_{\boldsymbol{x}} \boldsymbol{v}(\boldsymbol{x}, t)]^{\mathrm{T}} = \dot{\boldsymbol{F}} \cdot \boldsymbol{F}^{-1}$$

$\boldsymbol{L}$ is a second-order tensor, which can be split additively into symmetric and antisymmetric tensors

$$\begin{aligned} \boldsymbol{L} &= \frac{1}{2}\left(\boldsymbol{L} + \boldsymbol{L}^{\mathrm{T}}\right) + \frac{1}{2}\left(\boldsymbol{L} - \boldsymbol{L}^{\mathrm{T}}\right) \\ &= \frac{1}{2}\left[(\nabla_{\boldsymbol{x}} \boldsymbol{v})^{\mathrm{T}} + \nabla_{\boldsymbol{x}} \boldsymbol{v}\right] + \frac{1}{2}\left[(\nabla_{\boldsymbol{x}} \boldsymbol{v})^{\mathrm{T}} - \nabla_{\boldsymbol{x}} \boldsymbol{v}\right] = \boldsymbol{D} + \boldsymbol{W} \end{aligned}$$

Definition 1.10 (Strain rate tensor) The symmetric part of $\boldsymbol{L}$

$$\boldsymbol{D} = \frac{1}{2}\left(\boldsymbol{L} + \boldsymbol{L}^{\mathrm{T}}\right)$$

is the strain rate tensor.

Definition 1.11 (Vorticity tensor) The antisymmetric part of $\boldsymbol{L}$

$$\boldsymbol{W} = \frac{1}{2}\left(\boldsymbol{L} - \boldsymbol{L}^{\mathrm{T}}\right)$$

is the vorticity tensor.

1.6 Displacements, Displacement Gradient, Linearizations

Let us express the kinematical variables by the displacement vector and the displacement gradient tensor

- displacement vector in the reference configuration

$$\boldsymbol{u}(\boldsymbol{X}, t) = \boldsymbol{x}(\boldsymbol{X}, t) - \boldsymbol{X},$$

- displacement vector in the actual configuration

$$\boldsymbol{u}(\boldsymbol{x}, t) = \boldsymbol{x} - \boldsymbol{X}(\boldsymbol{x}, t),$$

- displacement gradient tensor

$$\begin{aligned} P_0(\boldsymbol{X}) \Longrightarrow P(\boldsymbol{x}): &\quad \boldsymbol{x} = \boldsymbol{X} + \boldsymbol{u}(\boldsymbol{X}, t), \\ Q_0(\boldsymbol{X} + \mathrm{d}\boldsymbol{X}) \Longrightarrow Q(\boldsymbol{x} + \mathrm{d}\boldsymbol{x}): &\quad \boldsymbol{x} + \mathrm{d}\boldsymbol{x} = \boldsymbol{X} + \mathrm{d}\boldsymbol{X} + \boldsymbol{u}(\boldsymbol{X} + \mathrm{d}\boldsymbol{X}, t) \end{aligned} \tag{19}$$

From Eq. (19) one gets

$$\mathrm{d}\boldsymbol{x} = \mathrm{d}\boldsymbol{X} + \boldsymbol{u}(\boldsymbol{X} + \mathrm{d}\boldsymbol{X}, t) - \boldsymbol{u}(\boldsymbol{X}, t)$$

and finally

$$\mathrm{d}\boldsymbol{x} = \mathrm{d}\boldsymbol{X} + (\nabla_X \boldsymbol{u})^{\mathrm{T}} \cdot \mathrm{d}\boldsymbol{X} = (\boldsymbol{I} + \boldsymbol{J}) \cdot \mathrm{d}\boldsymbol{X} \tag{20}$$

Definition 1.12 (Displacement gradient tensor) The material displacement gradient tensor is defined as

$$[\nabla_X \boldsymbol{u}(\boldsymbol{X}, t)]^{\mathrm{T}} \equiv \boldsymbol{J}$$

The spatial displacement gradient tensor is defined as

$$[\nabla_x \boldsymbol{u}(\boldsymbol{x}, t)]^{\mathrm{T}} \equiv \boldsymbol{K}$$

With

$$\begin{aligned} \boldsymbol{u}(\boldsymbol{X}, t) = \boldsymbol{x}(\boldsymbol{X}, t) - \boldsymbol{X} &\Longrightarrow (\nabla_X \boldsymbol{u})^{\mathrm{T}} = (\nabla_X \boldsymbol{x})^{\mathrm{T}} - \boldsymbol{I}, \qquad \boldsymbol{J} = \boldsymbol{F} - \boldsymbol{I}, \\ \boldsymbol{u}(\boldsymbol{x}, t) = \boldsymbol{x} - \boldsymbol{X}(\boldsymbol{x}, t) &\Longrightarrow (\nabla_x \boldsymbol{u})^{\mathrm{T}} = \boldsymbol{I} - (\nabla_x \boldsymbol{X})^{\mathrm{T}}, \qquad \boldsymbol{K} = \boldsymbol{I} - \boldsymbol{F}^{-1}, \end{aligned}$$

all kinematical tensors can be expressed by $\boldsymbol{u}$ and $\boldsymbol{J}$ or $\boldsymbol{K}$, for example

$$\begin{aligned} \boldsymbol{F} &= \boldsymbol{I} + \boldsymbol{J}, \qquad \boldsymbol{F}^{-1} = \boldsymbol{I} - \boldsymbol{K}, \\ \boldsymbol{C} &= (\boldsymbol{I} + \boldsymbol{J})^{\mathrm{T}} \cdot (\boldsymbol{I} + \boldsymbol{J}) = \boldsymbol{I} + \boldsymbol{J} + \boldsymbol{J}^{\mathrm{T}} + \boldsymbol{J}^{\mathrm{T}} \cdot \boldsymbol{J}, \\ \boldsymbol{C}^{-1} &= (\boldsymbol{I} - \boldsymbol{K}) \cdot (\boldsymbol{I} - \boldsymbol{K})^{\mathrm{T}} = \boldsymbol{I} - \boldsymbol{K} - \boldsymbol{K}^{\mathrm{T}} + \boldsymbol{K} \cdot \boldsymbol{K}^{\mathrm{T}}, \\ \boldsymbol{B} &= (\boldsymbol{I} + \boldsymbol{J}) \cdot (\boldsymbol{I} + \boldsymbol{J})^{\mathrm{T}} = \boldsymbol{I} + \boldsymbol{J} + \boldsymbol{J}^{\mathrm{T}} + \boldsymbol{J} \cdot \boldsymbol{J}^{\mathrm{T}}, \\ \boldsymbol{B}^{-1} &= (\boldsymbol{I} - \boldsymbol{K})^{\mathrm{T}} \cdot (\boldsymbol{I} - \boldsymbol{K}) = \boldsymbol{I} - \boldsymbol{K} - \boldsymbol{K}^{\mathrm{T}} + \boldsymbol{K}^{\mathrm{T}} \cdot \boldsymbol{K}, \\ \boldsymbol{G} &= \frac{1}{2}(\boldsymbol{C} - \boldsymbol{I}) = \frac{1}{2}(\boldsymbol{J} + \boldsymbol{J}^{\mathrm{T}} + \boldsymbol{J} \cdot \boldsymbol{J}^{\mathrm{T}}), \\ \boldsymbol{A} &= \frac{1}{2}(\boldsymbol{I} - \boldsymbol{B}^{-1}) = \frac{1}{2}(\boldsymbol{K} + \boldsymbol{K}^{\mathrm{T}} - \boldsymbol{K}^{\mathrm{T}} \cdot \boldsymbol{K}) \end{aligned}$$

It is easy to show that $\boldsymbol{G}$ and $\boldsymbol{A}$ are nonlinear

$$\boldsymbol{G} = \frac{1}{2}\left[(\nabla_X \boldsymbol{u})^{\mathrm{T}} + (\nabla_X \boldsymbol{u}) + (\nabla_X \boldsymbol{u}) \cdot (\nabla_X \boldsymbol{u})^{\mathrm{T}}\right] = G_{ij}\boldsymbol{e}_i\boldsymbol{e}_j,$$

$$\boldsymbol{A} = \frac{1}{2}\left[(\nabla_x \boldsymbol{u})^{\mathrm{T}} + (\nabla_x \boldsymbol{u}) - (\nabla_x \boldsymbol{u}) \cdot (\nabla_x \boldsymbol{u})^{\mathrm{T}}\right] = A_{ij}\boldsymbol{e}_i\boldsymbol{e}_j$$

Obviously, $\boldsymbol{G}$ and $\boldsymbol{A}$ contain quadratic terms w.r.t. $\boldsymbol{u}$. This is the so-called geometrical nonlinearity. It is easy to deduce the consistent geometrical linear relations (see, for example, [1]).

2 Stress State

2.1 Classification of External Actions

The actions on a body can by classified as volume or surface actions. The following actions are known: pure mechanical, thermal, electromagnetic, etc. Here we focus our attention on mechanical actions, which can split into forces and moments as known from general mechanics (part statics). Then we can introduce:

- mass or volume forces and moments and
- surface forces and moments.

In general, the actions are continuously defined in the volume or on the surface functions. They are introduced as models since they cannot be observed directly (only the response of the actions can be measured). It is easy to show that line and concentrated single actions are limit cases of the volume and surface actions. These limit cases are the result of the different order of the three spatial dimensions or of the two dimensions of the surface.

Any material body is characterized by a continuous mass density distribution $\rho(\boldsymbol{x})$. The mass or volume actions are also continuous functions applied to any material point of the body. Examples of volume forces are the gravitational force, the force of inertia and the Coriolis force among others. The sources of these forces are out of the body, that means they are external volume forces. By analogy one can introduce sources for volume moments.

Volume forces are related to volume or mass. By $\boldsymbol{k}^{\mathrm{V}}$ the volume force density and by $\boldsymbol{k}^{\mathrm{m}} \equiv \boldsymbol{k}$ the mass force density are denoted (in what is following $\boldsymbol{k}$ is used instead of $\boldsymbol{k}^{\mathrm{m}}$). It holds

$$\rho(\boldsymbol{x}, t)\boldsymbol{k}(\boldsymbol{x}, t) = \boldsymbol{k}^{\mathrm{V}} \tag{21}$$

with the mass force density $\boldsymbol{k}(\boldsymbol{x}, t)$, the volume force density $\boldsymbol{k}^{\mathrm{V}}(\boldsymbol{x}, t)$ and the mass density $\rho(\boldsymbol{x}, t)$. Examples of the volume force density are the weight, the centrifugal force or in general the potential forces:

- the weight
$$\rho\boldsymbol{k} = -\rho g \boldsymbol{e}_3,$$
where g is the gravity acceleration, $\boldsymbol{e}_3$ is the basis vector in the opposite direction to the gravity acceleration.
- the centrifugal force
$$\rho\boldsymbol{k} = -\rho\boldsymbol{\omega} \times (\boldsymbol{\omega} \times \boldsymbol{x}),$$
where $\boldsymbol{\omega}$ is the angular velocity
- general potential force
$$\rho\boldsymbol{k} = -\rho\nabla_x \Pi$$

The force potential Π in the case of weight or centrifugal force can be expressed as

$$\Pi = \boldsymbol{e}_3 \cdot \boldsymbol{x} g \qquad \text{or} \qquad \Pi = -\frac{1}{2}|\boldsymbol{\omega} \times \boldsymbol{x}|^2$$

In the case of volume moments the analogous equation is valid

$$\rho(\boldsymbol{x}, t)\boldsymbol{l}^{\mathrm{m}}(\boldsymbol{x}, t) = \rho(\boldsymbol{x}, t)\boldsymbol{l}(\boldsymbol{x}, t) = \boldsymbol{l}^{\mathrm{V}} \tag{22}$$

with $\boldsymbol{l}^{\mathrm{m}}(\boldsymbol{x}, t)$ as the mass moment density and $\boldsymbol{l}^{\mathrm{V}}(\boldsymbol{x}, t)$ as the volume moment density.

External surface loads are acting on surfaces. Such loads are named contact loads. The surface can be the surface of a material body with a volume $A(V)$, but also common interfaces between the parts of the body or between two different bodies. External surface loads are existing also between solids and fluids, for example, the hydrostatic pressure of the fluid on a solid surrounded by the fluid. The surface loads can be split again in surface forces and surface moments. The surface forces are related to the surface and result in the stress vector $\boldsymbol{t}$, the surface moments by analogy result in moment stress vectors $\boldsymbol{\mu}$. The following limits can be introduced [2, 3]

$$\boldsymbol{t} = \lim_{\triangle A \to 0} \frac{\triangle \boldsymbol{f}}{\triangle A}, \qquad \boldsymbol{\mu} = \lim_{\triangle A \to 0} \frac{\triangle \boldsymbol{m}}{\triangle A}, \tag{23}$$

where $\triangle \boldsymbol{f}$ and $\triangle \boldsymbol{m}$ are the resulting force and moment vectors acting on the surface $\triangle A$. Note that $\triangle A$ is oriented that means $\triangle \boldsymbol{A} = \boldsymbol{n}\triangle A$. The vectors depend on the position on the surface and the orientation of the surface ($\mathrm{d}\boldsymbol{A} = \boldsymbol{n}\mathrm{d}A$)

$$\boldsymbol{t} = \boldsymbol{t}(\boldsymbol{x}, \boldsymbol{n}, t), \qquad \boldsymbol{\mu} = \boldsymbol{\mu}(\boldsymbol{x}, \boldsymbol{n}, t) \tag{24}$$

One gets by integration of the external volume and surface forces the resulting external force $\boldsymbol{f}^{\mathrm{R}}$ acting on the body

$$\boldsymbol{f}^{\mathrm{R}} = \int_V \rho \boldsymbol{k}\, \mathrm{d}V + \int_A \boldsymbol{t}\, \mathrm{d}A \tag{25}$$

The resulting external moment can be introduced in the same manner

$$\boldsymbol{m}_0^{\mathrm{R}} = \int_V \rho(\boldsymbol{l} + \boldsymbol{x} \times \boldsymbol{k})\, \mathrm{d}V + \int_A (\boldsymbol{\mu} + \boldsymbol{x} \times \boldsymbol{t})\, \mathrm{d}A \tag{26}$$

In the classical mechanics the moment vectors are ignored and we get the following expression

$$\boldsymbol{m}_0^{\mathrm{R}} = \int_V \rho(\boldsymbol{x} \times \boldsymbol{k})\, \mathrm{d}V + \int_A (\boldsymbol{x} \times \boldsymbol{t})\, \mathrm{d}A \tag{27}$$

2.2 Cauchy's Stress Vector and Tensor

As the result of the external action on the body one obtains stresses in the body. Let us introduce the Euler-Cauchy stress principle.

Definition 1.13 (Euler-Cauchy stress principle) The external forces result in a vector field of stress vectors $\boldsymbol{t}(\boldsymbol{x}, \boldsymbol{n}, t)$ acting on a surface A with the normal $\boldsymbol{n}(\boldsymbol{x}, t)$. In the case that the surface is the body surface the stress vectors are resulting from the surface forces and named traction.

The stresses in the body can be defined using the method of sections (Fig. 2). From the Statics follows that we have some actions in the body (Fig. 3). $\mathrm{d}\boldsymbol{f}$ is the resulting force vector and $\mathrm{d}\boldsymbol{m}$ is the resulting moment vector on the surface element $\mathrm{d}A$, $\boldsymbol{n}$ is the unit normal vector on the surface. With respect to Eq. (23) one gets the stress and the couple stress vector

$$\boldsymbol{t}(\boldsymbol{x}, \boldsymbol{n}, t) = \frac{\mathrm{d}\boldsymbol{f}}{\mathrm{d}A}, \qquad \boldsymbol{\mu}(\boldsymbol{x}, \boldsymbol{n}, t) = \frac{\mathrm{d}\boldsymbol{m}}{\mathrm{d}A}$$

Ignoring the inner moments, we get the classical continuum (Fig. 4). In this case we have no surface moments and $\boldsymbol{\mu} \equiv \boldsymbol{0}$ is valid. For polar continua that is when $\boldsymbol{\mu} \neq \boldsymbol{0}$ we refer to [4–8].

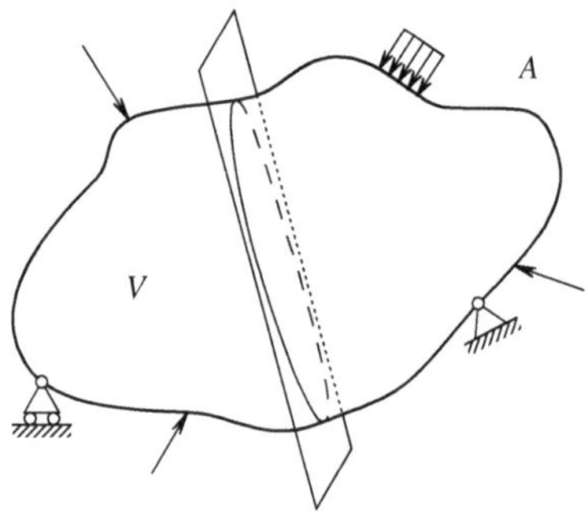

Fig. 2 Method of sections (arbitrary section in body under external loadings)

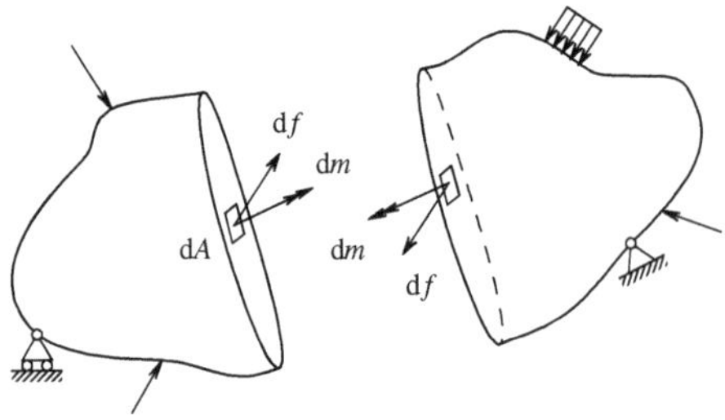

Fig. 3 Actions in the body (polar continuum)

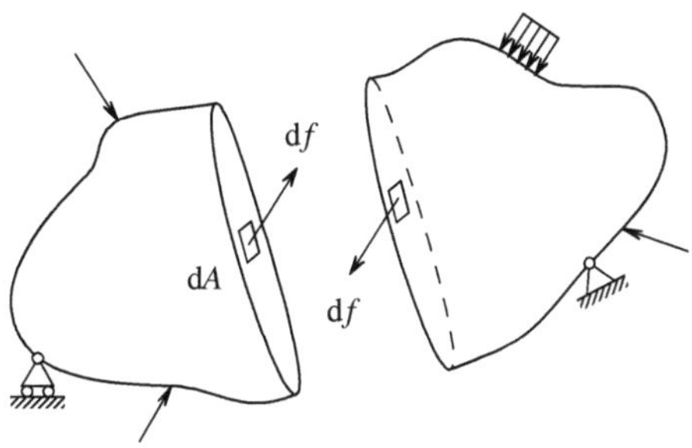

Fig. 4 Actions in the body (classical continuum)

Note 1.1 As a measure of the inner force in point P of the body we introduce the stress vector

$$\boldsymbol{t}(\boldsymbol{x}, \boldsymbol{n}, t) = \frac{\mathrm{d}\boldsymbol{f}}{\mathrm{d}A}$$

In general, $\boldsymbol{t}$ depends on the position, time and orientation of the surface. In each point of the body $\boldsymbol{t}(\boldsymbol{n}) = -\boldsymbol{t}(-\boldsymbol{n})$ holds (Cauchy's lemma).

Definition 1.14 (Stress state) All possible stress vectors in a material point P define the stress state in this point.

From material's testing we know two definitions for the stresses.

Definition 1.15 (Engineering stresses) The acting force is related to the surface in the reference configuration.

Definition 1.16 (True stresses) The acting force is related to the surface in the actual configuration.

In continuum mechanics we have more possibilities since we can define the force vector in both configurations, the surface orientation in two configurations and in addition we can introduce intermediate configurations.

Definition 1.17 (Cauchy's stress vector) The Cauchy stress vector is called true stress vector. The actual force is related to the actual section.

The stress tensor follows as

$$\boldsymbol{t}(\boldsymbol{x}, \boldsymbol{n}, t) = \boldsymbol{n} \cdot \boldsymbol{T}(\boldsymbol{x}, t)$$

with the Cauchy stress vector $\boldsymbol{t}(\boldsymbol{x}, \boldsymbol{n}, t)$, the normal $\boldsymbol{n}$ and the Cauchy stress tensor $\boldsymbol{T}(\boldsymbol{x}, t)$.

2.3 Equilibrium Equations, Equations of Motion

For the body under surface forces $\boldsymbol{t}\mathrm{d}A$ and volume forces $\rho\boldsymbol{k}\mathrm{d}V$, which is in an equilibrium state, the following equations considering Eqs. (25) and (27) hold

$$\int\limits_V \rho\boldsymbol{k}\,\mathrm{d}V + \int\limits_A \boldsymbol{t}\,\mathrm{d}A = \boldsymbol{0}, \qquad \int\limits_V (\boldsymbol{x}\times\rho\boldsymbol{k})\,\mathrm{d}V + \int\limits_A (\boldsymbol{x}\times\boldsymbol{t})\,\mathrm{d}A = \boldsymbol{0} \tag{28}$$

With

$$\boldsymbol{t} = \boldsymbol{n}\cdot\boldsymbol{T}$$

and applying the divergence theorem it follows

$$\int\limits_A \boldsymbol{t}\,\mathrm{d}A = \int\limits_A \boldsymbol{n}\cdot\boldsymbol{T}\,\mathrm{d}A = \int\limits_V \nabla_x\cdot\boldsymbol{T}\,\mathrm{d}V \tag{29}$$

and

$$\int\limits_V (\rho\boldsymbol{k} + \nabla_x\cdot\boldsymbol{T})\,\mathrm{d}V = \boldsymbol{0} \tag{30}$$

This is the integral equilibrium. If the volume is arbitrary and all fields are smooth the local equilibrium can be expressed as

$$\nabla_x\cdot\boldsymbol{T} + \rho\boldsymbol{k} = \boldsymbol{0} \tag{31}$$

Adding the inertial force $-\ddot{\boldsymbol{x}}\mathrm{d}M = -\ddot{\boldsymbol{x}}\rho\mathrm{d}V$ in the sense of the Newton/d'Alembert principle, one obtains

$$\int\limits_V \rho\boldsymbol{k}\,\mathrm{d}V + \int\limits_A \boldsymbol{T}\,\mathrm{d}A - \int\limits_V \ddot{\boldsymbol{x}}\rho\,\mathrm{d}V = \boldsymbol{0}$$

or

$$\int\limits_V (\rho\boldsymbol{k} + \nabla_x\cdot\boldsymbol{T} - \rho\ddot{\boldsymbol{x}})\,\mathrm{d}V = \boldsymbol{0}, \tag{32}$$

The local equation of motion holds

$$\rho\ddot{\boldsymbol{x}} = \nabla_x\cdot\boldsymbol{T} + \rho\boldsymbol{k} \tag{33}$$

The second equation of (28) results in the symmetry of the stress tensor ($\boldsymbol{T} = \boldsymbol{T}^{\mathrm{T}}$).

The stress tensor $\boldsymbol{T}$ can be split in a spherical and a deviatoric part

$$\boldsymbol{T} = \frac{1}{3}\boldsymbol{T} \cdot\cdot \boldsymbol{II} + \left(\boldsymbol{T} - \frac{1}{3}\boldsymbol{T} \cdot\cdot \boldsymbol{II}\right) = \frac{1}{3}(\mathrm{tr}\boldsymbol{T})\boldsymbol{I} + \left[\boldsymbol{T} - \frac{1}{3}(\mathrm{tr}\boldsymbol{T})\boldsymbol{I}\right] = \boldsymbol{T}^{\mathrm{K}} + \boldsymbol{T}^{\mathrm{D}}$$

2.4 Stress Vectors and Tensors After Piola-Kirchhoff

Up to now the stress vector and the stress tensor are presented in the Eulerian coordinates. The Cauchy tensor of true stresses is defined by the actual force vector and the actual oriented surface element. In many applications it is better to use the Lagrangian description and as the minimum to relate the volume and surface elements to the reference configuration. The transforms (8) and (9) are valid.

Definition 1.18 (First Piola-Kirchhoff stress tensor) If we relate the actual force vector $\mathrm{d}\boldsymbol{f}$ to the oriented surface element $\mathrm{d}\boldsymbol{A}_0 = \boldsymbol{n}_0 \mathrm{d}A_0$ in the reference configuration we get the following stress vector

$$^{I}\boldsymbol{t} = \frac{\mathrm{d}\boldsymbol{f}}{\mathrm{d}A_0}$$

The respective tensor $^{I}\boldsymbol{P}$, which describes the stress state in a material point of the reference configuration is named first Piola-Kirchhoff or Lagrangian stress tensor.

With

$$\boldsymbol{t} = \frac{\mathrm{d}\boldsymbol{f}}{\mathrm{d}A}, \qquad ^{I}\boldsymbol{t} = \frac{\mathrm{d}\boldsymbol{f}}{\mathrm{d}A_0}$$

one obtains

$$\boldsymbol{t}\mathrm{d}A =^{I} \boldsymbol{t}\mathrm{d}A_0 = \mathrm{d}\boldsymbol{f}, \qquad \boldsymbol{t} =^{I} \boldsymbol{t} \cdot (\mathrm{det}\boldsymbol{F})^{-1}\boldsymbol{F}^{\mathrm{T}}, \qquad ^{I}\boldsymbol{t} = \boldsymbol{t} \cdot \mathrm{det}\,\boldsymbol{F}\left(\boldsymbol{F}^{-1}\right)^{\mathrm{T}} \tag{34}$$

$^{I}\boldsymbol{P}$ is in general a non-symmetric tensor. With

$$\mathrm{d}\boldsymbol{A}_0 = (\mathrm{det}\,\boldsymbol{F})^{-1}\boldsymbol{F}^{\mathrm{T}} \cdot \mathrm{d}\boldsymbol{A}$$

one gets the relations between the Cauchy and the first Piola-Kirchhoff stress tensors

$$\boldsymbol{t} = \boldsymbol{n} \cdot \boldsymbol{T}, \qquad ^{I}\boldsymbol{t} = \boldsymbol{n}_0 \cdot^{I} \boldsymbol{P}, \tag{35}$$

$$\boldsymbol{T} - (\mathrm{det}\,\boldsymbol{F})^{-1}\boldsymbol{F} \cdot^{I} \boldsymbol{P}, \qquad ^{I}\boldsymbol{P} = (\mathrm{det}\,\boldsymbol{F})\boldsymbol{F}^{-1} \cdot \boldsymbol{T} \tag{36}$$

Assuming the force and the moment equilibrium in the reference configuration, we get the equations of motion in Lagrangian coordinates

$$\int_{A_0} \boldsymbol{n}_0 \cdot {}^I\boldsymbol{P}\,\mathrm{d}A_0 + \int_{V_0} \rho_0 \boldsymbol{k}\,\mathrm{d}V_0 - \int_{V_0} \ddot{\boldsymbol{x}}\rho_0\,\mathrm{d}V_0 = \boldsymbol{0}, \tag{37}$$

$$\int_{A_0} \left[\boldsymbol{x} \times (\boldsymbol{n}_0 \cdot {}^I\boldsymbol{P})\right] \mathrm{d}A_0 + \int_{V_0} (\boldsymbol{x} \times \rho_0 \boldsymbol{k})\,\mathrm{d}V_0 = \boldsymbol{0}, \tag{38}$$

$$\ddot{\boldsymbol{x}}\rho_0 = \nabla_{\boldsymbol{X}} \cdot {}^I\boldsymbol{P} + \rho \boldsymbol{k}, \qquad {}^I\boldsymbol{P} \cdot \boldsymbol{F}^{\mathrm{T}} = \boldsymbol{F} \cdot {}^I\boldsymbol{P}^{\mathrm{T}} \tag{39}$$

From the last equation results that ${}^I\boldsymbol{P}$ is a non-symmetrical tensor.

The non-symmetric stress tensor ${}^I\boldsymbol{P}$ is not convenient for the formulation of the constitutive equations. ${}^I\boldsymbol{P}$ should be modified in such a manner that we get a symmetric tensor again for the reference configuration. Let us introduce a "fictitious force vector"

$$\mathrm{d}\boldsymbol{f}_0 = \boldsymbol{F}^{-1} \cdot \mathrm{d}\boldsymbol{f} \tag{40}$$

Since

$$d\boldsymbol{f}_0 = {}^{II}\boldsymbol{P}^{\mathrm{T}} \cdot d\boldsymbol{A}_0 \tag{41}$$

the stress tensor ${}^{II}\boldsymbol{P}$ can be introduced.

Definition 1.19 (Second Piola-Kirchhoff stress tensor) If we relate the force vector $\mathrm{d}\boldsymbol{f}_0 = \boldsymbol{F}^{-1} \cdot \mathrm{d}\boldsymbol{f}$ to the oriented surface element $\mathrm{d}\boldsymbol{A}_0$ of the reference configuration, one gets the pseudo stress vector ${}^{II}\boldsymbol{t}$ with the respective pseudo stress tensor

$${}^{II}\boldsymbol{P} = {}^I\boldsymbol{P} \cdot \left(\boldsymbol{F}^{-1}\right)^{\mathrm{T}}$$

This is the second Piola-Kirchhoff stress tensor. $\mathrm{d}\boldsymbol{f} = \boldsymbol{T} \cdot \mathrm{d}\boldsymbol{A}$ is the analogous relation in the actual configuration to $\mathrm{d}\boldsymbol{f}_0 = {}^{II}\boldsymbol{P} \cdot \mathrm{d}\boldsymbol{A}_0$ in the reference configuration. ${}^{II}\boldsymbol{P}$ is a symmetric tensor since

$${}^{II}\boldsymbol{P} = (\det\boldsymbol{F})\boldsymbol{F}^{-1} \cdot \boldsymbol{T} \cdot \boldsymbol{F}^{-\mathrm{T}}$$

3 Balance Equations

3.1 General Formulation of Balance Equations

Here we present a brief description of the balance equations. They are material independent. We focus our attention to smooth fields. For further discussions and extensions we recommend the special literature, for example, [1, 9–12].

The problems of two- and one-dimensional or generalized continua are presented in [13–19] among others.

The task of continuum mechanics is the estimation the density $\rho = \rho(\boldsymbol{X}, t)$, motion $\boldsymbol{x} = \boldsymbol{x}(\boldsymbol{X}, t)$ and, if thermodynamics is taken into account, the temperature $\vartheta = \vartheta(\boldsymbol{X}, t)$ for all material points $\boldsymbol{X}$ as a function of the time t. It can be shown that the material independent statements can be given by balance equations.

Definition 1.20 (Balance equation) Balance equations are empirical judgements basic in the continuum mechanics, which express the relationships between the variables describing the state of the continuum and the external loadings on the body.

The general structure of a balance equation can be given as follows. $\Psi(\boldsymbol{x}, t)$ and $\Psi_0(\boldsymbol{X}, t)$ are distributions of a scalar mechanical variable w.r.t. the volume elements $\mathrm{d}V$ and $\mathrm{d}V_0$ in the actual and the reference configuration. The integration over the volume results in an additive (extensive) variable $Y(t)$

$$Y(t) = \int\limits_V \Psi(\boldsymbol{x}, t)\,\mathrm{d}V = \int\limits_{V_0} \Psi_0(\boldsymbol{X}, t)\,\mathrm{d}V_0 \tag{42}$$

With $\mathrm{d}V = (\det \boldsymbol{F})\mathrm{d}V_0$ holds $\Psi_0(\boldsymbol{X}, t) = (\det \boldsymbol{F})\Psi(\boldsymbol{x}, t)$. The material time derivative of $Y(t)$ is the rate of changes of the state of the system expressed by $\Psi(\boldsymbol{x}, t)$. This rate is balanced with the action of the surrounding on the body. In the actual configuration we have

$$\frac{\mathrm{d}}{\mathrm{d}t}Y(t) = \frac{\mathrm{d}}{\mathrm{d}t}\int\limits_V \Psi(\boldsymbol{x}, t)\,\mathrm{d}V = \int\limits_A \Phi(\boldsymbol{x}, t)\,\mathrm{d}A + \int\limits_V \Xi(\boldsymbol{x}, t)\,\mathrm{d}V \tag{43}$$

and in the reference configuration

$$\frac{\mathrm{d}}{\mathrm{d}t}Y(t) = \frac{\mathrm{d}}{\mathrm{d}t}\int\limits_{V_0} \Psi_0(\boldsymbol{X}, t)\,\mathrm{d}V_0 = \int\limits_{A_0} \Phi_0(\boldsymbol{X}, t)\,\mathrm{d}A_0 + \int\limits_{V_0} \Xi_0(\boldsymbol{X}, t)\,\mathrm{d}V_0 \tag{44}$$

Φ and Φ_0 are the external fluxes through the surface in both configurations, Ξ and Ξ_0 are the source (production) terms in the volume. Such balance equation for scalar fields can be extended to vectorial and tensorial fields.

Remark 1.1 Φ in the actual configuration is a function of $\boldsymbol{x}$ and t, but also of the orientation of the surface element $d\boldsymbol{A} = \boldsymbol{n}(\boldsymbol{x}, t)\mathrm{d}A$ (with other words dependence of the normal $\boldsymbol{n}$). This statement is valid for arbitrary tensor fields ${}^{(n)}\boldsymbol{\Phi} = {}^{(n)}\boldsymbol{\Phi}(\boldsymbol{x}, \boldsymbol{n}, t)$ of order $n \geq 0$. For the reference configuration we get ${}^{(n)}\boldsymbol{\Phi}_0 = {}^{(n)}\boldsymbol{\Phi}_0(\boldsymbol{X}, \boldsymbol{n}_0, t)$ and $d\boldsymbol{A}_0 = \boldsymbol{n}_0(\boldsymbol{X}, t)\mathrm{d}A_0$.

Remark 1.2 For the $\boldsymbol{n}-$ or $\boldsymbol{n}_0$ dependency of the surface functions $\boldsymbol{\Phi}$ or $\boldsymbol{\Phi}_0$ the Cauchy's lemma is valid

$$ {}^{(n)}\boldsymbol{\Phi}(\boldsymbol{x},\boldsymbol{n},t) = \boldsymbol{n} \cdot {}^{(n+1)}\tilde{\boldsymbol{\Phi}}(\boldsymbol{x},t), \quad {}^{(n)}\boldsymbol{\Phi}_0(\boldsymbol{X},\boldsymbol{n}_0,t) = \boldsymbol{n}_0 \cdot {}^{(n+1)}\tilde{\boldsymbol{\Phi}}_0(\boldsymbol{X},t), \tag{45} $$

Remark 1.3 For the fluxes the third Newton's law (actio = reactio) is valid. Two fluxes acting on the surface in a common material point and characterized by the normals $\boldsymbol{n}$ and $-\boldsymbol{n}$ or $\boldsymbol{n}_0$ and $-\boldsymbol{n}_0$ have the same value, but opposite sign

$$ \boldsymbol{\Phi}(\boldsymbol{n}) = -\boldsymbol{\Phi}(-\boldsymbol{n}), \qquad \boldsymbol{\Phi}_0(\boldsymbol{n}_0) = -\boldsymbol{\Phi}_0(-\boldsymbol{n}_0) \tag{46} $$

The balance equations stating the equilibrium between the changes of the state of the body and the fluxes on the surface and the production in the volume have in the actual configuration the following structure

$$ \frac{\mathrm{d}}{\mathrm{d}t}\int\limits_V {}^{(n)}\boldsymbol{\Psi}(\boldsymbol{x},t)\,\mathrm{d}V = \int\limits_A \boldsymbol{n}(\boldsymbol{x},t) \cdot {}^{(n+1)}\boldsymbol{\Phi}(\boldsymbol{x},t)\,\mathrm{d}A + \int\limits_V {}^{(n)}\boldsymbol{\Xi}(\boldsymbol{x},t)\,\mathrm{d}V \tag{47} $$

In the reference configuration one has

$$ \begin{aligned} \frac{\mathrm{d}}{\mathrm{d}t}\int\limits_{V_0} {}^{(n)}\boldsymbol{\Psi}_0(\boldsymbol{X},t)\,\mathrm{d}V_0 &\equiv \frac{\partial}{\partial t}\int\limits_{V_0} {}^{(n)}\boldsymbol{\Psi}_0(\boldsymbol{X},t)\,\mathrm{d}V_0 \\ &= \int\limits_{A_0} \boldsymbol{n}_0(\boldsymbol{X},t) \cdot {}^{(n+1)}\boldsymbol{\Phi}_0(\boldsymbol{X},t)\,\mathrm{d}A_0 + \int\limits_{V_0} {}^{(n)}\boldsymbol{\Xi}_0(\boldsymbol{X},t)\,\mathrm{d}V_0 \end{aligned} \tag{48} $$

${}^{(n)}\boldsymbol{\Psi}$ and ${}^{(n)}\boldsymbol{\Psi}_0$ or ${}^{(n)}\boldsymbol{\Xi}$ and ${}^{(n)}\boldsymbol{\Xi}_0$ are tensorial fields of the order n ($n \geq 0$), ${}^{(n+1)}\boldsymbol{\Phi}$ and ${}^{(n+1)}\boldsymbol{\Phi}_0$ are tensorial fields of order $(n+1)$.

Using the following transforms

$$ \begin{aligned} \boldsymbol{n} = \det\boldsymbol{F}\frac{\mathrm{d}A_0}{\mathrm{d}A}\left(\boldsymbol{F}^{-1}\right)^{\mathrm{T}} \cdot \boldsymbol{n}_0 &\Longleftrightarrow \boldsymbol{n}_0 = (\det\boldsymbol{F})^{-1}\frac{\mathrm{d}A}{\mathrm{d}A_0}\boldsymbol{F}^{\mathrm{T}} \cdot \boldsymbol{n}, \\ \mathrm{d}\boldsymbol{A} = \det\boldsymbol{F}\left(\boldsymbol{F}^{-1}\right)^{\mathrm{T}} \cdot \mathrm{d}\boldsymbol{A}_0 &\Longleftrightarrow \mathrm{d}\boldsymbol{A}_0 = (\det\boldsymbol{F})^{-1}\boldsymbol{F}^{\mathrm{T}} \cdot \mathrm{d}\boldsymbol{A}, \\ \mathrm{d}V = \det\boldsymbol{F}\mathrm{d}V_0 &\Longleftrightarrow \mathrm{d}V_0 = (\det\boldsymbol{F})^{-1}\mathrm{d}V \end{aligned} \tag{49} $$

we get for example from

$$ \boldsymbol{\Phi}_0 \cdot \mathrm{d}\boldsymbol{A}_0 = \boldsymbol{\Phi} \cdot \mathrm{d}\boldsymbol{A} = \boldsymbol{\Phi} \cdot \det\boldsymbol{F}\left(\boldsymbol{F}^{-1}\right)^{\mathrm{T}} \cdot \mathrm{d}\boldsymbol{A}_0 \tag{50} $$

the relationship between $\boldsymbol{\Phi}_0$ and $\boldsymbol{\Phi}$

$$ \boldsymbol{\Phi}_0 = (\det\boldsymbol{F})\boldsymbol{\Phi} \cdot \left(\boldsymbol{F}^{-1}\right)^{\mathrm{T}}, \qquad \boldsymbol{\Phi}_0 = \boldsymbol{\Phi}_0(\boldsymbol{X},\boldsymbol{n}_0,t), \qquad \boldsymbol{\Phi} = \boldsymbol{\Phi}(\boldsymbol{x},\boldsymbol{n},t) \tag{51} $$

and from $\boldsymbol{\Xi}_0\mathrm{d}V_0 = \boldsymbol{\Xi}\mathrm{d}V = \boldsymbol{\Xi}(\det\boldsymbol{F})\mathrm{d}V_0$ the relationship between $\boldsymbol{\Xi}_0$ and $\boldsymbol{\Xi}$

$$\boldsymbol{\Xi}_0 = (\det \boldsymbol{F})\boldsymbol{\Xi}, \qquad \boldsymbol{\Xi}_0 = \boldsymbol{\Xi}_0(\boldsymbol{X}, t), \qquad \boldsymbol{\Xi} = \boldsymbol{\Xi}(\boldsymbol{x}, t) \tag{52}$$

Sometimes it is more useful to apply the mass integrals. Using the same variables for the distribution functions $\boldsymbol{\Psi}$ and $\boldsymbol{\Xi}$i the global mechanical balance equation in the actual configuration can be written

$$\frac{\mathrm{d}}{\mathrm{d}t}\int\limits_m \boldsymbol{\Psi}(\boldsymbol{x}, t)\,\mathrm{d}m \equiv \frac{\mathrm{d}}{\mathrm{d}t}\int\limits_V \boldsymbol{\Psi}(\boldsymbol{x}, t)\rho\,\mathrm{d}V = \int\limits_A \boldsymbol{n}\cdot\boldsymbol{\Phi}(\boldsymbol{x}, t)\,\mathrm{d}A + \int\limits_V \boldsymbol{\Xi}(\boldsymbol{x}, t)\rho\,\mathrm{d}V \tag{53}$$

In Eq. (53) $\boldsymbol{\Psi}(\boldsymbol{x}, t)$, $\boldsymbol{\Xi}(\boldsymbol{x}, t)$ are tensorial fields of the same order n ($n \geq 0$), $\boldsymbol{\Phi}(\boldsymbol{x}, t)$ is a tensorial field of the order $(n + 1)$, $\boldsymbol{n}(\boldsymbol{x}, t)$ is the outer normal on A, $m(\boldsymbol{x}, t)$ is the mass as a continues function of the volume.

If the continuity in the sense of the divergence theorem is fulfilled for $\boldsymbol{\Psi}$ after application of the theorem to Eq. (53) one gets

$$\frac{\mathrm{d}}{\mathrm{d}t}\int\limits_V \boldsymbol{\Psi}(\boldsymbol{x}, t)\rho\,\mathrm{d}V = \int\limits_V \nabla_{\boldsymbol{x}}\cdot\boldsymbol{\Phi}(\boldsymbol{x}, t)\,\mathrm{d}V + \int\limits_V \boldsymbol{\Xi}(\boldsymbol{x}, t)\rho\,\mathrm{d}V \tag{54}$$

and with $\mathrm{d}V \to 0$ the local formulation of the general balance equation

$$\rho\frac{\mathrm{d}}{\mathrm{d}t}[\boldsymbol{\Psi}(\boldsymbol{x}, t)] = \nabla_{\boldsymbol{x}}\cdot\boldsymbol{\Phi}(\boldsymbol{x}, t) + \boldsymbol{\Xi}(\boldsymbol{x}, t)\rho \tag{55}$$

If we transform the equations into the reference configuration one gets

$$\begin{aligned}\frac{\partial}{\partial t}\int\limits_{V_0} \boldsymbol{\Psi}_0(\boldsymbol{X}, t)\rho_0\,\mathrm{d}V_0 &= \int\limits_{A_0} \boldsymbol{n}_0\cdot\boldsymbol{\Phi}_0(\boldsymbol{X}, t)\,\mathrm{d}A_0 + \int\limits_{V_0} \boldsymbol{\Xi}_0(\boldsymbol{X}, t)\rho_0\,\mathrm{d}V_0\\ &= \int\limits_{V_0} [\nabla_{\boldsymbol{X}}\cdot\boldsymbol{\Phi}_0(\boldsymbol{X}, t) + \boldsymbol{\Xi}_0(\boldsymbol{X}, t)\rho_0]\,\mathrm{d}V_0\end{aligned}$$

or locally

$$\rho_0\frac{\partial}{\partial t}[\boldsymbol{\Psi}_0(\boldsymbol{X}, t)] = \nabla_{\boldsymbol{X}}\cdot\boldsymbol{\Phi}_0(\boldsymbol{X}, t) + \boldsymbol{\Xi}(\boldsymbol{X}, t)\rho_0 \tag{56}$$

3.2 Mass Balance and Mass Conservation

The mass is one of the main characteristics of a material body. The mass of the body can be estimated as the volume integral over the density field

$$m = \int_V \rho(\boldsymbol{x}, t)\,\mathrm{d}V = \int_{V_0} \rho_0(\boldsymbol{X})\,\mathrm{d}V_0 \tag{57}$$

This equation contains the global law of mass conservation.

Definition 1.21 **(Mass conservation)** If we have no mass exchange through the surface and no mass production in the volume the mass of the body is always constant that means independent of the time.

$\rho \mathrm{d}V$ and $\rho_0 \mathrm{d}V_0$ are the mass of a material point before and after the deformation. If they are the same one gets $\rho \det \boldsymbol{F} = \rho_0$ and finally

$$\frac{\rho_0}{\rho} = \det \boldsymbol{F},$$

The mass conservation is valid locally.

Theorem 1.2 **(Law of mass conservation)** *The mass* $\mathrm{d}m = \rho(\boldsymbol{x}, t)\mathrm{d}V$ *of a material volume* $\mathrm{d}V$ *is always constant*

$$\mathrm{d}m = \rho(\boldsymbol{x}, t)\mathrm{d}V = \rho_0(\boldsymbol{X})\mathrm{d}V_0 = const$$

The law of mass conservation is equivalent to the continuity of the mass distribution for the continuum with continuous placement of the material points.

Applying the general balance Eq. (53) with $\boldsymbol{\Psi} \to 1$ (scalar), $\boldsymbol{\Phi} = 0$ (no mass exchange through the surface A) and $\boldsymbol{\Xi} \to 0$ (no mass production in the volume) follows

$$\frac{\mathrm{d}m}{\mathrm{d}t} = \frac{\mathrm{d}}{\mathrm{d}t}\int_V \rho(\boldsymbol{x}, t)\,\mathrm{d}V = \frac{\partial}{\partial t}\int_{V_0} \rho_0(\boldsymbol{X})\,\mathrm{d}V_0 = 0 \tag{58}$$

or locally

$$\frac{\mathrm{d}}{\mathrm{d}t}(dm) = \frac{\mathrm{d}}{\mathrm{d}t}(\rho \mathrm{d}V) = \frac{\partial}{\partial t}(\rho_0 \mathrm{d}V_0) = 0 \tag{59}$$

The global law of mass conservation in the Eulerian formulation can be given as

$$\begin{aligned}\frac{\mathrm{d}}{\mathrm{d}t}\int_V \rho(\boldsymbol{x}, t)\,\mathrm{d}V &= \int_V [\dot{\rho}(\boldsymbol{x}, t) + \rho(\boldsymbol{x}, t)\nabla_x \cdot \boldsymbol{v}]\,\mathrm{d}V \\ &= \int_V \left\{\frac{\partial}{\partial t}\rho(\boldsymbol{x}, t) + \nabla_x \cdot [\rho(\boldsymbol{x}, t)\boldsymbol{v}]\right\}\,\mathrm{d}V\end{aligned} \tag{60}$$

The local conservation law is

$$\frac{\mathrm{d}}{\mathrm{d}t}\rho(\boldsymbol{x}, t) + \rho(\boldsymbol{x}, t)\nabla_x \cdot \boldsymbol{v}(\boldsymbol{x}, t) = \frac{\partial}{\partial t}\rho(\boldsymbol{x}, t) + \nabla_x \cdot [\rho(\boldsymbol{x}, t)\boldsymbol{v}] = 0 \tag{61}$$

The equation

$$\frac{d\rho}{dt} + \rho \nabla_x \cdot \boldsymbol{v} = 0 \tag{62}$$

is the continuity equation, which can be given also as

$$\frac{\partial \rho}{\partial t} + \nabla_x \cdot (\rho \boldsymbol{v}) = 0 \tag{63}$$

From this equation can be made the following conclusions. If $\rho dV = \rho_0 dV_0$ one gets

$$\frac{d\rho_0}{dt} = \frac{1}{dV_0}(\rho dV)^{\cdot} = \frac{1}{dV_0}\left[\dot{\rho} dV + \rho (dV)^{\cdot}\right] = \frac{1}{dV_0}\left[\dot{\rho} + \rho \nabla_x \cdot \boldsymbol{v}\right] dV = 0$$

For $\rho_0 = \rho \det \boldsymbol{F}$ we can compute $(\rho \det \boldsymbol{F})^{\cdot} = 0$. $\dot{\rho} = 0$ results in div $\boldsymbol{v} = \nabla_x \cdot \boldsymbol{v} = 0$. In Eq. (61) the volume integral over $\nabla_x \cdot (\rho \boldsymbol{v})$ can be transformed into a surface integral. The global mass balance is in this case

$$\int\limits_V \frac{\partial \rho}{\partial t} \, dV + \int\limits_A \boldsymbol{n} \cdot (\rho \boldsymbol{v}) \, dA = 0 \tag{64}$$

3.2.1 Balance of Momentum

The momentum vector $\boldsymbol{p}$ of the body is defined by

$$\boldsymbol{p}(\boldsymbol{x}, t) = \int\limits_m \boldsymbol{v}(\boldsymbol{x}, t) \, dm = \int\limits_V \boldsymbol{v}(\boldsymbol{x}, t) \rho(\boldsymbol{x}, t) \, dV \tag{65}$$

The global balance of momentum is named first Euler-Cauchy law of motion and can be related to the second Newton's axiom specified for continua.

Theorem 1.3 (Balance of momentum) *The rate of changes of the momentum* $\boldsymbol{p}(\boldsymbol{x}, t)$ *during the deformation of the body is equal to the sum of all on the acting surface and volume forces.*

The balance of momentum in the Eulerian description is

$$\frac{d}{dt} \int\limits_V \boldsymbol{v}(\boldsymbol{x}, t) \rho(\boldsymbol{x}, t) \, dV = \int\limits_A \boldsymbol{t}(\boldsymbol{x}, \boldsymbol{n}, t) \, dA + \int\limits_V \boldsymbol{k}(\boldsymbol{x}, t) \rho(\boldsymbol{x}, t) \, dV \tag{66}$$

Equation (66) follows from the general balance Eq. (53) with $\boldsymbol{\Psi} = \boldsymbol{v}$, $\boldsymbol{\Phi} = \boldsymbol{T}$ and $\boldsymbol{\Xi} = \boldsymbol{k}$

$$\frac{\mathrm{d}}{\mathrm{d}t}\int_V \boldsymbol{v}\rho \,\mathrm{d}V = \int_A \boldsymbol{n}\cdot\boldsymbol{T}\,\mathrm{d}A + \int_V \boldsymbol{k}\rho\,\mathrm{d}V \tag{67}$$

In the reference configuration we have

$$\frac{\partial}{\partial t}\int_{V_0} \boldsymbol{v}(\boldsymbol{X},t)\rho_0(\boldsymbol{X})\,\mathrm{d}V_0 = \int_{A_0} {}^I\boldsymbol{t}(\boldsymbol{X},\boldsymbol{n}_0,t)\,\mathrm{d}A_0 + \int_{V_0} \boldsymbol{k}(\boldsymbol{X},t)\rho_0(\boldsymbol{X})\,\mathrm{d}V_0, \tag{68}$$

and with

$${}^I\boldsymbol{t} = \boldsymbol{n}_0 \cdot {}^I\boldsymbol{P}$$

one gets

$$\frac{\partial}{\partial t}\int_{V_0} \boldsymbol{v}(\boldsymbol{X},t)\rho_0(\boldsymbol{X})\,\mathrm{d}V_0 = \int_{A_0} \boldsymbol{n}_0(\boldsymbol{X})\cdot{}^I\boldsymbol{P}(\boldsymbol{X},t)\,\mathrm{d}A_0 + \int_{V_0} \boldsymbol{k}(\boldsymbol{X},t)\rho_0(\boldsymbol{X})\,\mathrm{d}V_0 \tag{69}$$

Applying the divergence theorem to Eqs. (67) and (69) we get

$$\frac{\mathrm{d}}{Dt}\int_V \boldsymbol{v}(\boldsymbol{x},t)\rho(\boldsymbol{x},t)\,\mathrm{d}V = \int_V \left[\nabla_x\cdot\boldsymbol{T}(\boldsymbol{x},t) + \boldsymbol{k}(\boldsymbol{x},t)\rho(\boldsymbol{x},t)\right]\mathrm{d}V, \tag{70}$$

$$\frac{\partial}{\partial t}\int_{V_0} \boldsymbol{v}(\boldsymbol{X},t)\rho_0(\boldsymbol{X})\,\mathrm{d}V_0 = \int_{V_0} [\nabla_X\cdot{}^I\boldsymbol{P}(\boldsymbol{X},t) + \boldsymbol{k}(\boldsymbol{X},t)\rho_0(\boldsymbol{X})]\,\mathrm{d}V_0 \tag{71}$$

The local formulations are

$$\nabla_x\cdot\boldsymbol{T}(\boldsymbol{x},t) + \rho(\boldsymbol{x},t)\boldsymbol{k}(\boldsymbol{x},t) = \rho(\boldsymbol{x},t)\frac{D\boldsymbol{v}(\boldsymbol{x},t)}{Dt}, \tag{72}$$

$$\nabla_X\cdot{}^I\boldsymbol{P}(\boldsymbol{X},t) + \rho_0(\boldsymbol{X})\boldsymbol{k}(\boldsymbol{X},t) = \rho_0(\boldsymbol{X})\frac{\partial\boldsymbol{v}(\boldsymbol{X},t)}{\partial t} \tag{73}$$

3.3 Balance of Moment of Momentum

Let us introduce the global vector of the moment of momentum

$$\boldsymbol{l}_O(\boldsymbol{x},t) = \int_V \boldsymbol{x}\times\rho(\boldsymbol{x},t)\boldsymbol{v}(\boldsymbol{x},t)\,\mathrm{d}V \tag{74}$$

The respective balance equations result in the second Euler-Cauchy equation of motion.

Theorem 1.4 (Balance of moment of momentum) *The rate of changes of the moment of momentum of a body $\boldsymbol{l}_O(\boldsymbol{x}, t)$ w.r.t. to the arbitrary point O is equal to the moments of all on the body acting surface and volume forces.*

The Eulerian balance of the moment of momentum is given by

$$\frac{\mathrm{d}}{\mathrm{d}t}\int\limits_V [\boldsymbol{x} \times \rho(\boldsymbol{x}, t)\boldsymbol{v}(\boldsymbol{x}, t)]\,\mathrm{d}V = \int\limits_V [\boldsymbol{x} \times \rho(\boldsymbol{x}, t)\boldsymbol{k}(\boldsymbol{x}, t)]\,\mathrm{d}V + \int\limits_A [\boldsymbol{x} \times \boldsymbol{t}(\boldsymbol{x}, \boldsymbol{n}, t)]\,\mathrm{d}A \tag{75}$$

With $\boldsymbol{t} = \boldsymbol{n} \cdot \boldsymbol{T}$ and $\boldsymbol{x} \times \boldsymbol{n} \cdot \boldsymbol{T} = -\boldsymbol{n} \cdot \boldsymbol{T} \times \boldsymbol{x}$ one gets

$$\frac{\mathrm{d}}{\mathrm{d}t}\int\limits_V \boldsymbol{x} \times \rho\boldsymbol{v}\,\mathrm{d}V = -\int\limits_A \boldsymbol{n} \cdot \boldsymbol{T} \times \boldsymbol{x}\,\mathrm{d}A + \int\limits_V \boldsymbol{x} \times \rho\boldsymbol{k}\,\mathrm{d}V \tag{76}$$

Equation (76) results from the general balance Eq. (53), if $\boldsymbol{\Psi} = (\boldsymbol{x} \times \boldsymbol{v})$, $\boldsymbol{\Phi} = -(\boldsymbol{T} \times \boldsymbol{x})$ and $\boldsymbol{\Xi} = (\boldsymbol{x} \times \boldsymbol{k})$. After some algebra it can be shown that for classical continua the balance of moment of momentum results in the symmetry condition for the Cauchy stress tensor. It can be shown that we have for the first and second Piola-Kirchhoff tensors different conclusions. The symmetry for the first Piola-Kirchhoff tensor cannot be established since we obtain

$${}^{I}\boldsymbol{P} \cdot \boldsymbol{F}^{\mathrm{T}} = \boldsymbol{F} \cdot {}^{I}\boldsymbol{P}^{\mathrm{T}}$$

3.4 Balance of Energy

Now we are prescribing two properties in each material point: the density ρ and the temperature θ. Both are non-negative ($\rho \geq 0, \theta \geq 0$). The foundation of any thermomechanical analysis is given by the first and the second law of thermodynamics. The first law is the balance of energy, the second one of entropy.

Within the framework of continuum thermodynamics we have to introduce at first suitable variables for the description of the macroscopic properties of the continuum. Let us assume macroscopic measurable, independent of each other parameters, which describe the state of the continuum in a unique manner. These parameters are named state variables. We have to distinguish extensive (additive) and intensive variables. The additive variables are proportional to the amount of continuum, expressed for example by the mass. The inner energy of the system is a typical extensive state variable. It depends only on the kinematical variables and the temperature: $\mathscr{U} = \mathscr{U}$ (kinematic variables, θ). If we divide a homogeneous system with the mass m into n homogeneous subsystems with the masses m_i the following statement is valid

$$\mathscr{U}_i = \left(\frac{m_i}{m}\right)\mathscr{U}, \quad i = 1, \ldots, n, \qquad \sum_{i=1}^{n} \mathscr{U}_i = \mathscr{U}, \quad \sum_{i=1}^{n} m_i = m \tag{77}$$

Intensive variables are independent of the amount of the continuum. If we divide now the system which is in an equilibrium state, into n subsystems, than the intensive state variable in each subsystem has the same value. Examples are the density and temperature.

Let us introduce the following limitations for the further developments:

- The continuum is assumed to be homogeneous that means in each material point we have the same properties.
- We assume that there is no mass exchange with the surrounding. The mass conservation is valid.
- We are assuming only mechanical and thermal actions.

Let us introduce the first law of thermodynamics.

Definition 1.22 (Thermomechanical balance of energy) The rate of changes of the total energy W within the volume is equal to the sum of the rate of the external heat supply Q and the power of all external forces P_a

$$\frac{\mathrm{d}}{\mathrm{d}t}\mathscr{W} = \mathscr{P}_\mathrm{a} + \mathscr{Q} \tag{78}$$

The total energy $\mathscr{W}$ consists of the inner energy $\mathscr{U}$ and the kinetic energy $\mathscr{K}$

$$\mathscr{W} = \mathscr{U} + \mathscr{K} \tag{79}$$

The kinetic energy is given as

$$\mathscr{K} = \frac{1}{2}\int\limits_V \boldsymbol{v} \cdot \boldsymbol{v}\rho \,\mathrm{d}V$$

The inner energy is an additive function of the mass

$$\mathscr{U} = \int\limits_m u \,\mathrm{d}m = \int\limits_V \rho u \,\mathrm{d}V$$

with u as the inner energy density. The power of the external forces $\mathscr{P}_\mathrm{a}$ is based on the introduced volume and surface forces

$$\mathscr{P}_\mathrm{a} = \int\limits_A \boldsymbol{t} \cdot \boldsymbol{v} \,\mathrm{d}A + \int\limits_V \boldsymbol{k} \cdot \boldsymbol{v}\rho \,\mathrm{d}V \tag{80}$$

The rate of heat supply consists of two parts: contribution of the heat sources r in the volume and the heat flux through A

$$\mathscr{Q} = \int\limits_V \rho r \,\mathrm{d}V - \int\limits_A \boldsymbol{n} \cdot \boldsymbol{h} \,\mathrm{d}A \tag{81}$$

$\boldsymbol{h}$ is the heat flux vector.

Now we have instead of Eq. (78)

$$\dot{\mathscr{U}} + \dot{\mathscr{K}} = \mathscr{P}_\mathrm{a} + \mathscr{Q} \tag{82}$$

or

$$\frac{\mathrm{d}}{\mathrm{d}t}\int\limits_V \left(u + \frac{1}{2}\boldsymbol{v}\cdot\boldsymbol{v}\right)\rho \,\mathrm{d}V = \int\limits_A \boldsymbol{t}\cdot\boldsymbol{v}\,\mathrm{d}A + \int\limits_V \boldsymbol{k}\cdot\boldsymbol{v}\rho\,\mathrm{d}V - \int\limits_A \boldsymbol{n}\cdot\boldsymbol{h}\,\mathrm{d}A + \int\limits_V r\rho\,\mathrm{d}V \tag{83}$$

Taking into account $\boldsymbol{t} = \boldsymbol{n}\cdot\boldsymbol{T}$ the first law of thermodynamics can be obtained from the general balance Eq. (53) with

$$\boldsymbol{\Psi} \to u + \frac{1}{2}\boldsymbol{v}\cdot\boldsymbol{v}, \quad \boldsymbol{\Phi} = \boldsymbol{T}\cdot\boldsymbol{v} - \boldsymbol{h}, \quad \Xi \to \boldsymbol{k}\cdot\boldsymbol{v} + r$$

In the reference configuration one gets

$$\begin{aligned}\frac{\partial}{\partial t}\int\limits_{V_0} \left(u + \frac{1}{2}\boldsymbol{v}\cdot\boldsymbol{v}\right)\rho_0 \,\mathrm{d}V_0 &= \int\limits_{A_0} {}^I\boldsymbol{t}\cdot\boldsymbol{v}\,\mathrm{d}A_0 + \int\limits_{V_0} \boldsymbol{k}\cdot\boldsymbol{v}\rho_0\,\mathrm{d}V_0 \\ &\quad - \int\limits_{A_0} \boldsymbol{n}_0\cdot\boldsymbol{h}_0\,\mathrm{d}A_0 + \int\limits_{V_0} r\rho_0\,\mathrm{d}V_0\end{aligned} \tag{84}$$

With respect to Eq. (34) for the relationship between $\boldsymbol{t}$ and ${}^I\boldsymbol{t}$ we have the relationship between $\boldsymbol{h}$ and $\boldsymbol{h}_0$

$$\boldsymbol{h}_0 = (\det\boldsymbol{F})\boldsymbol{F}^{-1}\cdot\boldsymbol{h}, \quad \boldsymbol{h} = (\det\boldsymbol{F})^{-1}\boldsymbol{F}\cdot\boldsymbol{h}_0 \tag{85}$$

With

$$\frac{\mathrm{d}}{\mathrm{d}t}\left(\frac{1}{2}\boldsymbol{v}\cdot\boldsymbol{v}\right) = \frac{1}{2}\dot{\boldsymbol{v}}\cdot\boldsymbol{v} + \frac{1}{2}\boldsymbol{v}\cdot\dot{\boldsymbol{v}} = \dot{\boldsymbol{v}}\cdot\boldsymbol{v},$$

$$\int\limits_A \boldsymbol{n}\cdot(\boldsymbol{T}\cdot\boldsymbol{v} - \boldsymbol{h})\,\mathrm{d}A = \int\limits_V [\nabla_{\boldsymbol{x}}\cdot(\boldsymbol{T}\cdot\boldsymbol{v}) - \nabla_{\boldsymbol{x}}\cdot\boldsymbol{h}]\,\mathrm{d}V$$

$$\nabla_{\boldsymbol{x}}\cdot(\boldsymbol{T}\cdot\boldsymbol{v}) = (\nabla_{\boldsymbol{x}}\cdot\boldsymbol{T})\cdot\boldsymbol{v} + \boldsymbol{T}\cdot\cdot(\nabla_{\boldsymbol{x}}\boldsymbol{v})^{\mathrm{T}} = (\nabla_{\boldsymbol{x}}\cdot\boldsymbol{T})\cdot\boldsymbol{v} + \boldsymbol{T}\cdot\cdot\boldsymbol{D}$$

from Eq. (83) it follows

$$\int\limits_V \left(\frac{\mathrm{d}u}{\mathrm{d}t} + \underline{\dot{\boldsymbol{v}} \cdot \boldsymbol{v}}\right) \rho \,\mathrm{d}V = \int\limits_V (\boldsymbol{T} \cdot\cdot \boldsymbol{D} - \nabla_x \cdot \boldsymbol{h} + \rho r) \,\mathrm{d}V + \int\limits_V \underline{[(\nabla_x \cdot \boldsymbol{T}) \cdot \boldsymbol{v} + \rho \boldsymbol{k} \cdot \boldsymbol{v}]} \,\mathrm{d}V \qquad (86)$$

The underlined terms are the balance of momentum and we can simplify Eq. (86)

$$\int\limits_V (\rho \dot{u} - \boldsymbol{T} \cdot\cdot \boldsymbol{D} + \nabla_x \cdot \boldsymbol{h} - \rho r) \,\mathrm{d}V = 0 \qquad (87)$$

The local form holds

$$\rho \dot{u} = \boldsymbol{T} \cdot\cdot \boldsymbol{D} - \nabla_x \cdot \boldsymbol{h} + \rho r \qquad (88)$$

The balance of energy in the reference configuration is

$$\int\limits_{V_0} \rho_0 \frac{\partial u}{\partial t} \,\mathrm{d}V_0 = \int\limits_{V_0} ({}^{II}\boldsymbol{P} \cdot\cdot \dot{\boldsymbol{G}} - \nabla_X \cdot \boldsymbol{h}_0 + \rho_0 r) \,\mathrm{d}V_0 \qquad (89)$$

or in the local form

$$\rho_0 \frac{\partial u}{\partial t} = {}^{II}\boldsymbol{P} \cdot\cdot \dot{\boldsymbol{G}} - \nabla_X \cdot \boldsymbol{h}_0 + \rho_0 r \qquad (90)$$

In Eq. (90) the conjugated pair $({}^{II}\boldsymbol{P}, \dot{\boldsymbol{G}})$ can be substituted by $({}^{I}\boldsymbol{P}, \dot{\boldsymbol{F}})$.

3.5 Balance of Entropy

One of the possible formulations of the second law of thermodynamics is the following.

Theorem 1.5 (Balance of entropy) *The rate of changes of the entropy S within the volume is not less than the rate of the external entropy supply.*

The entropy is an additive function that means

$$\mathscr{S} = \int\limits_m s \,\mathrm{d}m = \int\limits_V \rho s \,\mathrm{d}V \qquad (91)$$

with s as the inner entropy density. Now the second law of thermodynamics can be stated in the global form

$$\frac{\mathrm{d}}{\mathrm{d}t}\int\limits_V \rho s\,\mathrm{d}V \geq \int\limits_V \frac{r}{\theta}\rho\,\mathrm{d}V - \int\limits_A \frac{\boldsymbol{n}\cdot\boldsymbol{h}}{\theta}\,\mathrm{d}A \tag{92}$$

For all real processes (92) is an inequality (>) or with other words real processes are always irreversible.

With the transform

$$\int\limits_A \frac{\boldsymbol{n}\cdot\boldsymbol{h}}{\theta}\,\mathrm{d}A = \int\limits_V \boldsymbol{\nabla}_{\boldsymbol{x}}\cdot\left(\frac{\boldsymbol{h}}{\theta}\right)\mathrm{d}V = \int\limits_V \left(\frac{\boldsymbol{\nabla}_{\boldsymbol{x}}\cdot\boldsymbol{h}}{\theta} - \frac{\boldsymbol{h}\cdot\boldsymbol{\nabla}_{\boldsymbol{x}}\theta}{\theta^2}\right)\mathrm{d}V, \tag{93}$$

we get the local formulation

$$\rho\theta\dot{s} \geq \rho r - \boldsymbol{\nabla}_{\boldsymbol{x}}\cdot\boldsymbol{h} + \frac{1}{\theta}\boldsymbol{h}\cdot\boldsymbol{\nabla}_{\boldsymbol{x}}\theta \tag{94}$$

or

$$\rho\theta\dot{s} - (\rho r - \boldsymbol{\nabla}_{\boldsymbol{x}}\cdot\boldsymbol{h}) - \boldsymbol{h}\cdot\boldsymbol{\nabla}_{\boldsymbol{x}}\ln\theta \geq 0 \tag{95}$$

The term in brackets can be substituted by Eq. (88)

$$\rho\theta\dot{s} + \boldsymbol{T}\cdot\cdot\boldsymbol{D} - \rho\dot{u} - \boldsymbol{h}\cdot\boldsymbol{\nabla}_{\boldsymbol{x}}\ln\theta \geq 0 \tag{96}$$

and with

$$\rho\theta\dot{s} = \rho(\theta s)^{\cdot} - \rho s\dot{\theta}$$

it follows

$$\rho\frac{\mathrm{d}}{\mathrm{d}t}(\theta s - u) - \rho s\frac{\mathrm{d}\theta}{\mathrm{d}t} + \boldsymbol{T}\cdot\cdot\boldsymbol{D} - \boldsymbol{h}\cdot\boldsymbol{\nabla}_{\boldsymbol{x}}\ln\theta \geq 0 \tag{97}$$

The term

$$(u - \theta s) = f \tag{98}$$

is the Helmholtz free energy.

All entropy balances can be presented in the reference configuration, for example the global balance

$$\frac{\partial}{\partial t}\int\limits_{V_0} \rho_0 s\,\mathrm{d}V_0 \geq \int\limits_{V_0} \frac{r}{\theta}\rho_0\,\mathrm{d}V_0 - \int\limits_{A_0} \frac{\boldsymbol{n}_0\cdot\boldsymbol{h}_0}{\theta}\,\mathrm{d}A_0$$

or the local one

$$\rho_0\theta\frac{\partial s}{\partial t} \geq (\rho_0 r - \nabla_{\boldsymbol{X}}\cdot\boldsymbol{h}_0) + \frac{1}{\theta}\boldsymbol{h}_0\cdot\nabla_{\boldsymbol{X}}\theta$$

4 Constitutive Modeling

Equations describing the specific behavior of the continua are named constitutive, physical or state equations. Let us introduce according to [20] the following definition.

Definition 1.23 (Constitutive equations) Constitutive equations link all phenomenological variables describing the macroscopic behavior of the continuum.

Among such variables are: stresses, strains, temperature, heat flux, etc.

The number of necessary constitutive equations depends on the problem. In the case of thermomechanical problems we have the following balance equations: mass (1 scalar equation), momentum (1 vectorial equation or for three-dimensional problems 3 scalar equations), moment of momentum (1 vectorial or 3 scalar equations), energy (1 scalar equation) and entropy (1 scalar inequality). Since the entropy inequality yields only in the determination of the process direction we have only 8 scalar equations to estimate 19 variables: density ρ (1 scalar variable), displacements $\boldsymbol{u}$ or velocities $\boldsymbol{v}$ (1 vector or its 3 coordinates), stress tensor $\boldsymbol{T}$ (1 second-order tensor or 9 coordinates, inner energy u (1 variable), entropy s (1 variable), temperature θ (1 variable) and heat flux $\boldsymbol{h}$ (1 vector or 3 variables). In this case we have to add 11 constitutive equations otherwise the the system of governing equations is underestimated and we cannot solve problems.

Before we start the formulation of constitutive equations let us introduce some restrictions and definitions. We limit ourselves by the consideration of classical continua. Other continua are discussed in the literature, for example, micropolar continua are presented in [4]. In addition, we assume that we have only simple materials of first order [21].

Definition 1.24 (Simple material of first order) A simple material of first order defined by constitutive equations which contains only local variables, for example the local strain tensor and the locale heat flux vector with the local stress tensor and the local temperature gradient. All conclusions are made for the same material point and its differential surrounding of first order.

Since the constitutive equations present the individual response of the material universal constitutive equations cannot be established [22]. That means we are discussing always special cases. The constitutive equations can be formulated using the top-down approach, the bottom-up approach or rheological models. Here we present only the top-down approach that we start with some mathematical and physical statements. At first we assume that the balances are valid. The mathematical structure of the constitutive equations can be established with the axioms of the theory of materials.

Let us start with ordering the variables which are included in the set of constitutive equations. We assume that all variables depend on the position ($\boldsymbol{x}$ or $\boldsymbol{X}$) and the time t. The material behavior can be presented by functional relationships between the constitutive variables and the constitutive parameters. For example, within the framework of thermomechanics the temperature θ is a constitutive parameter and the heat flux vector $\boldsymbol{h}$ is a constitutive variable. The choice of the constitutive parameters and the constitutive variables is arbitrary.

If we want to describe the processes realized in the material points of the continuum history of changes of the constitutive parameters should be given.

Definition 1.25 (Process) The time changes of the constitutive parameters in the material points is named process.

Definition 1.26 (Constitutive variable) The behavior of the continuum in each material point can be presented by a set of constitutive variables. They can be time operators of the processes in the points.

The respective functional relationships are the constitutive equations. Below we discuss only constitutive equations for solids. With respect to the second axiom of rheology [23, 24] all real bodies show solid and fluid properties. So we consider that the solid behavior is dominant.

Definition 1.27 (Solid) The acting on the body loads results in the stress deviator to non-zero elements that means it resists to the changes of the shape of the body.

In this sense we distinguish time-independent and time-dependent material behavior [1]. Elastic and plastic materials belong to the first group, viscoelastic and viscoplastic materials—to the second one.

4.1 Basics of Material Theory

Let us discuss some basic tools in material modeling based on the top-down approach. As usual one should focus on the attention to the following three questions [1]:

- formulation of constitutive equations,
- consideration of symmetries of the material behavior and
- consideration of kinematic constraints.

The last one is important w.r.t. models for compressible and incompressible materials.

The systematical deduction of constitutive equations can be realized with the help of some principal axioms (constitutive principles). The main axioms are:

- causality,
- determinism,
- equipresence,
- materielle objectivity,
- locale action,

- memory and
- physical consistency.

A detailed discussion of the physical meaning and the mathematical consequences for the aforementioned axiom is given in [25].

4.2 General Constitutive Equations of Thermo-Mechanical Materials

The thermodynamical state of the continuum is defined by the motion $\boldsymbol{x} = \boldsymbol{x}(\boldsymbol{X}, t)$ and the temperature $\theta = \theta(\boldsymbol{X}, t)$ of the material points $\boldsymbol{X}$ at the instant time t. $\boldsymbol{x}$ and θ are independent variables. As dependent variables (constitutive variables) we postulate the stress tensor, the heat flux vector, the free energy and the entropy.

For general models of the material behavior we should assume that the actual state not only depends on the actual loading, but also from the loading history $t_0 < \tau \leq t$. In addition, the behavior of the given material point $\boldsymbol{X}$ also depends on the behavior of all other points of the body $\tilde{\boldsymbol{X}}$. If we assume for the functions $\boldsymbol{x}(\tilde{\boldsymbol{X}}, \tau)$ and $\theta(\tilde{\boldsymbol{X}}, \tau)$ the continuity w.r.t. $\tilde{\boldsymbol{X}}$ and τ, we can represent the behavior by Taylor series for the points $\boldsymbol{X}$ by powers of $(\tilde{\boldsymbol{X}} - \boldsymbol{X})$ and for the time τ by powers $(\tau - t)$. Applying the axiom of local action and memory axiom (in special situations fading memory the power series for $\boldsymbol{x}(\tilde{\boldsymbol{X}}, \tau)$ and $\theta(\tilde{\boldsymbol{X}}, \tau)$ can be limited by the first derivative with respect to $\tilde{\boldsymbol{X}}$ and τ, respectively. The constitutive variables depend in this case on $\boldsymbol{X}$ and θ, but also from $\nabla_X \boldsymbol{x}$, $\nabla_X \theta$ and $\dot{\theta}$. Other models take into account second gradients (see [26–29] among others).

Assuming the axiom of material objectivity there is no explicit dependency of $\boldsymbol{x}$ or $\dot{\boldsymbol{x}}$ since only the strains or the strain rates and not the rigid body motions define the material behavior. For many real materials the gradients $\nabla_X \boldsymbol{x}$ and $\nabla_X \theta$ are influenced by the loading history. Taking into account the first derivatives only in this case $\nabla_X \dot{\boldsymbol{x}}$ and $\nabla_X \dot{\theta}$ act as constitutive variables.

The constitutive equations of a simple thermomechanical material take the form

$$\begin{aligned}
\boldsymbol{P}(\boldsymbol{X}, t) &= \boldsymbol{P}\left\{\boldsymbol{X}, \theta(\boldsymbol{X}, t), \dot{\theta}(\boldsymbol{X}, t), \nabla_X \theta(\boldsymbol{X}, t), \nabla_X \dot{\theta}(\boldsymbol{X}, t), \boldsymbol{\Gamma}(\boldsymbol{X}, t)\right\}, \\
\boldsymbol{h}_0(\boldsymbol{X}, t) &= \boldsymbol{h}_0\left\{\boldsymbol{X}, \theta(\boldsymbol{X}, t), \dot{\theta}(\boldsymbol{X}, t), \nabla_X \theta(\boldsymbol{X}, t), \nabla_X \dot{\theta}(\boldsymbol{X}, t), \boldsymbol{\Gamma}(\boldsymbol{X}, t)\right\}, \\
f(\boldsymbol{X}, t) &= f\left\{\boldsymbol{X}, \theta(\boldsymbol{X}, t), \dot{\theta}(\boldsymbol{X}, t), \nabla_X \theta(\boldsymbol{X}, t), \nabla_X \dot{\theta}(\boldsymbol{X}, t), \boldsymbol{\Gamma}(\boldsymbol{X}, t)\right\}, \\
s(\boldsymbol{X}, t) &= s\left\{\boldsymbol{X}, \theta(\boldsymbol{X}, t), \dot{\theta}(\boldsymbol{X}, t), \nabla_X \theta(\boldsymbol{X}, t), \nabla_X \dot{\theta}(\boldsymbol{X}, t), \boldsymbol{\Gamma}(\boldsymbol{X}, t)\right\}
\end{aligned} \tag{99}$$

The set $\boldsymbol{\Gamma}$ includes all mechanical variables $\nabla_X \boldsymbol{x}(\boldsymbol{X}, t)$, $\nabla_X \dot{\boldsymbol{x}}(\boldsymbol{X}, t)$ characterizing the deformation. The stress tensor $\boldsymbol{P}$ can be the first or the second Piola-Kirchhoff tensor. The explicit dependency of the constitutive equations on $\boldsymbol{X}$ is equivalent to the statement that in each point of the body one can have different kinds of material behavior otherwise the body is homogeneous. In general the constitutive equations are functionals (operators). This is shown by the symbol $\{\ldots\}$. If the prehistory has no influence on the actual material behavior, the constitutive equations are functions and the symbol (...) is used.

The axiom of physical consistency axiom allows a further specification of Eq. (99). In the case of simple thermoviscoelastic materials (the time dependency is related to the initial state, but not to the prehistory) the elastic strains and the strain rates can be expressed by $\boldsymbol{C}, \dot{\boldsymbol{C}}$. If an independent relation to parameters $\rho, \dot{\rho}$ cannot be assumed, these parameters can be ignored and we have for simple thermoviscoelastic solids (no influence of the prehistory)

$$\begin{aligned}
\boldsymbol{P}(\boldsymbol{X},t) &= \boldsymbol{P}\,\left\{\boldsymbol{X},\theta,\dot{\theta},\nabla_{\boldsymbol{X}}\theta,\nabla_{\boldsymbol{X}}\dot{\theta},\boldsymbol{C},\dot{\boldsymbol{C}}\right\},\\
\boldsymbol{h}_0(\boldsymbol{X},t) &= \boldsymbol{h}\,\left\{\boldsymbol{X},\theta,\dot{\theta},\nabla_{\boldsymbol{X}}\theta,\nabla_{\boldsymbol{X}}\dot{\theta},\boldsymbol{C},\dot{\boldsymbol{C}}\right\},\\
f(\boldsymbol{X},t) &= f\,\left\{\boldsymbol{X},\theta,\dot{\theta},\nabla_{\boldsymbol{X}}\theta,\nabla_{\boldsymbol{X}}\dot{\theta},\boldsymbol{C},\dot{\boldsymbol{C}}\right\},\\
s(\boldsymbol{X},t) &= s\,\left\{\boldsymbol{X},\theta,\dot{\theta},\nabla_{\boldsymbol{X}}\theta,\nabla_{\boldsymbol{X}}\dot{\theta},\boldsymbol{C},\dot{\boldsymbol{C}}\right\}
\end{aligned} \tag{100}$$

For pure thermoelastische solids the time derivatives can be ignored

$$\begin{aligned}
\boldsymbol{P}(\boldsymbol{X},t) &= \boldsymbol{P}\,\{\boldsymbol{X},\theta,\nabla_{\boldsymbol{X}}\theta,\boldsymbol{C}\},\\
\boldsymbol{h}_0(\boldsymbol{X},t) &= \boldsymbol{h}\,\{\boldsymbol{X},\theta,\nabla_{\boldsymbol{X}}\theta,\boldsymbol{C}\},\\
f(\boldsymbol{X},t) &= f\,\{\boldsymbol{X},\theta,\nabla_{\boldsymbol{X}}\theta,\boldsymbol{C}\},\\
s(\boldsymbol{X},t) &= s\,\{\boldsymbol{X},\theta,\nabla_{\boldsymbol{X}}\theta,\boldsymbol{C}\}
\end{aligned} \tag{101}$$

4.3 Elastic Simple Material

As a first example let us discuss the ideal-elastic material behavior. In this case we have constitutive equations which are functions. In the pure mechanical case one has to present a relation between the stress and the strain tensors. Considering the material objectivity in this relation the following equations are valid

$$^{II}\boldsymbol{P}(\boldsymbol{X},t) = \boldsymbol{f}(\boldsymbol{C},\boldsymbol{X},t) \qquad \text{or} \qquad ^{II}\boldsymbol{P}(\boldsymbol{X},t) = \boldsymbol{g}(\boldsymbol{G},\boldsymbol{X},t)$$

Here the assumption of simple material is included since only the gradient of $\boldsymbol{x}(\boldsymbol{X},t)$ (deformation gradient tensor $\boldsymbol{F}(\boldsymbol{X},t)$) is taken into account.

The specific elementary work in the reference configuration is the starting point for the further discussions. The constitutive Eq. (101) have a very simple form

$$^{I}\boldsymbol{P}(\boldsymbol{X},t) = {}^{I}\boldsymbol{P}(\boldsymbol{F}),$$

and the variation of the elementary work W_i gives

$$\delta W_i = \frac{1}{\rho_0}\,{}^{I}\boldsymbol{P}\cdot\cdot\,\delta\boldsymbol{F}^{\mathrm{T}}$$

The work depends on the strains at current time t. This work is stored as the strain energy of the body

$$\delta W_i = \delta \mathscr{U} = \frac{1}{\rho_0}\, {}^I\boldsymbol{P} \cdot\cdot\, \delta \boldsymbol{F}$$

with $u = u(\boldsymbol{F})$ as the specific strain energy density function. This function should not be influenced by rigid motions of the body (material objectivity axiom). From this it follows for arbitrary rotations $\boldsymbol{Q}$

$$u(\boldsymbol{F}) = u(\boldsymbol{Q}\cdot\boldsymbol{F}) = u\left(\sqrt{(\boldsymbol{Q}\cdot\boldsymbol{F})^{\mathrm{T}}\cdot(\boldsymbol{Q}\cdot\boldsymbol{F})}\right) = u\left(\sqrt{\boldsymbol{F}^{\mathrm{T}}\cdot\boldsymbol{F}}\right) = u(\boldsymbol{U}) \qquad (102)$$

With $\boldsymbol{U}^2 = \boldsymbol{C}$ and $\boldsymbol{G} = \frac{1}{2}(\boldsymbol{C} - \boldsymbol{I})$ we have $u(\boldsymbol{U}) = \hat{u}(\boldsymbol{C})$ or $u(\boldsymbol{U}) = \check{u}(\boldsymbol{G})$. From

$$\delta u(\boldsymbol{F}) = \frac{\partial u(\boldsymbol{F})}{\partial \boldsymbol{F}} \cdot\cdot\, \delta \boldsymbol{F}^{\mathrm{T}} = \left[\frac{\partial u(\boldsymbol{F})}{\partial \boldsymbol{F}}\right] \cdot\cdot\, \delta \boldsymbol{F} = \left[u(\boldsymbol{F})_{,\boldsymbol{F}}\right]^{\mathrm{T}} \cdot\cdot\, \delta \boldsymbol{F}$$

one gets

$$\frac{1}{\rho_0}\, {}^I\boldsymbol{P} = \left[\frac{\partial u(\boldsymbol{F})}{\partial \boldsymbol{F}}\right] = \left[\frac{\partial \hat{u}(\boldsymbol{C})}{\partial \boldsymbol{F}}\right] \qquad (103)$$

or after some manipulations

$${}^I\boldsymbol{P} = 2\rho_0 \boldsymbol{F} \cdot \frac{\partial \hat{u}(\boldsymbol{C})}{\partial \boldsymbol{C}} \qquad (104)$$

With respect to the transform rules

$$\boldsymbol{T} = (\det \boldsymbol{F})^{-1}\, {}^I\boldsymbol{P} \cdot \boldsymbol{F}^{\mathrm{T}}, \qquad {}^{II}\boldsymbol{P} = \boldsymbol{F}^{-1} \cdot {}^I\boldsymbol{P}$$

we can formulate the constitutive equation for the Cauchy stress tensor $\boldsymbol{T}$ and the second Piola-Kirchhoff stress tensor ${}^{II}\boldsymbol{P}$

$$\boldsymbol{T} = 2\rho \boldsymbol{F} \cdot \frac{\partial \hat{u}}{\partial \boldsymbol{C}} \cdot \boldsymbol{F}^{\mathrm{T}},$$

$${}^{II}\boldsymbol{P} = 2\rho_0 \frac{\partial \hat{u}(\boldsymbol{C})}{\partial \boldsymbol{C}} = \boldsymbol{f}(\boldsymbol{C}) \qquad \text{or} \qquad {}^{II}\boldsymbol{P} = 2\rho_0 \frac{\partial \check{u}(\boldsymbol{G})}{\partial \boldsymbol{G}} = \boldsymbol{g}(\boldsymbol{G})$$

This is the general constitutive equation in the non-linear elasticity assuming large deformations and isothermal behavior. If we have special cases of anisotropy than further simplifications are possible. In the simplest case (isotropy) the energy density function $u = u(\boldsymbol{C})$ depends only on the invariants of the tensor $\boldsymbol{C}$

$$u = \hat{u}(\boldsymbol{C}) = \hat{u}[I_1(\boldsymbol{C}), I_2(\boldsymbol{C}), I_3(\boldsymbol{C})]$$

At first we consider the chain rule

$$\frac{\partial \hat{u}(\boldsymbol{C})}{\partial \boldsymbol{C}} = \frac{\partial \hat{u}}{\partial I_1}\frac{\partial I_1}{\partial \boldsymbol{C}} + \frac{\partial \hat{u}}{\partial I_2}\frac{\partial I_2}{\partial \boldsymbol{C}} + \frac{\partial \hat{u}}{\partial I_3}\frac{\partial I_3}{\partial \boldsymbol{C}}$$

With

$$\begin{aligned}
I_1(\boldsymbol{C}) &= \operatorname{tr}\boldsymbol{C}, \\
I_2(\boldsymbol{C}) &= \frac{1}{2}\left[I_1^2(\boldsymbol{C}) - I_1(\boldsymbol{C}^2)\right], \\
I_3(\boldsymbol{C}) &= \frac{1}{3}\left[I_1(\boldsymbol{C}^3) + 3I_1(\boldsymbol{C})I_2(\boldsymbol{C}) - I_1^3(\boldsymbol{C})\right]
\end{aligned}$$

one gets

$$\begin{aligned}
\frac{\partial I_1}{\partial \boldsymbol{C}} &= \boldsymbol{I}, \\
\frac{\partial I_2}{\partial \boldsymbol{C}} &= I_1\boldsymbol{I} - \boldsymbol{C}, \\
\frac{\partial I_3}{\partial \boldsymbol{C}} &= \boldsymbol{C}^2 + \boldsymbol{I}I_2(\boldsymbol{C}) + I_1(\boldsymbol{C})[I_1(\boldsymbol{C})\boldsymbol{I} - \boldsymbol{C}] - I_1^2(\boldsymbol{C})\boldsymbol{C}
\end{aligned}$$

and

$$\begin{aligned}
\hat{u}_{,\boldsymbol{C}} &= \left(\frac{\partial \hat{u}}{\partial I_1} + I_1\frac{\partial \hat{u}}{\partial I_2} + I_2\frac{\partial \hat{u}}{\partial I_3}\right)\boldsymbol{I} - \left(\frac{\partial \hat{u}}{\partial I_2} + I_1\frac{\partial \hat{u}}{\partial I_3}\right)\boldsymbol{C} + \frac{\partial \hat{u}}{\partial I_3}\boldsymbol{C}^2 \\
&= \phi_0\boldsymbol{I} + \phi_1\boldsymbol{C} + \phi_2\boldsymbol{C}^2,
\end{aligned}$$

Here $\phi_i = \phi_i(I_1, I_2, I_3)$. Finally we obtain

$${}^{I}\boldsymbol{P} = 2\rho_0\boldsymbol{F}\cdot(\phi_0\boldsymbol{I} + \phi_1\boldsymbol{C} + \phi_2\boldsymbol{C}^2)$$

For an arbitrary isotropic tensor function $\boldsymbol{f}(\boldsymbol{A})$ one gets in the case of orthogonal tensors $\boldsymbol{Q}$

$$\boldsymbol{Q}\cdot\boldsymbol{f}(A)\cdot\boldsymbol{Q}^{\mathrm{T}} = \boldsymbol{f}(\boldsymbol{Q}\cdot A\cdot\boldsymbol{Q}^{\mathrm{T}})$$

and the following representation

$$\boldsymbol{f}(\boldsymbol{A}) = \phi_0\boldsymbol{I} + \phi_1\boldsymbol{A} + \phi_2\boldsymbol{A}^2$$

Then for the isotropic elastic body we can write the constitutive equation

$${}^{II}\boldsymbol{P} = \psi_0\boldsymbol{I} + \psi_1\boldsymbol{G} + \psi_2\boldsymbol{G}^2,$$

where ψ_i is now a function of the invariants of $\boldsymbol{G}$.

Introducing kinematic restrictions another special case can be deduced. For example, for the incompressibility (isochoric motions) the following equations are valid

$$\det \boldsymbol{F} = 1 \qquad \det \boldsymbol{C} = 1 \qquad \text{or} \qquad \sqrt{\det(2\boldsymbol{G} - \boldsymbol{I})} - 1 = 0,$$

The kinematic constraint can be expressed as follows

$$\lambda(\boldsymbol{C}) = \det \boldsymbol{C} - 1 = 0$$

That means $I_3(\boldsymbol{C}) = 1$ and instead of $\hat{u} = \hat{u}(I_1, I_2, I_3)$ we have $\hat{u} = \hat{u}(I_1, I_2)$ or

$$\frac{\partial \hat{u}}{\partial I_3} = 0$$

With the help of the method of Lagrangian multipliers one gets the constitutive equation

$${}^{I}\boldsymbol{P} = 2\rho_0 \boldsymbol{F} \cdot \left[\left(\frac{\partial \hat{u}}{\partial I_1} + I_1 \frac{\partial \hat{u}}{\partial I_2} \right) \boldsymbol{I} - \frac{\partial \hat{u}}{\partial I_2} \boldsymbol{C} \right] - p\boldsymbol{F}^{-1},$$

where p is an a priori unknown function (called hydrostatic pressure), which can be estimated from the equilibrium/motion equation.

4.4 Models with Internal Variables

Dissipative effects can be modeled with the help of different concepts and we have various possibilities to formulate the constitutive equations. One possibility is the introduction of viscose stresses depending on the strain rates. Another possibility is given using the fading memory principle. Here we apply a third approach: at first some inner variables are postulated, which have an influence on the free energy. In addition, for the inner variable we need evolution equations. Examples of inner variables are creep strains, plastic strains or damage variables.

The starting point are the balances and the constitutive equations for homogeneous materials. The last one depends now additionally on $\boldsymbol{\Upsilon}_i(\boldsymbol{X}, t)$ $(i = 1, \ldots, n)$—the inner variables. The inner variables are tensor-valued variables of different order. For example, in the case of isotropic damage a scalar is used, for isotropic hardening—scalar, kinematic hardening-second-order tensor or anisotropic damage-fourth-order tensor. The evolution equations can be postulated in the following form

$$\frac{D\boldsymbol{\Upsilon}_i}{Dt} = \boldsymbol{Y}_i(\theta, \nabla_{\boldsymbol{x}}\theta, \boldsymbol{g}, \boldsymbol{\Upsilon}_1, \ldots, \boldsymbol{\Upsilon}_n) \qquad (105)$$

Finally, dissipative materials can be presented by the following set of constitutive and evolution equations

$$
\begin{aligned}
{}^{I}\boldsymbol{P}(\boldsymbol{X},t) &= {}^{I}\boldsymbol{P}\ (\theta, \nabla_{x}\theta, \boldsymbol{g}, \boldsymbol{\Upsilon}_i),\\
\boldsymbol{h}_0(\boldsymbol{X},t) &= \boldsymbol{h}_0\ (\theta, \nabla_{x}\theta, \boldsymbol{g}, \boldsymbol{\Upsilon}_i),\\
f(\boldsymbol{X},t) &= f\ (\theta, \nabla_{x}\theta, \boldsymbol{g}, \boldsymbol{\Upsilon}_i),\\
s(\boldsymbol{X},t) &= s\ (\theta, \nabla_{x}\theta, \boldsymbol{g}, \boldsymbol{\Upsilon}_i),\\
\dot{\boldsymbol{\Upsilon}}_i(\boldsymbol{X},t) &= \boldsymbol{Y}_i\ (\theta, \nabla_{x}\theta, \boldsymbol{g}, \boldsymbol{\Upsilon}_1, \ldots, \boldsymbol{\Upsilon}_n)
\end{aligned}
\tag{106}
$$

They should be completed by initial conditions

$$
\boldsymbol{\Upsilon}_i(\boldsymbol{X},t_0) = \boldsymbol{\Upsilon}_i^0(\boldsymbol{X}) \tag{107}
$$

The problems of the formulation of constitutive equations incorporating inner variables and large deformations are widely discussed in the literature, see [30] among others. Assuming geometrical linearity it will be much more simpler. Now we present the strains by $\boldsymbol{\varepsilon}$ (Cauchy or small strain tensor) and the stresses by $\boldsymbol{\sigma}$. It is not necessary to distinguish two configurations. In addition, we assume that in analogy to Sect. 4.3 the heat flux vector is given by the anisotropic Fourier's law. For the other constitutive equations we suggest the independence of the temperature gradient. Finally we have the constitutive and evolution Eq. (106)

$$
\begin{aligned}
\boldsymbol{\sigma} &= \boldsymbol{\sigma}(\theta, \boldsymbol{\varepsilon}, \boldsymbol{\Upsilon}_i),\\
\boldsymbol{h} &= -\boldsymbol{\kappa}\cdot\nabla\theta,\\
f &= f(\theta, \boldsymbol{\varepsilon}, \boldsymbol{\Upsilon}_i),\\
s &= s(\theta, \boldsymbol{\varepsilon}, \boldsymbol{\Upsilon}_i),\\
\dot{\boldsymbol{\Upsilon}}_i &= \boldsymbol{Y}_i(\theta, \nabla_X\theta, \boldsymbol{\varepsilon}, \boldsymbol{\Upsilon}_i)
\end{aligned}
\tag{108}
$$

The analysis will be performed as shown in Sect. 4.3. The staring point is the free energy f. Note that the strains consist of an elastic and an inelastic part

$$
\boldsymbol{\varepsilon} = \boldsymbol{\varepsilon}^{\mathrm{el}} + \boldsymbol{\varepsilon}^{\mathrm{inel}} = \boldsymbol{\varepsilon}^{\mathrm{el}} + \boldsymbol{\varepsilon}^{\mathrm{pl}} \tag{109}
$$

with $\boldsymbol{\varepsilon}^{\mathrm{el}}$ as thermoelastic strains, $\boldsymbol{\varepsilon}^{\mathrm{inel}}$ as inelastic strains and $\boldsymbol{\varepsilon}^{\mathrm{pl}}$ as plastic strains.

For the free energy w.r.t. Eq. (109) we can assume

$$
f = f(\theta, \boldsymbol{\varepsilon}, \boldsymbol{\varepsilon}^{\mathrm{el}}, \boldsymbol{\varepsilon}^{\mathrm{pl}}, \boldsymbol{\Upsilon}_i) \tag{110}
$$

The total strains $\boldsymbol{\varepsilon}$, the elastic strains $\boldsymbol{\varepsilon}^{\mathrm{el}}$ and the plastic strains $\boldsymbol{\varepsilon}^{\mathrm{pl}}$ are connected by one equation. Following [31] the form of the free energy considering the decomposition of the total strains can be obtained

$$
f = f(\theta, \boldsymbol{\varepsilon} - \boldsymbol{\varepsilon}^{\mathrm{pl}}, \boldsymbol{\Upsilon}_i) = f(\theta, \boldsymbol{\varepsilon}^{\mathrm{el}}, \boldsymbol{\Upsilon}_i)
$$

The derivative of the free energy is

$$
\dot{f} = \frac{\partial f}{\partial \boldsymbol{\varepsilon}^{\mathrm{el}}} \cdot\cdot\, \dot{\boldsymbol{\varepsilon}}^{\mathrm{el}} + \frac{\partial f}{\partial \theta}\dot{\theta} + \frac{\partial f}{\partial \boldsymbol{\Upsilon}_i} \odot \dot{\boldsymbol{\Upsilon}}_i
$$

and finally we get the following dissipative inequality

$$\underline{\left(\boldsymbol{\sigma} - \rho \frac{\partial f}{\partial \boldsymbol{\varepsilon}^{\text{el}}}\right) \cdot\cdot\, \dot{\boldsymbol{\varepsilon}}^{\text{el}}} + \boldsymbol{\sigma} \cdot\cdot\, \dot{\boldsymbol{\varepsilon}}^{\text{pl}} - \underline{\rho \left(s + \frac{\partial f}{\partial \theta}\right) \dot{\theta}} - \rho \frac{\partial f}{\partial \boldsymbol{\Upsilon}_i} \odot \dot{\boldsymbol{\Upsilon}}_i + \frac{1}{\theta} (\boldsymbol{\kappa} \cdot \nabla \theta) \cdot \nabla \theta \geq 0 \tag{111}$$

$\odot$ is the scalar product of tensors of an arbitrary order. The underlined terms in the inequality (111) are discussed in Sect. 4.3. If the thermoelastic strains are independent, for the stresses can be assumed

$$\boldsymbol{\sigma} = \rho \frac{\partial f}{\partial \boldsymbol{\varepsilon}^{\text{el}}} \tag{112}$$

This yields

$$s = -\frac{\partial f}{\partial \theta} \tag{113}$$

Equations (112) and (113) describe the thermoelastic state of the materials. This state is free from dissipation. From the dissipative inequality follows

$$\boldsymbol{\sigma} \cdot\cdot\, \dot{\boldsymbol{\varepsilon}}^{\text{pl}} - \rho \frac{\partial f}{\partial \boldsymbol{\Upsilon}_i} \odot \dot{\boldsymbol{\Upsilon}}_i + \frac{1}{\theta} (\boldsymbol{\kappa} \cdot \nabla \theta) \cdot \nabla \theta \geq 0 \tag{114}$$

The first two terms describe the mechanical dissipation, the last one—the thermal.

Let us assume the existence of a scalar dissipation potential and the thermal and mechanical dissipation can be decoupled. We suggest that the dissipation potential must be convex. For the mechanical dissipation we obtain

$$\chi = \chi(\dot{\boldsymbol{\varepsilon}}^{\text{pl}}, \dot{\boldsymbol{\Upsilon}}_i)$$

If we can assume an associated law (normality rule)

$$\boldsymbol{\sigma} = \frac{\partial \chi}{\partial \dot{\boldsymbol{\varepsilon}}^{\text{pl}}}$$

and

$$\boldsymbol{\Lambda}_i = \frac{\partial \chi}{\partial \dot{\boldsymbol{\Upsilon}}_i}$$

$\boldsymbol{\Lambda}_i$ are the associated to the inner variables functions.

5 Governing Equations of Mechanics of Hyperelastic Materials

A material is named hyperelastic, if an elastic potential u exists and u is a scalar function of the strain tensor. This energy is preserved as the result of the loading process. If we have only elastic behavior (we ignore, for example, viscoelastic behavior, creep or damage) this energy is reservable. In addition in the further developments we ignore the influence of the temperature. For rubber-like materials the elastic strains are sometimes more than 1,000 %. Limiting ourselves by the isotropy assumption, the nonlinear elasticity theory should be applied.

Note that in the case of finite strains we must distinguish the reference and the actual configurations. That means the constitutive equations can be presented in both configurations. Below we present all equations and variables in the reference configuration (Lagrangian description).

The starting point for the kinematics are the considerations of Sect. 1. The deformable body is a part of the three-dimensional space. The position of the material points in the reference configuration is given by the radius-vector $\boldsymbol{X}$, in the actual configuration by the radius-vector $\boldsymbol{x}$. The displacement vector can be presented by $\boldsymbol{u} = \boldsymbol{x} - \boldsymbol{X}$. The deformation gradient $\boldsymbol{F}$ is defined by Eq. (7). In the nonlinear elasticity the right Cauchy-Green tensor $\boldsymbol{C}$ (15) can be used as strain measure and the respective strain tensor $\boldsymbol{G}$ is the Green-Lagrange tensor (16).

In the case of rubber-like materials the incompressibility assumption often is applied. The physical meaning of this assumption is that the volumetric strains can be neglected in comparison with the strains responsible for the shape changes. The incompressibility condition is equivalent to the following equation

$$J \equiv \det \boldsymbol{F} = 1. \tag{115}$$

Let us introduce the following invariants

$$I_1 = \operatorname{tr} \boldsymbol{C}, \quad I_2 = \frac{1}{2}\left[\operatorname{tr}^2 \boldsymbol{C} - \operatorname{tr} \boldsymbol{C}^2\right], \quad I_3 = \det \boldsymbol{C} = J^2. \tag{116}$$

For incompressible materials we get $I_3 = 1$. The invariants I_k can be expressed with the help of the eigen-values λ_i^2 of the Cauchy-Green strain measure $\boldsymbol{C}$

$$I_1 = \lambda_1^2 + \lambda_2^2 + \lambda_3^2, \quad I_2 = \lambda_1^2\lambda_2^2 + \lambda_2^2\lambda_3^2 + \lambda_1^2\lambda_3^2, \quad I_3 = \lambda_1^2\lambda_2^2\lambda_3^2. \tag{117}$$

In linear elasticity the split of the strain tensor into spherical and deviatoric parts is used. The first one is responsible for the volumetric strains, the second one—for the deviatoric (isochoric) strains, which have no relations to the changes of the volume. In non-linear elasticity if we have finite strains the isochoric strains can be presented by the normalized strain gradient $\bar{\boldsymbol{F}} = J^{-1/3}\boldsymbol{F}$, which results in $\det \bar{\boldsymbol{F}} = 1$. In addition, one can introduce the normalized Cauchy-Green strain measure

$$\bar{\boldsymbol{C}} = \bar{\boldsymbol{F}}^{\mathrm{T}} \cdot \bar{\boldsymbol{F}}. \tag{118}$$

and instead of the invariants I_1, I_2, I_3-the normalized invariants of $\bar{\boldsymbol{F}}$

$$\bar{I}_1 = J^{-1/3} I_1, \quad \bar{I}_2 = J^{-2/3} I_2, \quad J. \tag{119}$$

The invariants (119) are used in formulation of constitutive equations for weak compressible materials. They can be expressed by the eigen-values $\bar{\lambda}_i^2$ of the strain measure of $\bar{\boldsymbol{C}}$

$$\bar{I}_1 = \bar{\lambda}_1^2 + \bar{\lambda}_2^2 + \bar{\lambda}_3^2, \quad I_2 = \bar{\lambda}_1^2\bar{\lambda}_2^2 + \bar{\lambda}_2^2\bar{\lambda}_3^2 + \bar{\lambda}_1^2\bar{\lambda}_3^2. \tag{120}$$

The constitutive equations of a non-linear elastic body can be introduced with the help of the specific strain energy (related to the volume in the reference configuration) W as a function of the Cauchy-Green strain measure or tensor

$$u = u(\boldsymbol{C}) = u(\boldsymbol{G}) \tag{121}$$

For finite strains the $\boldsymbol{C}$ is more used than $\boldsymbol{G}$. In the case of isotropic materials W is a function of the invariants of $\boldsymbol{C}$ and we have the representation by I_k or λ_k

$$u = u(I_1, I_2, I_3) = u(\lambda_1, \lambda_2, \lambda_3) = u(\bar{I}_1, \bar{I}_2, J) \tag{122}$$

Introducing u the stress tensors can be computed as

- the Piola stress tensor

$$\boldsymbol{P} = 2\boldsymbol{F} \cdot \frac{\partial u}{\partial \boldsymbol{C}}, \tag{123}$$

- or the Cauchy stress tensor (tensor of true stresses)

$$\boldsymbol{T} = 2J^{-1}\boldsymbol{F} \cdot \frac{\partial u}{\partial \boldsymbol{C}} \cdot \boldsymbol{F}^{\mathrm{T}}. \tag{124}$$

In the case of incompressible materials the stress tensors can be estimated from the strain energy only for terms which are related to isochoric strains. That means the following expressions are valid for

- the Piola stress tensor

$$\boldsymbol{P} = -p\boldsymbol{F}^{-1} + 2\boldsymbol{F} \cdot \frac{\partial u}{\partial \boldsymbol{C}}, \tag{125}$$

- for the Cauchy stress tensor

$$\boldsymbol{T} = -p\boldsymbol{I} + 2\boldsymbol{F} \cdot \frac{\partial u}{\partial \boldsymbol{C}} \cdot \boldsymbol{F}^{\mathrm{T}}. \tag{126}$$

Here $\boldsymbol{I}$ is the unit tensor and p is a unknown function. The physical meaning of the last one is similar to the pressure in an incompressible fluid. From the point of view of mathematics this is a Lagrangian multiplier related to Eq. (115).

Taking into account the equation for the derivatives of the invariants I_k

$$\frac{\partial I_1}{\partial \boldsymbol{C}} = \boldsymbol{I}, \quad \frac{\partial I_2}{\partial \boldsymbol{C}} = \boldsymbol{I} I_1 - \boldsymbol{C}, \quad \frac{\partial I_3}{\partial \boldsymbol{C}} = I_3\boldsymbol{C}^{-1}, \tag{127}$$

we get

$$\frac{\partial u}{\partial \boldsymbol{C}} = (u_{,1} + I_1 u_{,2})\boldsymbol{I} - u_{,2}\boldsymbol{C} + I_3 u_{,3}\boldsymbol{C}^{-1}, \quad u_{,k} \equiv \frac{\partial u}{\partial I_k} \tag{128}$$

Finally, for the isotropic incompressible material the Cauchy stress tensor (124) can be expressed by

$$\boldsymbol{T} = 2J^{-1}\left[(u_{,1} + I_1 u_{,2})\boldsymbol{B} - u_{,2}\boldsymbol{B}^2 + I_3 u_{,3}\boldsymbol{I}\right], \tag{129}$$

where $\boldsymbol{B} = \boldsymbol{F}^{\mathrm{T}} \cdot \boldsymbol{F}$ is the left Cauchy-Green tensor or Finger strain measure (15).

It can be shown [32] that the principal stresses σ_k (eigen-values of the Cauchy stress tensor $\boldsymbol{T}$) for isotropic materials can be estimated by derivatives of u w.r.t. λ_k

$$\sigma_k = J^{-1}\lambda_k \frac{\partial u}{\partial \lambda_k}. \tag{130}$$

Note that sometimes the principal forces t_k are applied. They are related to σ_k by the following equation

$$t_k = \frac{J}{\lambda_k}\sigma_k \tag{131}$$

and we get a very simple relation between t_k and u

$$t_k = \frac{\partial u}{\partial \lambda_k} \tag{132}$$

The physical meaning of σ_k and t_k are as follows. Let us introduce a infinitesimal cube in the volume, which is oriented in such a manner that the normals to surfaces have the same direction as the eigen-vectors of $\boldsymbol{T}$ (principal axes). Than σ_k are forces related to the surfaces in the actual configuration and t_k are forces related to surfaces in the reference configuration. In the case of finite strains both quantities can differ significantly. In the case of isotropic incompressible materials the Cauchy stress tensor (126) can be computed as

$$\boldsymbol{T} = -p\boldsymbol{I} + 2(u_{,1} + I_1 u_{,2})\boldsymbol{B} - 2u_{,2}\boldsymbol{B}^2, \tag{133}$$

the principal stresses σ_k and principal forces t_k as

$$\sigma_k = -p + \lambda_k \frac{\partial u}{\partial \lambda_k}, \quad t_k = -\frac{p}{\lambda_k} + \frac{\partial u}{\partial \lambda_k}. \tag{134}$$

6 Constitutive Equations of Incompressible Materials

The incompressibility assumption is often used in modeling of highly elastic materials. The strain should be a function of the first and the second invariant of the strain measure $\boldsymbol{C}$

$$u = u(I_1, I_2). \tag{135}$$

6.1 Polynomial Approximation

This model suggested in [33] is based on the Taylor series expansion of the strain energy u in the neighborhood of the undeformed state

$$u = \sum_{m=0}^{M} \sum_{n=0}^{N} C_{mn}(I_1 - 3)^m (I_2 - 3)^n \tag{136}$$

C_{mn} are material parameters with $C_{00} = 0$.

6.2 Treloar (Neo-Hookean) Material

The Treloar material [34] is an example of Eq. (136). This model is based on the Gaussian kinetic theory for rubber-like materials. The following constitutive equation is suggested

$$u = C_1(I_1 - 3), \quad C_1 > 0. \tag{137}$$

The model is very simple and some vulcanized rubbers with organic fillers can be presented at moderate strains.

6.3 Mooney Material

The model of the Mooney material [35] can be deduced from (136) if one preserves only the linear term

$$u = C_1(I_1 - 3) + C_2(I_2 - 3), \quad C_1 > 0, \quad C_2 \geq 0. \tag{138}$$

It is easy to show that the Treloar model is a special case of the Mooney material. Equation (138) allows to describe the behavior of some natural and vulcanized rubbers in a wide range. But in the case of strains higher 400–500 % the description is dissatisfying. Together with the Ogden material the Mooney material is mostly used in applications. Some generalizations of Eq. (138) are known.

6.4 Rivlin-Saunders Material

Rivlin and Saunders [36] generalized the Mooney material

$$u = C_1(I_1 - 3) + F(I_2 - 3), \tag{139}$$

where $F(I_2 - 3)$ is an arbitrary function, which should be estimated for the given material. As usual for $F(I_2 - 3)$ are proposed polynomial approximations, for example, Klosner and Segal introduced a cubic polynomial for $(I_2 - 3)$

$$u = C_1(I_1 - 3) + C_2(I_2 - 3) + C_3(I_2 - 3)^2 + C_4(I_2 - 3)^3. \tag{140}$$

6.5 Biderman Model

Biderman [37] suggested the following function for the strain energy for gray rubber

$$u = C_1(I_1 - 3) + B_1(I_1 - 3)^2 + B_2(I_1 - 3)^3 + C_2(I_2 - 3). \tag{141}$$

6.6 Non-Polynomial Approximation

Not in all notations for u the function $F(I_2 - 3)$ in (139) is included as a polynomial of $(I_2 - 3)$. For example, in [38] an exponential-hyperbolic constitutive equation is suggested

$$u = C\left[\int e^{k_1(I_1-3)^2} \mathrm{d}I_1 + k_2 \ln \frac{I_2}{3}\right]. \tag{142}$$

Alexander introduced [39]

$$u = C_1(I_1 - 3) + C_2(I_2 - 3) + C_3 \ln \frac{I_2 - 3 + k}{k} \tag{143}$$

and

$$u = C_1 \int e^{k_1(I_1-3)^2} \mathrm{d}I_1 + C_2(I_2 - 3) + C_3 \ln \frac{I_2 - 3 + k_2}{k_2} \tag{144}$$

with $C, C_1, C_2, C_3, k, k_1, k_2$ as material parameters. Equation (143) is equivalent to the Rivlin-Saunders Eq. (139) and (144) summarizes elements of (139) and (142). The last one describes the behavior of natural rubber. The last example is the strain function proposed by Hutchinson, Becker and Landel [40]

$$u = C_1(I_1 - 3) + B_1(I_1 - 3)^2 + B_2(1 - e^{k_1(I_2-3)}) + B_3(1 - e^{k_2(I_2-3)}). \quad (145)$$

describing the behavior of silicon rubber in one-and two-axial tests.

6.7 Incompressible Ogden Material

Ogden [41] suggested a constitutive equation for the strain energy u, based on λ_1, λ_2, λ_3

$$u = \frac{2\mu}{\alpha}(\lambda_1^\alpha + \lambda_2^\alpha + \lambda_3^\alpha - 3) = \frac{2\mu}{\alpha}\left[\operatorname{tr} \boldsymbol{C}^{\alpha/2} - 3\right]. \quad (146)$$

The generalization of this two-parameter model (146)

$$u = \sum_{n=1}^{N} \frac{2\mu_n}{\alpha_n}(\lambda_1^{\alpha_n} + \lambda_2^{\alpha_n} + \lambda_3^{\alpha_n} - 3) = \sum_{n=1}^{N} \frac{2\mu_n}{\alpha_n}(\operatorname{tr} \boldsymbol{C}^{\alpha_n/2} - 3). \quad (147)$$

is also called Ogden material. Similar to the polynomial approximation (136) model (147) contains enough material parameters for the fitting of experimental data. At the same time it is difficult to perform the large number of tests.

6.8 Chernykh-Shubina Material

Independent from the Ogden material model the two-parameter model of Chernykh and Shubina [42, 43] was created

$$\begin{aligned} u &= \mu\left[(1+\beta)(\lambda_1 + \lambda_2 + \lambda_3 - 3) + (1-\beta)(\lambda_1^{-1} + \lambda_2^{-1} + \lambda_3^{-1} - 3)\right] \\ &= \mu\left[(1+\beta)(\operatorname{tr} \boldsymbol{C}^{1/2} - 3) + (1-\beta)(\operatorname{tr} \boldsymbol{C}^{-1/2} - 3)\right]. \end{aligned} \quad (148)$$

6.9 Bartenev-Khazanovich Material

In the mechanics of polymers the one-parameter model of Bartenev-Khazanovich [44], which is a special case of the Ogden material or the Chernykh-Shubina model, is established

$$u = 2\mu(\lambda_1 + \lambda_2 + \lambda_3 - 3) = 2\mu(\operatorname{tr}\boldsymbol{C}^{-1/2} - 3). \tag{149}$$

In the literature the constitutive Eq. (149) is sometimes named Varga material [45].

7 Constitutive Equations of Compressible Materials

The incompressibility condition is often assumed in modeling of rubber-like materials. In some cases this approximation is not acceptable. For example, the deformation problem of high-elastic damping material between two rigid plates can be solved only if the weak compressibility of the material is taken into account. In the literature such models for weak compressible materials are introduced which are based on the aforementioned constitutive equations for incompressible materials adding terms accounting the changes of the specific volume of the material. In addition, there are models for compressible materials based on experimental data and not using the hypothesis on weak compressibility.

7.1 Compressible Neo-Hookean Material

This model for weak compressible materials is a generalization of the Treloar material. The constitutive equation has the form

$$u = C_1(\bar{I}_1 - 3) + \frac{1}{d}(J - 1)^2, \tag{150}$$

where d is a material parameter characterizing the compressibility. In the case of small strains the modulus of compressibility k for the material (150) is related to d by the equation $k = 2/d$. For incompressible materials $d = 0$ is valid and the last term in (150) is the penalty, presenting the contribution of the volumetric strains to the strain energy.

7.2 Mooney-Rivlin Material

The polynomial approximation (136) can be generalized for the case of materials with small compressibility, for example elastomers. The generalization is given as additional terms depending on J in the strain energy

$$u = \sum_{m=0}^{M} \sum_{n=0}^{N} C_{mn} (\bar{I}_1 - 3)^m (\bar{I}_2 - 3)^n + \sum_{n=1}^{N} \frac{1}{d_n} (J - 1)^{2n}. \quad (151)$$

The special case of (151) is the two-parameter Mooney-Rivlin material

$$u = C_{10}(\bar{I}_1 - 3) + C_{01}(\bar{I}_2 - 3) + \frac{1}{d}(J - 1)^2. \quad (152)$$

In FE codes polynomial approximations are implemented, which contain a big amount of material parameters, for example the three-parameter Mooney-Rivlin material

$$u = C_{10}(\bar{I}_1 - 3) + C_{01}(\bar{I}_2 - 3) + C_{11}(\bar{I}_1 - 3)(\bar{I}_2 - 3) + \frac{1}{d}(J - 1)^2, \quad (153)$$

the five-parameter Mooney-Rivlin material

$$\begin{aligned} u = {} & C_{10}(\bar{I}_1 - 3) + C_{01}(\bar{I}_2 - 3) + C_{20}(\bar{I}_1 - 3)^2 \\ & + C_{02}(\bar{I}_2 - 3)^2 + C_{11}(\bar{I}_1 - 3)(\bar{I}_2 - 3) + \tfrac{1}{d}(J - 1)^2, \end{aligned} \quad (154)$$

or the nine-parameter Mooney-Rivlin material

$$\begin{aligned} u = {} & C_{10}(\bar{I}_1 - 3) + C_{01}(\bar{I}_2 - 3) + C_{20}(\bar{I}_1 - 3)^2 + C_{02}(\bar{I}_2 - 3)^2 \\ & + C_{11}(\bar{I}_1 - 3)(\bar{I}_2 - 3) + C_{30}(\bar{I}_1 - 3)^3 + C_{21}(\bar{I}_1 - 3)^2(\bar{I}_2 - 3) \\ & + C_{12}(\bar{I}_1 - 3)(\bar{I}_2 - 3)^2 + C_{03}(\bar{I}_2 - 3)^3 + \frac{1}{d}(J - 1)^2 \end{aligned} \quad (155)$$

7.3 Compressible Ogden Material

Model (147) can be used for materials with small incompressibility

$$u = \sum_{n=1}^{N} \frac{2\mu_n}{\alpha_n} (\bar{\lambda}_1^{\alpha_n} + \bar{\lambda}_2^{\alpha_n} + \bar{\lambda}_3^{\alpha_n} - 3) + \sum_{n=1}^{N} \frac{1}{d_n} (J - 1)^{2n} \quad (156)$$

7.4 Arruda-Boyce Material

The potential in this case takes the form [46]

$$u = \mu\left[\frac{1}{2}(\bar{I}_1 - 3) + \frac{1}{20\lambda_L^2}(\bar{I}_1^2 - 9) + \frac{11}{1050\lambda_L^4}(\bar{I}_1^3 - 279)\right.$$
$$\left. + \frac{19}{7000\lambda_L^6}(\bar{I}_1^4 - 81) + \frac{519}{673750\lambda_L^8}(\bar{I}_1^5 - 243)\right] + \frac{1}{d}\left(\frac{J^2-1}{2} - \ln J\right) \tag{157}$$

Here μ, λ_L, d are the material parameters. With $\lambda_L \to \infty$ we get again the Neo-Hookean material.

7.5 Gent Material

For this material the following potential holds

$$u = -\frac{EI_m}{6}\log\left(1 - \frac{\bar{I}_1 - 3}{I_m}\right) \tag{158}$$

Here E, I_m are material parameters.

7.6 Blatz-Ko Material

This model was suggested by Blatz and Ko [47] for the case of modeling of several rubbers (porous materials—foams—made of polyurethane (PU)). The potential has the form

$$u = \frac{1}{2}\mu\beta\left[I_1 + \frac{1}{\alpha}(I_3^{-\alpha} - 1) - 3\right]\frac{1}{2}\mu(1-\beta)\left[I_2 I_3^{-1} + \frac{1}{\alpha}(I_3^{\alpha} - 1) - 3\right],$$
$$\alpha = \frac{\lambda}{2\mu} \tag{159}$$

This model contains 3 parameters: μ, α and β. In FE codes a simpler potential is implemented (159)

$$u = \frac{1}{2}\mu\left[\frac{I_2}{I_3} + 2\sqrt{I_3} - 5\right] \tag{160}$$

7.7 Other Material Models

In the literature there are given references to other compressible non-linear elastic materials (Seth, Signorini [48], Murnaghan [49], harmonic (semi-linear), etc.)

[32, 49]. The material parameters for these models are known. Applications of these models are related, for example to acoustics.

References

1. Haupt, P.: Continuum Mechanics and Theory of Materials, 2nd edn. Springer, Berlin (2002)
2. Altenbach, H. (ed.): Holzmann Meyer Schumpich Technische Mechanik Festigkeitslehre, 10th edn. Springer Vieweg, Stuttgart (2012)
3. Gross, D., Hauger, W., Schröder, J., Wall, W.A.: Technische Mechanik, vol. 2. Elastostatik, 9th edn. Springer, Berlin (2008)
4. Eremeyev, V., Lebedev, L., Altenbach, H.: Foundations of Micropolar Mechanics. Springer-Briefs in Applied Sciences and Technologies. Springer, Heidelberg (2013)
5. Eremeyev, V.A., Pietraszkiewicz, W.: Material symmetry group of the non-linear polar-elastic continuum. J. Solids Struct. **49**(14), 1993–2005 (2012)
6. Eringen, A.C.: Microcontinuum Field Theory, Foundations and Solids, vol. 2. Springer, New York (1999a)
7. Eringen, A.C.: Microcontinuum Field Theory, Fluent Media, vol. 2. Springer, New York (1999b)
8. Pietraszkiewicz, W., Eremeyev, V.A.: On natural strain measures of the non-linear micropolar continuum. Int. J. Solids Struct. **46**(3–4), 774–787 (2009)
9. Müller I (1973) Thermodynamik: die Grundlagen der Materialtheorie. Bertelsmann Universitätsverlag
10. Palmov, V.A.: Vibrations of Elasto-plastic Bodies. Foundations of Engineering Mechanics. Springer, Berlin (1998)
11. Salençon, J.: Handbbok of Continuum Mechanics. Springer, Berlin (2001)
12. Willner, K.: Kontinuums—und Kontaktmechanik: Synthetische und analytische Darstellung. Springer, Berlin (2003)
13. Altenbach, H., Eremeyev, V. (eds.): Generalized Continua— from the Theory to Engineering Applications, CISM International Centre for Mechanical Sciences, vol. 541. Springer, Wien (2013)
14. Altenbach, H., Naumenko, K., Zhilin, P.: A micro-polar theory for binary media with application to phase-transitional flow of fiber suspensions. Continuum Mech Thermodyn **15**, 539–570 (2003)
15. Altenbach, H., Maugin, G.A., Erofeev, V. (eds.): Mechanics of Generalized Continua, Advanced Structured Materials, vol. 7. Springer, Heidelberg (2011)
16. Altenbach, H., Forest, S., Krivtsov, A. (eds.): Generalized Continua as Models for Materials with Multi-scale Effects or Under Multi-field Actions, Advanced Structured Materials, vol. 22. Springer, Heidelberg (2013)
17. Maugin, G.A.: Continuum Mechanics Through the Twentieth Century. Springer, Heidelberg (2013)
18. Maugin, G.A., Metrikine, A. (eds.): Mechanics of Generalized Continua - One Hundred Years After the Cosserats, Advances in Mechanics and Mathematics 21. Springer, Berlin (2010)
19. Rubin, M.B.: Cosserat Theories: Shells Rods and Points. Kluwer, Dordrecht (2000)
20. Krawietz, A.: Materialtheorie. Springer, Berlin (1986)
21. Noll, W.: The Foundations of Mechanics and Thermodynamics. Springer, Berlin (1974)
22. Bertram, A.: What is the general constitutive equation? In: Beiträge Festschrift zum 65 . Geburtstag von Rudolf Trostel, TU Berlin. Berlin, pp. 28–37 (1994)
23. Giesekus, H.: Phänomenologische Rheologie : eine Einführung. Springer, Berlin (1994)
24. Reiner, M.: Rheologie. Fachbuchverlag, Leipzig (1968)
25. Truesdell, C.: A First Course in Rational Continuum Mechanics, Pure and Applied Mathematics, vol. 1. Academic Press, New York (1977)

26. Bertram, A., Forest, S.: The thermodynamics of gradient elastoplasticity. Continuum Mech. Thermodyn. **25**, 20 (2013)
27. dell'Isola F, Sciarra G, Vidoli S.: Generalized Hooke's law for isotropic second gradient materials. Proc. R Soc. A **495**, 2177–2196 (2009)
28. Forest, S., Trinh, D.K.: Generalized continua and non-homogeneous boundary conditions in homogenisation methods. ZAMM **91**(2), 90–109 (2011)
29. Podio-Guidugli, P., Vianello, M.: Hypertractions and hyperstresses convey the same mechanical information. Continuum Mech. Thermodyn. **22**, 163–176 (2010)
30. Backhaus, G.: Zum evolutionsgesetz der kinematischen verfestigung in objektiver darstellung. ZAMM **72**(9), 397–406 (1992)
31. Lemaitre, J., Chaboche, J.L.: Mechanics of Solid Materials. Cambridge University Press, Cambridge (1990)
32. Lurie, A.I.: Nonliner Theory of Elasticity. North-Holland, Amsterdam (1990)
33. Ogden, R.: Non-Linear Elastic Deformations. Ellis Horwood, Chichester (1984)
34. Treloar, L.: The Physics of Rubber Elasticity, 3rd edn. Oxford University Press, Oxford (1975)
35. Mooney, M.: A theory of large elastic deformations. J. Appl. Phys. **11**, 582–592 (1940)
36. Rivlin, R.S., Saunders, D.W.: Large elastic deformations of isotropic materials. Phil. Trans. Roy. Soc. A **243**, 251–288 (1951)
37. Biderman V.L.: On the calculation of rubber-like materials (in Russ.). In: Strengths Analysis (in Russ.), 3. Mashgiz, pp. 40–87 (1958)
38. Hart-Smith, L.J., Crisp, J.D.C.: Large elastic deformations in thin rubber membranes. Int. J. Eng. Sci. **5**(1), 1–24 (1967)
39. Alexander, H.: A constitutive relation for rubber-like materials. Int. J. Eng. Sci. **6**(9), 549–563 (1968)
40. Hutchinson, W.D., Becker, G.W., Landel, R.F. Determination of the stored energy function of rubber-like materials. In: Bull 4th Meeting Interagency Chemical Rocket Propulsion Group—Working Group Mechanical Behavior, vol 1. No. 94U, pp 141–152. CPIA
41. Ogden, R.W.: Large deformation isotropic elasticity I: on the correlation of theory and experiment for incompressible rubberlike solids. Proc. Roy. Soc. London A **326**, 565–584 (1972)
42. Chernykh, K.F., Shubina, I.M.: The laws of elasticity for isotropic compressible materials (in Russ.). In: Mekhanika Elastomerov. Krasnodar, vol **I**, 54–65 (1977)
43. FCK (1986) Nonlinear Theory of Elasticity in Engineering Calculations (in Russ.). Mashinostroenie, Moscow
44. Bartenev G.M, Khazanovich T.N: On a constitutive law for hyperelastic strans of cross-linked polymer (in russ.). Polym. Sci. **2**(1), 20–28 (1960)
45. Fu Y.B, Ogden R.W (eds.): Nonlinear elasticity. Theory and applications. No. 283 in Lecture Note Series. Cambridge University Press, Cambridge (2001)
46. Arruda, E., Boyce, M.: A three-dimensional constitutive model the large stretch behavior of rubber elastic materials. J. Mech. Phys. Solids **41**(2), 389–412 (1993)
47. Blatz, P.J., Ko, W.L.: Application of finite elasticity theory to the deformation of rubbery materials. Tran. Soc. Rheol. **6**, 223–251 (1962)
48. Signorini, A.: Solidi incompribili. Ann. Mat. Pura Appl. **39**, 147–201 (1955)
49. Murnaghan, F.D.: Finite Deformation of an Elastic Solid. Wiley, New York (1951)

Phenomenological Yield and Failure Criteria

Holm Altenbach, Alexandre Bolchoun and Vladimir A. Kolupaev

Abstract Models for isotropic materials based on the equivalent stress concept are discussed. At first, so-called classical models which are useful in the case of absolutely brittle or ideal ductile materials are presented. Tests for basic stress states are suggested. At second, standard models describing the intermediate range between the absolutely brittle and ideal-ductile behavior are introduced. Any criterion is expressed by various mathematical equations formulated, for example, in terms of invariants. At the same time the criteria can be visualized which simplifies the application. At third, in the main part pressure-insensitive, pressure-sensitive and combined models are separated. Fitting methods based on mathematical, physical and geometrical criteria are necessary. Finally, three examples (gray cast iron, poly(oxymethylene) (POM) and poly(vinyl chloride) (PVC) hard foam) demonstrates the application of different approaches in modeling certain limit behavior. Two appendices are necessary for a better understanding of this chapter: in Chap. 2 applied invariants are briefly introduced and a table of discussed in this chapter criteria with references is given.

Keywords Strength criteria · Yield criteria · Equivalent stress · Pressure-insensitive criteria · Pressure-sensitive criteria · Limit surfaces

H. Altenbach (✉)
Lehrstuhl für Technische Mechanik, Institut für Mechanik, Fakultät für Maschinenbau, Otto-von-Guericke-Universtät Magdeburg, Universitätsplatz 2, 39106 Magdeburg, Germany
e-mail: holm.altenbach@ovgu.de

A. Bolchoun
Abteilung Werkstoffe und Bauteile, Fraunhofer-Institut für Betriebsfestigkeit und Systemzuverlässigkeit LBF, Bartningstr. 47, 64289 Darmstadt, Germany
e-mail: alexandre.bolchoun@lbf.fraunhofer.de

V. A. Kolupaev
Deutsches Kunststoff-Institut (DKI), Schloßgartenstr. 6, 64289 Darmstadt, Germany
e-mail: VKolupaev@dki.tu-darmstadt.de

H. Altenbach and A. Öchsner (eds.), *Plasticity of Pressure-Sensitive Materials*, Engineering Materials, DOI: 10.1007/978-3-642-40945-5_2,

1 Need of Criteria

The strength theory assumes that the mechanical loading states can be characterized, for example, by stresses [62]. It is known that the stresses in each point of the material or structure are presented by the stress tensor $\boldsymbol{\sigma}$ [5]. For comparison purposes of various stress states the stress tensor cannot be applied that means a scalar quantity should be used. Let us introduce the following expression for such quantity

$$\sigma_{\mathrm{EQ}} = \sigma_{\mathrm{eq}}(\boldsymbol{\sigma}) + f(\nabla\boldsymbol{\sigma})\,R, \qquad R \geq 0. \tag{1}$$

∇ is the nabla operator, f denotes an arbitrary scalar-valued function and R is a structural parameter, which can be associated with the grain size in gray iron, with the cell size of a hard foam, with the particle size in nanomaterials, etc. This parameter represents the influence of the stress distribution expressed by the stress gradient $\nabla\boldsymbol{\sigma}$. The parameter R is positive-definit and bounded by the minimal dimension of the structural component, e. g. the plate or sheet thickness, cf. [214]. Equation (1) can be extended by introducing higher order stress gradients, however the application is limited. The additional material parameters should be estimated experimentally, but tests for their estimation are unknown.

Ignoring the microstructure influence Eq. (1) can be simplified

$$\sigma_{\mathrm{EQ}} = \sigma_{\mathrm{eq}}(\boldsymbol{\sigma}). \tag{2}$$

This implies that the stress state in each point can be described through the stresses at this point only [71]. This formulation has multiple limitations and must be applied with care if the calculation of stresses is performed for parts with significant stress gradients:

- stress concentration areas,
- load application areas,
- sharp corners, etc.

Nevertheless, the concept of the equivalent stress (2) is widely applicable. This concept allows to compare multi-axial stress states with material parameters, e. g. the tensile yield or failure stress σ_+

$$\sigma_{\mathrm{eq}} = \sigma_+. \tag{3}$$

Strength hypotheses and yield criteria for isotropic materials can be formulated using principal stresses

$$\Phi(\sigma_{\mathrm{I}}, \sigma_{\mathrm{II}}, \sigma_{\mathrm{III}}, \sigma_{\mathrm{eq}}) = 0, \tag{4}$$

or other invariants, e. g. axiatoric-deviatoric invariants

$$\Phi(I_1, I_2', I_3', \sigma_{\mathrm{eq}}) = 0, \tag{5}$$

or cylindrical invariants (Novozhilov's invariants)

$$\Phi(I_1, I_2', \theta, \sigma_{\mathrm{eq}}) = 0. \tag{6}$$

The invariants are named in accordance to [234] and detailed presented in Appendix 15. The three formulations (4), (5) and (6) are equivalent. Note that in the case of incompressible material behavior it can be shown that the first invariant I_1 has no influence on Φ

$$\Phi(I_2', I_3', \sigma_{\mathrm{eq}}) = 0, \qquad \Phi(I_2', \theta, \sigma_{\mathrm{eq}}) = 0. \tag{7}$$

The remaining invariants are named deviatoric invariants.

The equivalent stress concept (2) allows to formulate the material response under multi-axial loading in a compact form using only few parameters. Such formulations are used in

- elasticity theory (elastic potential) [8, 9, 130, 212],
- plasticity theory (plastic potential, yield criterion) [8, 139, 163, 234],
- creep theory (creep potential) [8, 125],
- strength of materials (strength hypothesis or criterion) [8, 35, 91, 157, 220, 229],
- low cyclic fatigue [8, 126] and
- phase transformation conditions [158, 165].

Phenomenological yield and failure criteria are widely discussed in the literature. Some reviews are given in [8, 16, 45, 71, 74, 157, 174, 204, 221, 228, 229, 234] among others.

2 Classical Strength Theories

The dimensioning of structural members is usually carried out under the assumption, that materials behave either brittle or ductile. The following hypotheses (sometimes named theories), which correspond to one of the two assumptions, are often used for strength or yield evaluation [65, 74, 132, 174].

The three classical models (normal stress hypothesis, Tresca, von Mises), which are presented as usual in textbooks on strength of materials and implemented in commercial finite element codes as a standard tool, and the model of Schmidt-Ishlinsky represent particular cases of material behavior and are sometimes unable to describe the behavior of materials properly. Because of their simplicity they are used in the engineering practice. For applied problems the computations can be performed using these models, if no information on the particular material properties is available. The normal stress hypothesis (Fig. 1) describes the "absolutely brittle" material behavior, the models of Tresca, von Mises and Schmidt-Ishlinsky—the "ideal ductile" behavior.

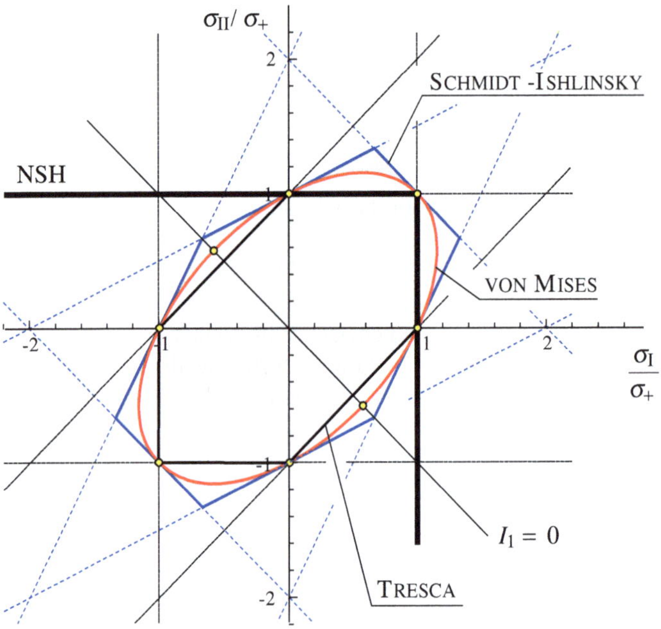

Fig. 1 Models for incompressible "ideal ductile" material behavior (von Mises, Tresca and Schmidt-Ishlinsky) and the normal stress hypothesis (NSH) for "absolutely brittle" material behavior in the plane $\sigma_{\mathrm{I}}-\sigma_{\mathrm{II}}$, $\sigma_{\mathrm{III}} = 0$ (after [93])

2.1 Normal Stress Hypothesis

The normal stress hypothesis (Clapeyron, Galilei, Leibniz, Lamé, Maxwell, Navier, Rankine), Fig. 2, i.e. the maximum tensile stress is responsible for the failure [75, 94, 166, 175], can be expressed as it as follows

$$\sigma_{\mathrm{eq}} = \max(\sigma_{\mathrm{I}},\ \sigma_{\mathrm{II}},\ \sigma_{\mathrm{III}}). \tag{8}$$

Another formulation is

$$(\sigma_{\mathrm{I}} - \sigma_{\mathrm{eq}})\,(\sigma_{\mathrm{II}} - \sigma_{\mathrm{eq}})\,(\sigma_{\mathrm{III}} - \sigma_{\mathrm{eq}}) = 0. \tag{9}$$

Equation (9) is a cubic equation with respect to σ_{eq}. With the help of a parameter identification this equation can be transformed into a third order polynomial of I_1^3, $I_1^2\,\sigma_{\mathrm{eq}}$, $I_1\,\sigma_{\mathrm{eq}}^2$, σ_{eq}^3, $I_2'\,\sigma_{\mathrm{eq}}$ and I_3'. It can be obtained using the model [178]

$$\frac{3\,I_2'\,\sigma_{\mathrm{eq}} + c_3\,I_3'}{1 + 2\,c_3/3^3} = \sigma_{\mathrm{eq}}^3, \tag{10}$$

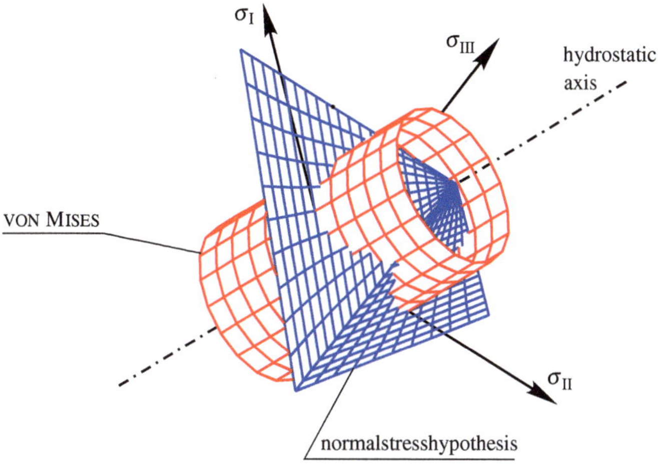

Fig. 2 Normal stress hypothesis (8) and cylinder of von Mises (16) in the principal stress space $(\sigma_{\rm I}, \sigma_{\rm II}, \sigma_{\rm III})$

and the substitution [103, 178]

$$\sigma_{\rm eq} \to \frac{\sigma_{\rm eq} - \gamma_1 I_1}{1 - \gamma_1} \tag{11}$$

with the parameter values

$$c_3 = \frac{3^2}{2}, \qquad \gamma_1 = \frac{1}{3} \tag{12}$$

for the better analysis, unified visualization techniques and systematization.

2.2 *Tresca Hypothesis*

The shear stress hypothesis (Coulomb, Guest, Mohr, Saint Venant, Tresca), i.e. the maximum difference of the principal stresses is relevant for the failure [48, 163, 209], can be written as follows (Fig. 1)

$$\tau_{\rm max} = \frac{1}{2} \max(|\sigma_{\rm I} - \sigma_{\rm II}|, |\sigma_{\rm II} - \sigma_{\rm III}|, |\sigma_{\rm III} - \sigma_{\rm I}|). \tag{13}$$

The equivalent stress can be expressed in this case as

$$\sigma_{\rm eq} = 2\,\tau_{\rm max}. \tag{14}$$

In analogy to Eq. (9) one can write

$$(\sigma_{\text{eq}} - |\sigma_{\text{I}} - \sigma_{\text{II}}|)\,(\sigma_{\text{eq}} - |\sigma_{\text{II}} - \sigma_{\text{III}}|)\,(\sigma_{\text{eq}} - |\sigma_{\text{III}} - \sigma_{\text{I}}|) = 0.$$

This hypothesis (often called Tresca hypothesis) can be also expressed by the deviatoric invariants [163, 171]

$$\left(I_2' - \sigma_{\text{eq}}^2\right)^2 \left(2^2\, I_2' - \sigma_{\text{eq}}^2\right) - 3^3\, I_3'^{\,2} = 0. \tag{15}$$

2.3 Huber-von Mises-Hencky Hypothesis

The distortion energy hypothesis (Huber, von Mises, Hencky),[1] Fig. 2, has different interpretations among them that the failure occurs if a critical amount of accumulated distortion energy is achieved [85, 91, 139, 194, 234]

$$\sigma_{\text{eq}}^2 = \frac{1}{2}\left[(\sigma_{\text{I}} - \sigma_{\text{II}})^2 + (\sigma_{\text{II}} - \sigma_{\text{III}})^2 + (\sigma_{\text{III}} - \sigma_{\text{I}})^2\right] = 3\, I_2'. \tag{16}$$

This hypothesis is often called von Mises hypothesis.

2.4 Schmidt-Ishlinsky Hypothesis

The criterion of the maximum deviatoric stress (Burzyński, Schmidt, Ishlinsky, Hill, Haythornthwaite), i. e. the failure occurs if a critical value of deviatoric components of the stress tensor is achieved [35, 84, 87, 92, 182] (Fig. 1)

$$\max\left[\left|\sigma_{\text{I}} - \frac{1}{3}\, I_1\right|,\ \left|\sigma_{\text{II}} - \frac{1}{3}\, I_1\right|,\ \left|\sigma_{\text{III}} - \frac{1}{3}\, I_1\right|\right] = \frac{2}{3}\,\sigma_{\text{eq}} \tag{17}$$

or in analogy to Eq. (9)

$$\left[\sigma_{\text{eq}} - \left|\sigma_{\text{I}} - \frac{1}{2}(\sigma_{\text{II}} + \sigma_{\text{III}})\right|\right]\left[\sigma_{\text{eq}} - \left|\sigma_{\text{II}} - \frac{1}{2}(\sigma_{\text{III}} + \sigma_{\text{I}})\right|\right]\left[\sigma_{\text{eq}} - \left|\sigma_{\text{III}} - \frac{1}{2}(\sigma_{\text{I}} + \sigma_{\text{II}})\right|\right] = 0. \tag{18}$$

This model can be expressed with the deviatoric invariants [11, 222, 224, 225]

$$\left[\frac{3^3}{2^3}\, I_3' + \frac{3^2}{2^2}\, I_2'\,\sigma_{\text{eq}} - \sigma_{\text{eq}}^3\right]\left[\frac{3^3}{2^3}\, I_3' - \frac{3^2}{2^2}\, I_2'\,\sigma_{\text{eq}} + \sigma_{\text{eq}}^3\right] = 0. \tag{19}$$

The naming Schmidt-Ishlinsky hypothesis has become established.

---

[1] This criterion was also formulated 1865 in a letter of Maxwell to Lord Kelvin [204].

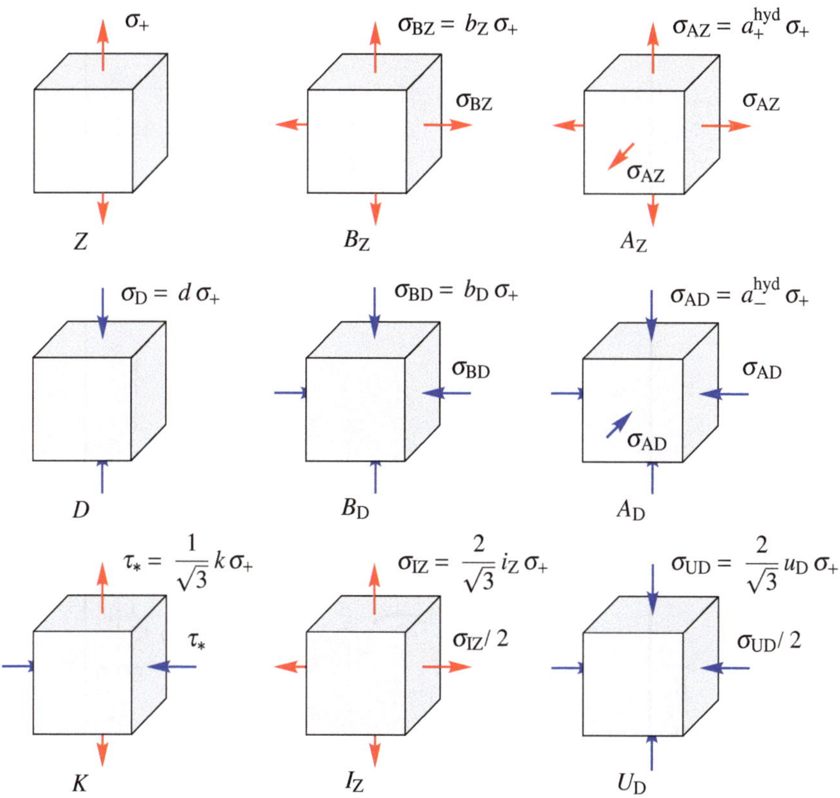

Fig. 3 *Nine* basic tests. The stresses, values and labels of loading are given in Table 1

3 Basic Stress States

All criteria can be visualized as a limit surface Φ. Nine tests (Fig. 3) are chosen for the analysis and comparison of the surfaces Φ:

- two loadings corresponding to one-dimensional stress states (tension, compression),
- five loadings corresponding to plane stress states (torsion, two balanced plane states, two thin-walled tube specimens with closed ends under inner and outer pressure) and
- two loadings corresponding to hydrostatic (3D balanced) tension and compression.

The relevant stresses are listed in Table 1. All these loading cases have approved verbal formulations and can be considered as basic tests.[2]

[2] Note that in material testing another definition of basic tests is given [32].

Table 1 Basic stress states, relations, normalized coordinates in the principal stress space, normalized axiatoric-deviatoric invariants (Appendix A.3), and the stress angle θ (224) [107]

Loading	Stress	Label	Relation	$\left(\frac{\sigma_{\mathrm{I}}}{\sigma_+}, \frac{\sigma_{\mathrm{II}}}{\sigma_+}, \frac{\sigma_{\mathrm{III}}}{\sigma_+}\right)$	$\dfrac{I_1}{\sigma_+}$	$\dfrac{\sqrt{3I_2'}}{\sigma_+}$	$\dfrac{3\sqrt[3]{I_3'/2}}{\sigma_+}$	θ
Basic stress states								
Tension	σ_+	Z	1	$(1,\ 0,\ 0)$	1	1	1	0
Compression	σ_-	D	d	$(-d,\ 0,\ 0)$	$-d$	d	d	$\dfrac{\pi}{3}$
Torsion	τ_*	K	k	$\left(\dfrac{k}{\sqrt{3}},\ -\dfrac{k}{\sqrt{3}},\ 0\right)$	0	k	0	$\dfrac{\pi}{6}$
Biaxial tension	σ_{BZ}	B_{Z}	b_{Z}	$(b_{\mathrm{Z}},\ b_{\mathrm{Z}},\ 0)$	$2\,b_{\mathrm{Z}}$	b_{Z}	$-b_{\mathrm{Z}}$	$\dfrac{\pi}{3}$
Biaxial compression	σ_{BD}	B_{D}	b_{D}	$(-b_{\mathrm{D}},\ -b_{\mathrm{D}},\ 0)$	$-2\,b_{\mathrm{D}}$	b_{D}	b_{D}	0
Inner pressure	σ_{IZ}	I_{Z}	i_{Z}	$\left(\dfrac{2}{\sqrt{3}}\,i_{\mathrm{Z}},\ \dfrac{1}{\sqrt{3}}\,i_{\mathrm{Z}},\ 0\right)$	$\sqrt{3}\,i_{\mathrm{Z}}$	i_{Z}	0	$\dfrac{\pi}{6}$
Outer pressure	σ_{UD}	U_{D}	u_{D}	$\left(-\dfrac{2}{\sqrt{3}}\,u_{\mathrm{D}},\ -\dfrac{1}{\sqrt{3}}\,u_{\mathrm{D}},\ 0\right)$	$-\sqrt{3}\,u_{\mathrm{D}}$	u_{D}	0	$\dfrac{\pi}{6}$
Hydrostatic tension	σ_{AZ}	A_{Z}	a_+^{hyd}	$\left(a_+^{\mathrm{hyd}},\ a_+^{\mathrm{hyd}},\ a_+^{\mathrm{hyd}}\right)$	$3\,a_+^{\mathrm{hyd}}$	0	0	–
Hydrostatic compression	σ_{AD}	A_{D}	a_-^{hyd}	$\left(-a_-^{\mathrm{hyd}},\ -a_-^{\mathrm{hyd}},\ -a_-^{\mathrm{hyd}}\right)$	$-3\,a_-^{\mathrm{hyd}}$	0	0	–
Additional tests								
Tension-torsion	σ_{H}	H	h	$\left(\sqrt{\dfrac{2}{3}}\,h,\ \sqrt{\dfrac{2}{3}}\,\dfrac{1-\sqrt{3}}{2}\,h,\ 0\right)$	$\sqrt{2-\sqrt{3}}\,h$	h	$\dfrac{h}{2^{1/6}}$	$\dfrac{\pi}{12}$
Compression-torsion	σ_{Q}	Q	q	$\left(-\sqrt{\dfrac{2}{3}}\,q,\ \sqrt{\dfrac{2}{3}}\,\dfrac{\sqrt{3}-1}{2}\,q,\ 0\right)$	$-\sqrt{2-\sqrt{3}}\,q$	q	$-\dfrac{q}{2^{1/6}}$	$\dfrac{\pi}{4}$

These nine tests are sufficient for the comparison of the most important features of surfaces. Their selection is however not unambiguous and can be expanded according to the available equipment, expected phenomena and requested precision, see e. g. loading cases labeled by Q and H (Table 1 and Sect. 9.2). Further considerations for the choice of loadings are discussed in [8, 36, 37].

The values (Table 1) relating the respective stresses to σ_+ are introduced in order to obtain

$$k = d = i_{\mathrm{Z}} = u_{\mathrm{D}} = b_{\mathrm{Z}} = b_{\mathrm{D}} = 1 \qquad \text{and} \qquad a_-^{\mathrm{hyd}},\ a_+^{\mathrm{hyd}} \to \infty \tag{20}$$

for the von Mises hypotheses (16).

For the models of incompressible material behavior the values on the angle $\theta = 0$, $\frac{\pi}{6}$ and $\frac{\pi}{3}$ are computed to [16, 103, 234]

$$b_{\mathrm{D}} = 1, \qquad k = i_{\mathrm{Z}} = u_{\mathrm{D}} \qquad \text{and} \qquad d = b_{\mathrm{Z}}. \tag{21}$$

For the classical models (Sect. 2) it follows [113]

$$b_{\mathrm{Z}} = 1, \qquad b_{\mathrm{D}} = d. \tag{22}$$

The models for incompressible behavior can be compared in the d–k-diagram (Fig. 4) [104, 105, 110]. In this diagram the models of Haythornthwaite and Sayir II (Sect. 9.1) limit the convex shapes of the surface Φ in the π-plane [33]. For the models of compressible material behavior (Sect. 10) the $\frac{1}{d}$–k-diagram, which allows to represent the properties $d \to \infty$, $k = \sqrt{3}$ of the normal stress hypothesis among others, is recommended (Fig. 5) [122, 156, 157]. In this diagram the areas of validity of all criterions and various ideas of generalization can be visualized.

The measurements σ_+, σ_- and τ_* for some materials are presented in [18, 41, 124, 157, 229]. Examples of experimental data for some polymers are given in Table 2.

The data are taken from various sources and they are related to different manufacturers. They can be used as first estimates only. Note that the experiments were performed for specimens with different geometries and using different techniques. The relations d and k of the materials can be represented in Figs. 4 and 5 together with the models in order to simplify the choice of the suitable model [104].

4 Inelastic Poisson's Ratio

In the linear theory of elasticity the Poisson's ratio is defined as the negative ratio of the strain $\varepsilon_{\mathrm{II}} = \varepsilon_{\mathrm{III}}$ in the direction orthogonal to the applied load and the strain ε_{I} in the tension direction [20, 89]:

$$\nu^{\mathrm{el}} = -\frac{\varepsilon_{\mathrm{II}}}{\varepsilon_{\mathrm{I}}} = -\frac{\varepsilon_{\mathrm{III}}}{\varepsilon_{\mathrm{I}}}, \qquad \nu^{\mathrm{el}} \in \left]-1, \frac{1}{2}\right[. \tag{23}$$

Table 2 Parameters for brittle and ductile failure at room temperature collected in [18, 184, 185] on the base of [13, 19, 60, 137, 164, 183, 184, 189, 190, 198, 203] and elastic Poisson's ratios

Polymer	σ_+ [MPa]	d [−]	k [−]	ν_+^{el}, [10] [−]	ν_+^{el}, [119] [−]	ν_+^{el}, [59] [−]	ν_+^{el}, [58] [−]	Remarks
Failure as the result of yielding								
PS	72.67	1.33	1.18	0.33…0.35	0.32…0.34	0.38	0.33	
PVC	57.08	1.17	1.25	–	$0.32\ldots0.35^1$	–	–	^{1}PVC unplasticized
PVC	53.94	1.30	1.18	–	$0.32\ldots0.35^1$	–	–	^{1}PVC unplasticized
PVC	32.36	1.33	1.11	–	$0.32\ldots0.35^1$	–	–	^{1}PVC unplasticized
PMMA	–	1.00	–	–	0.30…0.35	–	0.33	
PMMA	37.27	1.30	–	–	–	–	–	at 80°C
PC	45.11	1.20	1.06	0.37…0.38	0.30…0.35	–	0.42	
PC	58.84	1.22	1.13	0.37…0.38	0.32…0.35	–	0.42	
CAB	33.15	1.00	1.00	–	–	–	–	
PE	10.69	1.34	1.49*	$0.40\ldots0.44^2$	$0.35\ldots0.40^2$	–	0.38^2	^{2}high density
					$0.40\ldots0.45^3$	0.49^3	0.45^3	^{3}low density
PP	32.26	1.32	1.44*	0.37…0.38	0.35…0.40	–	0.4	
PA	66.29	0.92	1.04	–	0.32…0.40	0.44^4	–	^{4}PA6.6
							0.33^5	^{5}PA6, PA6.6 (dried)
							0.45^6	^{6}PA6, PA6.6 (dried) at 100°C

(continued)

Table 2 (continued)

Polymer	σ_+ [MPa]	d [−]	k [−]	ν_+^{el}, [10] [−]	ν_+^{el}, [119] [−]	ν_+^{el}, [59] [−]	ν_+^{el}, [58] [−]	Remarks
ABS	44.62	0.95	0.94	0.35…0.36	0.32…0.35	–	–	
Brittle fracture								
PMMA	62.08	1.50	1.20	–	0.30…0.35	–	0.33	
PMMA	59.04	1.40	1.16	–	0.30…0.35	–	0.33	
PMMA	–	1.00	–	–	0.30…0.35	–	0.33	
CAB	33.64	0.91	1.30	–	–	–	–	
CA	40.50	1.23	1.70*	–	–	–	–	
PVCA	65.90	1.29	1.80*	–	–	–	–	
EP resin	81.40	1.45	1.20	–	–	–	0.4	Araldit CY232
EP resin	78.45	1.33	1.13	–	–	–	0.4	Legutherm K57

* These values contradict our own experiences [104]

$^-$ Molar mass of the thermoplastic polymers was not specified

$^-$ Crystallinity of PE, PP and PA is unknown

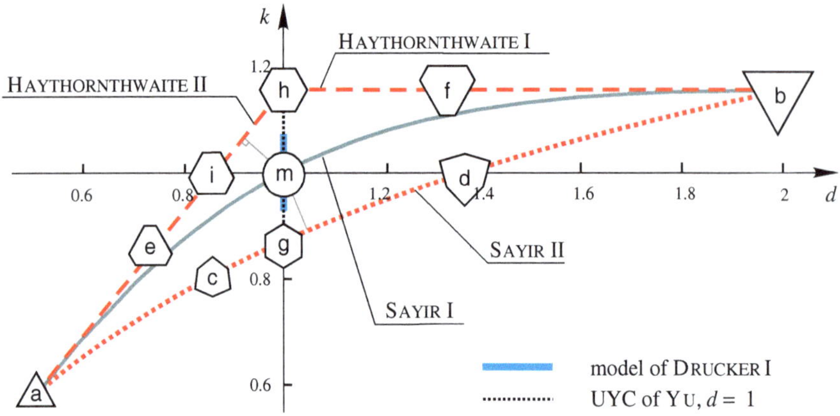

Fig. 4 Diagram d–k for convex models of incompressible material behavior compared with the hypothesis of von Mises [110]. Certain cross sections in the π-plane are visualized in order to achieve a better understanding:
$a.\ k = 1/\sqrt{3},\ d = 1/2$ $b.\ k = 2/\sqrt{3},\ d = 2$
$c.\ k = 3\,(2-\sqrt{3}),\ d = \sqrt{3}/2$ $d.\ k = 1,\ d = (1+\sqrt{3})/2$
$e.\ k = \sqrt{3}/2,\ d = 3/4$ $f.\ k = 2/\sqrt{3},\ d = 4/3$
$g.\ k = \sqrt{3}/2,\ d = 1$(Tresca) $h.\ k = 2/\sqrt{3},\ d = 1$(Schmidt – Ishlinsky)
$i.\ k = 1,\ d = \sqrt{3}/2$ $m.\ k = 1,\ d = 1$(von Mises)

If the elastic law is formulated on the base of the potential Φ, one can write

$$\nu^{\text{el}} = -\left(\frac{\partial\,\Phi}{\partial\,\sigma_{\text{II}}}\Big/\frac{\partial\,\Phi}{\partial\,\sigma_{\text{I}}}\right)\Bigg|_{\sigma_{\text{II}}=\sigma_{\text{III}}=0} = -\left(\frac{\partial\,\Phi}{\partial\,\sigma_{\text{III}}}\Big/\frac{\partial\,\Phi}{\partial\,\sigma_{\text{I}}}\right)\Bigg|_{\sigma_{\text{II}}=\sigma_{\text{III}}=0}. \tag{24}$$

The classical theory of elasticity [20, 64, 205] makes no difference between Poisson's ratio at tension and compression

$$\nu_+^{\text{el}} = \nu_-^{\text{el}}. \tag{25}$$

The Poisson's ratio for yield criteria in the isotropic case can be computed in a similar way as relations of the strain or strain rates

$$\nu^{\text{in}} = -\left(\frac{\partial\,\Phi}{\partial\,\sigma_{\text{II}}}\Big/\frac{\partial\,\Phi}{\partial\,\sigma_{\text{I}}}\right)\Bigg|_{\sigma_{\text{II}}=\sigma_{\text{III}}=0} = -\left(\frac{\partial\,\Phi}{\partial\,\sigma_{\text{III}}}\Big/\frac{\partial\,\Phi}{\partial\,\sigma_{\text{I}}}\right)\Bigg|_{\sigma_{\text{II}}=\sigma_{\text{III}}=0} \tag{26}$$

using the flow rule [126, 167, 234]

$$\dot{\varepsilon}_{ij} = \dot{\lambda}\,\frac{\partial\,\Phi}{\partial\,\sigma_{ij}}, \qquad \dot{\lambda} > 0 \tag{27}$$

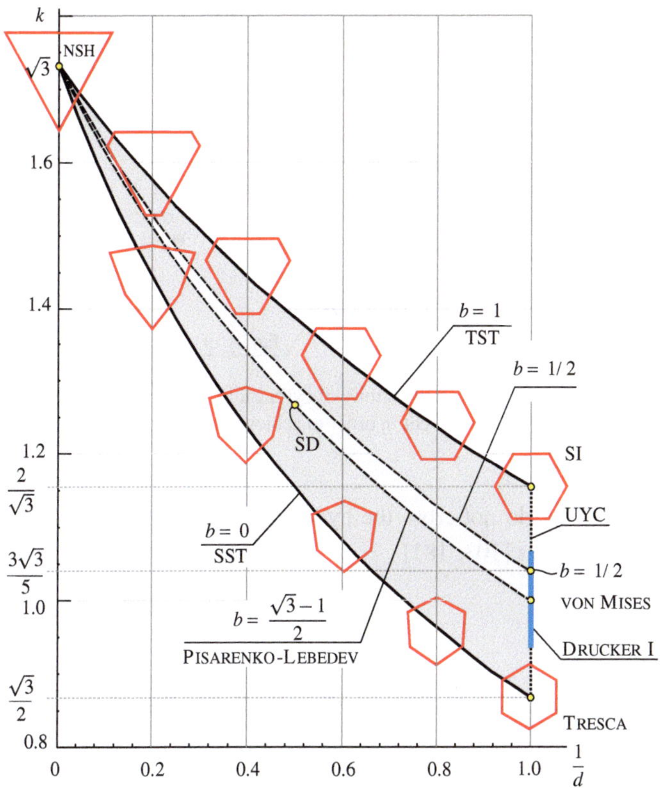

Fig. 5 $1/d$ versus k for the classical models (Sect. 2) and for the Unified Strength Theory of Yu (Sect. 10.2) as a function of $d \geq 1$ and $b \in [0,\ 1]$:
$d \geq 1, k \in \left[1,\ \sqrt{3}\right]$—Pisarenko-Lebedev model (48),
$d = 1, k \in \left[(2/3)^{1/6},\ (3/2)^{1/6}\right]$—model of Drucker I (Sect. 9.2),
$d = 1, k \in \left[\sqrt{3}/2,\ 2/\sqrt{3}\right]$—Unified Yield Criterion of Yu with $b \in [0,\ 1]$ (Sect. 9.2),
$d \to \infty, k = \sqrt{3}$—normal stress hypothesis (Sect. 2.1).
SI—model of Schmidt-Ishlinsky (Sect. 2.4), SD—model of Sdobirev with $d = 2$, $k = 3 - \sqrt{3} \approx 1.27$ (Sect. 6.3), UYC—Unified Yield Criterion (Sect. 9.2), SST—Single-Shear-Theory of Yu (model of Mohr-Coulomb), Sect. 10.2, TST—Twin-Shear-Theory of Yu. The cross-sections in the π-plane on $I_1 = \sigma_{eq}$ are provided for better understanding [104, 105, 115]

for $\dot{\varepsilon}_{II}$ and $\dot{\varepsilon}_{I}$. Further definitions of ν^{in} are given in [109]. There is a difference between the inelastic Poisson's ratios at tension ν^{in}_{+} with $\sigma_I = \sigma_+$ and compression ν^{in}_{-} with $\sigma_I = -d\,\sigma_+$, $d \geq 0$.

The convexity condition (necessary condition) for the meridian with the angle $\theta = 0$ and the associated point Z (tension) yields [105]

$$\nu^{in}_{+} \in \left] -1,\ \frac{1}{2}\right]. \tag{28}$$

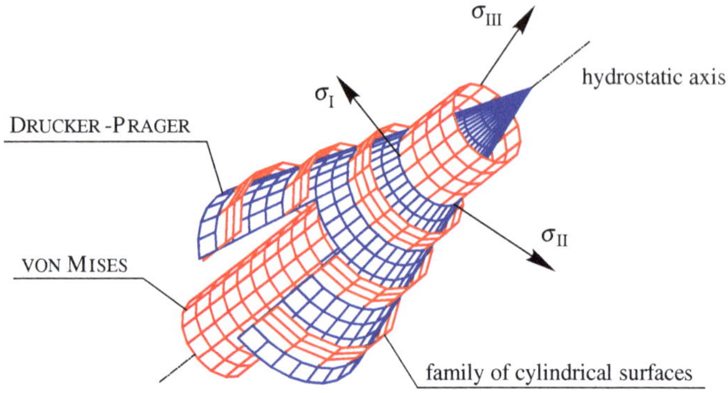

Fig. 6 *Cone* of Drucker-Prager with the family of cylindric surfaces and cylinder of von Mises in the principal stress space (the *cone* is cut in order to achieve better visualization)

For the surfaces that do not cross the hydrostatic axis in the compression region ($a_{-}^{\text{hyd}} \to \infty$), it follows [103, 181]

$$\nu_{-}^{\text{in}} \geq \frac{1}{2}. \tag{29}$$

For those surfaces (e. g. Drucker-Prager, Mohr-Coulomb, Pisarenko-Lebedev) a non-associated flow rule with

$$\nu_{-}^{\text{in}} = \frac{1}{2} \tag{30}$$

is often used [103]. The model of Drucker-Prager can be used as an example (Fig. 6). In this figure for each stress state in the region $I_1 \leq 0$ a cylindrical surface is defined. This results in a “family” of rings, which define the incompressible behavior for each stress state in the compression region.

For closed surfaces in the principal stress space will be assumed [105, 216]

$$\nu_{-}^{\text{in}} \in \left] -1, \ \frac{1}{2} \right]. \tag{31}$$

The restriction can be clarified in the $\left(\sqrt{3\,I_2'}, \ I_1\right)$-plane (Sect. 8.2): the maximum of a meridian lies in the region

$$\frac{I_1}{\sigma_+} \in [-d, \ 1].$$

Using (26) the inclination of the tangent line at the points Z (tension) and D (compression) of the surface Φ in the principal stress space with respect to the hydrostatic axis can be characterized:

- from the inclination $\psi = 0$ (tangent line parallel to the hydrostatic axis) follows $\nu^{\text{in}} = 1/2$ (Fig. 2, model of von Mises) and
- the inclination $\psi = \pi/2$ (tangent line is orthogonal to the hydrostatic axis) yields $\nu^{\text{in}} = -1$.

The Poisson's ratio $\nu_{+}^{\text{in}} = 0$ (Fig. 2, normal stress hypothesis (8)) yields[3] the inclination angle with

$$\tan \psi = \sqrt{2} \left[\frac{3}{2\,(1 + \nu_{+}^{\text{in}})} - 1 \right] \tag{32}$$

to $\psi = \arctan\left(\sqrt{2}/2\right) \approx 35.26°$ in the principal stress space (Fig. 7).

This geometrical interpretation of (26) can be used for description of the limit surface Φ. It completes the relations (Table 1) for the analytical comparison of different surfaces by fitting of measurements.

The following estimates are available for evaluation of the quality of the model:

- For ductile materials the experience-based inequality

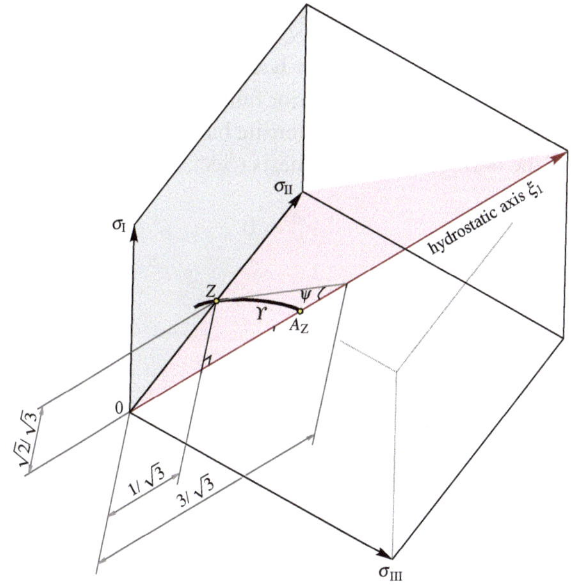

Fig. 7 Principal stress space ($\sigma_{\text{I}}, \sigma_{\text{II}}, \sigma_{\text{III}}$): inclination of the tangent line $\psi \approx 35.26°$ at the point Z (tension) of the surface Φ correlates with the Poisson's ratio at tension $\nu_{+}^{\text{in}} = 0$, Υ—intersection of the surface Φ with the surface ($\xi_1, 0, \sigma_{\text{II}}$) or meridian with $\theta = 0$ of the surface Φ

---

[3] Here the substitution $I_3' = \frac{2\sqrt{3}}{3^2}\,(I_2')^{3/2}$ is used, which corresponds to the meridian with the angle $\theta = 0$ (Sect. 8.2). The point Z (tension) belongs to this meridian (Table 1).

$$\nu_+^{\text{in}} \in \left[\nu_+^{\text{el}}, \frac{1}{2}\right] \tag{33}$$

can be used in order to check the quality of fitting to the measured data.

- The term "slight compressibility", see [140], can estimated by [110]

$$\nu_+^{\text{in}} \in \left[0.48, \frac{1}{2}\right]. \tag{34}$$

This range is recommended for yield criteria.

- For "very ductile" behavior [126] it can be required additionally

$$\nu_+^{\text{in}} \to \frac{1}{2}, \tag{35}$$

and one gets the desired parameters of the yield criterion in the fitting (Sect. 12).

- For brittle material behavior the following constraints can be formally written down

$$\nu_+^{\text{in}} \in]-1,\ \nu_+{}^{\text{el}}], \tag{36}$$

cf. the maximum strain hypothesis (Sect. 6.1) for the upper bound and the strain hypotheses with $\sigma_{\text{eq}} = I_1$ for the lower bound.

- For "absolutely" brittle material behavior failure occurs without plastic deformations in the cross sectional area of the tensile bar [206, 208]. In this case it can be assumed from the normal stress hypothesis (Sect. 2)

$$\nu_+^{\text{in}} \approx 0. \tag{37}$$

5 Ratios for a Torsion Bar

In addition to value k, one can define other values relating to torsion using the yield condition (27). By analogy to the Poisson's ratio the elongation/contraction ratio for a torsion bar or tubes can be established, cf. [8]:

$$\chi = \frac{\partial \Phi}{\partial \sigma_{11}} \Big/ \frac{\partial \Phi}{\partial \sigma_{12}} \tag{38}$$

with $\sigma_{12} = k\,\sigma_{\text{eq}}/\sqrt{3}$ and $\sigma_{11} = \sigma_{22} = \sigma_{33} = \sigma_{13} = \sigma_{23} = 0$. With the help of this ratio the Poynting-effect and the Poynting-Swift-effect [160–162, 200] can be described (Figs. 8 and 9). These effects are discussed in the literature, see [3, 8, 16, 27, 29, 69, 81, 168, 206, 215] among others. The material behavior of the von Mises-type results in $\chi = 0$. In the case of the application of the normal stress hypothesis (Sect. 2) one gets $\chi = 1/2$ [6, 110].

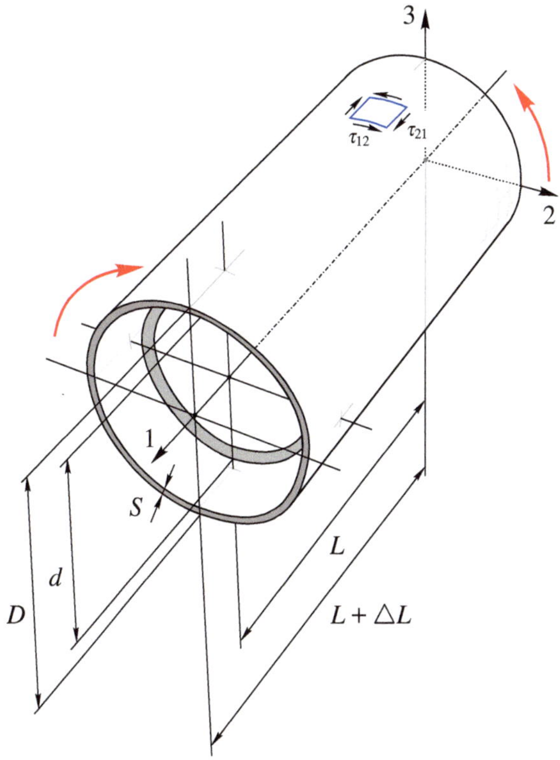

Fig. 8 Change of the geometry of a tube clamped on the right side at torsion

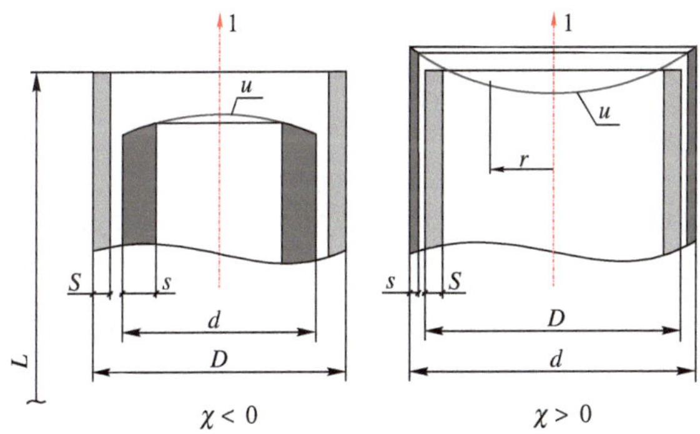

Fig. 9 Change of the geometry of a tube at torsion. D and d—outer diameter of the tube before and after loading, S and s—wall thickness before and after loading, $u(r)$—distribution law for the displacements in the direction 1

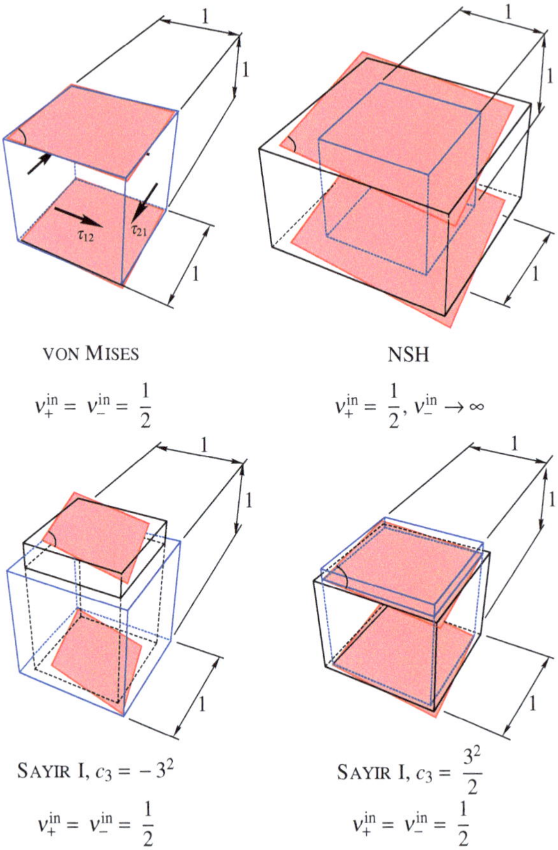

Fig. 10 Change of the geometry and the shape of a cube with the size $1 \times 1 \times 1$ under stress action τ_{12} for four settings of the Sayir I model (Table 3)

Furthermore, the volume strain caused by torsion (volume dilatation, see Kelvin-effect [8, 16]) can be computed as follows:

$$\varepsilon_{V\tau} = \left(\frac{\partial \Phi}{\partial \sigma_{11}} + \frac{\partial \Phi}{\partial \sigma_{22}} + \frac{\partial \Phi}{\partial \sigma_{33}}\right) \Big/ \frac{\partial \Phi}{\partial \sigma_{12}}. \tag{39}$$

In addition, a transverse contraction ratio can be obtained [110]:

$$\nu_\tau^{in} = -\frac{\partial \Phi}{\partial \sigma_{33}} \Big/ \frac{\partial \Phi}{\partial \sigma_{11}}. \tag{40}$$

By this way one gets more information about the material behavior from the torsion test. Some special cases are analyzed in Fig. 10 and Table 3 on the base of model of

Table 3 Effects computed at torsion considering the model of Sayir I (10) with the linear substitution (11)

Cross section in the π-plane (Fig. 4)	a	m			b	
$c_3 \in \left[-3^2,\ \dfrac{3^2}{2}\right]$	-3^2	0			$\dfrac{3^2}{2}$	
$\gamma_1 \in [0,\ 1[$			0			$\dfrac{1}{3}$
$\nu_+^{\text{in}} = \dfrac{1}{2}\,(1 - 3\,\gamma_1)$			$\dfrac{1}{2}$			0
Classical model	–		von Mises	–		NSH
$k = \dfrac{1}{1-\gamma_1}\sqrt{1 + \dfrac{2}{3^3}\,c_3}$	$\dfrac{1}{\sqrt{3}}$	1		$\dfrac{2}{\sqrt{3}}$		$\sqrt{3}$
d	$\dfrac{1}{2}$	1		2		∞
$\chi = \sqrt{3}\sqrt{1 + \dfrac{2}{3^3}\,c_3}\,\dfrac{c_3\,(1-\gamma_1) + 2\cdot 3^2\,\gamma_1}{2\cdot 3^3\,(1-\gamma_1)}$	$-\dfrac{1}{6}$	0		$\dfrac{1}{6}$		$\dfrac{1}{2}$
$\nu_\tau^{\text{in}} = 2\,\dfrac{c_3\,(1-\gamma_1) - 3^2\,\gamma_1}{c_3\,(1-\gamma_1) + 2\cdot 3^2\,\gamma_1}$	2	$\dfrac{0}{0}$		2		0
$\varepsilon V_\tau = \sqrt{3}\,\dfrac{\gamma_1}{1-\gamma_1}\sqrt{1 + \dfrac{2}{3^3}\,c_3}$			0			1

The von Mises hypothesis describes the torsion of the tubular specimen without elongation $\chi = 0$. The hypotheses for incompressible material behavior (von Mises and Sayir I with $c_3 = -3^2$ und $c_3 = 3^2/2$) reflect no volume change. The hypotheses of von Mises and of normal stress do not result in changes of the wall thickness $\partial\,\Phi/\partial\,\sigma_{33} = 0$, cf. the Mohr-Coulomb hypothesis [73, 79] and Pisarenko-Lebedev hypothesis [157].

Sayir I (10). It is obvious that the influence of I_1 and I_3' cannot be separated by these measurements.

For rotational-symmetric models with $a_-^{\text{hyd}} \to \infty$ one obtains $\chi \geq 0$. For various materials the cross-sections of the surface Φ in the π-plane result in the part g − b − h of the d–k-diagram (Fig. 4, $d \geq 1$) yielding $\chi > 0$. From this it follows that the length and the diameter increase at torsion is more realistic, cf. [25, 168, 206]. This effect is significantly influenced by:

- defects in the material,
- material anisotropy,
- loading-induced anisotropy,
- technological characteristics,
- deviation of the specimen shape from the ideal geometry,
- non-coaxial fixation of the specimen in the testing machine,
- temperature changing during the test, etc.

This is the reason for different suggestions to describe such behavior in the literature. The ratios χ, εV_τ and ν_τ^{in} are properties of the model.

If the torsion test is carried out with $\varepsilon_{11} = 0$ (strain is constrained), the axial force can be computed using the flow rule [172]. The sign of this force gives clues about the shape of the surface Φ.

6 Standard Criteria

The standard models (strain hypothesis, model of Mohr-Coulomb, model of Pisarenko-Lebedev and model of Burzyński-Yagn [157, 194, 234]) are frequently used models for first approximations of measurements: they are easy to handle, can be used to describe different material types (brittle-ductile range) and their parameters can be obtained using simple tests.

6.1 Strain Criterion

Strain model (strain hypothesis) is obtained assuming Hooke's law [20, 64, 205]

$$\sigma_{\mathrm{I}} - \nu_+^{\mathrm{in}}\,(\sigma_{\mathrm{II}} + \sigma_{\mathrm{III}}) - \sigma_{\mathrm{eq}} = 0. \tag{41}$$

The other two equations are obtained by cyclic permutation of indices. The model (10) together with the substitution (11) and the parameter values

$$c_3 = \frac{3^2}{2}, \qquad \gamma_1 = \frac{1}{3}\left(1 - 2\,\nu_+^{\mathrm{in}}\right). \tag{42}$$

gives rise to the representation in invariants.

The Poisson's ratio for the strain hypothesis follows, cf. [88, 201]

$$\nu_-^{\mathrm{in}} = \frac{1}{2}\,(d-1)\,, \qquad \nu_+^{\mathrm{in}} = \frac{1}{d}. \tag{43}$$

Further values are

$$d \geq 2, \qquad k = \frac{\sqrt{3}}{1+\nu_+^{\mathrm{in}}}, \qquad a_+^{\mathrm{hyd}} = \frac{1}{1-2\,\nu_+^{\mathrm{in}}}. \tag{44}$$

This model contains

- the normal stress hypothesis with $\nu_+^{\mathrm{in}} \doteq 0$ (Sect. 2),
- the maximum strain hypothesis[4] (Mariotte [134], Navier [146], St. Venant [175], Poncelet [159], Grashof [77], Resal [170] or Bach [15]) with $\nu_+^{\mathrm{el}} = \nu_+^{\mathrm{in}}$.

The limit cases are the following surfaces Φ

- triangular prism in the principal stress space with $\nu_-^{\mathrm{in}} = \nu_+^{\mathrm{in}} = 1/2$, [8, 31, 66, 97, 174, 177, 192] and
- plane through point Z orthogonal to hydrostatic axis with $\nu_+^{\mathrm{in}} = -1$ [97, 102].

[4] This hypothesis is analyzed in [38, 65, 94, 220, 221]. It does not reflect the experimental results [22, 50, 78, 157, 213] and is used in combinations of various hypotheses (Sect. 11).

6.2 Mohr-Coulomb Criterion

The model is introduced on the basis of geometrical considerations [76, 127, 141–143]. It arises as equations, which are obtained by permutation of indices in

$$\left[\sigma_{\mathrm{I}} - \frac{1}{d}\,\sigma_{\mathrm{II}} - \sigma_{\mathrm{eq}}\right]\left[\sigma_{\mathrm{I}} - \frac{1}{d}\,\sigma_{\mathrm{III}} - \sigma_{\mathrm{eq}}\right]\left[\sigma_{\mathrm{II}} - \frac{1}{d}\,\sigma_{\mathrm{III}} - \sigma_{\mathrm{eq}}\right] = 0. \tag{45}$$

The formulation in invariants is given in Sect. 10.2. The model leads to values

$$\nu_{-}^{\mathrm{in}} = \frac{d}{2}, \qquad \nu_{+}^{\mathrm{in}} = \frac{1}{2\,d}, \qquad d \geq 1, \tag{46}$$

and [129]

$$k = \sqrt{3}\,\frac{d}{d+1}, \qquad a_{+}^{\mathrm{hyd}} = \frac{d}{d-1}. \tag{47}$$

This model yields (Fig. 5, SST)

- with $d \to \infty$ to the normal stress hypothesis and
- with $d = 1$ to the model of Tresca.

The relation $d \geq 2$ is recommended if computations involving so called fatigue limits should be performed [30].

6.3 Pisarenko-Lebedev Criterion

The model is presented by

$$(1-\xi)\sqrt{\frac{1}{2}\left[(\sigma_{\mathrm{I}} - \sigma_{\mathrm{II}})^2 + (\sigma_{\mathrm{II}} - \sigma_{\mathrm{III}})^2 + (\sigma_{\mathrm{III}} - \sigma_{\mathrm{I}})^2\right]} + \xi\, max[\sigma_{\mathrm{I}},\, \sigma_{\mathrm{II}},\, \sigma_{\mathrm{III}}] = \sigma_{\mathrm{eq}} \tag{48}$$

with $\xi \in [0,\ 1]$ [82, 120, 121, 156, 157]. This is a linear combination of the equivalent stresses after the normal stress hypothesis ($\xi = 1$) and the von Mises hypothesis ($\xi = 0$). The relations compute to [123]

$$d = \frac{1}{1-\xi}, \qquad k = \frac{3}{3 + \left(-3 + \sqrt{3}\right)\xi}, \qquad a_{+}^{\mathrm{hyd}} = \frac{1}{\xi}, \tag{49}$$

and the Poisson's ratio

$$\nu_{+}^{\mathrm{in}} = \frac{1-\xi}{2}, \qquad \nu_{-}^{\mathrm{in}} = \frac{1}{2} + \frac{\xi}{1-\xi}. \tag{50}$$

The model of Sdobirev [191] follows with $\xi = 1/2$. The relations are $d = 2$, $k = 3 - \sqrt{3} \approx 1.27$, $a_{+}^{\mathrm{hyd}} = 2$, $\nu_{+}^{\mathrm{in}} = 1/4$ and $\nu_{-}^{\mathrm{in}} = 3/2$ (Fig. 5). The average values of the parameter ξ for some materials are given in [124].

6.4 Burzyński-Yagn Criterion

The rotationally symmetric model evolves the energy consideration and is a function of two parameters [16, 36, 37, 220]

$$3\, I_2' = \frac{\sigma_{\mathrm{eq}} - \gamma_1\, I_1}{1 - \gamma_1} \frac{\sigma_{\mathrm{eq}} - \gamma_2\, I_1}{1 - \gamma_2}. \tag{51}$$

The values k and d compute to

$$d = \frac{1}{1 - \gamma_1 - \gamma_2}, \qquad k^2 = \frac{1}{(1 - \gamma_1)(1 - \gamma_2)}. \tag{52}$$

The position of the hydrostatic nodes one gets from

$$\left(1 - 3\, \gamma_1\, a_{+}^{\mathrm{hyd}}\right) \left(1 - 3\, \gamma_2\, a^{\mathrm{hyd}}\right) = 0. \tag{53}$$

The Poisson's ratios at tension and compression are obtained using

$$\nu_{+}^{\mathrm{in}} = \frac{-1 + 2\,(\gamma_1 + \gamma_2) - 3\, \gamma_1\, \gamma_2}{-2 + \gamma_1 + \gamma_2}, \tag{54}$$

and

$$\nu_{-}^{\mathrm{in}} = -\frac{-1 + \gamma_1^2 + \gamma_2^2 - \gamma_1\, \gamma_2}{(-2 + \gamma_1 + \gamma_2)\,(-1 + \gamma_1 + \gamma_2)}. \tag{55}$$

The model (51) represents the general equation of a second order surface of revolution about the hydrostatic axis in the principal stress space. In dependence of the parameter values γ_1 and γ_2 one gets:

- cone of Drucker-Prager [56], Mirolyubov [138] (Fig. 6) with equal parameters $\gamma_1 = \gamma_2 \in [0,\ 1]$;
- paraboloid of Balandin [17], Burzyński-Torre [35–37, 207, 235] (Fig. 11) with $\gamma_1 \in [0,\ 1]$, $\gamma_2 = 0$;
- ellipsoid of Beltrami (strain energy hypothesis) [21], see Fig. 12, assuming $\gamma_1 = -\gamma_2 \in [0,\ 1]$ or with equal Poisson's ratios $\nu_{-}^{\mathrm{in}} = \nu_{+}^{\mathrm{in}} \in]-1,\ 1/2]$;
- ellipsoid of Schleicher [179, 180] with $\gamma_1 \in [0,\ 1]$, $\gamma_2 \in [-\infty,\ 0]$. The restriction (31) yields $\nu_{-}^{\mathrm{in}} \in]-1,\ 1/2]$, $\nu_{+}^{\mathrm{in}} \in [-1,\ 1/2]$;
- hyperboloid of Burzyński-Yagn [39, 221] with $\gamma_1 \in [0,\ 1]$ and $\gamma_2 \in [0,\ \gamma_1]$;

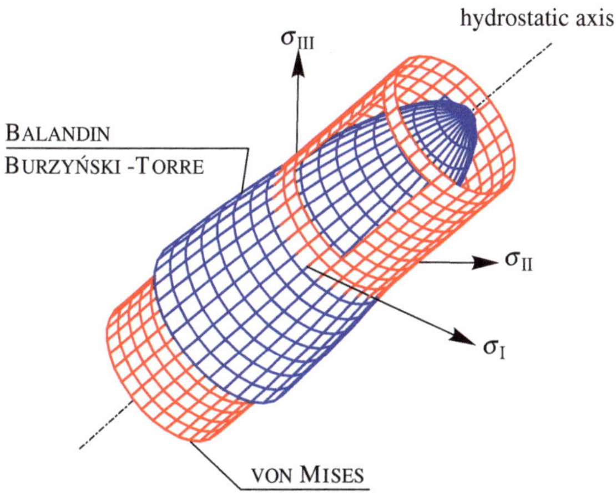

Fig. 11 Paraboloid of Balandin, Burzyński-Torre and the cylinder of von Mises in the principal stress space (the cylinder is cut for better understanding) [103]

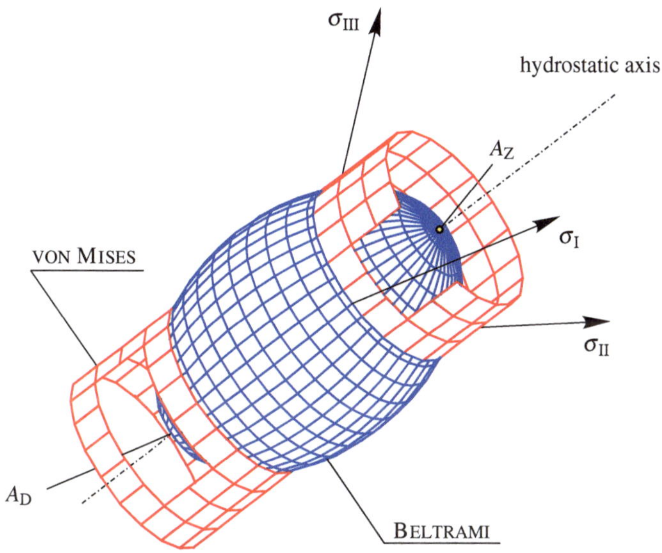

Fig. 12 Ellipsoid of Beltrami and cylinder of von Mises in the principal stress space (the cylinder is cut for better understanding) [103]

- hyperboloid of one sheet with complex conjugate γ_1 and γ_2 which is not convex (see Sect. 10.1); this model is used as partial surface in combined models (Sect. 11.2.1).

The models differ by the symmetry type in the π-plane (Fig. 4) and by the power of stresses n. The rotationally symmetric model (51) has the stress power $n = 2$. The strain hypothesis and the model of Pisarenko-Lebedev has the stress power $n = 3$

and the model of Mohr-Coulomb—$n = 6$. The last three models have a trigonal symmetry in the in the π-plane.

The models can be characterized by the shape of the meridian line. The strain hypothesis, the hypothesis of Mohr-Coulomb, the model of Pisarenko-Lebedev, the models of von Mises and of Drucker-Prager have a straight line as the meridian. The model (51) has additional to the straight line curvilinear meridians: parabola, hyperbola and ellipse.

The cross-sections of models (41) and (51) in the π-plane are unchangable. The form in the π-plane and the inclination of the meridian line of the models of Mohr-Coulomb and Pisarenko-Lebedev are controlled by a single parameter. This limits the capabilities of the models to be fitted to measured data.

There are no theoretical or experimental evidence known to support the application of models with a straight line as the meridian line. Models with further shapes of the meridian lines and independent from the shapes in the π-plane (Fig. 4) are required.

7 Mathematical Formulations

In the case of phenomenological models some mathematical framework is often applied for the formulation. The aim is to establish some general equation which includes classical models as special cases. The following three formulation ideas are known:

7.1 Criterion of Altenbach-Zolochevsky I

The criterion [7, 8]

$$\sigma_{\text{eq}} = \sqrt{3\, I_2'}\ (\lambda_1\ \sin\varphi + \lambda_2\ \cos\varphi + \lambda_3) + I_1\ (\lambda_4 + \lambda_5\ \sin\varphi + \lambda_6\ \cos\varphi) \qquad (56)$$

is a combination of the first invariant of the stress tensor, the second invariant of the stress deviator and the stress angle.[5] Various special cases can be deduced by different settings of λ_i (Table 4). It should be mentioned that the systematization of models can be based on the number of these parameters.

In the formulation of this model the following relations between the principal stresses σ_{I}, σ_{II}, σ_{III} and the invariants (Appendix 15) [42, 43, 147, 149, 211, 233] are used:

---

[5] In the original papers the following definition of the stress angle is used

$$\sin 3\varphi = -\frac{3\sqrt{3}}{2}\frac{I_3'(\boldsymbol{s})}{I_2'(\boldsymbol{s})^{3/2}}, \qquad |\varphi| \leq \frac{\pi}{6}.$$

Table 4 Settings of the criterion of Altenbach-Zolochevsky I (56)

<table>
<tr><th>Model</th><th>λ_1</th><th>λ_2</th><th>λ_3</th><th>λ_4</th><th>λ_5</th><th>λ_6</th><th>Reference /Equation</th></tr>
<tr><td>von Mises</td><td colspan="2">0</td><td>1</td><td colspan="3">0</td><td>(16)</td></tr>
<tr><td>Tresca</td><td>0</td><td>$\dfrac{2\sqrt{3}}{3}$</td><td colspan="4">0</td><td>(13)</td></tr>
<tr><td>SH, $\nu_+^{\text{in}} = \dfrac{1}{2}$</td><td>$-\dfrac{1}{2}$</td><td>$\dfrac{\sqrt{3}}{2}$</td><td colspan="4">$0$</td><td>(41)</td></tr>
<tr><td>SH, $\nu_+^{\text{in}} = \nu_+^{\text{el}}$</td><td>$-\dfrac{1+\nu_+^{\text{el}}}{3}$</td><td>$(1+\nu_+^{\text{el}})\ \dfrac{\sqrt{3}}{3}$</td><td>$0$</td><td>$\dfrac{1-2\,\nu_+^{\text{el}}}{3}$</td><td colspan="2">$0$</td><td>(41)</td></tr>
<tr><td>NSH</td><td>$-\dfrac{1}{3}$</td><td>$\dfrac{\sqrt{3}}{3}$</td><td>0</td><td>$\dfrac{1}{3}$</td><td colspan="2">0</td><td>(8)</td></tr>
<tr><td>Sdobirev</td><td>$-\dfrac{1}{6}$</td><td>$\dfrac{\sqrt{3}}{6}$</td><td>$\dfrac{1}{2}$</td><td>$\dfrac{1}{6}$</td><td colspan="2">0</td><td>(48)</td></tr>
<tr><td>Mohr-Coulomb</td><td>$\dfrac{1}{3}\left(\dfrac{1}{d}-1\right)$</td><td>$\dfrac{\sqrt{3}}{3}\left(\dfrac{1}{d}+1\right)$</td><td>$0$</td><td>$\dfrac{1}{3}\left(1-\dfrac{1}{d}\right)$</td><td colspan="2">$0$</td><td>(45)</td></tr>
<tr><td>Drucker-Prager</td><td colspan="2">0</td><td>$\dfrac{1}{2}\left(1+\dfrac{1}{d}\right)$</td><td>$\dfrac{1}{2}\left(1-\dfrac{1}{d}\right)$</td><td colspan="2">$0$</td><td>(51)</td></tr>
<tr><td>Pisarenko-Lebedev</td><td>$\dfrac{1}{3}\left(\dfrac{1}{d}-1\right)$</td><td>$\dfrac{\sqrt{3}}{3}\left(1-\dfrac{1}{d}\right)$</td><td>$\dfrac{1}{d}$</td><td>$\dfrac{1}{3}\left(1-\dfrac{1}{d}\right)$</td><td colspan="2">$0$</td><td>(48)</td></tr>
<tr><td>Sandel</td><td>0</td><td>$\dfrac{\sqrt{3}}{3}\left(1+\dfrac{1}{d}\right)$</td><td>$0$</td><td>$\dfrac{1}{2}\left(1-\dfrac{1}{d}\right)$</td><td colspan="2">$0$</td><td>(58)</td></tr>
<tr><td>Edelman-Drucker</td><td>0</td><td>$\dfrac{2\sqrt{3}}{3}\ (1-\chi)$</td><td>$\chi$</td><td colspan="3">$0$</td><td>(59)</td></tr>
<tr><td>Tsvelodub</td><td>λ_1</td><td>λ_2</td><td>λ_3</td><td colspan="3">0</td><td>[211]</td></tr>
<tr><td>Paul</td><td>$\dfrac{1}{3}\ (2\,a_2-a_1-a_3)$</td><td>$\dfrac{\sqrt{3}}{3}\ (a_1-a_3)$</td><td>$0$</td><td>$\dfrac{1}{3}\ (a_1+a_2+a_3)$</td><td colspan="2">$0$</td><td>(60)</td></tr>
<tr><td>Birger</td><td>$\dfrac{1}{3}\ (2\,a_2-a_1-a_3)$</td><td>$\dfrac{\sqrt{3}}{3}\ (a_1-a_3)$</td><td>$a_4$</td><td>$\dfrac{1}{3}\ (a_1+a_2+a_3)$</td><td colspan="2">$0$</td><td>(61)</td></tr>
<tr><td>Tarasenko</td><td>0</td><td>λ_2</td><td>0</td><td>λ_4</td><td>λ_5</td><td>λ_6</td><td>[201]</td></tr>
</table>

$$
\begin{aligned}
\sigma_{\mathrm{I}} &= \frac{1}{3}\left[2\sqrt{3\,I_2'}\,\sin\left(\varphi+\frac{2\,\pi}{3}\right)+I_1\right],\\
\sigma_{\mathrm{II}} &= \frac{1}{3}\left[2\sqrt{3\,I_2'}\,\sin(\varphi)+I_1\right],\\
\sigma_{\mathrm{III}} &= \frac{1}{3}\left[2\sqrt{3\,I_2'}\,\sin\left(\varphi+\frac{4\,\pi}{3}\right)+I_1\right],
\end{aligned}
\tag{57}
$$

and $\sigma_{\mathrm{I}} \geq \sigma_{\mathrm{II}} \geq \sigma_{\mathrm{III}}$ is assumed.

This model includes not only the standard models (Sect. 6). The following models can be obtained:

- criterion of Sandel [176]

$$
\sigma_{\mathrm{eq}} = \sigma_{\mathrm{I}} + \frac{1}{2}\left(1-\frac{1}{d}\right)\sigma_{\mathrm{II}} - \frac{1}{d}\sigma_{\mathrm{III}},
\tag{58}
$$

- convex combination of von Mises and Tresca models after Edelman-Drucker, Koval'chuk [57, 117]

$$
\chi\sqrt{3\,I_2'} + (1-\chi)\,(\sigma_{\mathrm{I}}-\sigma_{\mathrm{III}}) = \sigma_{\mathrm{eq}}, \qquad \chi \in [0,\ 1],
\tag{59}
$$

- criterion of Paul [154]

$$
a_1\,\sigma_{\mathrm{I}} + a_2\,\sigma_{\mathrm{II}} + a_3\,\sigma_{\mathrm{III}} = \sigma_{\mathrm{eq}},
\tag{60}
$$

- criterion of Birger [30]

$$
a_1\,\sigma_{\mathrm{I}} + a_2\,\sigma_{\mathrm{II}} + a_3\,\sigma_{\mathrm{III}} + a_4\sqrt{3\,I_2'} = \sigma_{\mathrm{eq}}.
\tag{61}
$$

A disadvantage of this model is the number of parameters, which should be identified by six independent tests.

7.2 Model in Terms of the Integrity Basis

This model results from the invariants I_1, $\left(I_2'\right)^{1/2}$ and I_3', cf. [23, 24, 51, 188]. The basic idea is the formulation of scalar valued functions of a given order:

$$
\begin{aligned}
S_1 &= a_1\,I_1 + b_1\,\left(I_2'\right)^{1/2},\\
S_2 &= a_2\,I_1^2 + b_2\,I_2',\\
S_3 &= a_3\,I_1^3 + b_3\,\left(I_2'\right)^{3/2} + c_3\,I_3 + d_3\,I_1\,I_2 + e_3\,I_1^2\,\left(I_2'\right)^{1/2},\\
&\dots
\end{aligned}
\tag{62}
$$

The sum of the S_i with the same power n yields

$$(S_1)^n + (S_2)^{n/2} + (S_3)^{n/3} + \ldots = \sigma_{eq}^n. \tag{63}$$

The choice of integer exponents n/i is recommended for the terms S_i. In [103] the following exponents are suggested

$$n = 1,\ 2,\ 3,\ 6,\ 9 \ \text{and}\ 12. \tag{64}$$

The model of Altenbach-Zolochevsky II follows with $n = 1$ for S_1, S_2 and S_3 [2, 8, 103]. The parameters of the model are related to the scaling (3). The advantage of this model is that the equivalent stress σ_{eq} can be expressed explicitly.

Another modification one gets if the weight σ_{eq}^{n-i} for S_i is introduced

$$\sigma_{eq}^{n-1} S_1 + \sigma_{eq}^{n-2} S_2 + \sigma_{eq}^{n-3} S_3 + \ldots + \sigma_{eq} S_{n-1} + S_n = \sigma_{eq}^n. \tag{65}$$

By this way we get the same power of the stresses in each term [103]. The exponent $n > 1$ and the terms in (65) can be selected in such a manner that an analytical solution is possible with respect to σ_{eq}. Equation (51) is an example of a quadratic equation, models which are given cubic, bi-cubic and tri-quadratic equations are

$$S_1 \sigma_{eq}^2 + S_2 \sigma_{eq} + S_3 = \sigma_{eq}^3, \tag{66}$$

$$S_2 \sigma_{eq}^4 + S_4 \sigma_{eq}^2 + S_6 = \sigma_{eq}^6, \tag{67}$$

$$S_3 \sigma_{eq}^3 + S_6 = \sigma_{eq}^6. \tag{68}$$

More examples are presented in [103]. Disadvantages of this approach can be summarized as follows:

- increasing number of parameters,
- difficult convexity limits for the parameters and
- missing geometrical interpretation of the parameters.

7.3 *Models Based on the Stress Deviator*

The functions of the invariants of the stress deviator can be defined as follows

$$\begin{aligned}
S_2' &= b_2 I_2', \\
S_3' &= b_3 \left(I_2'\right)^{3/2} + c_3 I_3, \\
S_4' &= b_4 \left(I_2'\right)^2 + f_4 \left(I_2'\right)^{1/2} I_3, \\
&\ldots
\end{aligned} \tag{69}$$

The sum of S_i' with the same power results in

$$\left(S_2'\right)^{n/2} + \left(S_3'\right)^{n/3} + \left(S_4'\right)^{n/4} + \ldots = \sigma_{eq}^n \tag{70}$$

and

$$\sigma_{eq}^{n-2}\, S_2' + \sigma_{eq}^{n-3}\, S_3' + \ldots + \sigma_{eq}\, S_{n-1}' + S_n' = \sigma_{eq}^n, \tag{71}$$

cf. Eqs. (63) and (65). Another possibility is [106]

$$\sigma_{eq}^{n-2\,m_2}\,\left(S_2'\right)^{m_2} + \sigma_{eq}^{n-3\,m_3}\,\left(S_3'\right)^{m_3} + \ldots + \sigma_{eq}\,\left(S_{n-1}'\right) + \left(S_n'\right) = \sigma_{eq}^n. \tag{72}$$

The formulations (70), (71) and (72) yield in the models of incompressible material behavior (Sect. 9): Sayir I, Drucker I, Dodd-Naruse, TQM, BCM, Spitzig, Iyer, Freudenthal [16, 68, 90] and Maitra [131, 226].

Multiplicative combinations of various S_i' are possible, for example,

$$\left(S_2'\right)^{(n-j)/2}\, S_j' = \sigma_{eq}^n. \tag{73}$$

This equation results in the geometrical-mechanical model (Sect. 9.1.6).

The formulation of the models with the deviatoric basis (69) should be preferred since they are simpler in comparison with models on the basis of Eq. (62). The compressible generalization can be performed using the substitution presented in Sect. 10.1. In the case of rational functions of I_3' (functions of I_3' with integer power) one gets convex shapes in the π-plane.

8 Visualization Methods

Several possibilities of the visualization of the limit surface Φ are presented in the literature. In this section main approaches are briefly discussed and examples are given.

8.1 Spatial Representation of the Limit Surface

Strength hypotheses and flow criteria can be represented in the principal stress space (σ_{I}, σ_{II}, σ_{III}) [43, 219, 234], which is also known as the Haigh-Westergaard space [80, 218]. By means of an orthogonal transform the decomposition of the stress tensor in the hydrostatic and deviatoric components can be carried out. For this purpose the coordinates (ξ_1, ξ_2, ξ_3) are introduced (Fig. 13), which are related to the coordinates (σ_{I}, σ_{II}, σ_{III}) as follows [23, 123, 186, 187, 210]:

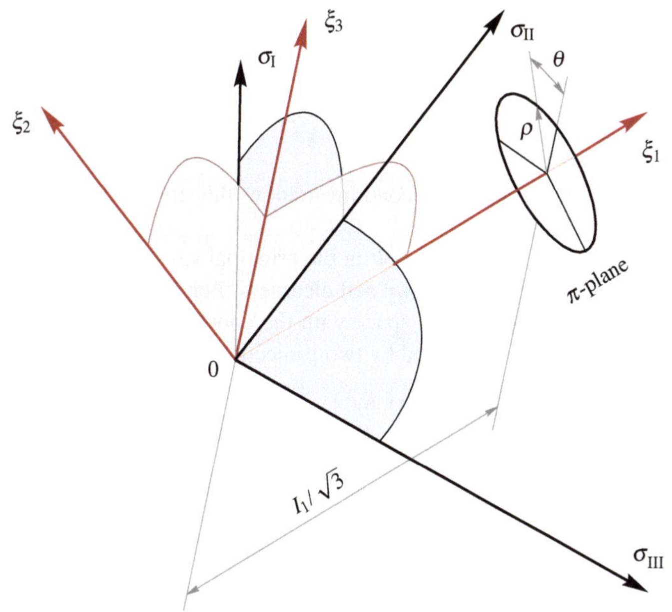

Fig. 13 Principal stress space (σ_{I}, σ_{II}, σ_{III}), coordinates (ξ_1, ξ_2, ξ_3) and (ξ_1, ρ, θ)

$$\begin{pmatrix} \sigma_{\mathrm{I}} \\ \sigma_{\mathrm{II}} \\ \sigma_{\mathrm{III}} \end{pmatrix} = \begin{pmatrix} \frac{1}{\sqrt{3}} & \frac{1}{\sqrt{2}} & -\frac{1}{\sqrt{6}} \\ \frac{1}{\sqrt{3}} & 0 & \frac{2}{\sqrt{6}} \\ \frac{1}{\sqrt{3}} & -\frac{1}{\sqrt{2}} & -\frac{1}{\sqrt{6}} \end{pmatrix} \begin{pmatrix} \xi_1 \\ \xi_2 \\ \xi_3 \end{pmatrix}. \tag{74}$$

In these coordinates

$$\xi_1 = \frac{1}{\sqrt{3}} (\sigma_{\mathrm{I}} + \sigma_{\mathrm{II}} + \sigma_{\mathrm{III}}) = \frac{1}{\sqrt{3}} I_1 \tag{75}$$

is the hydrostatic axis ($\sigma_{\mathrm{I}} = \sigma_{\mathrm{II}} = \sigma_{\mathrm{III}}$). The axis

$$\xi_3 = \frac{1}{\sqrt{6}} (-\sigma_{\mathrm{I}} + 2\,\sigma_{\mathrm{II}} - \sigma_{\mathrm{III}}) \tag{76}$$

lies in the plane $\xi_1 - \sigma_{\mathrm{II}}$. The axis

$$\xi_2 = \frac{1}{\sqrt{2}} (\sigma_{\mathrm{I}} - \sigma_{\mathrm{III}}) \tag{77}$$

constitutes together with the axes ξ_1 and ξ_3 an orthogonal coordinate system (Fig. 13).

The surface Φ can be formulated in cylindrical coordinates or the Haigh-Westergaard coordinates $(\xi_1,\ \rho,\ \theta)$ [234]. The value of the radius ρ is computed to [23, 43]

$$\rho = \sqrt{2\, I_2'} = \sqrt{\xi_2^2 + \xi_3^2} \tag{78}$$

and represents a function of the second invariant of the stress deviator. The angle θ is given by Eq. (224).

The representation of the surface Φ in the principal stress space $(\sigma_{\mathrm{I}},\ \sigma_{\mathrm{II}},\ \sigma_{\mathrm{III}})$ is widely used because of its simplicity and clearness. For the analysis of the surface properties the Haigh-Westergaard space with the coordinates $(\xi_1,\ \xi_2,\ \xi_3)$ is better. Such a surface can be characterized by two projections:

- the meridian cross section $(\xi_1,\ \xi_3)$ and
- the planes with the cuts $\xi_1 = \text{const.}$ (π-plane with the coordinates $(\xi_2,\ \xi_3)$).

8.2 *Burzyński-Plane*

Instead of the meridian cross section $(\xi_1,\ \xi_3)$ the Burzyński-plane is often used. The Burzyński-plane is introduced for the rotationally symmetric models [35, 39]

$$\Phi\left(I_1,\ I_2',\ \sigma_{\mathrm{eq}}\right) = 0. \tag{79}$$

For these models the surface is represented by a line in the upper half-plane in the diagram $(\xi_1,\ \rho)$ (Fig. 14, model of von Mises).

The whole surface Φ is obtained by the rotation of this line about the axis ξ_1 [234]. For a better clearness the coordinates $\left(I_1,\ \sqrt{3\, I_2'}\right)$ are used [103]. This scaling is due to the relation

$$I_1^2 = 3\, I_2' \tag{80}$$

at tension and compression. It simplifies the comparison of the presented models with the model of von Mises (16) and leads to a geometrical interpretation of the relations (Table 1). Other scalings are used in [86, 95, 101, 133, 155, 174, 181, 194, 231, 234]. A normalization of axes with respect to the tensile stress

$$\left(\frac{I_1}{\sigma_+},\ \frac{\sqrt{3\, I_2'}}{\sigma_+}\right)$$

can be applied in order to compare the shape of the surfaces for different materials [105, 113].

The surface Φ (5) or (6) can be presented in the Burzyński-plane by the meridians defined using the stress angle [136, 145, 155]

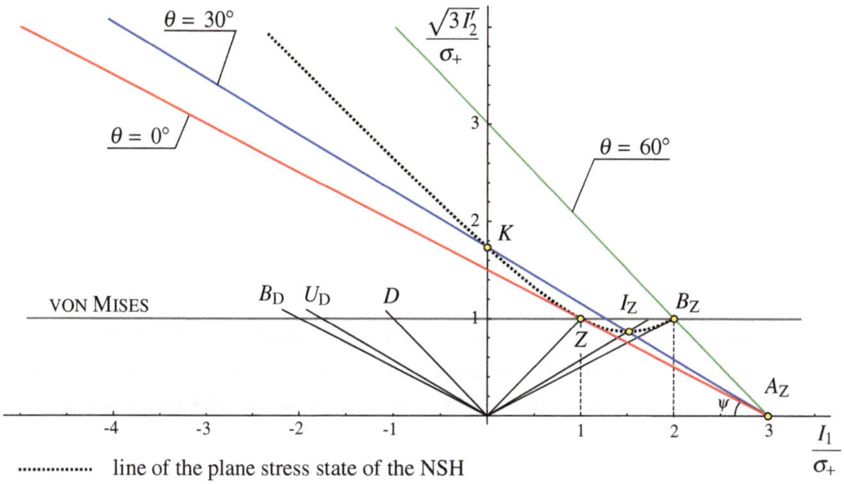

Fig. 14 Normal stress hypothesis (NSH) and the model of von Mises in the Burzyński-plane [105]. The inclination of the tangent line at the point Z (tension) of the surface Φ is $\psi = \arctan 1/2 \approx 27.56°$, cf. (Fig. 7)

- $\theta = 0 \Rightarrow I_3' = \dfrac{2\sqrt{3}}{3^2}\,(I_2')^{3/2}$,
- $\theta = \dfrac{\pi}{6} \Rightarrow I_3' = 0$,
- $\theta = \dfrac{\pi}{3} \Rightarrow I_3' = -\dfrac{2\sqrt{3}}{3^2}\,(I_2')^{3/2}$.

These meridians for the normal stress hypothesis (8) are shown in Fig. 14.

The three meridians are enough to display the most important properties (Table 1). The line of the plane stress state, which results from

$$\sigma_{\mathrm{I}}\,\sigma_{\mathrm{II}}\,\sigma_{\mathrm{III}} = 0, \tag{81}$$

is obtained in this diagram using the substitution [103]

$$I_3' = \frac{1}{3}\,I_1\,I_2' - \frac{1}{3^3}\,I_1^3. \tag{82}$$

This line contains the points B_{D}, U_{D}, D, K, Z, I_{Z} and B_{Z} (Figs. 15 and 16) and it is convex for axisymmetric models only, e. g. the model of Burzyński-Yagn (51), Fig. 17.

The introduced representation allows to show all the measurements on their respective meridians. Different extrapolations of the measurements to the point A_{Z} (hydrostatic tension) with the relation

$$a_+^{\text{hyd}} > \frac{1}{3} \quad \text{or} \quad \nu_+^{\text{in}} \in \left[-1, \frac{1}{2}\right]$$

can be easily evaluated.

8.3 π-Plane

The cross-section of the surface Φ with the cut $\xi_1 = \text{const.}$ (Fig. 13) is denoted as the π-plane [12, 43, 167, 234]. For incompressible material behavior these cross-sections do not depend on the coordinate ξ_1. For a compressible material it is important to consider the cross-sections, which contain certain points of the plane stress state for the most important loading cases, e. g.:

- point Z (tension): $\xi_1 = \dfrac{1}{\sqrt{3}} \sigma_{\text{eq}}$,
- point K (torsion): $\xi_1 = 0$ and
- point D (compression): $\xi_1 = -\dfrac{1}{\sqrt{3}} d\,\sigma_{\text{eq}}$.

The line of the plane stress state defined, for instance, by the condition $\sigma_{\text{III}} = 0$ can be projected onto the π-plane. The projection is given by the equality, see (74),

$$\xi_1 = \frac{1}{\sqrt{2}} (\sqrt{3}\, \xi_2 + \xi_3), \tag{83}$$

which must be substituted into the equation of the surface Φ.

In the most general case the surface Φ has a trigonal symmetry in the π-plane (Fig. 4). If only even powers of the third invariant of deviators are present, then the model has a hexagonal symmetry (e. g. models of Tresca and Schmidt-Ishlinsky, Fig. 4). The absence of the third invariant leads to a rotationally symmetric surface, e. g. Eq. (51) [105].

9 Pressure-Insensitive Criteria

In this section the most important models with the property

$$\nu_+^{\text{in}} = \nu_-^{\text{in}} = \frac{1}{2} \tag{84}$$

are discussed. These models are of the form

$$\Phi(I_2', I_3', \sigma_{\text{eq}}) = 0 \tag{85}$$

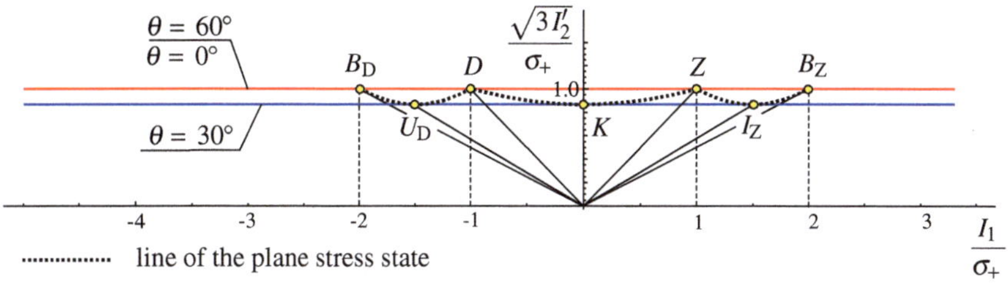

Fig. 15 Model of Tresca in the Burzyński-plane [103]

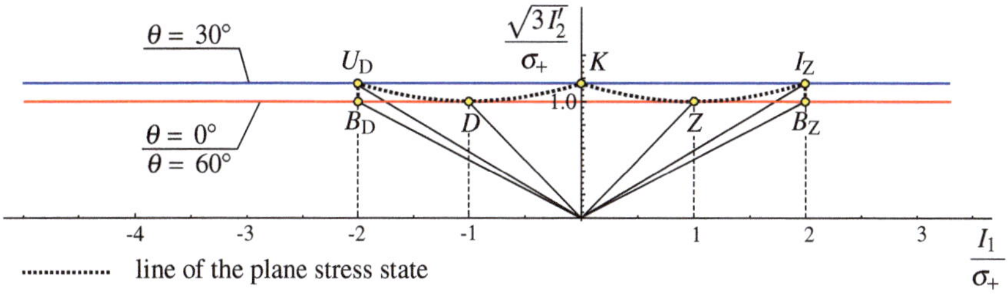

Fig. 16 Model of Schmidt-Ishlinsky in the Burzyński-plane

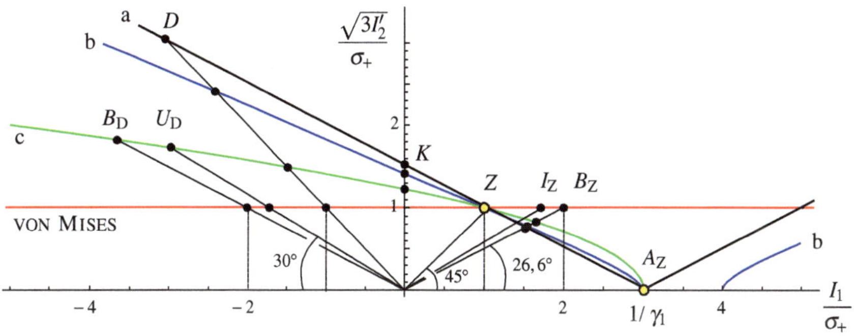

Fig. 17 Quadratic rotationally symmetric models (51) with the hydrostatic node A_Z ($\gamma_1 = 1/3$) in the Burzyński-plane [105]: *a* cone of Drucker-Prager: $\gamma_2 = 1/3$, $\nu_-^{\text{in}} = 2$, $\nu_+^{\text{in}} = 0$, $d = 3$, $k = 3/2$; *b* hyperboloid of Burzyński-Yagn: $\gamma_2 = 1/4$, $\nu_-^{\text{in}} = 1.54$, $\nu_+^{\text{in}} = 0.06$, $d = 2.4$, $k = 1.41$; *c* paraboloid of Balandin: $\gamma_2 = 0$, $\nu_-^{\text{in}} = 0.8$, $\nu_+^{\text{in}} = 0.2$, $d = 3/2$, $k = 1.22$

or

$$\Phi(I_2', \theta, \sigma_{\text{eq}}) = 0, \tag{86}$$

and hence are cylindric or prismatic surfaces aligned along the hydrostatic axis. They do not restrict the hydrostatic stresses. These models can only be used in the region $I_1 \leq 0$ in combined models (Sect. 11), cf. [139].

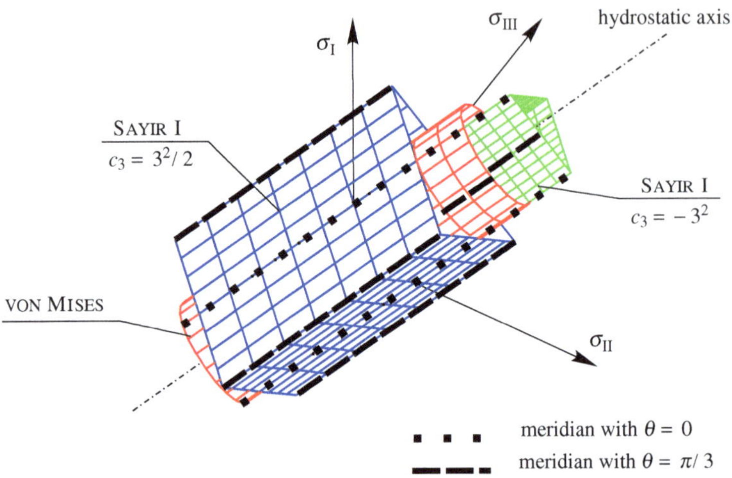

Fig. 18 *Triangular prisms* of Sayir I (87) with the limit convexity values of the parameter $c_3 \in \left[-3^2,\ 3^2/2\right]$ and the cylinder of von Mises (16) in the principal stress space [103]

9.1 Yield Surfaces with Trigonal Symmetry

Theoretical considerations allow conclusions about the symmetry of the yield surface Φ in the π-plane only. This surface shows trigonal, hexagonal or rotational symmetry. No suggestions can be made based on microstructure. The material behavior is described by neglecting the real structure with its microscopic defects and inhomogeneity. The effects of material behavior could be captured correctly only in average [8].

9.1.1 Model of Sayir I

The cylinder of Sayir I [178] is defined as follows

$$\frac{\sigma_{\text{eq}}\, 3\, I_2' + c_3\, I_3'}{1 + 2\, c_3/3^3} = \sigma_{\text{eq}}^3, \qquad c_3 \in \left[-3^2,\ \frac{3^2}{2}\right]. \tag{87}$$

The model has the structure of the reduced cubic equation with respect to σ_{eq}. For $c_3 = 0$ the cylinder of von Mises arises (Fig. 18). The relations k and d compute to

$$k^2 = 1 + \frac{2}{3^3}\, c_3, \qquad d = \frac{3^3 + 2\, c_3 - \sqrt{3\,(3^2 - 2\, c_3)\,(3^3 + 2\, c_3)}}{2^2\, c_3} \tag{88}$$

and shown in Fig. 4.

9.1.2 Model of Sayir II

The hexagonal prism of Sayir II [178] is defined by

$$\sigma_{\mathrm{I}} - \frac{1}{1+b_1}\,(b_1\,\sigma_{\mathrm{II}} + \sigma_{\mathrm{III}}) - \sigma_{\mathrm{eq}} = 0, \qquad b_1 \in \left[-\frac{1}{2},\, 1\right]. \tag{89}$$

Further equations result from the cyclic permutations of indices. The model is formulated in the deviatoric invariants

$$\begin{aligned}\Phi_{\mathrm{SAY}} &= \alpha_{41}\,\sigma_{\mathrm{eq}}^4\,I_2' + \alpha_{31}\,\sigma_{\mathrm{eq}}^3\,I_3' + \alpha_{21}\,\sigma_{\mathrm{eq}}^2\,(I_2')^2 \\ &\quad + \alpha_{11}\,\sigma_{\mathrm{eq}}\,I_2'\,I_3' + \beta_{21}\,(I_2')^3 + \beta_{31}\,(I_3')^2 - \sigma_{\mathrm{eq}}^6\end{aligned} \tag{90}$$

with

$$\begin{aligned}\alpha_{41} &= \frac{2\cdot 3\,(1+b_1+b_1^2)}{(1+b_1)^2}, & \alpha_{31} &= \frac{3^3\,b_1}{(1+b_1)^2}, \\ \alpha_{21} &= -3^2\,\frac{(1+b_1+b_1^2)^2}{(1+b_1)^4}, & \alpha_{11} &= -3^4\,b_1\,\frac{1+b_1+b_1^2}{(1+b_1)^4}, \\ \beta_{21} &= \frac{(-1+b_1)^2\,(2+b_1)^2\,(1+2b_1)^2}{(1+b_1)^6}, & \beta_{31} &= -\frac{3^3\,(1+b_1+b_1^2)^3}{(1+b_1)^6}.\end{aligned} \tag{91}$$

For $b_1 = 0$ the model of Tresca (15) arises (Figs. 1, 15). With $b_1 = -1/2$ and $b_1 = 1$ the model corresponds to the limit convexity cases of the model of Sayir I (87).

The relations are

$$k = \sqrt{3}\,\frac{1+b_1}{2+b_1}, \qquad d = 1 + b_1. \tag{92}$$

The model is representing the lower bound of the convexity region in the d–k-diagram (Fig. 4). For the model of Sayir II the point, which has the shortest distance to the point $M(1,\ 1)$ can be obtained from the equation

$$(d-1)^2 + (k-1)^2 \rightarrow \text{min.}, \tag{93}$$

which results in $b_1 \approx 0.0471$, $d \approx 1.05$ and $k \approx 0.89$.

9.1.3 Model of Haythornthwaite

The model of Haythornthwaite [40, 83, 107] consists of two overlapping triangles in the π-plane described with the model (87) with $c_3 = -3^2$ and $c_3 = 3^2/2$

$$\left[\frac{(2\,d\,\sigma_{\mathrm{eq}})\,3\,I_2' + (-3^2)\,I_3'}{1+2\,(-3^2)/3^3} - (2\,d\,\sigma_{\mathrm{eq}})^3\right]\left[\frac{\sigma_{\mathrm{eq}}\,3\,I_2' + (3^2/2)\,I_3'}{1+2\,(3^2/2)/3^3} - \sigma_{\mathrm{eq}}^3\right] = 0. \tag{94}$$

It is separated in two regions in the $d-k$ diagram (Fig. 4)

$$Haythornthwaite\,I: \quad k = \frac{2}{\sqrt{3}} \qquad \text{for} \quad d \in [1,\, 2], \tag{95}$$

$$Haythornthwaite\,II: \quad k = \frac{2d}{\sqrt{3}} \qquad \text{for} \quad d \in \left[\frac{1}{2},\, 1\right]. \tag{96}$$

The model is representing the upper bound of the convexity region of the $d-k$-diagram. The boundaries, where $d = 1/2$ and $d = 2$, correspond to the limit convexity cases of the models of Sayir I and Sayir II. For $d = 1$ the model of Schmidt-Ishlinsky (18) arises (Figs. 1, 16). The point, which has the shortest distance to the point $M(1,\, 1)$ (Fig. 4), can be obtained from Eq. (93). This results in $d = (3 + 2\sqrt{3})/7 \approx 0.9234$ and $k \approx 1.07$.

9.1.4 Convex π-Plane Model

The model of Haythornthwaite (94)

$$\begin{aligned}\Phi_{\mathrm{HAY}} = {} & \frac{3^6}{2^6}\frac{1}{d^3}\,(I_3')^2 - \frac{3^5}{2^5}\frac{d-1}{d^3}\,I_2'\,I_3'\,\sigma_{\mathrm{eq}} - \frac{3^4}{2^4}\frac{1}{d^2}\,(I_2')^2\,\sigma_{\mathrm{eq}}^2 \\ & - \frac{3^3}{2^3}\frac{1-d^3}{d^3}\,I_3'\,\sigma_{\mathrm{eq}}^3 + \frac{3^2}{2^2}\frac{1+d^2}{d^2}\,I_2'\,\sigma_{\mathrm{eq}}^4 - \sigma_{\mathrm{eq}}^6\end{aligned} \tag{97}$$

and the model of Sayir II (90) with the parameters

$$\begin{aligned} \alpha_{41} &= 6\,\frac{(d-1)^2+d}{d^2}, & \alpha_{31} &= 3^3\,\frac{d-1}{d^2}, \\ \alpha_{21} &= -3^2\,\frac{\left((d-1)^2+d\right)^2}{d^4}, & \alpha_{11} &= -3^4\,\frac{(d-1)\left((d-1)^2+d\right)}{d^4}, \\ \beta_{21} &= \frac{(1-2\,d)^2\,(d-2)^2\,(1+d)^2}{d^6}, & \beta_{31} &= -3^3\,\frac{\left((d-1)^2+d\right)^3}{d^6} \end{aligned} \tag{98}$$

are functions of $d \in [1/2,\, 2]$. With the linear combination [33]

$$\Phi_6 = \xi\,\Phi_{\mathrm{HAY}} + (1-\xi)\,\Phi_{\mathrm{SAY}}, \qquad \xi \in [0,\, 1] \tag{99}$$

one obtains the model with the power of stress $n = 6$ in each term. The resulting model describes with two parameters $(d,\, \xi)$ all points in the $d-k$-diagram (Fig. 4) with a convex form in the π-plane. A drawback is that an explicit solution of (99) with respect to σ_{eq} is not possible.

The model contains as special cases:

- the model of hexagonal symmetry in the π-plane with $d = 1$ (bicubic model, Sect. 9.2),
- the value $k = 1$ of the model of von Mises results with $d = 1$ in the parameter $\xi = \frac{2^6}{7\cdot 13} \approx 0.7033$ and
- the approximation of the model of Sayir I (87) with $\xi = \frac{2^6}{7\cdot 13}$ and $\xi \in [0,\ 1]$.

9.1.5 Radcig Model

The Radcig model[6] consists of two overlapping hexagonal prisms of Sayir II (89) [105, 115]. It is deduced from the Unified Yield Criterion of Yu (Sect. 9.2.1). The defining equations are:

$$\begin{cases} \sigma_{\mathrm{I}} - \dfrac{1}{1+b_1}(b_1\,\sigma_{\mathrm{II}} + \sigma_{\mathrm{III}}) - \sigma_{\mathrm{eq}} = 0, \\ \sigma_{\mathrm{I}} - \dfrac{1}{1+b_2}(b_2\,\sigma_{\mathrm{II}} + \sigma_{\mathrm{III}}) + \eta\,\sigma_{\mathrm{eq}} = 0. \end{cases} \tag{100}$$

Further equations are obtained by cyclic permutations of indices. The cross-section of this model in the π-plane is a dodecagon (twelve-sided figure). The parameters are bounded as follows

$$b_1 \in \left[-\frac{1}{2},\ 1\right], \qquad b_2 \in \left[-\frac{1}{2},\ 1\right], \qquad \eta \in \left[\frac{1}{1+b_2},\ 2\right]. \tag{101}$$

The model (100) can be formulated in invariants of the deviator [105]:

$$\begin{aligned} &\big(\alpha_{41}\,\sigma_{\mathrm{eq}}^4\,I_2' + \alpha_{31}\,\sigma_{\mathrm{eq}}^3\,I_3' + \alpha_{21}\sigma_{\mathrm{eq}}^2\,(I_2')^2 \\ &\qquad + \alpha_{11}\sigma_{\mathrm{eq}}\,I_2'\,I_3' + \beta_{21}\,(I_2')^3 + \beta_{31}\,(I_3')^2 - \sigma_{\mathrm{eq}}^6\big) \times \\ &\big[\alpha_{42}\,(\eta\,\sigma_{\mathrm{eq}})^4\,I_2' - \alpha_{32}\,(\eta\,\sigma_{\mathrm{eq}})^3\,I_3' + \alpha_{22}\,(\eta\,\sigma_{\mathrm{eq}})^2\,(I_2')^2 \\ &\qquad - \alpha_{12}\,(\eta\,\sigma_{\mathrm{eq}})\,I_2'\,I_3' + \beta_{22}\,(I_2')^3 + \beta_{32}\,(I_3')^2 - (\eta\,\sigma_{\mathrm{eq}})^6\big] = 0. \end{aligned} \tag{102}$$

This allows a representation of the model in the Burzyński-plane. The first part corresponds to the model of Sayir II (90) with parameters which are given by (91). The parameters of the second part are computed using the substitution b_2 by b_1. The Radcig model contains the following models:

- the model of Sayir II with $b_1 \in [-1/2,\ 1]$, $\eta = 2$ and for arbitrary $b_2 \in [-1/2,\ 1]$ or $b_1 = \eta - 1$, $b_2 = (1-\eta)/\eta$ and $\eta \in [1/2,\ 2]$ with the relations (92),
- the model of Haythornthwaite with $b_1 = 1$, $b_2 = 1$ and $\eta \in [1/2,\ 2]$ with relations (95) and (96),

[6] This model is dedicated to Jurij Antonovič Radcig (1900–1976), who was a professor at the Kazan State University of Technology (KAI), Kazan, Russia.

- the continuous approximation of the model of Sayir I with $\eta \in [1/2,\ 2]$ and

$$\begin{aligned} b_1 &= (1-\sqrt{3})\,\eta^2 + \frac{1}{2}\left(5\sqrt{3}-3\right)\,\eta - \sqrt{3}, \\ b_2 &= (2-\sqrt{3})\,\eta^2 + \left(\frac{5}{2}\sqrt{3}-6\right)\,\eta + \left(\frac{7}{2}-\sqrt{3}\right). \end{aligned} \tag{103}$$

- the UYC of Yu with $\eta = 1$ and $b = b_1 = b_2 \in [0,\ 1]$ (Sect. 9.2) with

$$k = \sqrt{3}\,\frac{1+b_1}{2+b_1}, \qquad d = 1. \tag{104}$$

Further discussion of the Radcig-model can be found in the Sect. 10.2.

9.1.6 Geometrical-Mechanical Model

This model is a function of the stress angle θ (224) [6, 33, 107]

$$(3\,I_2')^{\frac{n}{2}}\,\frac{1+c_3\cos 3\theta + c_6\cos^2 3\theta}{1+c_3+c_6} = \sigma_{\mathrm{eq}}^n. \tag{105}$$

The main idea is to include the influence of the odd and the even functions of I_3' separately and to give a geometrical meaning to the parameters with respect to the mechanical properties. Computation of the equivalent stress σ_{eq} can be performed directly.

The two parameters c_3 and c_6 determine the geometry of the model in the π-plane. With $c_3 = 0$ a model with hexagonal symmetry is obtained: there is no difference between tension and compression ($d = 1$). With $c_3 = c_6 = 0$ one gets the model of von Mises.

The values d and k compute to

$$d^n = \frac{1+c_3+c_6}{1-c_3+c_6}, \qquad k^n = 1+c_3+c_6. \tag{106}$$

This leads to the two inequalities

$$1-c_3+c_6 > 0, \qquad 1+c_3+c_6 \geq 0. \tag{107}$$

The recommended values for the exponent are $n = 2, 3$ and 6. The exponent $n = 2$ is suitable, if the modeling with rational compressible substitution (Sect. 10.1) involves energy considerations [212]. The values $n = 2$ and 3 allow to solve the equation given by (105) with respect to σ_{eq} analytically even if the rational compressible substitution (Sect. 10.1) is introduced. The convexity region of the geometrical-mechanical model (105) in the parameter space c_3–c_6 is shown in Figs. 19 and 20).

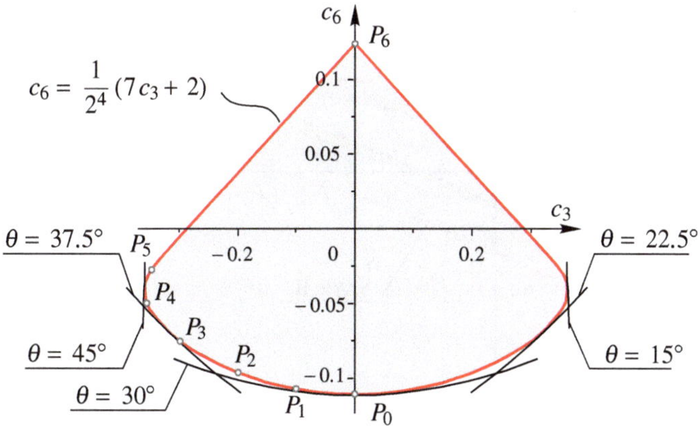

Fig. 19 Convexity region of the geometrical-mechanical model (105) with $n = 2$: with θ some curves are marked, which build the boundary of the convexity region, Coordinates at the points $P_0(0, -1/9)$, $P_1(-0.1, -0.1086)$, $P_2(-0.2, -0.1011)$, $P_3(-0.3, -0.0886)$, $P_4(-0.3560, -0.05)$, $P_5(-0.3478, -0.02717)$, $P_6(0, 1/2^3)$

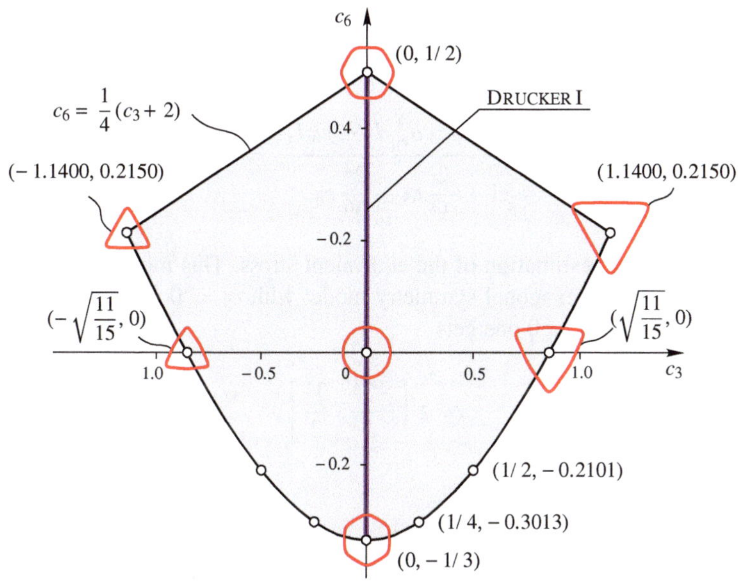

Fig. 20 Convexity region of the geometrical-mechanical model (105) with $n = 6$ (the cross-sections in the π-plane are shown for clarity)

With the exponent $n = 6$ a model is obtained, which has the largest convexity region in the d–k-diagram (Figs. 4 and 21).

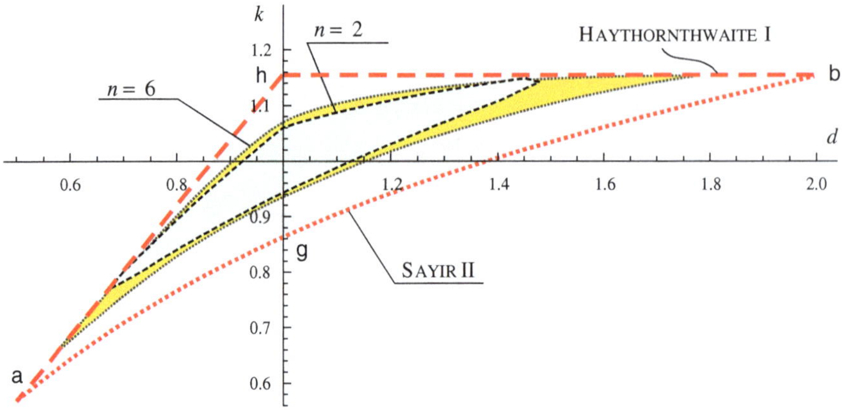

Fig. 21 Convexity condition of the geometrical-mechanical model (105) in the d–k-diagram together with the boundaries of the convexity region as presented in Fig. 4

9.1.7 Triquadratic Model

The triquadratic model is formulated as follows [103, 105, 107]

$$\frac{3^3\, I_2'^3 + c_3\, \sigma_{\mathrm{eq}}^3\, I_3' + c_6\, I_3'^2}{1 + \dfrac{2}{3^3}\, c_3 + \dfrac{2^2}{3^6}\, c_6} = \sigma_{\mathrm{eq}}^6. \tag{108}$$

It allows analytical estimation of the equivalent stress. This model with the power $n = 6$ contains the hexagonal symmetry model with $c_3 = 0$ (model of Drucker I, Sect. 9.2) and with $c_6 = 0$ one gets

$$c_3 \in \left[-\frac{3^3}{2^2}, \frac{3^3}{2}\right]. \tag{109}$$

The relations are

$$k^6 = 1 + \frac{2}{3^3} c_3 + \frac{2^2}{3^6} c_6, \qquad d^3 = \frac{k^6}{1 + \dfrac{2^2}{3^6}\, c_6}. \tag{110}$$

The idea for this model is similar to the model (105). Comparing with (105) we conclude that the triquadratic model (TQM) is more difficult to use. It should be noted that the parameters c_3, c_6 have no mechanical or geometrical meaning.

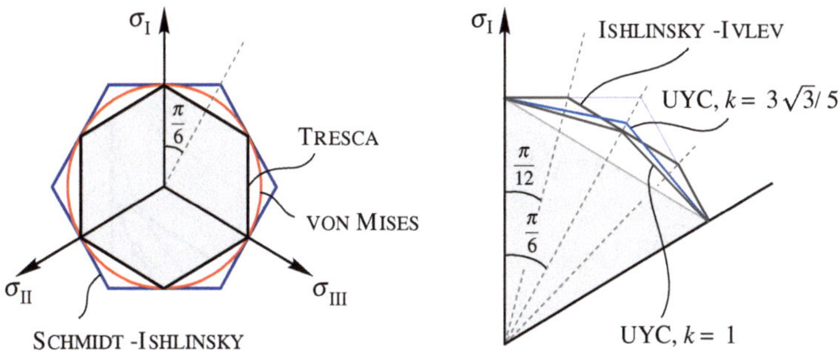

Fig. 22 Continuous surfaces with hexagonal symmetry and the model of von Mises (16) in the π-plane, incompressible material behavior, $d = 1$ [135]. On the *right hand side* an enlarged cross-section with $\theta \in [0, \pi/3]$ is presented [33]

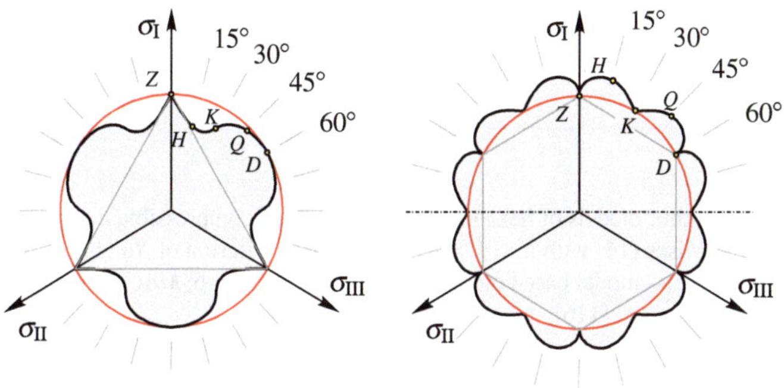

Fig. 23 Non-convex models for incompressible material behavior with $d = 1$ in the π-plane: *left hand side* model of triangular symmetry, *right hand side* model of hexagonal symmetry [96]. The rotationally symmetric model of von Mises is presented for comparison

9.2 Yield Surfaces with Hexagonal Symmetry

The models for incompressible material behavior with hexagonal symmetry have the properties $d = 1$ and $h = q$ (Table 1, Figs. 22 and 23). Such models are often used for the description of yield of ideal ductile materials in the theory of plasticity. Numerous problems are treated in the engineering practice using these criteria. These models are of the form

$$\Phi(I_2', (I_3')^2, \sigma_{\rm eq}) = 0, \qquad \Phi(I_2', \cos^2 2\theta, \sigma_{\rm eq}) = 0$$

The meridians with $\theta = 0$ and $\pi/3$ coincide in the Burzyński-plane (Figs. 15 and 16). Such models can be represented in the $h{-}k$-diagram (Fig. 24) and compared with von Mises model with $h = k = 1$.

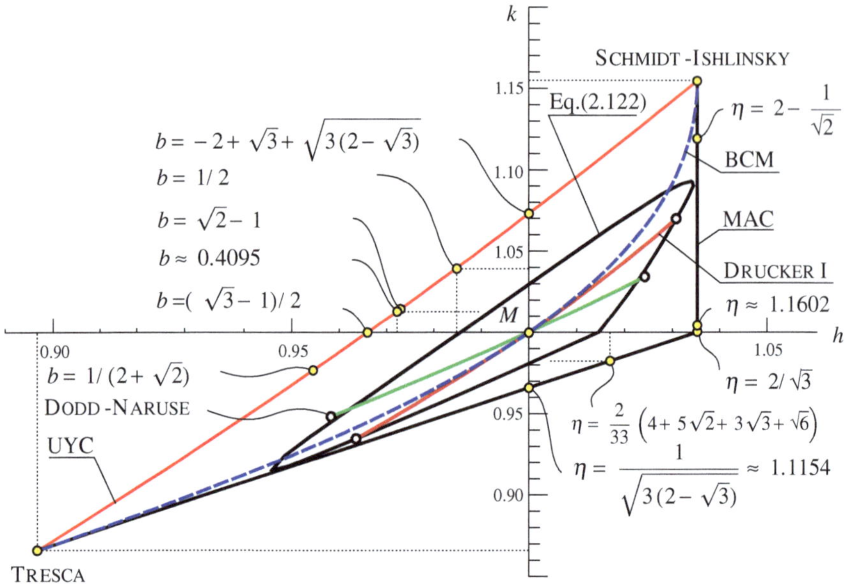

Fig. 24 $h - k$-diagram: models of hexagonal symmetry for incompressible material behavior: *M*—model of von Mises (16) with $h = k = 1$; *UYC*—yield criterion of Yu (111); *BCM*—bicubic model (113); Eq. (122)—model based on the stress angel with $n = 6$; *MAC*—multiplicative ansatz criterion, special points—s. [116]

9.2.1 Unified Yield Criterion of Yu

The criterion of Yu [222, 229] with the parameter $b \in [0, 1]$ can be expressed

$$\begin{cases} \sigma_{\text{I}} - \dfrac{1}{1+b}(b\,\sigma_{\text{II}} + \sigma_{\text{III}}) - \sigma_{\text{eq}} = 0, \\ \sigma_{\text{I}} - \dfrac{1}{1+b}(b\,\sigma_{\text{II}} + \sigma_{\text{III}}) + \sigma_{\text{eq}} = 0, \end{cases} \tag{111}$$

and using Eq. (102) formulated in terms of the invariants of the deviator [105]

$$\begin{aligned} \Phi_{\text{UYC}} = {} & (\alpha_{41}\,\sigma_{\text{eq}}^4\,I_2' + \alpha_{21}\,\sigma_{\text{eq}}^2\,I_2'^2 + \beta_{21}\,I_2'^3 + \beta_{31}\,I_3'^2 - \sigma_{\text{eq}}^6)^2 \\ & - (\alpha_{31}\,\sigma_{\text{eq}}^3\,I_3' + \alpha_{11}\,\sigma_{\text{eq}}\,I_2'\,I_3')^2 \end{aligned} \tag{112}$$

with the coefficients (91). In Table 5 some settings for special cases are presented.

The yield criterion of Yu defines the left convexity bound of the models with hexagonal symmetry in the $h - k$-diagram (Fig. 24, UYC).

Table 5 Settings of UYC (111) and the respective relations k and h (Fig. 24)

Model	b	k	h	Fig.
Tresca	0	$\frac{\sqrt{3}}{2} \approx 0.8660$	0.8966	1, 15, 22
–	$\frac{1}{2+\sqrt{2}} \approx 0.2929$	0.9767	0.9545	–
Sokolovsky[a]	$\frac{1}{2}\left(\sqrt{3}-1\right) \approx 0.3660$	1	0.9659	22
–	0.4095	1.0132	0.9723	–
–	$\sqrt{2}-1 \approx 0.4142$	$\sqrt{6}\,(\sqrt{2}-1) \approx 1.0146$	0.9729	–
–	$\frac{1}{2}$	$3\,\frac{\sqrt{3}}{5} \approx 1.0392$	0.9845	22,33
–	0.6286	1.0731	1	–
Schmidt-Ishlinsky	1	$\frac{2}{\sqrt{3}} \approx 1.1547$	1.0353	1, 16, 22

[a]The model is named after Sokolovsky following Pisarenko-Lebedev [157] "…it was attempted to introduce some intermediate criteria by replacing the hexagonal prism of Coulomb with a dodecagonal prism [195] (inscribed in the von Mises-cylinder) …". Further references to this models are [28, 222, 223, 226].

9.2.2 Bicubic Model

This model is obtained as a linear combination of the models of Tresca (15) and Schmidt-Ishlinsky (19) [33, 103]

$$\begin{aligned}&(1-\xi)\left[\left(I_2'-\sigma_{\mathrm{eq}}^2\right)^2\left(2^2\,I_2'-\sigma_{\mathrm{eq}}^2\right)-3^3\,I_3'^{\,2}\right]\\&+\xi\left[\frac{3^3}{2^3}\,I_3'+\frac{3^2}{2^2}\,I_2'\,\sigma_{\mathrm{eq}}-\sigma_{\mathrm{eq}}^3\right]\left[\frac{3^3}{2^3}\,I_3'-\frac{3^2}{2^2}\,I_2'\,\sigma_{\mathrm{eq}}+\sigma_{\mathrm{eq}}^3\right]=0.\end{aligned} \tag{113}$$

This model also results from (99) with $d = 1$.

The bicubic model divides the $h-k$-diagram into two areas. The models of Tresca and Schmidt-Ishlinsky are obtained with $\xi = 0$ and $\xi = 1$. The value $k = 1$ results in $\xi = 2^6/(7\cdot 13) \approx 0.7033$. This model is continuously differentiable (excluding the borders of ξ) and allows an explicit solution for σ_{eq}. For this reason, the BCM is appropriate for practical use. The parameters k and h are obtained from bicubic equations

$$2^4\cdot 3^3+2^3\cdot 3^3\,k^2\,(\xi-2^2)+2^6\,k^6\,(\xi-1)-3^3\,k^4\,(7\,\xi-2^4)=0, \tag{114}$$

$$2^5\cdot 3^3+2\cdot 3^3\,h^4\,(2^4-7\,\xi)+2^4\cdot 3^3\,h^2\,(\xi-2^2)+h^6\,(37\,\xi-2^6)=0 \tag{115}$$

as the lowest positive solutions. The analytical solution of (114) and (115) is complex and hence omitted.

Table 6 Settings of MAC (116) and the respective relations k and h (Fig. 24)

Model	η	k	h	Fig.
Tresca	1	$\dfrac{\sqrt{3}}{2} \approx 0.8660$	0.8966	1, 15, 22
–	$\dfrac{1}{\sqrt{3\,(2-\sqrt{3})}} \approx 1.1154$	$\dfrac{\sqrt{2+\sqrt{3}}}{2} \approx 0.9659$	1	–
–	1.1344	0.9824	1.0170	–
Ishlinsky-Ivlev[a]	$\dfrac{2}{\sqrt{3}} \approx 1.1547$	1	1.0353	22
–	$\dfrac{3}{14}\left(4+\sqrt{2}\right) \approx 1.1602$	1.0048	1.0353	–
–	$2-\dfrac{1}{\sqrt{2}} \approx 1.2929$	1.1197	1.0353	–
Schmidt-Ishlinsky	$\dfrac{4}{3}$	$\dfrac{2}{\sqrt{3}} \approx 1.1547$	1.0353	1, 16, 22

[a]The regular dodecagon in the π-plane is named after Ishlinsky-Ivlev [33, 105, 115], s. also [93, 98–100, 128, 193, 227].

9.2.3 Multiplicative Ansatz Criterion

Multiplicative combination of the models of Tresca (15) and Schmidt-Ishlinsky (19) [116] lies on the right boundary of the convexity region of the models of hexagonal symmetry (Fig. 24, MAC) [93, 98, 99, 193]. It is obtained as follows [105]

$$\begin{aligned}\Phi_{\mathrm{MAC}} = &\left[\left(I_2' - (\eta\,\sigma_{\mathrm{eq}})^2\right)^2 \left(2^2\,I_2' - (\eta\,\sigma_{\mathrm{eq}})^2\right) - 3^3\,I_3'^2\right]\\ &\times\left[\frac{3^3}{2^3}\,I_3' + \frac{3^2}{2^2}\,I_2'\,\sigma_{\mathrm{eq}} - \sigma_{\mathrm{eq}}^3\right]\left[\frac{3^3}{2^3}\,I_3' - \frac{3^2}{2^2}\,I_2'\,\sigma_{\mathrm{eq}} + \sigma_{\mathrm{eq}}^3\right].\end{aligned} \qquad (116)$$

The value η lies in the interval $\eta \in [1,\ 4/3]$. The parameters k and h compute to

$$k = \frac{\sqrt{3}}{2}\,\eta, \qquad h = \begin{cases} \eta\,\sqrt{3\,(2-\sqrt{3})}, & \eta \in \left[1,\ \dfrac{2}{\sqrt{3}}\right], \\ \sqrt{4\,(2-\sqrt{3})}, & \eta \in \left[\dfrac{2}{\sqrt{3}},\ \dfrac{4}{3}\right]. \end{cases} \qquad (117)$$

The models of Tresca and Schmidt-Ishlinsky are obtained with $\eta = 1$ and $\eta = 4/3$. With $\eta = \frac{2}{\sqrt{3}}$ one gets the regular dodecagon in the π-plane (Table 6).

For UYC and MAC the points, which have the shortest distance to the point $M(1,\ 1)$ (Fig. 24, model of von Mises), can be obtained from the equation

$$(h-1)^2 + (k-1)^2 \to \min. \qquad (118)$$

Using these points the model of von Mises can be approximated with the dodecagons of UYC with $b = 0.4095$ and the dodecagons of multiplicative ansatz criterion (MAC) with $\eta = 1.1344$.

9.2.4 Universal Model with Hexagonal Symmetry

The parameter $b \in [0,\ 1]$ of the UYC (111) can be replaced by the parameter $k \in \left[\sqrt{3}/2,\ 2/\sqrt{3}\right]$

$$b = \frac{\sqrt{3} - 2\,k}{k - \sqrt{3}}. \qquad (119)$$

The parameter $\eta \in [1,\ 4/3]$ in MAC (116) can be replaced by $k \in \left[\sqrt{3}/2,\ 2/\sqrt{3}\right]$ with (117)

$$\eta = \frac{2}{\sqrt{3}}\,k. \qquad (120)$$

With the linear (convex) combination of the two latter models [116]

$$\Phi_{12} = \xi\,\Phi_{\mathrm{MAC}} + (1 - \xi)\,\Phi_{\mathrm{UYC}}, \qquad \xi \in [0,\ 1] \qquad (121)$$

the model with the power of stress $n = 12$ is obtained, cf. Eq. (99). It covers all the convex forms in the h–k-diagram with two parameters $(k,\ \xi)$. The values $k = 1$ and $\xi = 0.3901$ result in $h = 1$, which corresponds to the model of von Mises (Fig. 24). With $\xi = 0.3901$ and $k \in \left[\sqrt{3}/2,\ 2/\sqrt{3}\right]$ one gets the approximation of BCM (113). With $k = 1$ and $\xi \in [0,\ 1]$ one obtains a model, which links the regular dodecagon of Sokolovsky and Ishlinsky-Ivlev: $h \in [0.9659,\ 1.0353]$. The major disadvantage is, that the model (121) is not analytically solvable with respect to σ_{eq}.

9.2.5 Model Based on the Stress Angle

Cosine ansatz to the power 2 and 4 is introduced in [33]

$$(3\,I_2')^{n/2}\,\frac{1 + c_6\cos^2 3\theta + c_{12}\,\cos^4 3\theta}{1 + c_6 + c_{12}} = \sigma_{\mathrm{eq}}^n, \qquad n = 1,\ 2\,\ldots \qquad (122)$$

with

$$k^n = 1 + c_6 + c_{12}, \qquad h^n = 2^2\,\frac{1 + c_6 + c_{12}}{2^2 + 2\,c_6 + c_{12}}. \qquad (123)$$

This model contains the following criteria:

- Drucker I [54, 154, 199] with $n = 6$, $c_6 \in [-1/3,\ 1/2]$, $c_{12} = 0$ and

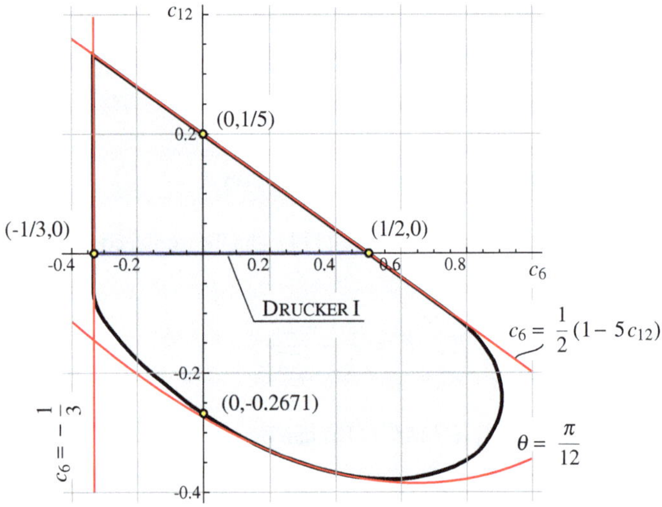

Fig. 25 Convexity region of the model (122) with $n = 6$ in the parameter space (c_6, c_{12}). The constraints at $\theta = 0$ with $c_6 = (1 - 5\,c_{12})/2$, $\theta = \pi/12$ and $\theta = \pi/6$ with $c_6 = -1/3$ are shown for clarity

- Dodd-Naruse [53, 229] with $n = 12$, $c_6 = 0$, $c_{12} \in \left[\frac{2}{3^3}\left(2\sqrt{11} - 13\right), \frac{1}{2}\right]$.

The boundaries of the parameters of the model (122) with $n = 6$, which result from the convexity conditions [33], are shown in Fig. 25.

10 Pressure-Sensitive Criteria

The behavior of real materials can be presented by the models (5) and (6). The first invariant of the stress tensor should be included in the pressure-insensitive criteria (Sect. 9) in such a way that the shape in the π-plane will be preserved.

10.1 Compressible Generalization

A compressible generalization of the models of incompressible material behavior (Sect. 9) is obtained by substitution [115]

$$\sigma_{\text{eq}} \to \sqrt[j+l+m]{\left(\frac{\sigma_{\text{eq}} - \gamma_1\, I_1}{1 - \gamma_1}\right)^j \left(\frac{\sigma_{\text{eq}} - \gamma_2\, I_1}{1 - \gamma_2}\right)^l \sigma_{\text{eq}}^m}. \tag{124}$$

The parameters γ_1 and γ_2 determine the position of the hydrostatic nodes A_Z and A_D on the hydrostatic axis (Figs. 2, 12). The powers j, l and m are chosen to be integer and positive. They are restricted by the following experience-based relation:

$$j + l + m \leq 6. \tag{125}$$

For materials, which do not fail under hydrostatic pressure (brass, plumb, steel), the surfaces Φ has a single hydrostatic node A_Z ($a_+^{\text{hyd}} > 1/3$, $a_-^{\text{hyd}} \to \infty$), e. g.:

- in order to obtain straight meridians the substitution with $l = m = 0$ is

$$\sigma_{\text{eq}} \to \frac{\sigma_{\text{eq}} - \gamma_1 I_1}{1 - \gamma_1}, \qquad +++\gamma_1 \in [0,\ 1[, \tag{126}$$

- for parabolic models it follows $l = 0$ and
- for hyperbolic meridians $\gamma_2 \in]0,\ \gamma_1[$; the second node does not belong to the relevant region of the surface and has no physical meaning, cf. [220]; due to this fact the hyperbolic surfaces are not recommended for applications.

For materials, which fail under hydrostatic compression (hard foams, ceramics, sintered and granular materials, etc.) the additional hydrostatic node A_D is needed. The parameters in (124) are then bounded as follows

$$\gamma_1 \in]0,\ 1[, \qquad \gamma_2 < 0. \tag{127}$$

For instance the closed surface with the substitution

$$\sigma_{\text{eq}} \to \sqrt[3]{\frac{\sigma_{\text{eq}} - \gamma_1 I_1}{1 - \gamma_1} \left(\frac{\sigma_{\text{eq}} - \gamma_2 I_1}{1 - \gamma_2} \right)^2} \tag{128}$$

and the shape b of the cross-section in the π-plane (Fig. 4) can be considered. The properties of the surface with $\gamma_1 = 1/3$, $\gamma_2 = -1/3$ are shown in Figs. 27, 28 and 26.

The closed surfaces with $j = l$ possess a symmetry plane orthogonal to the hydrostatic axis

$$\frac{I_1}{\sigma_+} = \frac{1}{2} \left(\frac{1}{\gamma_1} + \frac{1}{\gamma_2} \right). \tag{129}$$

The ellipsoid of Schleicher (Sect. 6) with this property is widely applied in modeling [114].

There is no method known, which allows to choose the powers in (124) analytically. Rational substitution, e. g. such transform where the root in (124) spared, is a simple possibility. For example a quadratic substitution

$$\sigma_{\text{eq}}^2 \to \frac{\sigma_{\text{eq}} - \gamma_1 I_1}{1 - \gamma_1} \frac{\sigma_{\text{eq}} - \gamma_2 I_1}{1 - \gamma_2}, \qquad \gamma_1 \in [0,\ 1[\tag{130}$$

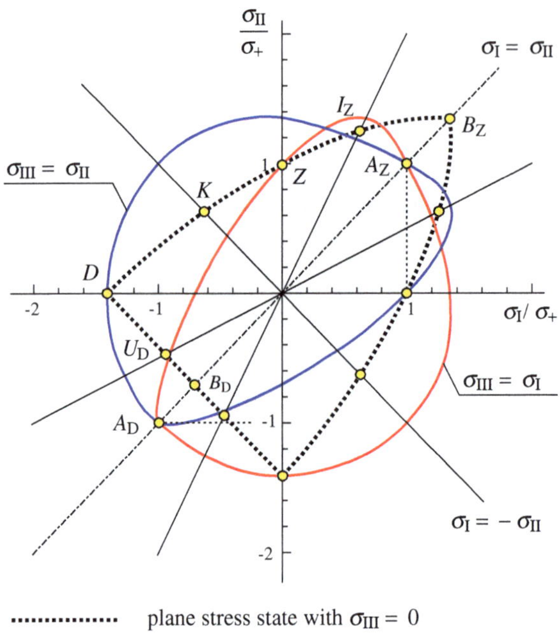

Fig. 26 Model of Sayir I (87) with $c_3 = 3^2/2$ and the substitution (128) with $\gamma_1 = 1/3$, $\gamma_2 = -1/3$ in the $\sigma_{\mathrm{I}}-\sigma_{\mathrm{II}}$-plane (s. Burzyński-plane (Fig. 27) and π-plane (Fig. 28)). The meridians with $\sigma_{\mathrm{III}} = \sigma_{\mathrm{II}}$ and $\sigma_{\mathrm{III}} = \sigma_{\mathrm{I}}$ are shown for clarity (spatial image)

can be applied for the model of von Mises (16), see rotationally symmetric model (51), and the models of hexagonal symmetry (Sect. 9.2) for fitting the available measured data.

The nonconvex surfaces in the meridian section are obtained with (124), if among the parameters γ_i there are two complex conjugated values. Figure 29 represents, as an example, a hyperboloid [63]. Such surfaces can be used as parts of combined models (Sect. 11.2).

10.2 Unified Strength Theory of Yu

The Unified Strength Theory (UST) is built up from two six-edge pyramids in the principal stress space [105, 228, 229]

$$\begin{cases} \sigma_{\mathrm{I}} - \dfrac{1}{d\,(1+b)}\,(b\,\sigma_{\mathrm{II}} + \sigma_{\mathrm{III}}) - \sigma_{\mathrm{eq}} = 0, \\[2ex] \dfrac{1}{d}\,\sigma_{\mathrm{I}} - \dfrac{1}{1+b}\,(b\,\sigma_{\mathrm{II}} + \sigma_{\mathrm{III}}) + \sigma_{\mathrm{eq}} = 0. \end{cases} \tag{131}$$

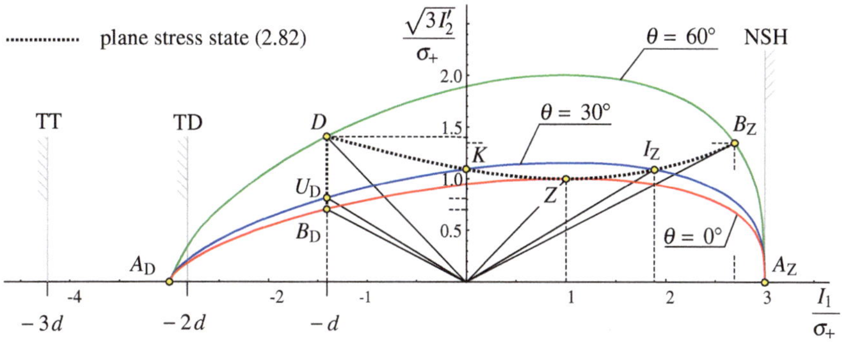

Fig. 27 Model of Sayir I (87) with $c_3 = 3^2/2$ and the substitution (128) with $\gamma_1 = 1/3, \gamma_2 = -1/3$ in the Burzyński-plane. Properties: $d = 1.41$, $k = 1.09$, $i_Z = 1.09$, $u_D = 0.81$, $b_Z = 1.35$, $b_D = 0.70$, $a_+^{\text{hyd}} = 1$, $a_-^{\text{hyd}} = 1$, $\nu_+^{\text{in}} = 1/2$, $\nu_-^{\text{in}} = -0.05$. The reference values for the hydrostatic nodes A_Z and A_D are: NSH—upper bound due to the normal stress hypothesis $a_+^{\text{hyd}} = 1$, (41), TT—lower bound for the point A_D with respect to the normal stress hypothesis as trigonal trapezohedron, $a_-^{\text{hyd}} = d$, TD—lower bound for the point A_D with respect to the normal stress hypothesis as triangular dipyramid, $a_-^{\text{hyd}} = 2\,d/3$

The faces of the first pyramid are obtained from the first equation with the cyclic permutations of indices. The faces of the second one are obtained in the same manner from the second equation.

The model (131) describes the compressible material behavior with the properties (22) using two parameters $d \geq 1$ and $b \in [0,\ 1]$. The value d corresponds to the relation d (Table 1)

$$d = \frac{|\sigma_-|}{\sigma_+}, \tag{132}$$

which simplifies the application of the model.

The analysis of the UST leads to the following special cases (Figs. 5 and 30):

- $b = 1$ results in the Twin-Shear Theory (TST) of Yu,
- with

$$b = \frac{\sqrt{3} - 1}{2} \tag{133}$$

  follows a continuous analogy of the model of Pisarenko-Lebedev (48),
- with $b = 0$ the model of Mohr-Coulomb (Single-Shear Theory of Yu), Eq. (45), is obtained,
- with $d \to \infty$, $b \in [0,\ 1]$ absolutely brittle materials can be described (normal stress hypothesis), Eq. (8),
- the Unified Yield Criterion (111) results from the UST with $d = 1$ and $b \in [0,\ 1]$.

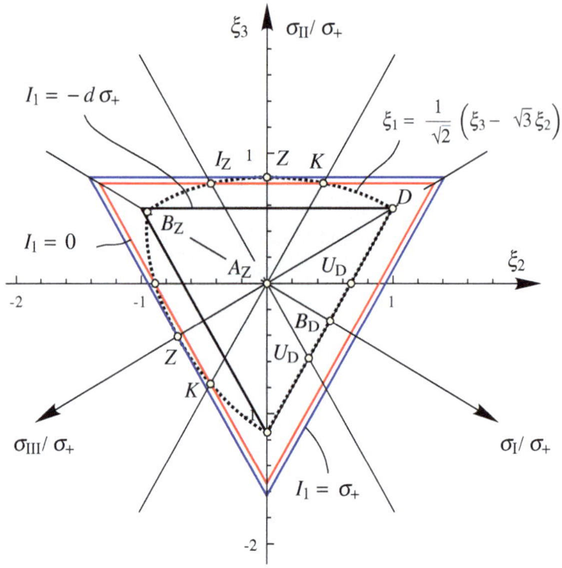

Fig. 28 Cross-sections $I_1 = \sigma_+$, $I_1 = 0$, $I_1 = -d\,\sigma_+$ and the line of the plane stress state of the model of Sayir I (87) with $c_3 = 3^2/2$ and the substitution (128) with $\gamma_1 = 1/3$, $\gamma_2 = -1/3$ in the π-plane, see Fig. 27

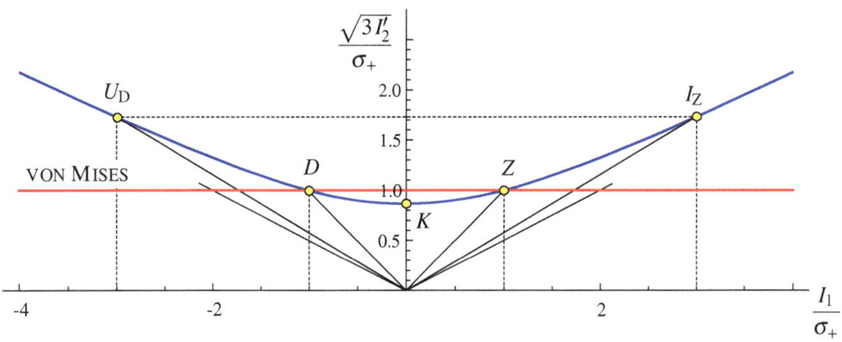

Fig. 29 Model of von Mises and the hyperboloid of one sheet (51) with $\gamma_1 = -\gamma_2 = 1/\sqrt{3}\,i$ in the Burzyński-plane with the properties: $d = 1$, $k = \sqrt{3}/2$, $i_Z = u_D = \sqrt{3}$, $\nu_+^{\text{in}} = \nu_-^{\text{in}} = 1$

The hydrostatic tensile stress computes to

$$a_+^{\text{hyd}} = \frac{1}{1 - \dfrac{1}{d}} = \frac{1}{1 - 2\,\nu_+^{\text{in}}}. \tag{134}$$

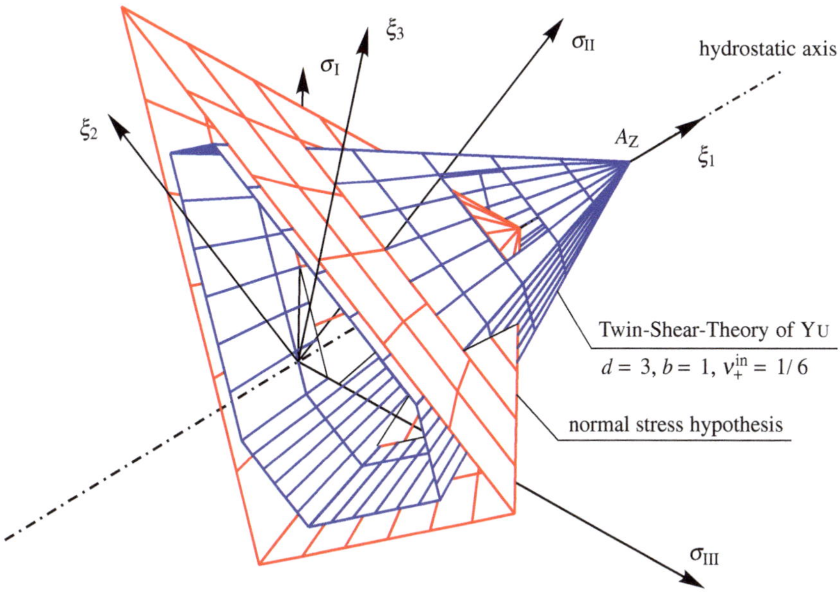

Fig. 30 Twin-Shear Theory ($d = 3, b = 1, \nu_+^{\text{in}} = 1/6$) and the normal stress hypothesis ($d \to \infty$, $b = 0 \ldots 1$, $\nu_+^{\text{in}} = 0$) in the principal stress space. The point A_Z of the normal stress hypothesis is shown for better understanding

The surface Φ of the UST is open in the hydrostatic compression direction ($I_1 < 0$):

$$a_-^{\text{hyd}} \to \infty. \tag{135}$$

The relation k equals to

$$k = \sqrt{3}\,\frac{1+b}{1+b+\dfrac{1}{d}} = \sqrt{3}\,\frac{1+b}{1+b+2\,\nu_+^{\text{in}}}. \tag{136}$$

For i_Z and u_D it follows

$$i_Z = \frac{\sqrt{3}\,(1+b)}{2-b\,(1/d-2)}, \qquad u_D = \frac{\sqrt{3}\,(1+b)}{(2+b)\,1/d}. \tag{137}$$

The relations b_Z and b_D are given by (22). The Poisson's ratios at tension and compression are

$$\nu_+^{\text{in}} = \frac{1}{2\,d}, \qquad \nu_-^{\text{in}} = \frac{d}{2}. \tag{138}$$

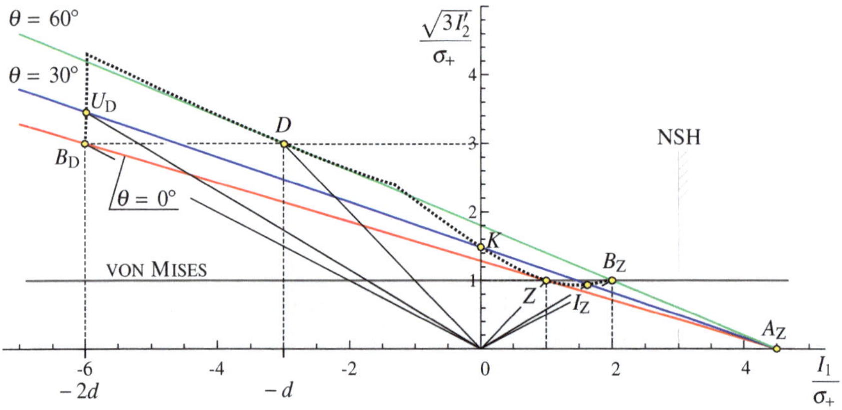

Fig. 31 UST with $d = 3$ and $b = 1$ with the properties $k = 6\sqrt{3}/7$, $i_Z = 6\sqrt{3}/11$, $u_D = 2\sqrt{3}$, $b_D = 3$, $b_Z = 1$, $a_+^{\text{hyd}} = 3/2$, $\nu_+^{\text{in}} = 1/6$, $\nu_-^{\text{in}} = 3/2$ and the model of von Mises in the Burzyński-plane, cf. [105]. NSH—reference value $a_+^{\text{hyd}} = 1$ from the normal stress hypothesis

The invariant representation of the UST (131) is given in [105, 115]. It results from the Radcig model (102) by the linear transform (126) and the parameters

$$\gamma_1 = \frac{1 - 1/d}{3}, \qquad \eta = \frac{2 + 1/d}{1 + 2/d}, \tag{139}$$

$$b_1 = \frac{1 + b - 1/d + 2\,b/d}{1 + b + 2/d - b/d}, \qquad b_2 = \frac{3(1 - b)}{b\,(1 - 1/d) - 2 - 1/d} + 1. \tag{140}$$

The meridians with $\theta = 0$, $\pi/6$ and $\pi/3$ of the UST are represented in the Burzyński-plane by straight lines through the point $A_Z(3\,a_+^{\text{hyd}},\ 0)$ and the points $Z\,(1,\ 1)$, $K\,(0,\ k)$ and $D\,(-d,\ d)$, respectively. The Twin Stress Theory (TST) with the parameter values $d = 3$ and $b = 1$ is shown in Figs. 30, 31 and 32.

10.3 *Models for Applications*

Because of their simplicity and versatility the following models can be recommended for various classes of isotropic materials.

10.3.1 Unified Strength Theory of Yu

The UST (131) is well-accepted and often used in computations and theoretical investigations. The linear relations of the model result in low computational complexity.

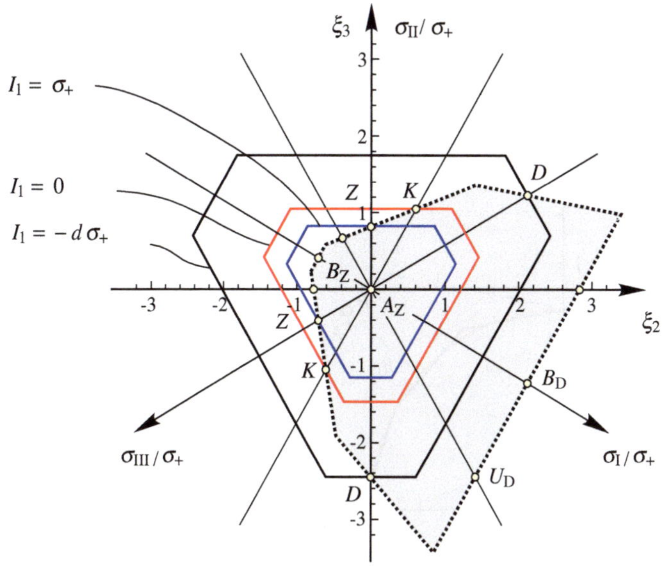

.... line of the plane stress state with $\xi_1 = \frac{1}{\sqrt{2}}\left(\xi_3 - \sqrt{3}\,\xi_2\right)$

Fig. 32 UST with $d = 3$ and $b = 1$: cross-sections $I_1 = \sigma_+$, $I_1 = 0$, $I_1 = -d\,\sigma_+$ and the line of the plane stress state in the π-plane, see principal stress space (Fig. 30) and Burzyński-plane (Fig. 31)

The model is restricted to materials with $a_-^{\text{hyd}} \to \infty$ (the hydrostatic compression is unbounded). The straight lines of the meridians of this model simplify the description of the material behavior, however they have no theoretical foundation. The model can be modified near the hydrostatic tension (Sect. 11.1) in order to reduce the relation a_+^{hyd} [229].

The UYC (UST with $d = 1$) cannot describe the SD-effect (no strength differential effect) and the Poynting-Swift-effect for incompressible material [105]. Because the intersections of the planes (131) with each other lie outside of the closed region, where the model is valid (Fig. 33), the approximation of the measurements using different optimization criteria (Sect. 12) becomes difficult.

If the UST is used as a plastic potential these intersections lead to singularities in the strain field.

10.3.2 Rotationally Symmetric Models

Depending on the number of experiments, the quality of the measurements and the required modeling precision, the function Φ can be simplified, if the influence

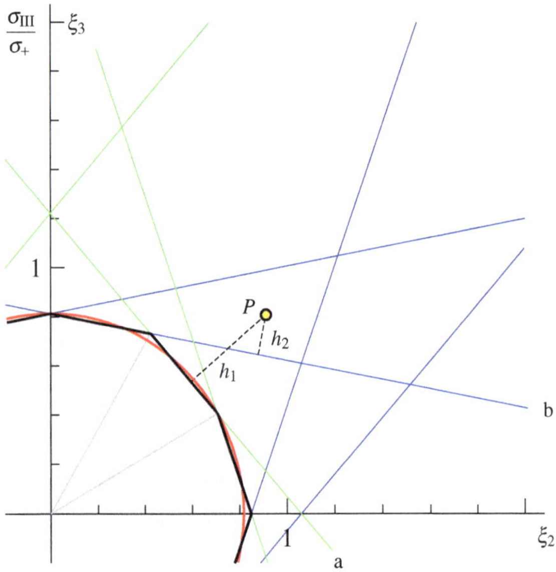

Fig. 33 UYC with $b = 1/2$ in the π-plane. P—measurement (schematic); h_1, h_2—normals to some planes of the model: $a\ \sigma_{eq} = \dfrac{1}{1+b}(\sigma_{II} + b\,\sigma_{III}) - \sigma_I$, $b\ \sigma_{eq} = \sigma_{II} - \dfrac{1}{1+b}(\sigma_I + b\,\sigma_{III})$. The model of von Mises is presented for comparison

of the third deviatoric invariant is neglected. These models contains the quadratic rotationally symmetric model (51), Figs. 11, 12 and 34.

Using the substitution (124) the rotationally symmetric model of the form

$$(3I_2')^3 = \left(\frac{\sigma_{eq} - \gamma_1 I_1}{1-\gamma_1}\right)^j \left(\frac{\sigma_{eq} - \gamma_2 I_1}{1-\gamma_2}\right)^l \sigma_{eq}^m, \quad \gamma_1 \in [0,\ 1[\qquad (141)$$

with

$$j + l + m = 6 \qquad (142)$$

can be introduced. The following combination of the parameters γ_i for materials with $a_-^{hyd} \to \infty$ can be used

- for a cone with $\gamma_1 = \gamma_2$, $m = 0$ and
- for a paraboloid with $l = 0$, $m = 1 \dots 5$.

For closed criteria, which restrict in addition to the hydrostatic tension the hydrostatic compression, the following values of parameters γ_i should be considered

- $\gamma_2 < 0$, $m = 0 \dots 3$, $j \neq l$ (Fig. 35) or
- $\gamma_2 < 0$, $m = 0,\ 2,\ 4$, $j = l$ a surface similar to the ellipsoid of Schleicher (Fig. 34) with the symmetry plane orthogonal to the hydrostatic axis with (129).

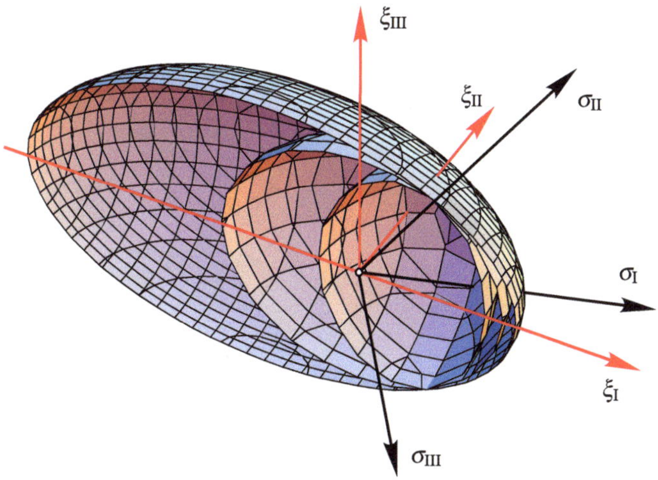

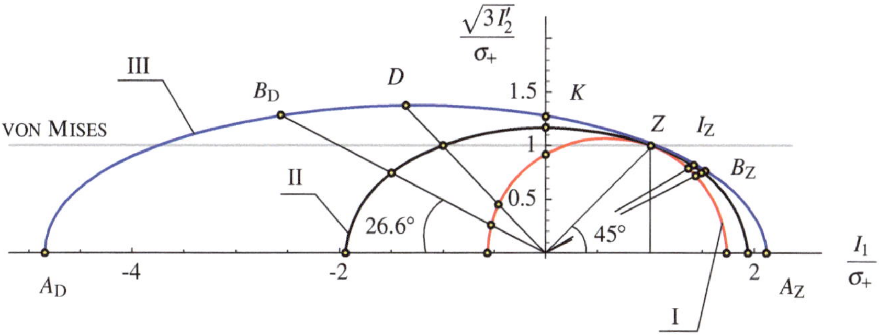

Fig. 34 Model of Schleicher (51) with the Poisson's ratio $\nu_+^{\text{in}} = \frac{1}{10}$, [103]: *top* in the principal stress space $\sigma_{\text{I}}, \sigma_{\text{II}}, \sigma_{\text{III}}$ (for clarity the surfaces *I, II* and *III* are cut $\xi_{\text{II}} \geq 0$) *bottom* in the Burzyński-plane *I* $\nu_-^{\text{in}} = -\dfrac{1}{2}$, $\gamma_1 = \dfrac{1}{15}(-9 + 2\sqrt{78})$, $\gamma_2 = \dfrac{1}{15}(-9 - 2\sqrt{78})$, $d = 0.45$, $k = 0.92$; *II* $\nu_-^{\text{in}} = \dfrac{1}{10}$, $\gamma_1 = \gamma_2 = \dfrac{2}{\sqrt{15}}$, $d = 1$, $k = 1.17$ (ellipsoid of Beltrami); *III* $\nu_-^{\text{in}} = \dfrac{1}{2}$, $\gamma_1 = \dfrac{1}{15}(2 + \sqrt{26})$, $\gamma_2 = \dfrac{1}{15}(2 - \sqrt{26})$, $d = 1.36$, $k = 1.25$

This model yields more possibilities for approximations in comparison to (51). If an analytical solution of the equation with respect to σ_{eq} is required, the model

$$(3I_2')^{3/2} = \left(\frac{\sigma_{\text{eq}} - \gamma_1\, I_1}{1 - \gamma_1}\right)^j \left(\frac{\sigma_{\text{eq}} - \gamma_2\, I_1}{1 - \gamma_2}\right)^l \sigma_{\text{eq}}^m, \quad \gamma_1 \in [0,\ 1[\tag{143}$$

with

$$j + l + m = 3 \tag{144}$$

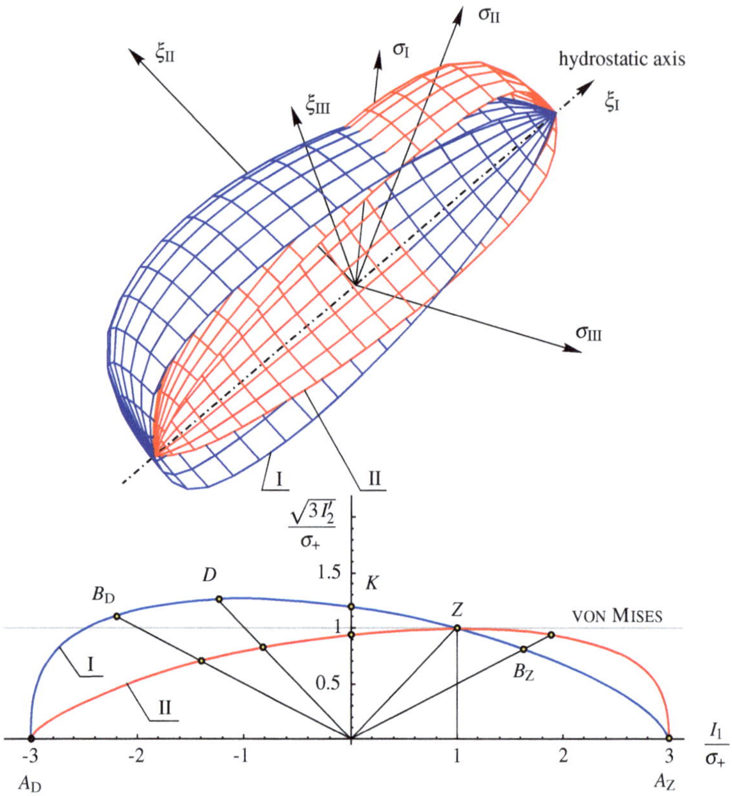

Fig. 35 Rotationally symmetric model (141) with $\gamma_1 = 1/3$, $\gamma_2 = -1/3$ or $a_+^{\text{hyd}} = 1$, $a_-^{\text{hyd}} = 1$ *I.* $j = 4$, $l = 2$, $d = 1.25$, $k = 1.19$, $b_Z = 0.82$, $b_D = 1.10$, $\nu_+^{\text{in}} = \frac{1}{5}$, $\nu_-^{\text{in}} = 0.44$; *II.* $j = 2$, $l = 4$, $d = 0.83$, $k = 0.94$, $b_Z = 0.94$, $b_D = 0.70$, $\nu_+^{\text{in}} = \frac{1}{2}$, $\nu_-^{\text{in}} = 0.27$ in the principal stress space and in the Burzyński-plane

can be used too. The number of the possible meridian shapes is lower in comparison with (141).

10.3.3 Geometrical-Mechanical Model

The model

$$3\, I_2' \, \frac{1 + c_3 \, \cos 3\theta + c_6 \, \cos^2 3\theta}{1 + c_3 + c_6} = \frac{\sigma_{\text{eq}} - \gamma_1 \, I_1}{1 - \gamma_1} \, \frac{\sigma_{\text{eq}} - \gamma_2 \, I_1}{1 - \gamma_2} \tag{145}$$

is a generalization of the rotationally symmetric model (51). The equivalent stress σ_{eq} can be computed analytically. The model can be used for describing of the multi-modular theory of elasticity.[7]

The relations compute to

$$d^2\,\frac{1-c_3+c_6}{1+c_3+c_6}=\frac{(1+d\,\gamma_1)\,(1+d\,\gamma_2)}{(1-\gamma_1)\,(1-\gamma_2)},\tag{146}$$

$$k^2=\frac{1+c_3+c_6}{(1-\gamma_1)\,(1-\gamma_2)}.\tag{147}$$

The Poisson's ratio at tension ν_+^{in} is the same as in (54). The Poisson's ratio at compression ν_-^{in} is not provided because of its complexity.

The model

$$(3I_2')^{3/2}\frac{1+c_3\cos 3\theta+c_6\cos^2 3\theta}{1+c_3+c_6}=\left(\frac{\sigma_{\text{eq}}-\gamma_1 I_1}{1-\gamma_1}\right)^{3-l-m}\left(\frac{\sigma_{\text{eq}}-\gamma_2 I_1}{1-\gamma_2}\right)^{l}\sigma_{\text{eq}}^{m}\tag{148}$$

with the substitution (124) and the adjustment (144) allows the analytical computation of the equivalent stress σ_{eq}. The number of possible shapes of the meridian line is however still not sufficient for a fitting of measurements. The relations compute to

$$d^3\,\frac{1-c_3+c_6}{1+c_3+c_6}=\left[\frac{1+d\,\gamma_1}{1-\gamma_1}\right]^{3-l-m}\left[\frac{1+d\,\gamma_2}{1-\gamma_2}\right]^{l},\tag{149}$$

$$k^3=\frac{1+c_3+c_6}{(1-\gamma_1)^{3-l-m}\,(1-\gamma_2)^{l}}.\tag{150}$$

As strength hypothesis it can be recommended to use the geometrical-mechanical model (105) with the substitution (124) and the adjustment (142)

$$(3I_2')^{3}\,\frac{1+c_3\cos 3\theta+c_6\cos^2 3\theta}{1+c_3+c_6}=\left(\frac{\sigma_{\text{eq}}-\gamma_1 I_1}{1-\gamma_1}\right)^{6-l-m}\left(\frac{\sigma_{\text{eq}}-\gamma_2 I_1}{1-\gamma_2}\right)^{l}\sigma_{\text{eq}}^{m}.\tag{151}$$

This model has a large region of convex forms in the π-plane and various settings for the meridian. The relations are obtained analogous to (149) and (150)

$$d^6\,\frac{1-c_3+c_6}{1+c_3+c_6}=\left[\frac{1+d\,\gamma_1}{1-\gamma_1}\right]^{6-l-m}\left[\frac{1+d\,\gamma_2}{1-\gamma_2}\right]^{l},\tag{152}$$

$$k^6=\frac{1+c_3+c_6}{(1-\gamma_1)^{6-l-m}\,(1-\gamma_2)^{l}}.\tag{153}$$

---

[7] Theory of elasticity with different Young's moduli $E_+\neq E_-$ and elastic Posisson's ratios $\nu_+^{\text{el}}\neq\nu_-^{\text{el}}$ at tension and compression [9, 212]

The models (145), (148) and (151) describe incompressible material behavior with $\gamma_1 = \gamma_2 = 0$ and for $c_3 = c_6 = 0$ become rotationally symmetric model (141). With $c_3 = 0$ one obtains the models of hexagonal symmetry. The convexity conditions in the parameter space $c_3 - c_6$ must be taken into account for these models.

10.3.4 Convex π-Plane Model

The model (99) with the substitution (124) and the adjustment (142) can be used for analysis in certain special situations. For instance, in order to check if the given measurements can be described by a convex model. The number of measurements must be sufficient in order to obtain reasonable approximations. This model contains applying the linear substitution (126) the pyramids of Sayir II [178] and due to Haythornthwaite. The model of Drucker II [55] and due to Schmidt-Ishlinsky are special cases of these models. This model incorporate various conditions, e.g. $\xi = 0$, $b_Z = 1$, $a_+^{\mathrm{hyd}} = 1$, to obtain special theories.

11 Combined Criteria

The mechanical behavior of modern materials can seldom be represented by a single surface Φ [157]. The extrapolated behavior at a hydrostatic tension (point A_Z) is in this case frequently overestimated. It occurs also that the Poisson's ratio at compression can admit incorrect values with $\nu_-^{\mathrm{in}} > 1/2$ for yield criteria (Sect. 4).

For a reliable description of the measured data a number of combined models is proposed: the standard hypotheses (strain hypothesis (41), Mohr-Coulomb model (45) and Burzyński-Yagn model (51)) are combined in a different manner. Further reasons for the development of the combined models are:

- a small number of well-recognized models (Sect. 2, 6),
- simple interpretation of the measurements, which for instance can be separated in the regions of the brittle and of the ductile failure based on the hydrostatic stress with e. g. $I_1 < \sigma_+$ and $I_1 \geq \sigma_+$. In this case the surfaces with hexagonal symmetry in the π-plane are often used for ductile material behavior (Tresca and Schmidt-Ishlinsky models) and the surfaces with the trigonal symmetry (Fig. 4, cross section b) for the brittle one (normal stress hypothesis),
- taking into account the incompressibility $\nu_-^{\mathrm{in}} = 1/2$ for loadings with $I_1 < 0$,
- the restriction of the hydrostatic stresses with, e. g., $a_+{}^{\mathrm{hyd}} \in \left[1/3,\ 1 \right]$ (Fig. 7),
- decrease of the power of the stresses in each part of the combined surface Φ to $n \leq 6$, which simplifies the analysis of the measurements and results in an increased computational stability.

Combined models containing a "cap" (cut-off), which bounds the hydrostatic stress at the point A_Z, and a "body". Models for incompressible material behavior (e. g. Schmidt-Ishlinsky, von Mises or Tresca) are usually chosen as the "body" in

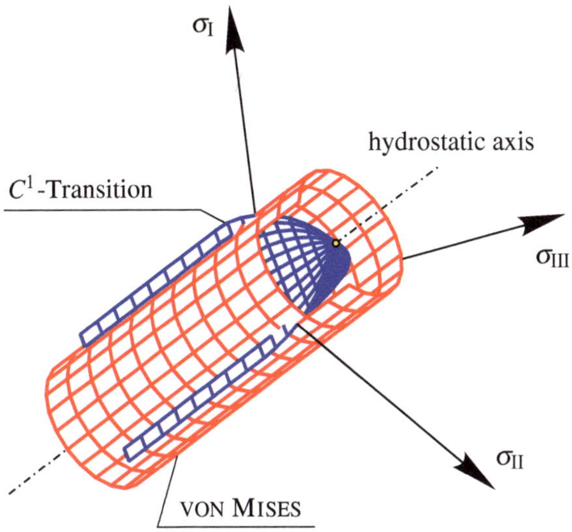

Fig. 36 Combined model with a C^1-transition in the cross-section $I_1 = 0$ built up from the model with the cross-section a in the π-plane (Fig. 4) in the principal stress space. The *cylinder* of von Mises is shown for comparison

the compression region ($I_1 < 0$). The influence of the first invariant in the model Φ changes with the transition from the "body" to the "cap" [107].

In order to reduce the number of possible combinations of surfaces additional plausibility conditions [105] are introduced

- the C^0-transition (continuous, not differentiable transition) follows for polyhedral surfaces (Sect. 11.1),
- for combinations of continuously differentiable surfaces it is a natural requirement the C^1-transition (Sect. 11.2),
- for combinations of surfaces, which have the same shape in the π-plane and are continuous, the C^1-transition in the meridian sections is recommended (Fig. 36).

The above mentioned conditions prohibit for instance the use of the model of Pelczyński [155, 234], which is built up from the normal stress hypothesis (8) and the model of von Mises (Figs. 2 and 37), some modifications, see [43, 45, 55, 173, 197, 217, 232]. The complicated shapes of the lines, resulting from the combination of the two surfaces, have no physical meaning.

11.1 Criteria with C^0-Transition

These models are built up from the Unified Strength Theory of Yu (Sect. 10.2) and the strain hypothesis (41). The normal stress hypothesis (8) is usually used as cut-off

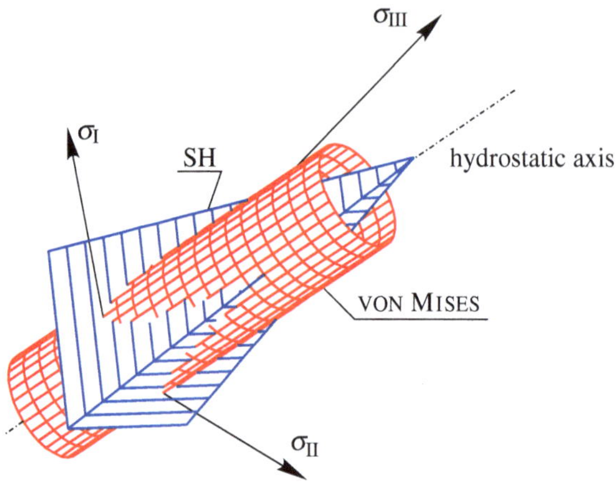

Fig. 37 "Pencil" of Pelczyński in the principal stress space with a C^0-transition between the surface of the strain hypothesis (SH), (41) and the *cylinder* of von Mises [107], cf. [234]

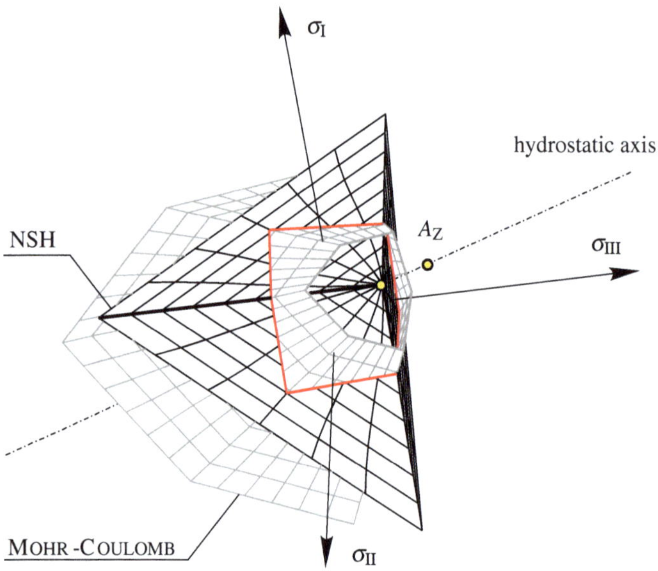

Fig. 38 Combined model with a C^0-transition in the principal stress space (model of Mohr-Coulomb with $d = 3$, $\nu_+^{\text{in}} = 1/6$ and the normal stress hypothesis $\nu_+^{\text{in}} = 0$ as a cut-off) [107]

(Figs. 30 and 38) instead of the strain hypothesis in order to reduce the number of parameters in the model [49, 70, 153, 154, 229].

The inclination of the meridian line of the angle $\theta = 0$ of the combined model, defined by the Poisson's ratio ν_+^{in}, has a jump at the point Z (Figs. 39 and 40)

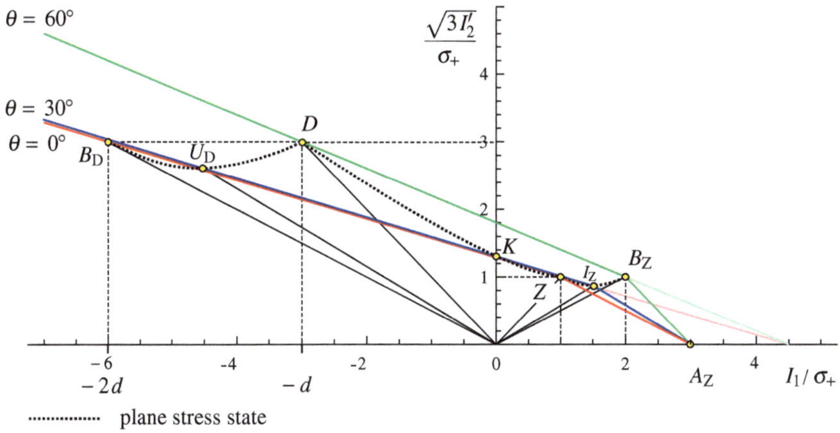

Fig. 39 Pyramid of the UST (131) with $b = 0$, $d = 3$, $k = 1.30$, $b_Z = 1$, $i_Z = 0.87$, $a_+^{\text{hyd}} = 3/2$, $\nu_+^{\text{in}} = 1/6$, $\nu_-^{\text{in}} = 3/2$ in the Burzyński-plane with the normal stress hypothesis as cut-off, $i_Z = 0.87$, $a_+^{\text{hyd}} = 1$, $\nu_+^{\text{in}} = 0$ (Fig. 38)

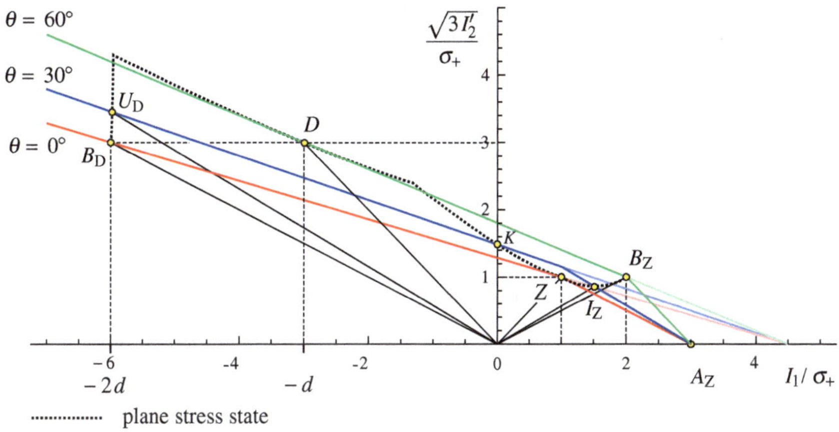

Fig. 40 Pyramid of the UST (131) with $b = 1$, $d = 3$, $k = 1.48$, $b_Z = 1$, $i_Z = 0.94$, $a_+^{\text{hyd}} = \frac{3}{2}$, $\nu_+^{\text{in}} = \frac{1}{6}$, $\nu_-^{\text{in}} = \frac{3}{2}$ in the Burzyński-plane with the normal stress hypothesis as cut-off, $i_Z = 0.87$, $a_+^{\text{hyd}} = 1$, $\nu_+^{\text{in}} = 0$ (Fig. 30), cf. [105]

$$\nu_+^{\text{in}} = \begin{cases} 0, & I_1 > \sigma_+, \text{ NSH as cut-off;} \\ 0 \ldots \dfrac{1}{2}, & I_1 \leq \sigma_+, \text{ UST of Yuas body.} \end{cases} \tag{154}$$

It is possible to combine two surfaces of UST with different parameter sets $(d,\ b)$ under the constraint

$$\nu_+^{\text{cut-off}} \in \left[0,\ \nu_+^{\text{body}}\right] \tag{155}$$

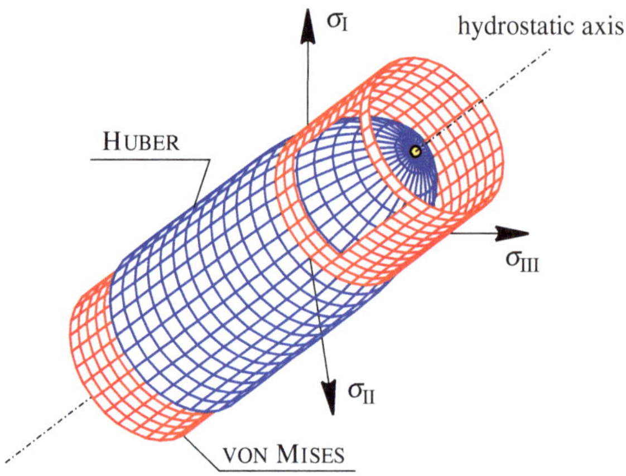

Fig. 41 Model of Huber with the C^1-transition in the cross section $I_1 = 0$ and the cylinder of von Mises in the principal stress space [107]

However, the resulting surface is too complex for applications. The existing information on the material behavior, which allow to deduce two different parameter sets (d, b) of the UST, can be as usual better taken into account with the help of C^1-criteria.

11.2 Criteria with C^1-Transition

Combined surfaces with C^1-transitions have the following advantages:

- unique computation of the strain rates for the yield surface Φ with the flow rule, e. g. with (27) and
- lower number of parameters compared to C^0-criteria.

11.2.1 Model of Huber

The first combined hypothesis was proposed in 1904 by Huber [36, 37, 67, 91, 103, 118]. The model consists of the ellipsoid of Beltrami (Sect. 6.4) and of a cylinder with the transition at the cross-section $I_1 = 0$ (Figs. 41 and 42)

$$3\, I_2' = \begin{cases} \dfrac{\sigma_{eq} - \gamma_1\, I_1}{1-\gamma_1}\, \dfrac{\sigma_{eq} + \gamma_1\, I_1}{1+\gamma_1}, & I_1 > 0, \text{ cap;} \\ \dfrac{\sigma_{eq}}{1-\gamma_1}\, \dfrac{\sigma_{eq}}{1+\gamma_1}, & I_1 \leq 0, \text{ body.} \end{cases} \tag{156}$$

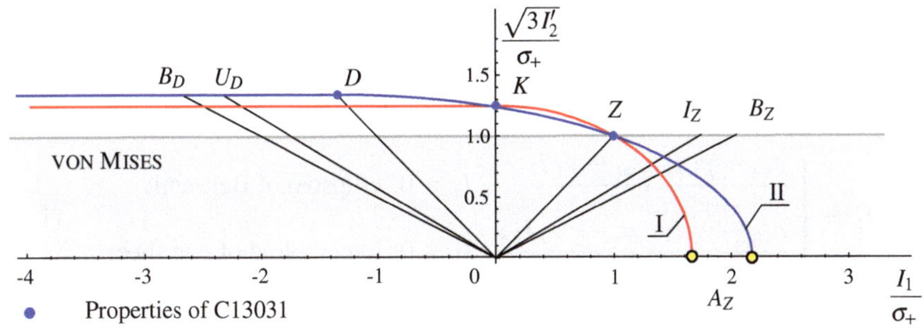

Fig. 42 Models fitted to the measured data for polyoxymethylene (POM) Hostaform C13031, Ticona GmbH, Sulzbach in the Burzyński-plane [113]: *I* model of Huber (156) with $k = 1.25$, ($d = 1.25$, $\nu_+^{\text{in}} = -0.04$, $\nu_-^{\text{in}} = 1/2$, $3\,a_+^{\text{hyd}} = 1.67$), *II* modified model of Huber (163) with $d = 1.34$, ($k = 1.24$, $\nu_+^{\text{in}} = 0, 20$, $\nu_-^{\text{in}} = 12$, $3\,a_+^{\text{hyd}} = 2.18$)

The transition between the surfaces in the cross-section $I_1 = 0$ is continuously differentiable. The model has the property $k = d \geq 1$. The parameter $\gamma_1 \in [0,\ 1[$ results from the relation (52)

$$k^2 = \frac{1}{(1-\gamma_1)(1+\gamma_1)}. \tag{157}$$

The Poisson's ratio at tension can be computed using (54) as follows

$$\nu_+^{\text{in}} = \frac{3}{2\,k^2} - 1. \tag{158}$$

Further values are

$$b_Z = \frac{1}{\sqrt{2\,\left(1-\nu_+^{\text{in}}\right)}}, \qquad a_+^{\text{hyd}} = \frac{1}{\sqrt{3\,\left(1-2\,\nu_+^{\text{in}}\right)}}. \tag{159}$$

The model is simple and represents the "classical view" with respect to the inelastic material behavior, cf. [140]:

- compressible properties for $I_1 > 0$ with $\nu_+{}^{\text{in}} \in \left]-1,\ 1/2\right]$ and
- incompressible properties for $I_1 < 0$, $\nu_-^{\text{in}} = 1/2$.

The model can be used as a yield surface with the empirical restriction (34), which leads to the relation $d = k \in [1,\ 1.007]$. The latter condition is rather restrictive, which makes the fitting of the model to the measurements harder. However, the model should be preferred to the von Mises-model (16), since one obtains safer results in the region $I_1 > \sigma_+$ with $b_Z \in [0.98,\ 1]$ and $a_+^{\text{hyd}} \geq 2.89$ in regions, where the information on these loading cases are missing.

The model of Huber (156) can be extended (Fig. 43).

The new model reflects the observation, that "the hydrostatic pressure improves the material strength" [118]. This model consists of the ellipsoid of Beltrami with $\gamma_1 \in [0,\ 1[$ for $I_1 > 0$ and of the hyperboloid (Sect. 10.1) for $I_1 \leq 0$:

$$3\,I_2' = \begin{cases} \dfrac{\sigma_{eq} - \gamma_1\,I_1}{1-\gamma_1}\,\dfrac{\sigma_{eq} + \gamma_1\,I_1}{1+\gamma_1}, & I_1 > 0,\ \text{ellipsoid of Beltrami;} \\ \dfrac{\sigma_{eq} - \gamma_0\,I_1\,i}{1-\gamma_1}\,\dfrac{\sigma_{eq} + \gamma_0\,I_1\,i}{1+\gamma_1}, & I_1 \leq 0,\ \text{hyperboloid of one sheet.} \end{cases} \tag{160}$$

The transition at the cross-section $I_1 = 0$ is continuously differentiable. The values compute to

$$\nu_+^{\text{in}} = \frac{1}{2}\left(1 - 3\,\gamma_1^2\right), \qquad \nu_-^{\text{in}} = \frac{1}{2} + \frac{3\,\gamma_0^2}{2\,\left(1 - \gamma_0^2 - \gamma_1^2\right)}, \tag{161}$$

$$k^2 = \frac{1}{(1-\gamma_1)\,(1+\gamma_1)}, \qquad d = \frac{1}{\sqrt{1 - \gamma_0^2 - \gamma_1^2}}, \qquad a_+^{\text{hyd}} = \frac{1}{3\,\gamma_1}. \tag{162}$$

With $\gamma_0 = 0$ the model of Huber (156) is obtained. The value $\gamma_0 \neq 0$ results in a non-convex model, cf. [144, 221, 231]. With the setting $\gamma_0 = \gamma_1$ the model of Kuhn [118] is obtained. A model built up of two ellipsoids with the transition at the point K results for purely complex γ_0. The tangent line at the point K is parallel to the hydrostatic axis, cf. [212].

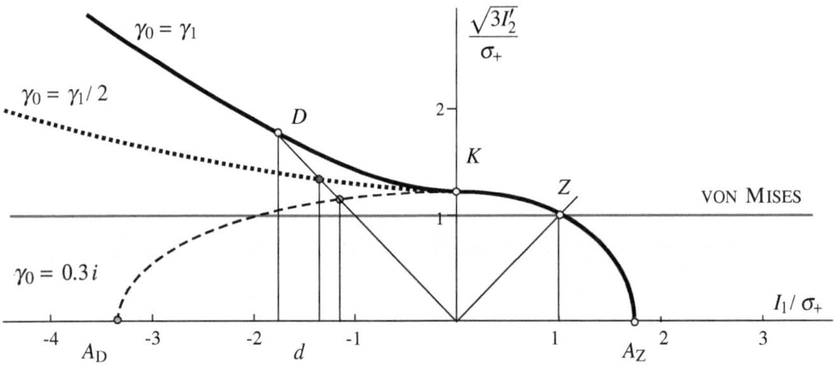

Fig. 43 Model of Kuhn with the C^1-transition at $I_1 = 0$ [107]: $\gamma_1 = 1/\sqrt{3}$, $\nu_+^{in} = 0$, $k = \sqrt{3/2}$, $a_+^{\text{hyd}} = 1/\sqrt{3}$ with; $\gamma_0 = \gamma_1$ ($d = \sqrt{3}$, $\nu_-^{\text{in}} = 2$), modification with $\gamma_0 = \gamma_1/2$ ($d = 2\sqrt{3/7}$, $\nu_-^{\text{in}} = 5/7$), combination of two ellipsoids with $\gamma_0 = 0.3\,i$ ($d = 1.1496$, $\nu_-^{in} = 0.32$, $a_-^{\text{hyd}} = -10/9$)

11.2.2 Modification of the Model of Huber

An analogous approach results in the modified model of Huber (Fig. 42) for the materials with the relation $d \geq k \geq 1$ and incompressibility in the region $I_1 \leq -d\,\sigma_+$ [103, 107, 113]

$$3\,I_2' = \begin{cases} \dfrac{\sigma_{\text{eq}} - \gamma_1\,I_1}{1-\gamma_1}\dfrac{\sigma_{\text{eq}} - \gamma_2\,I_1}{1-\gamma_2}, & I_1 > -d\,\sigma_+, \text{ ellipsoid;} \\ \dfrac{\sigma_{\text{eq}}}{1-\gamma_1}\dfrac{\sigma_{\text{eq}}}{1-\gamma_2}, & I_1 \leq -d\,\sigma_+, \text{ cylinder.} \end{cases} \tag{163}$$

In contrast to the model of Huber (156) the C^1-transition between two surfaces at the cross-section $I_1 = -d\,\sigma_+$ is defined by the point D (compression). The model can be better fitted to the measurements, that belong to the region $I_1 \in \left[-d\,\sigma_+,\ \sigma_+\right]$, it means in the region $D - K - Z$.

Further we obtain with (52)

$$k = \frac{\sqrt{2}\,d}{\sqrt{1+d}} \qquad \text{or} \qquad d = \frac{2\,k}{\sqrt{2^3 + k^2} - k} \tag{164}$$

based on the relation

$$d + \frac{1}{\gamma_1} = -\left(d + \frac{1}{\gamma_2}\right). \tag{165}$$

This relation sets the symmetry plane of the ellipsoid in the cross-section with $I_1 = -d\,\sigma_+$, see (129). There are three equations: for d and k and the constraint ν_-^{in}. The parameters of the model $\gamma_1 \in [0,\ 1[$ and $\gamma_2 < 0$ are unknown and should be determined. There are two solutions

$$\begin{cases} \nu_-^{\text{in}} = -\dfrac{-1 + \gamma_1^2 + \gamma_2^2 - \gamma_1\,\gamma_2}{(-2+\gamma_1+\gamma_2)\,(-1+\gamma_1+\gamma_2)}, \\ d = \dfrac{1}{1-\gamma_1-\gamma_2} \end{cases} \tag{166}$$

and

$$\begin{cases} \nu_-^{\text{in}} = -\dfrac{-1 + \gamma_1^2 + \gamma_2^2 - \gamma_1\,\gamma_2}{(-2+\gamma_1+\gamma_2)\,(-1+\gamma_1+\gamma_2)}, \\ k^2 = \dfrac{1}{(1-\gamma_1)\,(1-\gamma_2)}, \end{cases} \tag{167}$$

which should be compared. The conservative solution will be chosen. The solution of the above equations with $\nu_-^{\text{in}} = 1/2$ defined through (165) leads to

$$\begin{cases} \gamma_1 = \dfrac{1}{2\,d}\left(d - 1 + \sqrt{d^2 - 1}\right), \\ \gamma_2 = \dfrac{1}{2\,d}\left(d - 1 - \sqrt{d^2 - 1}\right) \end{cases} \tag{168}$$

or

$$\begin{cases} \gamma_1 = \dfrac{1}{4\,k}\left(3\,k - \sqrt{2^3 + k^2} + \sqrt{2}\sqrt{k\,(k + \sqrt{2^3 + k^2}) - 4}\right), \\ \gamma_2 = \dfrac{1}{4\,k}\left(3\,k - \sqrt{2^3 + k^2} - \sqrt{2}\sqrt{k\,(k + \sqrt{2^3 + k^2}) - 4}\right), \end{cases} \tag{169}$$

respectively.

The setting $d = k = 1$, which yields $\gamma_1 = \gamma_2 = 0$, results in the model of von Mises. With $d = 3\left(\sqrt{17} - 1\right)/8 \approx 1.17$ or $k = \sqrt{\left(9\sqrt{17} - 27\right)/8} \approx 1.12$ the value $\gamma_1 = 1/3$ is obtained. This corresponds to the value $a_+^{\text{hyd}} = 1$ of the normal stress hypotheses. The Poisson's ratio at tension is computed as follows:

$$\nu_+^{\text{in}} = \frac{3}{2\,d} - 1 = \frac{3\sqrt{8 + k^2}}{4\,k} - \frac{7}{4}. \tag{170}$$

For this model the yield condition should be restricted by (34). This results in restrictions for the parameters $d \in [1,\ 1.014]$ and $k \in [1,\ 1.010]$. With

$$a_+^{\text{hyd}} = \frac{2}{3}\,\frac{d}{d - 1 + \sqrt{d^2 - 1}} \tag{171}$$

one gets $a_+^{\text{hyd}} \geq 3.79$. This model is more suitable than the model of von Mises (16), which results in underpredictions in the region $I_1 > \sigma_+$.

The model can be applied instead of the paraboloid of Balandin (51). In this case incompressible yielding at compression can be obtained without the use of a non-associated flow rule (Sect. 4).

11.2.3 Combined Geometrical-Mechanical Model

A combined model can be built up based on the geometrical-mechanical model (105) with the transition at the cross-section $I_1 = 0$ (Fig. 44)

$$(3\,I_2')^3\,\frac{1 + c_3\cos 3\theta + c_6\cos^2 3\theta}{1 + c_3 + c_6} = \begin{cases} \left(\dfrac{\sigma_{\text{eq}} - \gamma_1\,I_1}{1 - \gamma_1}\,\dfrac{\sigma_{\text{eq}} + \gamma_1\,I_1}{1 + \gamma_1}\right)^3, & I_1 > 0; \\ \left(\dfrac{\sigma_{\text{eq}}}{1 - \gamma_1}\,\dfrac{\sigma_{\text{eq}}}{1 + \gamma_1}\right)^3, & I_1 \leq 0. \end{cases} \tag{172}$$

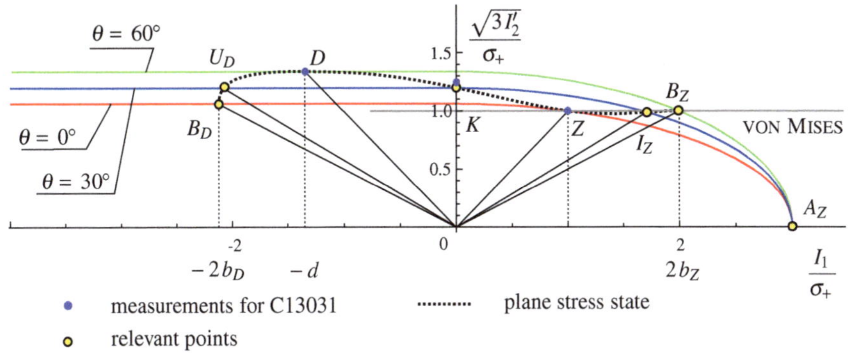

Fig. 44 Combined geometric mechanical model (172) for the polyoxymethylene (POM) Hostaform C13031, Ticona GmbH, Sulzbach, fitting starts from the values $d = 1.34$, $k = 1.25$ in the Burzyński-plane [113]: $k = 1.20$, $b_Z = 0.999$, $b_D = 1.06$, $\nu_+^{in} = 1/3$, $\nu_-^{in} = 1/2$, $a_+^{hyd} = 1$, $c_3 = 0.7885$, $c_6 = 0.3029$, $\gamma_1 = 1/3$

Since the necessary information on the material behavior under the hydrostatic tension (point A_Z) is almost always missing, the response under the hydrostatic tension can be defined by setting $\gamma_1 = 1/3$, which is based on the normal stress hypothesis (Sect. 2.1). It leads to the Poisson's ratio at tension with

$$\nu_+^{in} = \frac{1}{2}\left(1 - 3\,\gamma_1^2\right) \tag{173}$$

to $\nu_+^{in} = 1/3$. This setting can be corrected in dependence on the Poisson's ratio ν_+^{in}. For $\nu_+^{in} = 0.48$ one gets $\gamma_1 = 1/5\,\sqrt{3} \approx 0.1155$.

The values c_3 and c_6 can be computed from relations d and k, if the convexity restrictions (Fig. 20) are taken into account. These values result from the equations

$$k^6 = \frac{1 + c_3 + c_6}{(1 - \gamma_1^2)^3}, \qquad d^6 = \frac{1 + c_3 + c_6}{1 - c_3 + c_6}\,\frac{1}{(1 - \gamma_1)^3\,(1 + \gamma_1)^3}. \tag{174}$$

The model (172) can be recommended for many applications (Fig. 44).

A similar model with the C^1-transition at the cross-section $I_1 = -d\,\sigma_+$, cf. the modified model of Huber (163), can be formulated (Fig. 45).

The parameters c_3, c_6 and γ_2 of the model result from the formulas for d, k and $\nu_-^{in} = 1/2$. The last condition leads to the geometric relation (165)

$$\frac{1}{\gamma_2} = -\left(2\,d + \frac{1}{\gamma_1}\right). \tag{175}$$

This model has additional fitting possibilities in comparison to (172). Whether the model can be preferred over the model (172) it can be determined if further measurements, for instance at the points B_Z or B_D, are available. If only three measurements

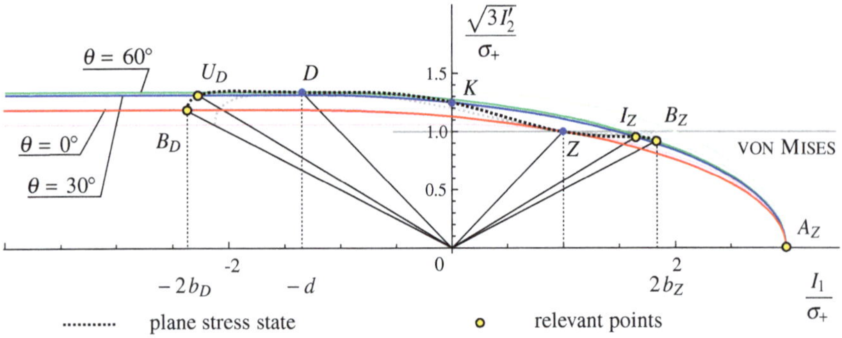

Fig. 45 Combined geometrical-mechanical model with the transition at $I_1 = -d\,\sigma_{eq}$ for the polyoxymethylene (POM) Hostaform C13031, Ticona GmbH, Sulzbach, with the values $d = 1.34$, $k = 1.25$ in the Burzyński-plane: $c_3 = 0.4743, c_6 = 0.3642, \gamma_2 = -0.1761, b_Z = 0.92, b_D = 1.19$, $\nu_+^{in} = 0.28$, $\nu_-^{in} = 1/2$, $a_+^{hyd} = 1$. The lines from Fig. 44 are shown for comparison [113]

at tension, compression and torsion (Z, K and D) exist, the application of the model with C^1-transition at $I_1 = -d\sigma_+$ is not meaningful.

12 Fitting

The objective function for fitting of the model to the measurements can be formulated in many ways, which lead to different results. The following three kinds of objective functions

- mathematical,
- physical and
- geometrical

can be considered [111, 114].

The mathematical objective functions are derived in a purely formal way, so that the fast convergence of the optimization routine can be achieved. Physical objective functions are based on a measurable value, which can be "related to mechanics". These conditions usually lead to complex implementations and slow computations. Geometrical criteria are based on the properties of the surface Φ.

12.1 Mathematical Criteria

This kind of objective functions will be presented using the geometrical-mechanical model (151) with the restriction (142) and the powers $j = 4$, $l = 2$, $m = 0$. The function Φ is rewritten in the form

$$\Omega = (3\,I_2')^3\,\frac{1 + c_3\,\cos 3\theta + c_6\,\cos^2 3\theta}{1 + c_3 + c_6} - \left(\frac{\sigma_{\mathrm{eq}} - \gamma_1\,I_1}{1 - \gamma_1}\right)^4 \left(\frac{\sigma_{\mathrm{eq}} - \gamma_2\,I_1}{1 - \gamma_2}\right)^2, \tag{176}$$

so that the surface is given by the equation $\Omega = 0$.

The n measurements are given e.g. in the principal stress space σ_{I}^i, σ_{II}^i, σ_{III}^i, $i = 1 \ldots n$. The objective functions can be formulated as follows

$$f = \frac{1}{n-1} \sum_{i=1}^{n} \left| \Omega(c_3, c_6, \gamma_1, \gamma_2, \sigma_{\mathrm{I}}^i, \sigma_{\mathrm{II}}^i, \sigma_{\mathrm{III}}^i) \right|^{m_1} \tag{177}$$

with $m_1 = 1,\ 2$ or

$$f_\infty = \max_{i=1\ldots n} \left| \Omega(c_3, c_6, \gamma_1, \gamma_2, \sigma_{\mathrm{I}}^i, \sigma_{\mathrm{II}}^i, \sigma_{\mathrm{III}}^i) \right|. \tag{178}$$

Other exponents m_1 can be used, however they do not lead to any significantly different results. In order to compare various fitting results the following value is considered

$$f_{m_1} = (f)^{1/m_1}. \tag{179}$$

The optimization problem is formulated as

$$\text{minimize } f(c_3, c_6, \gamma_1, \gamma_2) \tag{180}$$

for the chosen meridian shape through j, l and m. The solution is obtained in the form of the parameters of the models c_3, c_6, γ_1 and γ_2. Using these parameters the measurements σ_{I}^i, σ_{II}^i, σ_{III}^i, $i = 1 \ldots n$ are approximated. This optimization problem contains the constraints for the parameters of meridians γ_1, γ_2, parameters of the cross-section c_3, c_6 and for the Poisson's ratios ν_+^{in}, ν_-^{in} (Sect. 4).

The function Ω (176) can be modified, so that additional solutions of the optimization for comparisons become possible, e.g.:

$$\left[(3\,I_2')^3\,\frac{1 + c_3\,\cos 3\theta + c_6\,\cos^2 3\theta}{1 + c_3 + c_6}\right]^{l_1} - \left[\left(\frac{\sigma_{\mathrm{eq}} - \gamma_1\,I_1}{1 - \gamma_1}\right)^4 \left(\frac{\sigma_{\mathrm{eq}} - \gamma_2\,I_1}{1 - \gamma_2}\right)^2\right]^{l_1} = 0 \tag{181}$$

with the integer exponent $l_1 \geq 1$.

The formulation (176) is derived with $\sigma_+ = \sigma_{\mathrm{eq}}$. If there are "enough" measurements the equivalent stress σ_{eq} can also be seen as a parameter subject to optimization.

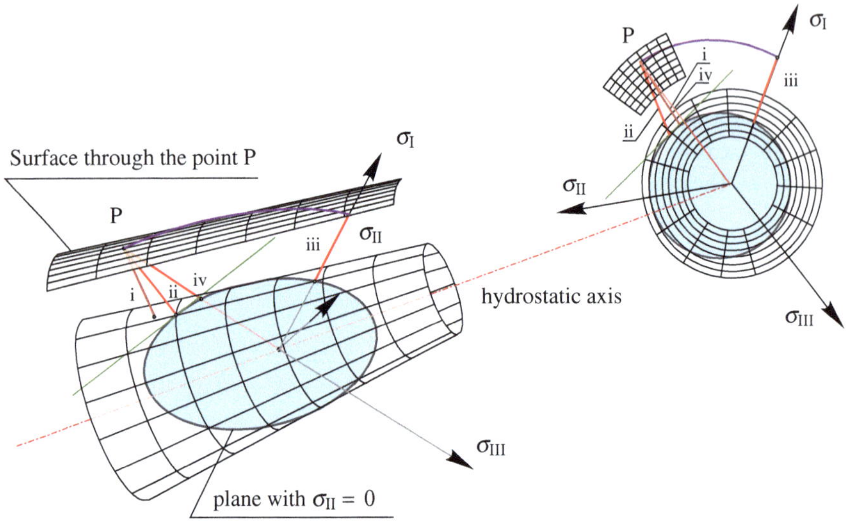

Fig. 46 Cone of Drucker-Prager (51) in the principal stress state (*left*) and in the π-plane (*right*). Comparison of the *four* physical optimization criteria

12.2 Physical Criteria

For a limit surface, which is defined implicitly in the form (5) or (6) and available measurements it is required to estimate the quality of fitting. Four physical criteria for estimation of the optimization quality are discussed (Fig. 46). In order to apply the criteria the measurements must be transformed into a principal stress state $(\sigma_I^i, \sigma_{II}^i, \sigma_{III}^i)$ for $i = 1 \dots n$. The four criteria can hardly be used in order to obtain the parameters of a model because of the high computational complexity. However, they can be used for comparison of different optimization results.

The criteria can be formulated as follows:

1. The regression quality in the principal stress space is evaluated, that is for each measurement the distance from the limit surface in the principal stress space $(\sigma_I, \sigma_{II}, \sigma_{III})$ is computed and then averaged over all measurements. Formally that means, we start with a set of measurements $(\sigma_I^i, \sigma_{II}^i, \sigma_{III}^i)$ and solve for each measurement the optimization problem

$$\begin{aligned} &\min\ (\sigma_I - \sigma_I^i)^2 + (\sigma_{II} - \sigma_{II}^i)^2 + (\sigma_{III} - \sigma_{III}^i)^2 \\ &\text{subject to } \Phi(\sigma_I, \sigma_{II}, \sigma_{III}) = 0. \end{aligned} \tag{182}$$

The solution is obtained using a Lagrange multiplier. For example the function

$$F(\sigma_I, \sigma_{II}, \sigma_{III}, \lambda) = (\sigma_I - \sigma_I^i)^2 + (\sigma_{II} - \sigma_{II}^i)^2 + (\sigma_{III} - \sigma_{III}^i)^2 - \lambda\, \Phi(\sigma_I, \sigma_{II}, \sigma_{III}) \tag{183}$$

is defined and the stationary points are obtained from the equation

$$\nabla F = 0. \tag{184}$$

Generally this equation has more than one solution, however a single point $(z^i_{\mathrm{I}},\, z^i_{\mathrm{II}},\, z^i_{\mathrm{III}})$ of minimal distance on the limit surface is to be determined. Since the number of solutions is small, the correct one can be chosen by trial-and-error. Finally, the value of the objective function $f_{3\mathrm{D}}$ computes to

$$f_{3\mathrm{D}} := \frac{1}{n-1}\sum_{i=1}^{n}\sqrt{(\sigma^i_{\mathrm{I}} - z^i_{\mathrm{I}})^2 + (\sigma^i_{\mathrm{II}} - z^i_{\mathrm{II}})^2 + (\sigma^i_{\mathrm{III}} - z^i_{\mathrm{III}})^2}. \tag{185}$$

2. The minimal distance can be computed not in the principal stress space but for a plane stress state. For example we put $\sigma_{\mathrm{III}} = 0$ and hence simplify the optimization problem (182)

$$\begin{array}{l} \min\ (\sigma^i_{\mathrm{I}} - \sigma_{\mathrm{I}})^2 + (\sigma^i_{\mathrm{II}} - \sigma_{\mathrm{II}})^2 \\ \text{subject to } \Phi(\sigma_{\mathrm{I}}, \sigma_{\mathrm{II}}, 0) = 0. \end{array} \tag{186}$$

A further computation is performed in analogy to the previous case. For each point $(\sigma^i_{\mathrm{I}},\, \sigma^i_{\mathrm{II}},\, 0)$ the point of minimal distance $(z^i_{\mathrm{I}},\, z^i_{\mathrm{II}},\, 0)$ on the curve

$$\Phi(\sigma_{\mathrm{I}},\, \sigma_{\mathrm{II}},\, 0) = 0 \tag{187}$$

is determined and the value of the objective function $f_{2\mathrm{D}}$ is estimated as follows

$$f_{2\mathrm{D}} := \frac{1}{n-1}\sum_{i=1}^{n}\sqrt{(\sigma^i_{\mathrm{I}} - z^i_{\mathrm{I}})^2 + (\sigma^i_{\mathrm{II}} - z^i_{\mathrm{II}})^2}. \tag{188}$$

3. If the model is based on the equivalent stress concept, that is

$$\Phi(\sigma_{\mathrm{I}},\, \sigma_{\mathrm{II}},\, \sigma_{\mathrm{III}},\, \sigma_{\mathrm{eq}}) = 0, \tag{189}$$

whereas $\sigma_{\mathrm{eq}} = \sigma_+$, a simple estimation for the quality of fitting can be proposed. The equivalent stress σ_{eq} is considered as a parameter as fitting is performed. The fitted equivalent stress is denoted by σ^*_{eq}. In order to estimate the quality of fitting for each point $(\sigma^i_{\mathrm{I}},\, \sigma^i_{\mathrm{II}},\, \sigma^i_{\mathrm{III}})$, $i = 1 \ldots n$ the value $\sigma_{\mathrm{eq}} = \sigma^i_{\mathrm{eq}}$ is computed so, that the point lies on the surface

$$\Phi(\sigma_{\mathrm{I}},\, \sigma_{\mathrm{II}},\, \sigma_{\mathrm{III}},\, \sigma^i_{\mathrm{eq}}) = 0, \tag{190}$$

i. e. the equation

$$\Phi(\sigma^i_{\mathrm{I}},\, \sigma^i_{\mathrm{II}},\, \sigma^i_{\mathrm{III}},\, \sigma^i_{\mathrm{eq}}) = 0 \tag{191}$$

must be solved for each $i = 1 \dots n$ with respect to σ^i_{eq}. The estimated value f_{eq} computes to

$$f_{\mathrm{eq}} := \frac{1}{n-1} \sum_{i=1}^{n} \left| \frac{\sigma^i_{\mathrm{eq}} - \sigma^*_{\mathrm{eq}}}{\sigma^*_{\mathrm{eq}}} \right| . \tag{192}$$

4. The distance between the experimental point and the surface of model is measured along the line connecting the point with the origin. The sum of all the distances normalized by $n-1$ computes to the value f_{ray}.

The criterion 1 is ubiquitous and can be used for an arbitrary set of measurements and an arbitrary surface. Often the measurements belong to a plane stress state, in this case the criterion 2 is of stronger physical relevance. Since it is a plausible assumption, that a measurement corresponding to a plane stress state is approximated by a point of the model, which also belong to a plane stress state. The criterion 3 is only suitable for the models based on the equivalent stress concept (2). It can be used too, if for instance torsion τ_* or compression σ_- are taken as the equivalent stress. An advantage of the criterion 4 is that it has a "mechanical background": Each measurement (e.g. torsion) is compared to the corresponding point on the surface (point K for torsion). It is clear that it can be easily applied and leads to relatively fast convergence of the optimization routine.

12.3 Geometrical Criteria

The principle of the conservative modeling can be stated as follows: among all best possible solutions the one is preferred, which represents the most conservative assumption about the material behavior [103]. Geometrical criteria allow to compare different optimal solutions. These criteria are listed below [108]:

- linear criteria:
  - the shortest length of the line in the meridian section of the closed surface Φ with the plane $\xi_2 = 0$ (Fig. 13, coordinates (ξ_1, ξ_2, ξ_3), Figs. 34 and 35),
  - the minimal length of the line of the plane stress state,
  - the distance between the hydrostatic nodes A_{Z} and A_{D} for foams, ceramics, etc.,

$$\text{minimize} \left(\chi\, a_-^{\mathrm{hyd}} + (1-\chi)\, a_+^{\mathrm{hyd}} \right), \qquad \chi \in [0,\ 1] \tag{193}$$

  and for materials with $a_-^{\mathrm{hyd}} \to \infty$

$$\text{minimize} \left(a_+^{\mathrm{hyd}} \right), \tag{194}$$

  - relation k: minimize (k),
  - relation d: minimize (d),

- square criteria:
  - the minimal area of the closed surface Φ in the principal stress space,
  - minimal area of the closed surface Φ in the meridian cross-section with the plane $\xi_2 = 0$,
  - minimal area circumscribed by the line of the plane stress state,
- cubic criteria, for instance the minimal volume circumscribed by the closed surface Φ in the principal stress space.

These criteria can be chosen in dependence on the measurements and the requirements imposed upon the model. A comparison of the criteria can be performed in the Pareto-diagram [44, 61, 152, 196]. The geometrical criteria allow to select a single point from the Pareto-solutions [114].

13 Applications

In order to illustrate the application of the models and fitting of the parameter, some measurements from the literature are analyzed. The stages of the analysis are visualized. The experimental data are normalized with respect to σ_+ for better comparison of the surfaces Φ.

13.1 Measurements of Coffin for Gray Cast Iron

29 measurements for the plane stress state for gray cast iron[8] are shown in Coffin-Schenectady [47] and approximated as follows:

- region Tension-Tension

$$\sigma_{\mathrm{eq}} = \sigma_{\mathrm{I}}, \qquad \sigma_{\mathrm{eq}} = \sigma_{\mathrm{II}} \tag{195}$$

  with $\sigma_{\mathrm{eq}} = 33 \times 10^3$ psi.
- region Tension-Compression

$$\begin{aligned} (\kappa\,\sigma_{\mathrm{eq}} + \sigma_0)^2 &= (k\,\sigma_{\mathrm{I}} + \sigma_0)^2 - \sigma_{\mathrm{II}}\,(k\,\sigma_{\mathrm{I}} + \sigma_0) + \sigma_{\mathrm{II}}^2, \\ (\kappa\,\sigma_{\mathrm{eq}} + \sigma_0)^2 &= (k\,\sigma_{\mathrm{II}} + \sigma_0)^2 - \sigma_{\mathrm{I}}\,(k\,\sigma_{\mathrm{II}} + \sigma_0) + \sigma_{\mathrm{I}}^2 \end{aligned} \tag{196}$$

  with $\kappa = 3$ and $\sigma_0 = 30$ psi.
- region Compression-Compression

$$\sqrt{\frac{1}{2}\left[(\sigma_{\mathrm{I}} - \sigma_{\mathrm{II}})^2 + \sigma_{\mathrm{I}}^2 + \sigma_{\mathrm{II}}^2\right]} + \frac{1}{3}\,\mu\,(\sigma_{\mathrm{I}} + \sigma_{\mathrm{II}}) = \vartheta \tag{197}$$

[8] 3.08 % total C, 2.04 % Si, 0.56 % Mn, 0.112 % S, 0.33 % P

with $\mu = 0.51$ and $\vartheta = 88 \times 10^3$ psi.

The measurements were digitalized from the diagram $\sigma_{\rm I} - \sigma_{\rm II}$ [47]. The agreement with the approximation presented in the paper results for $\kappa = 2.9050$, $\mu = 0.5307$ and $\vartheta = 89.0202 \times 10^3$ psi and is to be ascribed to the precision of the digitalization and the rounding error. The computation of stresses leads to $\sigma_{\rm BD} = 137.76 \times 10^3$ psi, $\sigma_{\rm UD} = 148.20 \times 10^3$ psi, $\sigma_- = 108.15 \cdot 10^3$ psi, $\tau_* = 27.49 \times 10^3$ psi and $\sigma_{\rm AZ} = 33 \cdot 10^3$ psi.

For further evaluations the measurements are normalized with respect to $\sigma_+ = 33 \times 10^3$ psi (Table 7). The relations of the evaluations are summarized in Table 8. The relations $a_{\rm Z}$ and $b_{\rm D}$ result from the extrapolations. The smallest value

$$b_{\rm D} = \frac{103.35}{33} = 3.13 \tag{198}$$

is obtained from the results of the test 11 with the convexity requirement

$$\frac{\sigma_{\rm I} - (-69.4)}{(-137.3) - (-69.4)} = \frac{\sigma_{\rm II} - (-137.3)}{(-69.4) - (-137.3)} \tag{199}$$

for $\sigma_{\rm I} = \sigma_{\rm II}$. If the classical material behavior (22) with $b_{\rm D} = d$ is assumed, the estimate

$$b_{\rm D} \in [3,\ 3.29] \tag{200}$$

is obtained from the measurements at axial compression.

The value $b_{\rm Z}$ can be estimated

$$b_{\rm Z} = \frac{28.52}{33} \dots \frac{30.22}{33} = 0.86 \dots 0.92. \tag{201}$$

The lower bound for $a_{\rm Z}$ is defined using the stress value $\sigma_{\rm BZ} = 30.22 \times 10^3$ psi and the convexity condition

$$2\sigma_{\rm BZ} < 3\sigma_{\rm AZ}. \tag{202}$$

It follows

$$a_+^{\rm hyd} > \frac{2}{3} \cdot \frac{30.22}{33} = 0.61. \tag{203}$$

13.1.1 Strain Hypothesis

In the first quadrant of the $\sigma_{\rm I}$–$\sigma_{\rm II}$-diagramm (region Tension-Tension, Fig. 47) the strain hypothesis (41) can be used instead of the normal stress hypothesis (8) in order to estimate relations in this region. For fitting of parameters and computation of the relations $i_{\rm Z}$, $b_{\rm Z}$ and $a_{\rm Z}$ the model is used in the form (10) with (42). The Poisson's ratio computes to $\nu_+^{\rm in} \in [-0.0870,\ -0.0521]$ which is equivalent to $\gamma_{\rm I} \in [0.3681,\ 0.3913]$. Relations $i_{\rm Z}$, $b_{\rm Z}$ and $a_{\rm Z}$ are specified in Table 8.

Table 7 Measurements by Coffin-Schenectady and their normalized values with respect to $\sigma_+ = 33 \cdot 10^3$ psi, the axiatoric-deviatoric invariants and the stress angle, plane stress state $\sigma_{III} = 0$

Test number	σ_I 10^3 [psi]	σ_{II} 10^3 [psi]	σ_I [-]	σ_{II} [-]	I_1 [-]	I'_2 [-]	I'_3 [-]	$\cos 3\theta$ [-]	θ [grad]
7	−77.60	−151.45	−2.35	−4.59	−6.94	5.2667	0.199442	0.0429	29. 2
11	−69.40	−137.30	−2.10	−4.16	−6.26	4.3278	0.065570	0.0189	29.6
12	−30.00	−116.95	−0.91	−3.54	−4.45	3.3881	−1.758632	−0.7327	45.7
23	−24.45	−120.52	−0.74	−3.65	−4.39	3.7268	−2.317541	−08369	48.9
	0	−108.67	0	−3.29	−3.29	3.6147	−2.645169	−1	60
	0	−101.78	0	−3.08	−3.08	3.1709	−2.173324	−1	60
46	0	−98.90	0	−3.00	−3.00	2.9939	−1.993946	−1	60
19	17.03	−67.68	0.52	−2.05	−1.53	1.8434	−0.809151	−0.8399	49.0
	23.28	−52.79	0.71	−1.60	−0.89	1.3948	−0.389280	−0.6140	42.6
18	25.35	−50.85	0.77	−1.54	−0.77	1.3827	−0.339070	−0.5418	40.9
	23.67	−47.65	0.72	−1.44	−0.73	1.2117	−0.279228	−0.5439	41.0
	32.60	−32.51	0.99	−0.99	0	0.9733	0.000814	0.0022	30
17	27.60	−27.47	0.84	−0.83	0	0.6962	0.000914	0.0041	29.9
	26.49	−26.49	0.80	−0.80	0	0.6442	−0.000002	0	30
	28.36	−14.52	0.86	−0.44	0.42	0.4368	0.058351	0.5251	19.4
9	30.25	−13.38	0.92	−0.41	0.51	0.4587	0.073234	0.6125	17.4
	30.58	0	0.93	0	0.93	0.2862	0.058921	1	0
6	32.55	0	0.99	0	0.99	0.3243	0.071085	1	0
	29.47	13.67	0.89	0.41	1.31	0.1997	0.004307	0.1254	27.6
	29.68	15.21	0.90	0.46	1.36	0.2023	−0.001518	−0.0434	30.8
	35.18	17.62	1.07	0.53	1.60	0.2841	−0.000188	−0.0032	30.1
	28.64	28.52	0.87	0.86	1.73	0.2501	−0.048125	−0.9999	59.8
	29.59	29.80	0.90	0.90	1.80	0.2699	−0.053974	−0.9998	59.7
4	30.22	30.22	0.92	0.92	1.83	0.2795	−0.056886	−1	60
	15.90	31.57	0.48	0.96	1.44	0.2288	−0.000551	−0.0131	30.3
	16.32	32.91	0.49	1	1.49	0.2487	0.000710	0.0149	29.7
	0	30.66	0	0.93	0.93	0.2877	0.059381	1	0
	0	33.28	0	1.01	1.01	0.3390	0.075957	1	0
1	0	34.18	0	1.04	1.04	0.3576	0.082308	1	0

The measurement number and the respective values for σ_I, σ_{II} are shown in Figs. 6, 7 and 8 in [47]
The values for the measurement 23 with $\sigma_I = -42.900 \times 10^3$ psi, $\sigma_{II} = -104,000 \times 10^3$ psi (Fig. 8 in [47]) are different from the digitalized values, which are used for evaluation

13.1.2 Burzyński-Yagn Model

The measurements can be described using the rotationally symmetric model (51). The best approximation is obtained with the hyperboloid (Table 9).

The position of the measurements in the Burzynski-plane suggests that a rotationally symmetric model is not suitable in this case. The material behavior in the region Tension-Tension is underestimated and overestimated near the point B_D (Table 8).

Table 8 Relations for the evaluations of the measurements by Coffin-Schenectady, $\sigma_+ = 33 \times 10^3$ psi with different models

Approximation	b_D	u_D	d	k	i_Z	b_Z	a_+^{hyd}	ν_+^{in}	ν_-^{in}
Coffin	4.17	3.89	3.38	1.44	0.87	1	1	≥ 0	–
SH in the region TT	–	–	–	–	0.84	0.95	0.91	−0.05	–
SH in the region TT	–	–	–	–	0.82	0.92	0.85	−0.09	–
Eq. (51), hyperboloid	4.39	3.91	2.79	1.66	0.73	0.66	0.53	−0.20	1.24
Eq. (51), cone	7.15	4.61	2.34	1.40	0.83	0.78	1.62	0.07	1.51
Eq. (51), paraboloid	4.29	3.88	2.82	1.68	0.73	0.66	0.52	−0.21	1.21
UST by Yu, $b = 0.35$ with cut-off	3.05	3.03	3.05	1.39	0.87	1	1	0…0.16	1.53
UST by Yu, $b = 1$ with cut-off	3.05	3.52	3.05	1.49	0.87	1	1	0…0.16	1.53
Bigoni-Piccolroaz GMM (for Bigoni), $l = 3, m = 1$	3.58	3.71	3.04	1.50	0.88	0.89	0.92	0.09	1.51
GMM, straight line $l = m = 0$	3.48	3.85	2.92	1.49	0.92	0.94	1.40	0.14	1.54
GMM, parabola $l = 0, m = 1$	3.48	3.84	2.94	1.53	0.90	0.90	1.10	0.10	1.47
CPM: straight line $l = m = 0$	3.31	3.82	2.50	1.50	0.94	0.91	1.43	0.15	1.37
CPM: straight line $l = m = 0$	3.34	3.82	3.00	1.50	0.93	0.97	1.43	0.15	1.55
CPM: parabola $l = 0, m = 1$	3.41	3.80	3.00	1.53	0.90	0.91	1.11	0.10	1.48

13.1.3 Unified Strength Theory of Yu

The values b_D (198) and d are similar. It follows that UST (Sect. 10.2) with the relationship $b_D = d$ can be used. In [230] Yu puts $d = 3.05$, in order to describe the Tension-Compression region. In the Tension-Tension region the normal stress hypothesis is used as a cut-off (Sect. 11.1) [228, 229]. In Table 8 the relations for the parameters $b = 0.35$ und $b = 1$ are presented. In the third quadrant of the $\sigma_I - \sigma_{II}$-diagramm (Compression-Compression region) the UST is not sufficient.

Table 9 Parameters of the models for the approximation of the measurements by Coffin-Schenectady, $\sigma_+ = 33 \times 10^3$ psi

Model	Eq.	Meridian	f_2	γ_1	γ_2	c_3	c_6
		straight line	1.1935	0.2867		–	–
Burzyński-Yagn	(51)	hyperbola	0.7763	0.6345	0.0073	–	–
		parabola	0.7800	0.6451	0	–	–
GMM, $l = m = 0$		straight line	53.7926	0.2376		0.8671	0.2832
GMM, $l = 0, m = 1$	(152)	parabola	62.6810	0.3033	0	0.8161	0.2960
(for Bigoni), $l = 3, m = 1$		hyperbola	84.6122	0.3617	0.2721	0.6969	0.0805

Values of the objective function f_2 (179) are to be compared for respective models only
(for Bigoni)—approximation by Bigoni-Piccolroaz reformulated using GMM

13.1.4 Geometrical-Mechanical Model

The best approximation for GMM (151) is obtained with a straight meridian (Figs. 47, 48, 49, 50). The value $b_D = 3.48$ lies outside of the bounds given by (200).

These approximations with GMM can be compared to the approach by Bigoni-Piccolroaz [26] with seven parameters (Table 8). This approximation can be obtained using GMM with $l = 3$, $m = 1$ (Table 9).

The experimental results with $\sigma_I = \sigma_{II}/4$ cannot be represented by GMM. A precise approximation is however possible using the continuously differentiable GMM (Sect. 11.2.3) . The switching occurs in the plane $I_1 = -d\,\sigma_+$ with $d = 3 \ldots 3.29$.

13.1.5 Convex π-Plane Model

The results of fitting using GMM (105) with $n = 6$ can be also represented by CPM (99). The parameter γ_1 of the linear substitution (126) is the same (Table 9): $\gamma_1 = 0.2376$. With the parameters $c_3 = 0.8672$ and $c_6 = 0.2832$ the relations (106) are computed to

$$d_{\text{inc}} = \sqrt[6]{\frac{1 + c_3 + c_6}{1 - c_3 + c_6}}, \qquad k_{\text{inc}} = \sqrt[6]{1 + c_3 + c_6}. \tag{204}$$

The parameter $d_{\text{inc}} = 1.3149$ corresponds to the same parameter of CPM and with $d_{\text{inc}} = 1.3149$, $k_{\text{inc}} = 1.1361$ the bridge-parameter $\xi = 0.8766$ is computed. These values can be used as a starting point for optimization:

$$f_2 = 7.6597: \quad d_{\text{inc}} = 1.3149, \quad \xi = 0.8766, \quad \gamma_1 = 0.2376. \tag{205}$$

The optimization results

$$f_2 = 6.0831: \quad d_{\text{inc}} = 1.2122, \quad \xi = 1, \quad \gamma_1 = 0.2325 \tag{206}$$

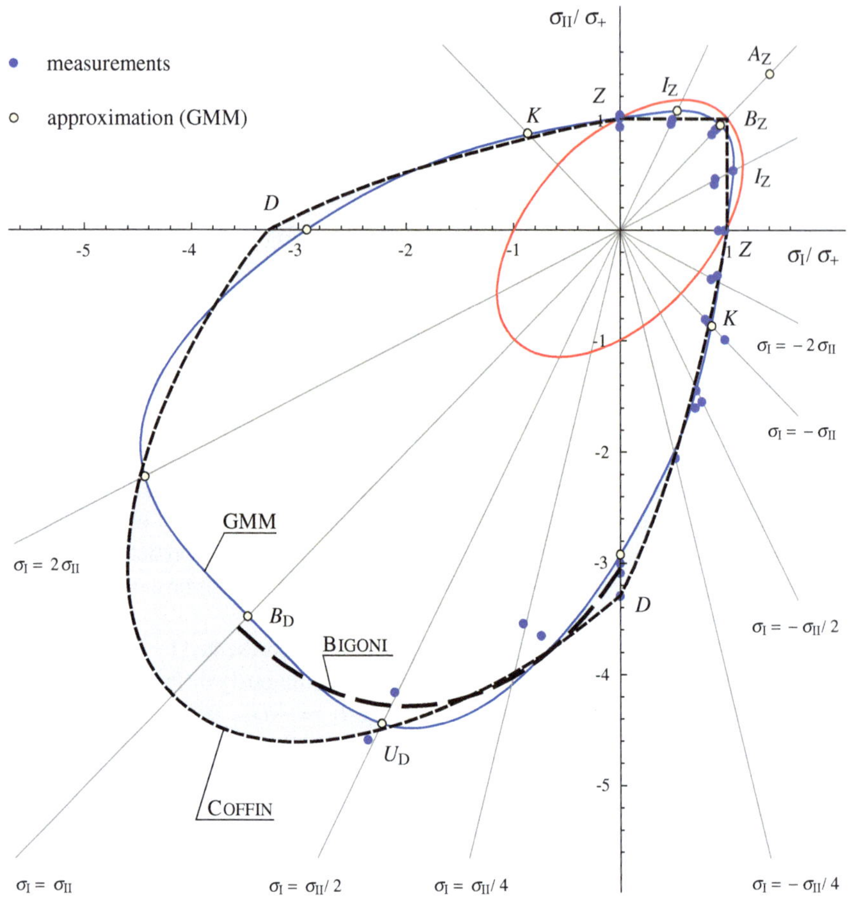

Fig. 47 Plane stress state $\sigma_{\mathrm{III}} = 0$: geometrical-mechanical model ($\sigma_+ = 33 \times 10^3$ psi, $\gamma_1 = \gamma_2 = 0.2375$, $c_3 = 0.8672$, $c_6 = 0.2832$, $l = m = 0$) with the values $b_{\mathrm{D}} = 3.48$, $d = 2.92$, $k = 1.49$ for gray cast iron (Table 8). The models of von Mises and the approximations of Coffin-Schenectady and Bigoni-Piccolroaz (between the points B_{D} and D) are shown for comparison

lead to the pyramid due to Haythornthwaite, which follows from the prism of Haythornthwaite (94) with the substitution (126) [107]. This approximation (Fig. 48, Model 1) underestimates the value at compression, it yields $d = 2.50$.

Additionally, it can be required, that the curve of the plane stress state contains the point D with the coordinates $\sigma_{\mathrm{I}} = 0$, $\sigma_{\mathrm{II}} = -98.90/33$ (relation $d = 3$), it follows (Fig. 48, Model 2)

$$f_2 = 8.1939: \quad d_{\mathrm{inc}} = 1.3522, \quad \xi = 0.9662, \quad \gamma_1 = 0.2335. \tag{207}$$

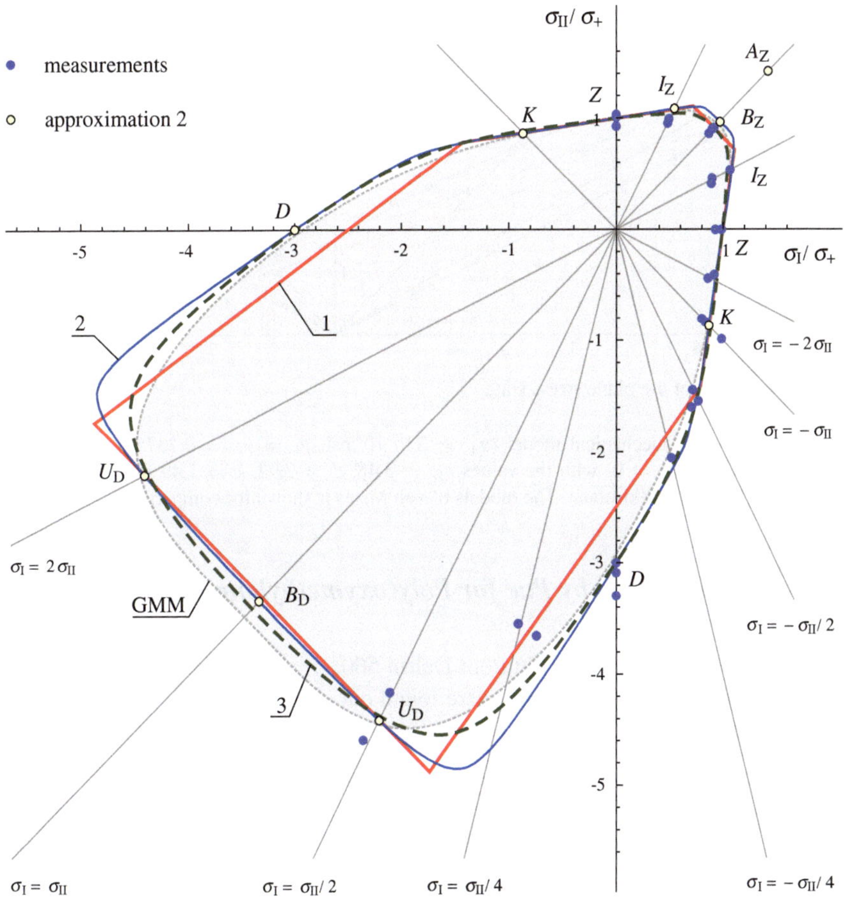

Fig. 48 Plane stress state $\sigma_{\rm III} = 0$: CPM ($\sigma_+ = 33 \times 10^3$ psi) for gray cast iron (Table 8): *1* CPM with $l = m = 0$, $f_2 = 6.0831$: $d_{\rm inc} = 1.2122$, $\xi = 1$, $\gamma_1 = 0.2325$, *2* CPM with $l = m = 0$ and $d \geq 3$, $f_2 = 8.1939$: $d_{\rm inc} = 1.3522$, $\xi = 0.9662$, $\gamma_1 = 0.2335$, *3* CPM with $l = 0$, $m = 1$ and $d \geq 3$, $f_2 = 8.8342$: $d_{\rm inc} = 1.3028$, $\xi = 0.8868$, $\gamma_1 = 0.3003$. The GMM (Fig. 47) is shown for comparison

Similar approximation (Fig. 48, Model 3) is obtained with the paraboloid $l = 0$, $m = 1$ using the substitution (124) and the restriction $d \geq 3$

$$f_2 = 8.8342: \qquad d_{\rm inc} = 1.3028, \qquad \xi = 0.8868, \qquad \gamma_1 = 0.3003. \tag{208}$$

The models with a straight meridian fitted with the parameters (206), (207) should be preferred here because of their simplicity. The setting (208) with the parabolic meridian $l = 0$, $m = 1$ yields a lower value $a_{\rm Z} = 1.11$. For further approximations see [1, 14, 34, 202].

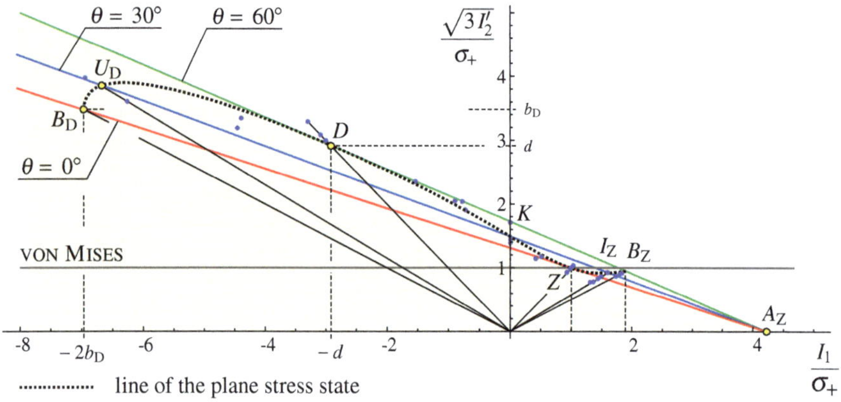

Fig. 49 Geometric-mechanical model ($\sigma_+ = 33 \cdot 10^3$ psi, $\gamma_1 = \gamma_2 = 0.2375$, $c_3 = 0.8672$, $c_6 = 0.2832$, $l = m = 0$) with the values $b_D = 3.48$, $d = 2.92$, $k = 1.49$ for gray cast iron (Table 8) in the Burzyński-plane. The models of von Mises is shown for comparison, s. (Fig. 47)

13.2 Measurements by Pae for Poly(oxymethylene) (POM)

The measurements for POM, du Pont Delrin 500, $\rho = 1.425\,\mathrm{g/cm^3}$ are provided in [151]. The following inaccuracies were found out after the analysis of the measurements from [151]:

- Molar mass and crystallinity were not specified;
- Table 1, hydrostatic pressure, (psi)$\times 10^{-3}$;
- Table 1, average experimental yield stresses, the meausrement 10.5 is shifted from the column "Shear" in to the column "Tension";
- Figure 3 (a), hydrostatic axis, the factor $1/\sqrt{3}$ for the first invariant I_1 was not taken into account as the measurements were represented in the principal stress space: $p = 7.3\,\mathrm{kbar}$, $I_1 = 3 \cdot 7.3 \cdot 14503.8 = 317.63 \times 10^3$ psi. From the normalization with respect to σ_+ it follows $\frac{I_1}{\sigma_+} = \frac{317.63 \times 10^3}{10.6 \cdot 10^3} = 29.97\,[-]$. The hydrostatic node A_Z should lie at $29.97/\sqrt{3} = 17.3\,[-]$, (or with $\frac{7.25 \cdot 10^3}{0.0230} \frac{1}{10.6 \times 10^3} \frac{1}{\sqrt{3}} = 17.2\,[-]$, cf. Eq. (3) with $I'_2 = 0$ and Table III in [151]). The difference results from the rounding error;
- Figures 3 (a) and (b), representation of the model in the π-plane, additionally to the cross-sections $I_1 = $ const. of the models of von Mises and Tresca a model with trigonal symmetry is shown, which is however not defined;
- The units in the Table III, [151] are not provided.

These measurements are visualized in the Burzyński-plane and approximated with a quadratic rotationally symmetric model (51), Fig. 51.

It can be seen in Fig. 51 that the points on the meridian $\theta = \pi/3$ are separated from the points on the meridians with $\theta = 0$ and $\theta = \pi/6$. So the trigonal symmetry of the surface can be assumed. The application of the geometrical-mechanical model

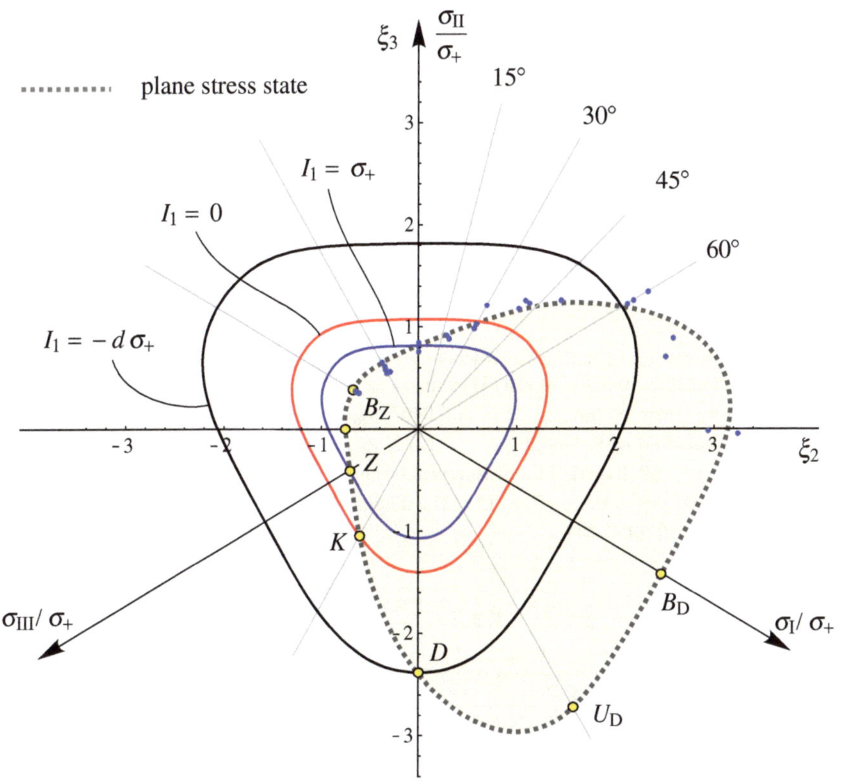

Fig. 50 Measurements by Coffin-Schenectady [47] in the π-plane approximated with the geometrical-mechanical model (151), s. Fig. 49, $d = 2.92$; line of the plane stress state with $\sigma_I = 0$. The cross-sections orthogonal to the hydrostatic axis with $I_1 =$ const. through the points Z, K and D are shown

(151) with the compressible substitution (146) can be recommended in this case. A possible approximation is shown in the Burzyński-plane (Fig. 52), in the plane $\sigma_I - \sigma_{II}$ (Fig. 53) and in the π-plane (Fig. 54). Further applications to POM can be found in [113].

13.3 Measurements of Cristensen for PVC Hard Foam

Closed-cell PVC (polyvinyl chloride) foam Divinycell H200 (DIAB International AB, Schweden) with the density $\rho = 200\,\text{kg/m}^3$ was tested in the region $D-K-Z$ (compression-torsion-tension) [46]. 25 measurements presented in the diagram σ_{11}-τ_{12} in [46] were digitalized (Fig. 55). The stresses at tension σ_+ corresponds to the data provided by the manufacturer [52] and indirectly to the measurements by

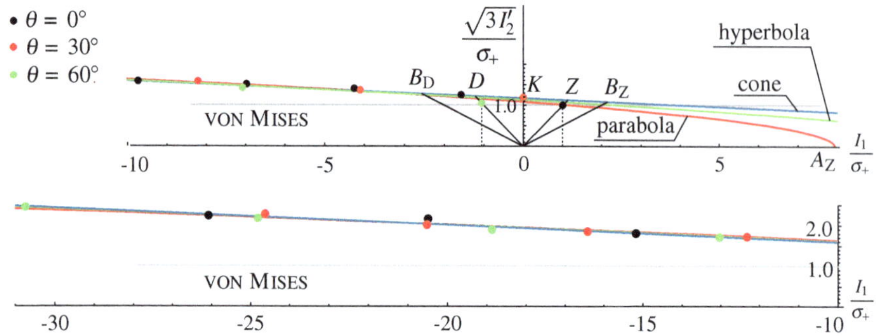

Fig. 51 Approximation of the measurements by Pae for POM in the Burzyński-plane with the quadratic rotationally symmetric model (51) with $\sigma_+ = 10.6 \times 10^3$ psi and σ_{eq} as a parameter: cone of Drucker-Prager (Mirolyubov): $\gamma_1 = \gamma_2 = 0.0403$, $\sigma_{eq} = 1.12$; $d = 1.09$, $k = 1.04$, $\nu_+^{in} = 0.44$, $\nu_-^{in} = 0.57$, paraboloid of Balandin: $\gamma_1 = 0.1277$, $\gamma_2 = 0$, $\sigma_{eq} = 1.01$; $d = 1.14$, $k = 1.07$, $\nu_+^{in} = 0.40$, $\nu_-^{in} = 0.60$, hyperboloid of Burzyński-Yagn: $\gamma_1 = 0.089$, $\gamma_2 = 0.0107$, $\sigma_{eq} = 1.07$; $d = 1.11$, $k = 1.05$, $\nu_+^{in} = 0.42$, $\nu_-^{in} = 0.58$. The diagram is divided in two regions $I_1 \in [-10,\ 8]$ and $I_1 \in [-31,\ -10]$ for clarity

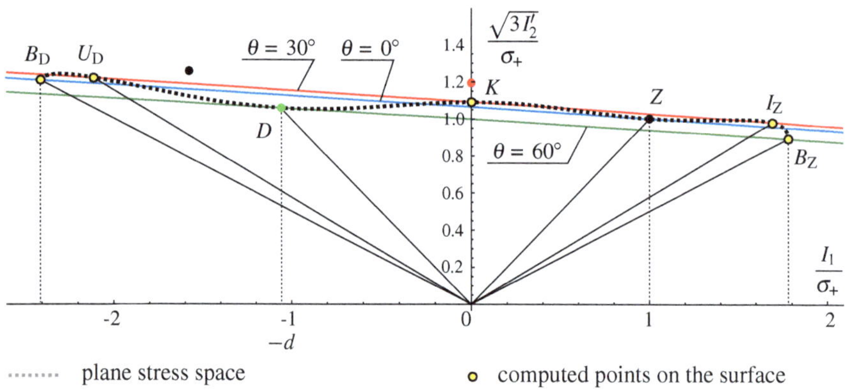

Fig. 52 POM-measurements by Pae in the Burzyński-plane approximated using the geometrical-mechanical model (151) with the parameters $\gamma_1 = 0.0869$, $\gamma_2 = 0$, $j = 4$, $l = 2$, $m = 0$, $c_3 = -0.2717$, $c_6 = 0.4314$ and $\sigma_{eq} = 1$: $d = \frac{11.2}{10.6} = 1.06$, $k = \sqrt{3}\,\frac{6.67}{10.6} = 1.09$, $\nu_+^{in} = 0.41$, $\nu_-^{in} = 0.63$. The region $I_1 \in [-2.5,\ 2]$ is shown for clarity

Gdoutos for the PVC-foam H 250 [72]. The stresses at compression σ_- are significantly lower compared to the data in [52, 72]. The value for torsion is also lower than the manufacturer's data and the measurements by Deshpande-Fleck and Gdoutos.

In order to present the models in the diagram σ_{11}–τ_{12} (Fig. 55) the invariants (Appendix 15) were reduced with $\sigma_{22} = \sigma_{33} = \tau_{13} = \tau_{23} = 0$ as follows

$$I_1 = \sigma_{11}, \qquad I_2' = \frac{1}{3}\,(\sigma_{11}^2 + 3\,\tau_{12}^2), \qquad I_3' = \frac{2}{3^3}\,\sigma_{11}^3 + \frac{1}{3}\,\sigma_{11}\,\tau_{12}^2. \tag{209}$$

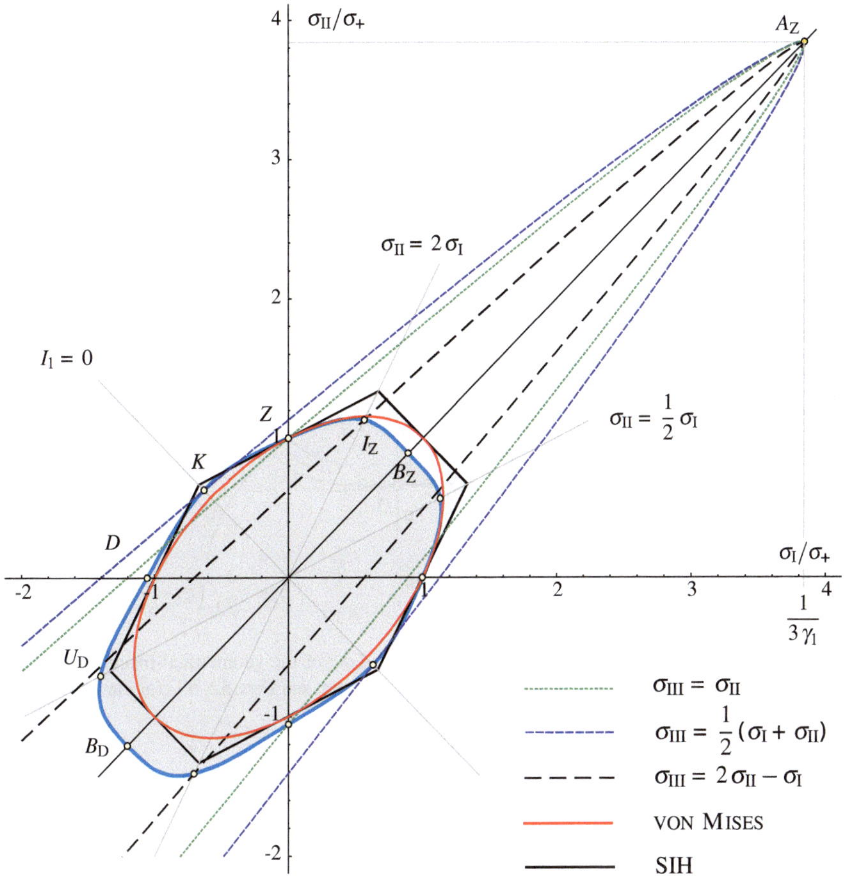

Fig. 53 Plane stress state $\sigma_{III} = 0$: geometrical-mechanical model for POM with the values $d = 1.06$, $k = 1.09$ (Fig. 52). The models of von Mises and Schmidt-Ishlinsky are shown for comparison

The restriction of the hydrostatic tension was set by Christensen according to the normal stress hypothesis ($a_+^{\text{hyd}} = 1$, $\gamma_1 = 1/3$) in order to obtain a closed limit surface in the tension region for the approximation with the paraboloid (open in the region $I_1 > 0$).

The combined model (paraboloid bounded by the normal stress hypothesis) presented by Christensen has singularities. A similar approximation can be obtained with the hyperboloid and the cone (51), Figs. 55 and 56. Further measurements can be described using the ellipsoid of Schleicher (51). These approximations are shown in the Burzyński-plane (Fig. 57), however they are not optimal:

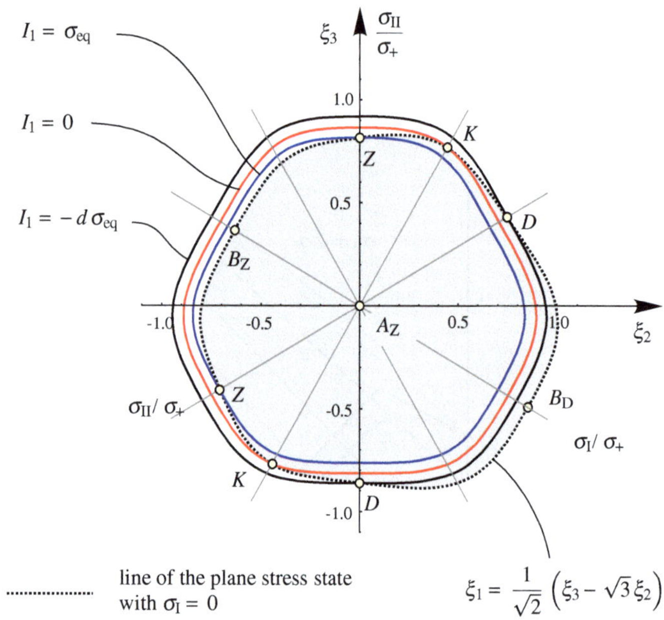

Fig. 54 Cross-sections $I_1 = \sigma_+$, $I_1 = 0$ and $I_1 = -d\,\sigma_+$ of the geometrical-mechanical model for POM with the values $d = 1.06$, $k = 1.09$ in the π-plane; see Figs. 52, 53 (for clarity only one section of the plane stress state with $\sigma_I = 0$ is shown)

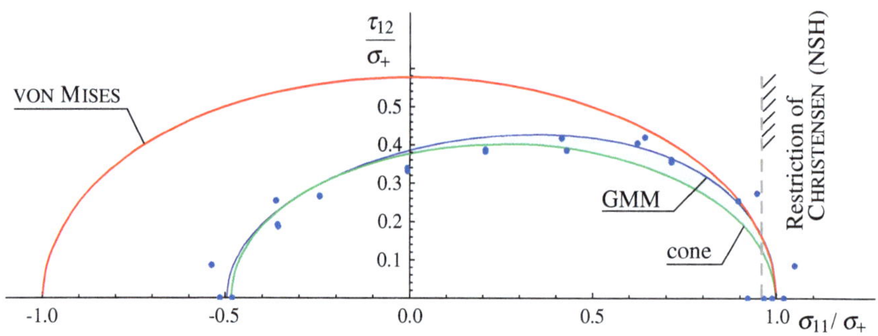

Fig. 55 Measurements by Christensen [46] in the diagram $\sigma_{11}-\tau_{12}$ with the normalization by $\sigma_+ = 7.12\,$MPa and the models: cylinder of von Mises, Eq. (16); cone of Drucker-Prager, Eq. (51): $\gamma_1 = \gamma_2 = -0.5325$, $\nu_+^{\text{in}} = 1.30$, $\nu_-^{\text{in}} = 0.11$, $d = 0.48$, $k = 0.65$; GMM Eq. (151), $j = 2$, $l = 4$, $m = 0$ with $c_3 = -1.0585$, $c_6 = 0.2354$, $\gamma_1 = 0.4219$, $\gamma_2 = -0.5747$, $\nu_+^{\text{in}} = 1/2$, $\nu_+^{\text{in}} = -0.24$, $d = 0.49$, $k = 0.66$

- The cone, the paraboloid and the hyperboloid are in this case open in the direction $I_1 > 0$. The hydrostatic tension is not constrained, however such constraint is required.

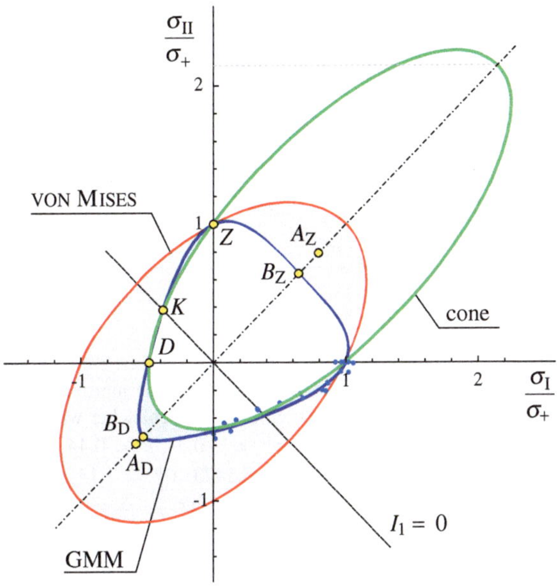

Fig. 56 Measurements by Christensen [46] in the diagram $\sigma_I - \sigma_{II}$, $\sigma_{III} = 0$ with the normalization by $\sigma_+ = 7.12\,\text{MPa}$ with the models from Fig. 55. The points A_Z and A_D of the GMM are shown for clarity

- The ellipsoid of Schleicher is fitted under the constraint $\nu_+^{\text{in}} \leq 1/2$. The quality of this approximation is not sufficient, the material strength in the region $D-K-Z$ is overestimated.
- All the approximations with the rotationally symmetric models can not describe the typical ∽-form of the curve of the plane stress state in the Burzyński-plane for the materials with $d < 1$ (cf. Figs. 58, 59 and 62).

In Figs. 57, 58 and 59 on the left and on the right hand side the constraints for the hydrostatic stresses at tension and compression are shown. These constraints result from the normal stress hypothesis and lead to a bounded region on the hydrostatic axis for approximation of hard foams.

13.3.1 Geometrical-Mechanical Model

In the first step the optimization $f_2 \to \min$ is performed for the geometric-mechanical model (151), (Fig. 58). Here the constraints are applied:

- $\gamma_1 = 1/3$ from the normal stress hypothesis and
- $\nu_+^{\text{in}} \leq 1/2$ as plausibility condition.

The line of the plane stress state approximates the measurements with a good quality, however the resulting extrapolation in the region $I_1 > \sigma_+$ is unconvincing:

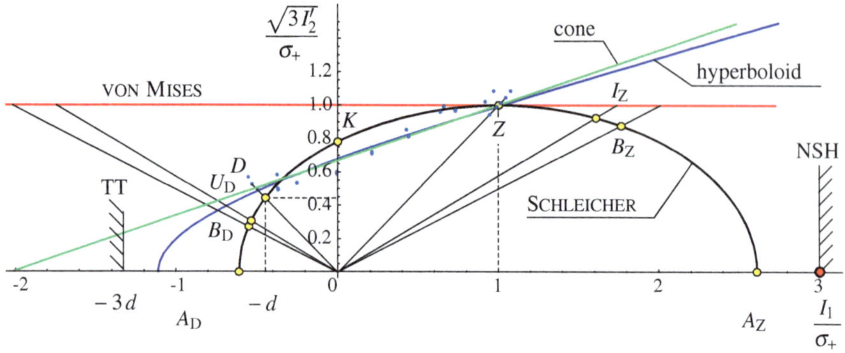

Fig. 57 Measurements by Christensen [46] in the Burzyński-plane approximated with the quadratic rotationally symmetric model (51), $\sigma_+ = 6.94$: ellipsoid of Schleicher with the constraints $\nu_+^{\text{in}} \leq 1/2$: $\gamma_1 = -1.6389$, $\gamma_2 = 0.3831$: $\nu_+^{\text{in}} = 1/2$, $\nu_-^{\text{in}} = -0.34$, $d = 0.44$, $k = 0.78$, $b_Z = 0.88$, $b_D = 0.27$; hyperboloid: $\gamma_1 = -0.9001$, $\gamma_2 = -0.1473$: $\nu_+^{\text{in}} = 1.14$, $\nu_-^{\text{in}} = 0.05$, $d = 0.49$, $k = 0.68$, $b_Z = 1.64$, $b_D = 0.37$; cone of Drucker-Prager: $\gamma_1 = \gamma_2 = -0.4956$: $\nu_+^{\text{in}} = 1.24$, $\nu_-^{\text{in}} = 0.13$, $d = 0.50$, $k = 0.67$, $b_Z = 1.98$, $b_D = 0.40$; NSH: Restriction from the normal stress hypothesis $\gamma_1 = 1/3$; TT: Restriction from the normal stress hypothesis as trigonal trapezohedron $3\,a_-^{\text{hyd}} = 3\,d$

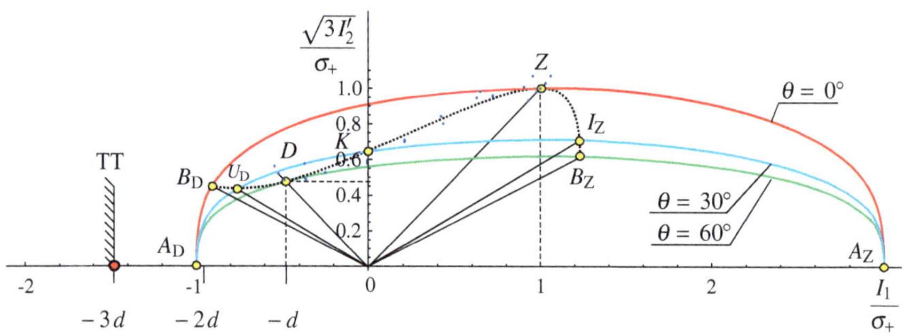

Fig. 58 Measurements by Christensen [46] in the Burzyński-plane approximated with the geometrical-mechanical model (151), $j = 1, l = 1, m = 4$ with $f_2 \to$min, ($f_2 = 0.321$): $c_3 = -1.0950$, $c_6 = 0.2263$, $\gamma_1 = 1/3$, $\gamma_2 = -1$, $\gamma_3 == 0$, $\sigma_+ = 7.04$: $\nu_+^{\text{in}} = 1/2$, $\nu_-^{\text{in}} = 0.29$, $d = 0.53$, $k = 0.68$, $b_Z = 0.62$, $b_D = 0.49$; the optimization constraints $\gamma = 1/3$ and $\nu_+^{\text{in}} \leq 1/2$

- the surface Φ has a symmetry plane $I_1 = \sigma_+$ and
- there are no measurements available in the region $I_1 > \sigma_+$, which confirm the setting $\gamma_1 = 1/3$.

In Figs. 59, 60 and 61 a further approximation with the GMM (151) and setting for the meridian $j = 2, l = 4, m = 0$ is presented. This representation shows that the restriction obtained from the normal stress hypothesis $-\frac{1}{\gamma_2} > 3\,d$ does not hold in this case. A further approximation can be obtained with the GMM (151), $j = 1$, $l = 5$, $m = 0$, $\gamma_1 \in [0,\ 1[$, $\gamma_2 < 0$ (Fig. 62). The point A_D is shifted to the left.

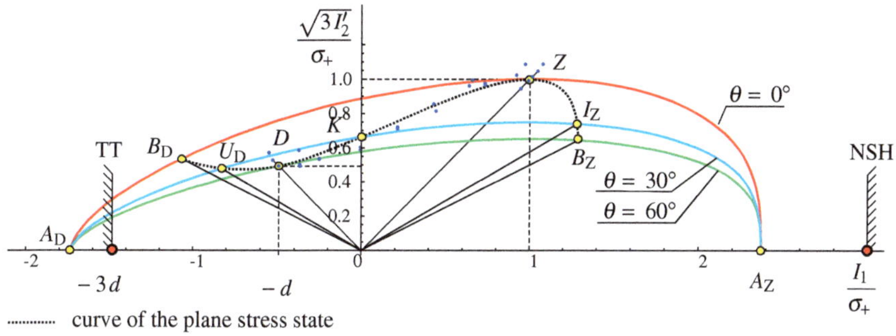

Fig. 59 Measurements by Christensen [46] in the Burzyński-plane approximated with the geometrical-mechanical model (151), $j = 2$, $l = 4$, $m = 0$, $f_2 \rightarrow$min, ($f_2 = 0.229$): $c_3 = -1.0585$, $c_6 = 0.2354$, $\gamma_1 = 0.4219$, $\gamma_2 = -0.5747$, $\gamma_3 = 0$, $\sigma_+ = 7.12$: $\nu_+^{\text{in}} = 1/2$, $\nu_-^{\text{in}} = -0.24$, $d = 0.49$, $k = 0.66$, $b_Z = 0.64$, $b_D = 0.53$; the optimization constraints $d \geq 0.49$ and $\nu_+^{\text{in}} \leq 1/2$

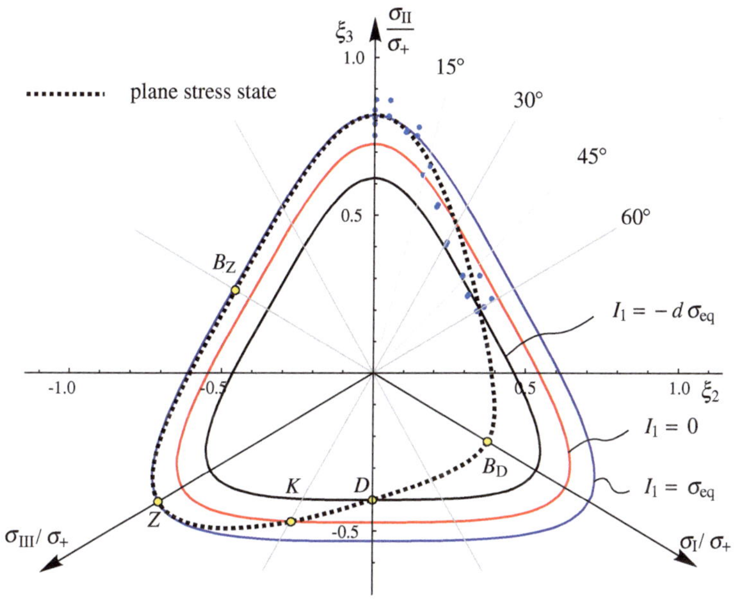

Fig. 60 Measurements by Christensen [46] in the π-plane approximated with the geometrical-mechanical model (151), see Fig. 59. The cross-sections orthogonal to the hydrostatic axis with $I_1 =$ const. through the points Z, K and D are shown

Both approximations with the constraints $d \geq 0.49$ and $\nu_+^{\text{in}} \leq 1/2$ lead to similar results for the points B_Z and B_D. A comparison of the Figs. 57, 58, 59 and 62 shows the differences of the approximations. For a more precise description of the plane stress state the loading points B_D and B_Z are necessary. The respective testing

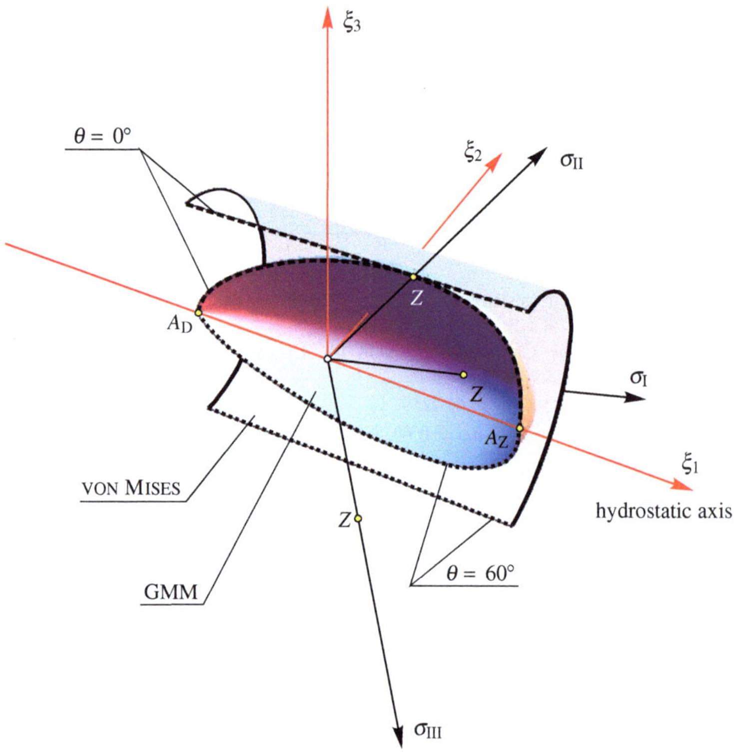

Fig. 61 Geometric-mechanical model (151) with the parameters from Fig. 59 and the model of von Mises in the principal stress space (σ_{I}, σ_{II}, σ_{III}), (the surfaces are cut for clarity; with $\theta = 0°$ and $\theta = 60°$ the meridians are labeled)

procedures for hard foams are described in [112]. The value at the point B_{Z} reduces the interval $a_{+}{}^{\text{hyd}} \in \left]\frac{1}{3},\ 1\right]$ to $a_{+}{}^{\text{hyd}} \in \left]\frac{2}{3}\,b_{\text{Z}},\ 1\right]$ because of the convexity condition.

The optimization constraint $\nu_{+}^{\text{in}} \leq 1/2$ allows to obtain plausible approximations as in Figs. 57, 58, 59 and 62 with $\nu_{+}^{\text{in}} = 1/2$. In order to justify this constraint the test B_{Z} is required as well.

13.3.2 Linear Geometrical Criterion

A further criterion based on a simple approach (Sect. 12.3) can be applied in order to compare the available approximations, e. g. the geometric criterion (193) with the equally weighted nodes $\chi = 1/2$. For the approximation in Fig. 62 it can be obtained:

$$\frac{1}{2}\left(\frac{1}{\gamma_1} - \frac{1}{\gamma_2}\right) = \frac{1}{2}\left(\frac{1}{0.6050} - \frac{1}{0.4415}\right) = 1.96. \tag{210}$$

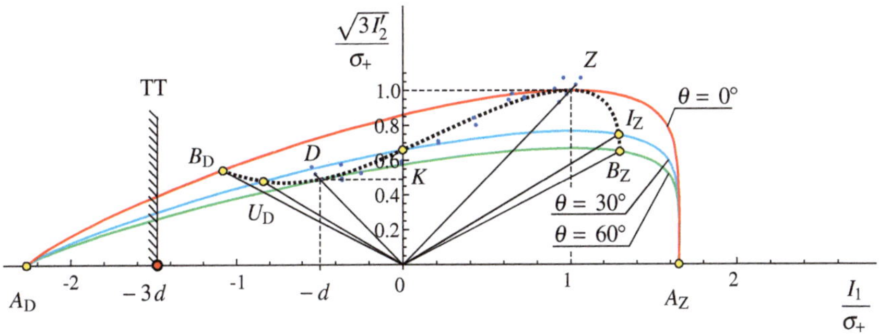

Fig. 62 Measurements by Christensen [46] in the Burzyński-plane approximated with the geometrical-mechanical model (151), $j = 1$, $l = 5$, $m = 0$, $f_2 \rightarrow$min, ($f_2 = 0.226$): $c_3 = -1.0380$, $c_6 = 0.2405$, $\gamma_1 = 0.6050$, $\gamma_2 = -0.4415$, $\sigma_+ = 7.22$: $\nu_+^{\text{in}} = 1/2$, $\nu_-^{\text{in}} = 0.26$, $d = 0.49$, $k = 0.66$, $b_Z = 0.65$, $b_D = 0.54$; the optimization constraints $d \geq 0.49$ and $\nu_+^{\text{in}} \leq 1/2$

The criterion leads to the values 2 and 2.06 for Figs. 58 and 59, respectively. It follows, that the approximation (Fig. 62) is to be preferred according to the criterion (210). The same result is obtained for $\chi = 0$.

13.3.3 Combined Geometrical-Mechanical Criterion

The position of the hydrostatic nodes A_D and A_Z can be adjusted, if the combined models (Sect. 11.2) are used and so a more conservative solution can be obtained. For this approximation of the measurements by Christensen a C^1-combination of two surfaces can be used. The surfaces are connected in the cross-section $I_1 = \sigma_+$, so that $\nu_+^{\text{in}} = 1/2$ holds. The second surface in the region $I_1 > \sigma_+$ with the same values of the parameters c_3, c_6 is continuously differentiable coupled with the first one.

The parameter $\gamma_1 \in [1/3,\ 1\ [$ of the right surface in the Burzyński-plane can be fitted according to the assumption regarding the position of the point A_Z, for instance based on the normal stress hypothesis ($a_+^{\text{hyd}} = 1$):

- $\gamma_1 = \frac{1}{3}$, $\gamma_2 = -1$, $j = 1$, $l = 1$, $m = 4$, cf. Fig. 58;
- $\gamma_1 = \frac{1}{3}$, $\gamma_2 = -\frac{1}{3}$, $j = 2$, $l = 4$, $m = 0$, cf. Fig. 59;
- $\gamma_1 = \frac{1}{3}$, $\gamma_2 = -\frac{1}{9}$, $j = 1$, $l = 5$, $m = 0$, cf. Fig. 62, etc.

These approximations result in similar values of b_Z. Contrary to $a_+^{\text{hyd}} = 1$ a more conservative value, for instance with $a_+^{\text{hyd}} = 1/2$ and $\nu_+^{\text{in}} = 1/2$ can be obtained. It follows then $\gamma_1 = 2/3$ (Fig. 63).

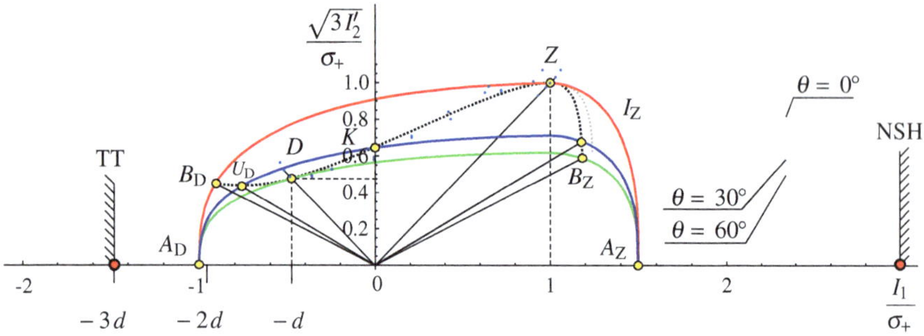

Fig. 63 Measurements by Christensen [46] in the Burzyński-plane approximated with the combined geometrical-mechanical model (151) with $j = l = 1, m = 4, c_3 = -1.0950, c_6 = 0.2263$: Region $I_1 \leq \sigma_{eq}$: $\gamma_1 = 1/3$, $\gamma_2 = -1$, $\nu_+^{in} = 1/2$, $\nu_-^{in} = 0.29$, $d = 0.53$, $k = 0.68$, $b_D = 0.49$ (Fig. 58), Region $I_1 > \sigma_+$: $\gamma_1 = 2/3$, $\gamma_2 = 2$, $\nu_+^{in} = 1/2$, $b_Z = 0.60$

14 Summary and Outlook

The modeling of the deformation and limit behavior of real materials is influenced by many phenomena. In the case of traditional materials and applications the number of influence factors taken into account is reduced. The corresponding models are simple and easy manageable. As usual they are based on a small number of material parameters. The experimental identification of the parameters and verification of the models is often simple. In many situations one needs only one basic test [8].

For high tech materials or applications with increasing safety requirements it is necessary to take into account non-classical effects since they have a significant influence on the deformation and limit behavior. Such phenomena are the different behavior at tension and compression—strength differential effect (SD-effect), the influence of the hydrostatic pressure, the Poynting-Swift-effect, the Kelvin-effect, etc. [8, 16, 229]. In contrast to the classical material behavior, which can be described by tensorial linear equations only, non-classical behavior partly should be presented by tensorial non-linear equations. The effects related to these equations are sometimes named second order effects [169]. They can be observed for elastic, plastic, creep behavior and in fluid mechanics [8, 169]. Therefore in this chapter several models allowing the description of complex material behavior are presented.

The models discussed in this chapter are limited by the assumption of isotropic material behavior. Further investigation should be directed on extension of these models to the case of anisotropic materials. In addition, the application of the suggested models to coated materials is not clear.

The concept of the equivalent stress is a simple and traditional engineering way to solve problems related to the strength prediction or material behavior modeling. The formulation and investigation of limit criteria will be in the focus of the scientific community in the future. The reason ist that one has new materials and particular application field [4].

The development of the concept of the equivalent stress is considered in conjunction with:

- materials science,
- thermo-dynamics,
- morphology, and
- advanced measurement programs.

This results in verification of new effects that result from the model, in the precise description of the failure with the consideration of the physical processes, taken into account the phase transitions for the definition of the limits of the parameters.

15 Invariants

Assuming isotropic material behavior the invariants of the stress tensor play an important role in the formulation of the equivalent stress expression. Here we define several sets of invariants mostly used in practical situations [8, 234].

15.1 Principal Invariants

Let us postulate that the stress state is defined by the stress tensor $\boldsymbol{\sigma}$. This tensor is a symmetric second rank tensor. The principal invariants are the solution of the eigenvalue problem

$$(\boldsymbol{\sigma} - \lambda \boldsymbol{I}) \cdot \boldsymbol{n} = \boldsymbol{0}. \tag{211}$$

$\boldsymbol{I}$ is the unit second rank tensor, $\boldsymbol{n}$ denotes the eigendirections and λ the eigenvalues of the problem. In our case the eigenvalues are named principal values or principal stresses. For the stress tensor it can be shown that the principal stresses are real-valued. Three different cases should be distinguished:

- three different solutions,
- one single and one double solution, and
- one triple solution

The necessary conditions for the solution of the problem (211) results in

$$\boldsymbol{n} = \boldsymbol{0} \qquad \text{or} \qquad |\boldsymbol{\sigma} - \lambda \boldsymbol{I}| = \det(\boldsymbol{\sigma} - \lambda \boldsymbol{I}) = 0. \tag{212}$$

The first Eq. (212) is a trivial solution. The second equation (212) is of greater importance since the principal stresses can be computed with help of this condition. The solution can be obtained from

$$\lambda^3 - I_1(\boldsymbol{\sigma})\lambda^2 + I_2(\boldsymbol{\sigma})\lambda - I_3(\boldsymbol{\sigma}) = 0, \tag{213}$$

where $I_i(\boldsymbol{\sigma})$, $(i = 1, 2, 3)$ are the invariants of the stress tensor

$$\begin{aligned}
I_1(\boldsymbol{\sigma}) &= \boldsymbol{I} \cdot\cdot\, \boldsymbol{\sigma}, \\
I_2(\boldsymbol{\sigma}) &= \frac{1}{2}\left[I_1^2(\boldsymbol{\sigma}) - I_1\left(\boldsymbol{\sigma}^2\right)\right], \\
I_3(\boldsymbol{\sigma}) &= |\boldsymbol{\sigma}| = \det\boldsymbol{\sigma} = \frac{1}{3}\left[I_1\left(\boldsymbol{\sigma}^3\right) + 3I_1(\boldsymbol{\sigma})I_2(\boldsymbol{\sigma}) - I_1^3(\boldsymbol{\sigma})\right].
\end{aligned} \tag{214}$$

The solutions of Eq. (213) are named principal invariants or principal stresses λ_i, $i = 1, 2, 3$. The following relations are valid after ordering the λ_i-values

$$\sigma_{\mathrm{I}} \geq \sigma_{\mathrm{II}} \geq \sigma_{\mathrm{III}}. \tag{215}$$

If the principal stresses are distinguished that means

$$\sigma_{\mathrm{I}} \neq \sigma_{\mathrm{II}} \neq \sigma_{\mathrm{III}}, \tag{216}$$

the following spectral decomposition holds

$$\boldsymbol{\sigma} = \sigma_{\mathrm{I}}\boldsymbol{n}_{\mathrm{I}}\boldsymbol{n}_{\mathrm{I}} + \sigma_{\mathrm{II}}\boldsymbol{n}_{\mathrm{II}}\boldsymbol{n}_{\mathrm{II}} + \sigma_{\mathrm{III}}\boldsymbol{n}_{\mathrm{III}}\boldsymbol{n}_{\mathrm{III}}, \tag{217}$$

where $\boldsymbol{n}_{\mathrm{I}}, \boldsymbol{n}_{\mathrm{II}}, \boldsymbol{n}_{\mathrm{III}}$ are the eigendirections (principal directions), which can be obtained from the solution of

$$(\boldsymbol{\sigma} - \sigma_i \boldsymbol{I}) \cdot \boldsymbol{n}_i = \boldsymbol{0}, \quad \boldsymbol{n}_i \cdot \boldsymbol{n}_j = \delta_{ij}. \tag{218}$$

δ_{ij} is the Kronecker symbol. The last equation in (218) is the orthogonality condition for the principal directions.

15.2 Irreducible Invariants

The stress tensor has three irreducible invariants:

- the linear invariant $J_1(\boldsymbol{\sigma}) = I_1(\boldsymbol{\sigma}) = \boldsymbol{I} \cdot\cdot\, \boldsymbol{\sigma}$,
- the quadratic invariant $J_2(\boldsymbol{\sigma}) = \boldsymbol{\sigma} \cdot\cdot\, \boldsymbol{\sigma}$,
- the cubic invariant $J_3(\boldsymbol{\sigma}) = (\boldsymbol{\sigma} \cdot \boldsymbol{\sigma}) \cdot\cdot\, \boldsymbol{\sigma}$.

The following representation is also possible

- the linear invariant $J_1(\boldsymbol{\sigma}) = I_1(\boldsymbol{\sigma}) = \boldsymbol{I} \cdot\cdot\, \boldsymbol{\sigma}$,
- the quadratic invariant $J_2(\boldsymbol{\sigma}) = \frac{1}{2}\boldsymbol{\sigma} \cdot\cdot\, \boldsymbol{\sigma}$,
- the cubic invariant $J_3(\boldsymbol{\sigma}) = \frac{1}{3}(\boldsymbol{\sigma} \cdot \boldsymbol{\sigma}) \cdot\cdot\, \boldsymbol{\sigma}$.

15.3 Axiatoric-Deviatoric Invariants

The axiatoric-deviatoric invariants are based on the principal invariants of the stress deviator, which can be computed from the eigenvalue problem for the deviator

$$\boldsymbol{s} = \boldsymbol{\sigma} - \frac{1}{3}\boldsymbol{\sigma} \cdot\cdot \boldsymbol{I} = \boldsymbol{\sigma} - \frac{1}{3} I_1(\boldsymbol{\sigma})\boldsymbol{I} \tag{219}$$

Principal deviatoric stresses follow from

$$|\boldsymbol{s} - \lambda \boldsymbol{I}| = \det(\boldsymbol{s} - \lambda \boldsymbol{I}) = 0 \tag{220}$$

or

$$\lambda^3 - I_1(\boldsymbol{s})\lambda^2 + I_2(\boldsymbol{s})\lambda - I_3(\boldsymbol{s}) = 0, \tag{221}$$

which can be simplified with respect to $I_1(\boldsymbol{s}) = 0$

$$\lambda^3 + I_2(\boldsymbol{s})\lambda - I_3(\boldsymbol{s}) = 0.$$

Here the second and the third invariants are equal to

$$I_2(\boldsymbol{s}) = -\frac{1}{2} J_2\left(\boldsymbol{\sigma}^2\right), \qquad I_3(\boldsymbol{s}) = |\boldsymbol{s}| = \det \boldsymbol{s} = \frac{1}{3} J_3\left(\boldsymbol{s}^3\right). \tag{222}$$

For a better distinguishing incompressible and compressible material behavior the second, the third deviatoric and the axiatoric invariant will be used

$$I_1(\boldsymbol{\sigma}) = \boldsymbol{\sigma} \cdot\cdot \boldsymbol{I}. \tag{223}$$

15.4 Cylindrical Invariants

There are other sets of invariants, for example, Novozhilov's invariants [148], which are defined as it follows

- the axiatoric invariant (223)
- the second invariant of the stress deviator (222)

$$I_2(\boldsymbol{s}) = -\frac{1}{2} J_2(\boldsymbol{s}),$$

- and the stress angle

Table 10 Main criteria

Abbrev.	Name	Reference	Section	Equation	BP	3D	π
Classical Strength Theories							
NSH	Normal Stress Hypothesis (Rankine)	[166, 175]	2.1	8	14	2	
	Tresca	[209]	2.2	13	15		22
	von Mises	[85, 91, 139]	2.3	16	14	2	22
SI	Schmidt-Ishlinsky	[92, 182]	2.4	17	15		22
Standard Models (Pressure-Sensitive Criteria)							
SH	Strain Hypothesis	[8, 174]	6.1	41			
MSH	Maximum Strain Hypothesis	[134]	6.1				
MC	Mohr-Coulomb	[141]	6.2	45			
PL	Pisarenko-Lebedev	[121]	6.3	48		38	
SD	Sdobirev	[191]	6.3				
	Burzyński-Yagn	[36, 220]	6.4	51	17		
	Drucker-Prager, Mirolyubov	[56, 138]	6.4		17	6	
	Balandin, Burzyński-Torre	[17, 36, 207]	6.4		17	11	
	Beltrami	[21]	6.4		17	12	
	Schleicher	[179]	6.4			34	
Mathematical Formulations (Pressure-Sensitive Criteria)							
	Altenbach-Zolochevsky I	[7]	7.1	56			
	Altenbach-Zolochevsky II	[8]	7.2	63			
Pressure-Insensitive Criteria with Trigonal Symmetry							
	Sayir I	[178]	9.1.1	87		18	
	Sayir II	[178]	9.1.2	89			
	Haythornthwaite	[83]	9.1.3	94			
CPM	Convex π-plane model	[33]	9.1.4	99			
	Radcig model	[105]	9.1.5	100			
GMM	Geometrical-Mechanical Model	[107]	9.1.6	105			
TQM	Triquadratic model	[103]	9.1.7	108			
Pressure-Insensitive Criteria with Hexagonal Symmetry							
UYC	Unified Yield Criterion of Yu	[222, 229]	9.2.1	111			22
	Sokolovsky	[157]	9.2.1				
BCM	Bicubic model	[103]	9.2.2	113			
MAC	Multiplicative Ansatz Criterion	[116]	9.2.3	116			
	Ishlinsky-Ivlev	[93, 98]	9.2.3				
	Universal model with hexagonal symmetry	[116]	9.2.4	121			
	Model based on the stress angle	[33]	9.2.5	122			
	Dodd-Naruse	[53]	9.2.5				
	Drucker I	[54]	9.2.5				
Pyramidal Criteria (Pressure-Sensitive Criteria)							
UST	Unified Strength Theory of Yu	[228]	10.2	131			
SST	Single-Shear-Theory of Yu	[228]	10.2				
TST	Twin-Shear-Theory of Yu	[228]	10.2			30	
	Drucker II	[55]	10.3.4				
	Haythornthwaite	[107]	10.3.4				
Combined Criteria							
	Pelczyński	[155]	11			37	
	Huber	[36, 91]	11.2.1	156	42	41	
	Kuhn	[118]	11.2.1	160	43		
	Modification of the model of Huber	[103]	11.2.2	163	42		
	Combined Geometrical-Mechanical Model	[103]	11.2.3	172	44		

$$\cos 3\theta = \frac{3\sqrt{3}}{2}\frac{\det \boldsymbol{s}}{(I_2'(\boldsymbol{s}))^{3/2}}, \qquad \theta \in \left[0, \frac{\pi}{3}\right], \tag{224}$$

see [43, 150, 234] among others.

16 Criteria of this Chapter 2

In Table 10 many of the discussed in this chapter criteria are summarized. The table is organized as it follows:

- In the first column (Abbrev.) some abbreviations are presented.
- The main name of the criteria is presented in column 2.
- The third column indicates the main reference(s).
- Column 4 presents the section, where the given criterion is discussed.
- In the column 5 the relevant equation is presented.
- In the last three columns are given the references to the figures (in the Burzyński-plane (BP), in the principal stress space (3D) and in the π-plane (π)).

References

1. Alpa, G.: On a statistical approach to brittle rupture for multiaxial states of stress. Eng. Fract. Mech. **19**(5), 881–901 (1984)
2. Altenbach, H.: Criterion with application to strength, yielding, and damage of isotropic materials. In: Lemaitre, J. (ed.) Handbook of Materials Behavior Models, pp. 175–186. Academic Press, San Diego (2001a)
3. Altenbach, H.: A nonclassical model for creep-damage processes. Mater. Phys. Mech. **3**, 25–35 (2001b)
4. Altenbach, H.: Strength hypotheses—a never ending story. Czasopismo Techniczne (Techn Trans) Politechniki Krakowkiej **107**(20), 5–15 (2010)
5. Altenbach, H.: Kontinuumsmechanik - Eine elementare Einführung in die materialunabhängigen und materialabhängigen Gleichungen. Springer Vieweg, Heidelberg (2012)
6. Altenbach, H., Kolupaev, V.A.: Fundamental forms of strength hypotheses. In: Indeitcev, D.A., Krivtsov, A.M. (eds.) Proceedings of XXXVI Summer School Advanced Problems in Mechanics, Institute for Problems in Mechanical Engineering, pp. 32–45. RAS, St. Petersburg (2009)
7. Altenbach, H., Lauschke, U., Zolochevsky, A.: Ein verallgemeinertes Versagenskriterium und seine Gegenüberstellung mit Versuchsergebnissen. ZAMM **73**(4–5), T 372-T 375 (1993)
8. Altenbach, H., Altenbach, J., Zolochevsky, A.: Erweiterte Deformationsmodelle und Versagenskriterien der Werkstoffmechanik. Deutscher Verlag für Grundstoffindustrie, Stuttgart (1995)
9. Ambarcumyan, S.A.: Multimodulus Elasticity Theory (in Russ.: Raznomodul'naja teorija uprugosti). Nauka, Moscow (1982)
10. Andrew, W.: The effect of creep and other time related factors on plastics and elastomers, vol. I. Book B, New York (1991)
11. Annin, B.D.: Theory of ideal plasticity with a singular yield surface. J. Appl. Mech. Tech. Phys. **40**(2), 347–353 (1999)

12. de Araújo, F.C.: Elasticidade e plasticidade. Imprensa Portuguesa, Porto (1961)
13. Argon, A., Andrews, R., Godrick, J., Whitney, W.: Plastic deformation bands in glassy polystyrene. J. Appl. Phys. **39**(3), 1899–1906 (1968)
14. Aubertin, M., Simon, R.: Un critère de rupture multiaxial pour matériaux fragiles. Can. J. Civ. Eng. **25**(2), 277–290 (1998)
15. Bach, C.: Elastizität und Festigkeit. Springer, New York (1902)
16. Backhaus, G.: Deformationsgesetze. Akademie-Verlag, Berlin (1983)
17. Balandin, P.P.: On the strength hypotheses (in Russ.: K voprosu o gipotezakh prochnosti). Vestnik inzhenerov i tekhnikov **1**, 19–24 (1937)
18. Bardenheier, R.: Mechanisches Versagen von Polymerwerkstoffen: Anstrengungsbewertung mehrachsialer Spannungszustände. Hanser, München (1982)
19. Bauwens, J.: Yield condition and propagation of Lüders' lines in tension-torsion experiments on poly (vinyl chloride). J. Polym. Sci., Part A-2: Polym. Phys. **8**(6), 893–901 (1970)
20. Becker, W., Gross, D.: Mechanik elastischer Körper und Strukturen. Springer, Berlin (2002)
21. Beltrami, E.: Sulle condizioni di resistenza dei corpi elastici. Il Nuovo Cimento **18**(1), 145–155 (1885)
22. Belyaev, N.M.: Strength of Materials. Mir Publ, Moscow (1979)
23. Betten, J.: Kontinuumsmechanik, 2nd edn. Springer, Berlin (2001)
24. Betten, J.: Creep Mechanics. Springer Verlag, Berlin (2008)
25. Betten, J., Borrmann, M.: Der Poynting-Effekt als Ursache einer werkstoffbedingten Anisotropie. Forsch. Ingenieurwes. **54**(1), 16–18 (1988)
26. Bigoni, D., Piccolroaz, A.: Yield criteria for quasibrittle and frictional materials. Int. J. Solids Struct. **41**(11), 2855–2878 (2004)
27. Billington, E.W.: The Poynting-Swift effect in relation to initial and post-yield deformation. Int. J. Solids Struct. **21**(4), 355–371 (1985)
28. Billington, E.W.: Introduction to the Mechanics and Physics of Solids. Adam Hilger Ltd., Bristol (1986a)
29. Billington, E.W.: The Poynting-Swift effect. Acta Mech. **58**, 19–31 (1986b)
30. Birger, I.A.: On a criterion for fracture and plasticity (in Russ.: Ob odnom kriterii razrushenija i plastichnosti). Mechanika tverdogo tela, Izvestija Akademii Nauk SSSR **4**, 143–150 (1977)
31. Birger, I.A., Shopp, B.F., Iosilevich, G.B.: Strength Computations for Machine Components (in Russ.: Raschet na prochnost' detalej mashin, Spravochnik). Mashinostroenie, Moscow (1993)
32. Blumenauer, H.: Werkstoffprüfung. Dt. Verl. für Grundstoffindustrie, Leipzig (1996)
33. Bolchoun, A., Kolupaev, V.A., Altenbach, H.: Convex and non-convex flow surfaces (in German: Konvexe und nichtkonvexe Fließflächen). Forsch. Ingenieurwes. **75**(2), 73–92 (2011)
34. Brencich, A., Gambarotta, L.: Isotropic damage model with different tensile-compressive response for brittle materials. Int. J. Solids Struct. **38**(34), 5865–5892 (2001)
35. Burzyński, W.: Studjum nad hipotezami wytężenia. Akademia Nauk Technicznych, Lwów (1928)
36. Burzyński, W.: Über die Anstrengungshypothesen. Schweizerische Bauzeitung **94**(21), 259–262 (1929a)
37. Burzyński, W.: Über die Anstrengungshypothesen. Schweizerische Bauzeitung **95**(7), 87–88 (1929b)
38. Burzyński, W.: Theoretical foundations of the hypotheses of material effort. Eng. Trans. Pol. Acad. Sci. **56**(Special Issue), 9–45 (2008)
39. Burzyński, W.: Selected passages from Włodzimierz Burzyński's doctoral dissertation "Study on material effort hypotheses" printed in Polish by the Academy of Technical Sciences, Lwów, 1928, 1–192. Eng. Trans. Pol. Acad. Sci. **57**(3–4), 127–157 (2009)
40. Candland, C.T.: Implications of macroscopic failure criteria which are independent of hydrostatic stress. Int. J. Fract. **11**(3), 540–543 (1975)
41. Cardarelli, F.: Materials Handbook: A Concise Desktop Reference, 2nd edn. Springer, London (2008)
42. Chen, W.F., Han, D.J.: Plasticity for Structural Engineers. Springer, New York (1988)

43. Chen, W.F., Zhang, H.: Structural Plasticity—Theory, Problems, and CAE Software. Springer, New York (1991)
44. Chong, E.K.P., Zak, S.H.: An Introduction to Optimization. Wiley, Hoboken, NJ (2008)
45. Christensen, R.M.: The Theory of Materials Failure. University Press, Oxford (2013)
46. Christensen, R.M., Freeman, D.C., DeTeresa, S.J.: Failure criteria for isotropic materials, applications to low-density types. Int. J. Solids Struct. **39**(4), 973–982 (2002)
47. Coffin, L., Schenectady, N.: The flow and fracture of a brittle material. J. Appl. Mech. **17**, 233–248 (1950)
48. Coulomb, C.A.: Essai sur une application des regles des maximis et minimis a quelquels problemes de statique relatifs, a la architecture. Mem. Acad. Roy. Div. Sav. **7**, 343–387 (1776)
49. Cowan, H.J.: The strength of plain, reinforced and prestressed concrete under the action of combined stresses, with particular reference to the combined bending and torsion of rectangular sections. Mag. Concr. Res. **5**(14), 75–86 (1953)
50. Darkov, A., Shpiro, G.: Strength of Materials (in Russ.: Soprotivlenie materialov). Visshaja Shkola, Moscow (1965)
51. Desai, C.S.: A general basis for yield, failure and potential functions in plasticity. Int. J. Numer. Anal. Meth. Geomech. **4**(4), 361–375 (1980)
52. Diabgroup (2011) Technical manual, divinicell h. Tech. rep., www.diabgroup.com, Hemmingen
53. Dodd, B., Naruse, K.: Limitation on isotropic yield criteria. Int. J. Mech. Sci. **31**(7), 511–519 (1989)
54. Drucker, D.C.: Stress-strain relations for strain hardening materials: Discussion and proposed experiments. In: Reissner E., Prager W., Stoker R.R. (eds.) Non-Linear Problems in Mechanics of Continua. Brown University. Proceedings of the First Symposium in Applied Mathematics, American Mathematical Society, vol. 1, PP. 181–187. New York (1949)
55. Drucker, D.C.: Limit analysis of two and three dimensional soil mechanics problems. J. Mech. Phys. Solids **1**(4), 217–226 (1953)
56. Drucker, D.C., Prager, W.: Soil mechanics and plastic analysis or limit design. Q. Appl. Math. **10**, 157–165 (1952)
57. Edelman, F., Drucker, D.C.: Some extensions of elementary plasticity theory. J. Franklin Inst. **251**(6), 581–605 (1951)
58. Ehrenstein, G.W.: Mit Kunststoffen konstruieren. Hanser, München (1995)
59. Elias, H.G.: Makromoleküle: Physikalische Strukturen und Eigenschaften, vol. 2. Wiley-VCH, Weinheim (2001)
60. Ely, R.: Biaxial stress testing of acrylic tube specimens. Polym. Eng. Sci. **7**(1), 40–44 (1967)
61. Eschenauer, H., Olhoff, N., Schnell, W.: Applied Structural Mechanics: Fundamentals of Elasticity, Load-bearing Structures, Structural Optimization. Springer, Berlin (1997)
62. Feodosiev, V.I.: Ten Lectures-Discussions of the Strength Theory (in Russ.: Desjat' lekzij-besed po soprotivleniju materialov). Nauka, Moscow (1975)
63. Filin, A.P.: Applied Mechanics of Solid Deformable Bodies (in Russ.: Prikladnaja mechanika tverdogo deformiruemogo tela), vol. 1. Nauka, Moscow (1975)
64. Filonenko-Borodich, M.M.: Theory of Elasticity. P. Noordhoff W. N, Groningen (1960)
65. Filonenko-Boroditsch, M.M.: Festigkeitslehre, vol. 1. Technik, Berlin (1960)
66. Finnie, I., Heller, W.R.: Creep of Engineering Materials. McGraw-Hill, New York (1959)
67. Föppl, A., Föppl, L.: Drang und Zwang: Eine höhrere Festigkeitslehre für Ingenieure. R. Oldenbourg, München (1920)
68. Freudenthal, A., Gou, R.: Second order effects in the theory of plasticity. Acta Mech. **8**(1), 34–52 (1969)
69. Freudenthal, A.M.: Constitutive equations of rock with shear dilatancy. Tech. Rep. AD-AOll 402, DTIC Document (1975)
70. Fridman, Y.B.: Unified Theory of Strength (in Russ.: Edinaya teorija prochnosti). Oborongiz, Moscow (1943)

71. Fromm, H.: Grenzen des elastischen Verhaltens beanspruchter Stoffe. In: Auerbach F, Hort W. (eds.) Statik und Dynamik elastischer Körper nebst Anwendungsgebieten. II. Teil. Zum Gebrauch für Ingenieure, Physiker und Mathematiker, vol. 4, pp. 359–435. Verlag von Barth, J.A., Leipzig (1931)
72. Gdoutos, E.E.: Failure of cellular foams under multiaxial loading. Compos. A **33**, 163–176 (2002)
73. Geniev, G.A., Kissjuk, V.N., Tjupin, G.A.: Plasticity Theory for Concrete and Reinforced Concrete (in Russ.: Teorija plastichnosti betona i zhelezhobetona). Strojizdat, Moscow (1974)
74. Gol'denblat, I.I., Kopnov, V.A.: Yield and Strength Criteria for Structural Materials (in Russ.: Kriterii prochnosti i plastichnosti konstrukzionnych materialov). Mashinostroenie, Moscow (1968)
75. Göldner, H., Holzweißig, F.: Leitfaden der Technischen Mechanik: Statik, Festigkeitslehre, Kinematik. Dynamik, Fachbuchverlag, Leipzig (1989)
76. Gollub, W.: Grenzen und Möglichkeiten der Mohr-Coulombschen Bruchbedingung, 18: Mechanik/Bruchmechanik, vol. 68. VDI, Düsseldorf (1989)
77. Grashof, F.: Theorie der Elasticität und Festigkeit. Gaertner, Berlin (1878)
78. Gross, D., Seelig, T.: Fracture Mechanics: with an Introduction to Micromechanics. Springer, Berlin (2011)
79. Gummert, P., Reckling. K.A.: Computation of the Bearing Capacity of Structures by the Method of Limiting Equilibrium (in Russ.: Raschet nesushej sposobnosti konstrukzij po metody predel'nogo ravnovesija). Strohizdat, Moscow (1949)
80. Haigh, B.P.: The strain-energy function and the elastic limit. Engineering **109**, 158–160 (1920)
81. Häusler, O., Tsakmakis, C.: Torsion eines Kreiszylinders bei großen Deformationen und inkompressiblem Materialverhalten. Forschungszentrum Karlsruhe GmbH, Karlsruhe (1995)
82. Hayhurst, D.R.: Creep rupture under multi-axial states of stress. J. Mech. Phys. Solids **20**(6), 381–390 (1972)
83. Haythornthwaite, R.M.: Mechanics of triaxial tests for soil. Proc. ASCE J. Soil Mech. Found. Div. **86**(SM5), 35–62 (1960)
84. Haythornthwaite, R.M.: Range of yield condition in ideal plasticity. Proc. ASCE J. Eng. Mech. Div. **87**(EM6), 117–133 (1961)
85. Hencky, H.: Zur Theorie plastischer Deformationen und der hierdurch im Material hervorgerufenen Nachspannungen. ZAMM **4**(4), 323–334 (1924)
86. Hencky, H.: Ermüdung, Bruch, Plastizität. Stahlbau **16**(23/24), 95–97 (1943)
87. Hill, R.: On the inhomogeneous deformation of a plastic lamina in a compression test. Phil. Mag. Series 7 **41**(319), 733–744 (1950)
88. Hoffman, O., Sachs, G.: Introduction to the Theory of Plasticity for Engineers. McGraw-Hill, New York (1953)
89. Holzmann, G., Meyer, H., Schumpich, G.: Technische Mechanik Festigkeitslehre. Springer, Vieweg, Wiesbaden (2012)
90. Hu, W., Wang, Z.R.: Multiple-factor dependence of the yielding behavior to isotropic ductile materials. Comput. Mater. Sci. **32**(1), 31–46 (2005)
91. Huber, M.T.: Specific strain work as a measurment of material effort (in Polish: Właściwa praca odkształcenia jako miara wytężenia materyału). Czasopismo Techniczne **22**, 34–40, 49–50, 61–62, 80–81 (1904)
92. Ishlinsky, A.Y.: Hypothesis of strength of shape change (in Russ.: Gipoteza prochnosti formoizmenenija). Uchebnye Zapiski Moskovskogo Universiteta, Mekhanika **46**, 104–114 (1940)
93. Ishlinsky, A.Y., Ivlev, D.D.: Mathematical Theory of Plasticity (in Russ.: Matematicheskaja teorija plastichnosti). Fizmatlit, Moscow (2003)
94. Ismar, H., Mahrenholtz, O.: Über Beanspruchungshypothesen für metallische Werkstoffe. Konstruktion **34**, 305–310 (1982)
95. Issler, L., Ruoß, H., Häfele, P.: Festigkeitslehre - Grundlagen. Springer, Berlin (2006)
96. Ivlev, D.D.: On the development of a theory of ideal plasticity. J. Appl. Math. Mech. **22**(6), 1221–1230 (1958)

97. Ivlev, D.D.: The theory of fracture of solids. J. Appl. Math. Mech. **23**(3), 884–895 (1959)
98. Ivlev, D.D.: On extremal properties of the yield criteria (in Russ.: Ob ekstremal'nych svojstvach uslovij plastichnosti). J. Appl. Math. Mech. **5**, 1439–1446 (1960)
99. Ivlev, D.D.: Theory of Ideal Plasticity (in Russ.: Teorija idealnoj plastichnosti). Nauka, Moscow (1966)
100. Ivlev, D.D.: Theory of Limit State and Ideal Plasticity (in Russ.: Teorija predel'nogo sostojanija i ideal'noj plastichnosti). Voronezhskij Gosudarstvennyj Universitet, Voronezh (2005)
101. Kłębowski, Z.: Obecny stan wytrzymałościowego obliczenia materiałów o własnościach uogólnionych; uogólnione obliczenie osiowo symetrycznego cienkościennego naczynia pod ciśnieniem. Przegląd Techniczny **11**, 7–31 (1934)
102. Ko, W.L.: Application of the finite elastic theory to the behavior of rubberlike materials. Ph.D. thesis, California Institute of Technology, Pasadena (1963)
103. Kolupaev, V.A.: 3D-Creep Behaviour of Parts Made of Non-Reinforced Thermoplastics (in German: Dreidimensionales Kriechverhalten von Bauteilen aus unverstärkten Thermoplasten). Ph.D. thesis, Martin-Luther-Universität Halle-Wittenberg, Halle (2006)
104. Kolupaev, V.A., Altenbach, H.: Application of the generalized model of Mao-Hong Yu to plastics (in German: Anwendung der Unified Strength Theory (UST) von Mao-Hong Yu auf unverstärkte Kunststoffe. In: Grellmann, W. (ed.) Martin-Luther-Universität Halle-Wittenberg, vol. 12, pp. 320–339. Merseburg, Tagung Deformations- und Bruchverhalten von Kunststoffen (2009)
105. Kolupaev, V.A., Altenbach, H.: Considerations on the unified strength theory due to Mao-Hong Yu (in German: Einige Überlegungen zur Unified Strength Theory von Mao-Hong Yu). Forsch. Ingenieurwes. **74**(3), 135–166 (2010)
106. Kolupaev, V.A., Altenbach, H.: Consistent view of generalised yield surfaces. In: Khan, A.S. (ed.) 18th International Symposium on Plasticity & Its Current Applications—Non-linear Response of Conventional & Advanced Materials, and Multi-scale Modeling, pp. 76–78. Neat Press, Fulton MD (2011)
107. Kolupaev, V.A., Bolchoun, A.: Combined yield and fracture criteria (in German: Kombinierte Fließ- und Grenzbedingungen). Forsch. Ingenieurwes. **72**(4), 209–232 (2008)
108. Kolupaev, V.A., Kraatz, A., Moneke, M., Bolchoun, A.: Description of the multiaxial creep for hard foams (in German: Beschreibung der mehraxialen Kriechphänomene bei Hartschaumstoffen). Kautschuk, Gummi, Kunststoffe **59**(1–2), 17–27 (2006)
109. Kolupaev, V.A., Bolchoun, A., Moneke, M.: Computation of the Poisson's ratio for the hardening of thermoplastics (in German: Ermittlung der Querkontraktionszahl bei der Verfestigung von Thermoplasten). In: Grellmann, W. (ed.) Martin-Luther-Universität Halle-Wittenberg, vol. 11, pp. 375–385. Merseburg, Tagung Deformations- und Bruchverhalten von Kunststoffen (2007)
110. Kolupaev, V.A., Bolchoun, A., Altenbach, H.: New trends in application of strength hypotheses (in German: Aktuelle Trends beim Einsatz von Festigkeitshypothesen). Konstruktion **5**, 59–66 (2009a)
111. Kolupaev, V.A., Bolchoun, A., Altenbach, H.: Unified representation and evaluation of the strength hypotheses. In: Elboujdaini M., Tyson B., Patnaik P. (eds.) 12th International Conference on Fracture ICF 12, p. 10. Ottawa (2009b)
112. Kolupaev, V.A., Bolchoun, A., Altenbach, H.: Testing of multi-axial strength behavior of hard foams. In: Brémand, F. (ed.) ICEM 14–14th International Conference on Experimental Mechanics, 4–9 July, EPJ Web of Conferences 6, 16003 (2010), 8 p. Poitiers, France (2010)
113. Kolupaev, V.A., Mohr-Matuschek, U., Altenbach. H.: Application of strength hypotheses for POM (in German: Anwendung von Festigkeitshypothesen an POM). In: Radusch H.J., Fiedler L. (eds.) 14. International Scientific Conference on Polymeric Materials P.2010, 15. - 17. September, Martin-Luther-Universität Halle-Wittenberg, p. 12. Halle (Saale) (2010b)
114. Kolupaev, V.A., Bolchoun, A., Altenbach. H.: Strength hypothesis applied to hard foams. Appl. Mech. Mater. (Advances in Experimental Mechanics VIII) **70**, 99–104 (2011)
115. Kolupaev, V.A., Yu, M.H., Altenbach, H.: Visualisation of the unified strength theory. Arch. Appl. Mech. (2013a). doi:10.1007/s00419-013-0735-8

116. Kolupaev, V.A., Yu, M.H., Altenbach, H.: Yield criteria of hexagonal symmetry in the π-plane. Acta Mech. (2013b). doi:10.1007/s00707-013-0830-5
117. Koval'chuk, B.I.: On the criterion for limit state of some hull steels under complex loading states at normal and above-normal temperatures (in Russ.: O kriterii predel'nogo sostojanija nekotorich korpusnich stalej v uslovijach slozhnogo naprjazhennogo sostojanija pri kompatnoj i povishennich temperaturach). Problemi Prochnosti **5**, 10–15 (1981)
118. Kuhn, P.: Grundzüge einer allgemeinen Festigkeitshypothese, Auszug aus Antrittsvorlesung des Verfassers vom 11. Juli, : Vom Konstrukteur und den Festigkeitshypothesen. Inst. für Maschinenkonstruktionslehre, Karlsruhe (1980)
119. Kunz, J., Michaeli, W., Herrlich, N., Land, W.: Kunststoffpraxis: Konstruktion. WEKA Media GmbH & Co, KG, Kissing (2002)
120. Lebedev, A.A.: Experimental study of long-term strength of chromium-nickel steel in biaxial tension (in Russ.: Eksperimental'nie issledovanija dlitel'noj prochnosti chromnikelevoj stali v uslovijakh dvustoronnego rastjazhenija). In: Thermal Strength of Materials and Structure Elements (in Russ.: Termoprochnost' materialov i konstrukcionnykh elementov), vol. 3, pp. 77–83 . Naukova Dumka, Kiev (1965a)
121. Lebedev, A.A.: Generalized criterion for the fatigue strength (in Russ.: Obobshennij kriterij dlitel'noj prochnosti). In: Thermal Strength of Materials and Structure Elements (in Russ.: Termoprochnost' materialov i konstrukcionnykh elementov), vol. 3, pp. 69–76. Naukova Dumka, Kiev (1965b)
122. Lebedev, A.A.: On a possible combination of a yield criterion with a criterion for brittle failure (in Russ.: O vozmozhnom sovmeshenii uslovij plastichnosti i khrupkogo razrushenija). Prikladnaja. Mechanika **4**(8), 85–93 (1968)
123. Lebedev, A.A., Panchin, V.V.: Geometrical interpretation of the generalized criterion for the fatigue strength (in Russ.: Geometricheskaja interpretazija obobshennogo kriterija dlitel'noj prochnosti). In: Thermal Strength of Materials and Structure Elements (in Russ.: Termoprochnost' materialov i konstrukcionnykh elementov), vol. 4, pp. 187–192. Naukova Dumka, Kiev (1967)
124. Lebedev, A.A., Koval'chuk, B.I., Giginjak, F.F., Lamashevsky, V.P.: Handbook of Mechanical Properties of Structural Materials at a Complex Stress State. Begell House, New York (2001)
125. Leckie, F.A., Hayhurst, D.R.: Constitutive equations for creep rupture. Acta Metall. **25**, 1059–1070 (1977)
126. Lemaitre, J., Chaboche, J.L.: Mechanics of Solid Materials. Cambridge University Press, Cambridge (1990)
127. Leon, A.: Über die Rolle des Trennbruches im Rahmen der Mohrschen Anstrengungshypothese. Der Bauingenieur **31**(32), 318–321 (1934)
128. Lequeu, P.H., Gilormini. P., Montheillet. F., Bacroix, B., Jonas, J.J.: Yield surfaces for textured polycrystals. Acta Metall. **35**(2), 439–451, 1159–1174 (1987)
129. Lüpfert, H.P.: Beurteilung der statischen Festigkeit und Dauerfestigkeit metallischer Werkstoffe bei mehrachsiger Beanspruchung. Deutscher Verlag für Grundstoffindustrie, Leipzig (1994)
130. Lurie, A.I.: Theory of Elasticity. Springer, Berlin (2005)
131. Maitra, M., Majumdar, K., Das, A.: Unified plastic yield criterion for ductile solids. AIAA Journal **11**(10), 1428–1429 (1973)
132. Mālmeisters, A., Tamužs, V., Teters, G.: Mechanik der Polymerwerkstoffe. Akademie-Verlag, Berlin (1977)
133. Marciniak, Z.: Graphical representation of states of stress and strain. Arch. Mech. **3**, 261–274 (1971)
134. Mariotte, M.: Traité du mouvement des eaux et des autres corps fluides. J. Jambert, Paris (1700)
135. Mendelson, A.: Plasticity: Theory and Application. Krieger, Malabar, Fla (1968)
136. Mendera, Z.: Wytężenie spoiny czołowej w interpretacji powierzchni granicznych. Przegląd Spawalnictwa SIMP **XVIII**(1), 6–13 (1966)

137. Miles, M., Mills, N.: The yield locus of polycarbonate. J. Polym. Sci.: Polym. Lett. Ed. **11**(9), 563–568 (1973)
138. Mirolyubov, I.N.: On the generalization of the strengt theory based on the octaedral stresses in the case of brittle materials (in Russ.: K voprosu ob obobshenii teorii prochnosti oktaedtich-eskikh kasatelnyh naprjazhenij na khrupkie materialy). Trudy Leningradskogo Technologich-eskogo Instituta pp 42–52 (1953)
139. von Mises, R.: Mechanik des festen Körpers im plastischen deformablen Zustand. Nachrichten der Königlichen Gesellschaft der Wissenschaften Göttingen, Mathematisch-physikalische Klasse pp. 589–592 (1913)
140. von Mises, R.: Mechanik der plastischen Formänderung von Kristallen. ZAMM **8**, 161–185 (1928)
141. Mohr, O.: Welche Umstände bedingen die Elastizitätsgrenze und den Bruch eines Materials. Z VDI **45**, 1524–1530 (1900a)
142. Mohr, O.: Welche Umstände bedingen die Elastizitätsgrenze und den Bruch eines Materials. Z VDI **46**, 1572–1577 (1900b)
143. Mohr, O.: Abhandlungen aus dem Gebiete der technischen Mechanik. Wilhelm & Sohn, Berlin (1914)
144. Müller, R.: Theoretische Untersuchung des Einflusses der Spannungskonzentration durch Kerben bei mehrachsiger Belastung. Ph.D., TH Karlsruhe, Karlsruhe (1987)
145. Murzewski, J., Mendera, Z.: Yield surface of steel determined by semi-empirical method. Bulletin de L'Academie Polonaise des Sciences, Serie des sciences, techniques**XI**(7), 35–42 (1963)
146. Navier, M.: De la resistance des corps solides. Dunand, Paris (1864)
147. Nayak, G., Zienkiewicz, O.: Convenient form of stress invariants for plasticity. In: Proceedings of the ASCE Journal of the Structural Division, vol. 98, pp. 949–953 (1972)
148. Novozhilov, V.: On the principles of the statical analysis of the experimental results for isotropic materials (in Russ.: O prinzipakh obrabotki rezultatov staticheskikh ispytanij izotrop-nykh materialov). Prikladnaja Matematika i Mechanika **XV**(6), 709–722 (1951a)
149. Novozhilov, V.V.: On the connection between stresses and strains in a nonlinear-elastic contin-uum (in Russ.: O svjazi mezhdu naprjazhenijami i deformazijami v nelinejno-uprugoj srede). Prikladnaja Matematika i Mechanika **XV**(2):183–194 (1951b)
150. Ottosen, N.S., Ristinmaa, M.: The Mechanics of Constitutive Modeling. Elsevier Science, London (2005)
151. Pae, K.D.: The macroscopic yielding behaviour of polymers in multiaxial stress fields. J. Mater. Sci. **12**, 1209–1214 (1977)
152. Papageorgiou, M.: Optimierung: statische, dynamische, stochastische Verfahren für die Anwendung. Oldenbourg, München (1996)
153. Paul, B.: A modification of the Coulomb-Mohr theory of fracture. J. Mater. Sci. Ser E **28**(2), 259–268 (1961)
154. Paul, B.: Macroscopic plastic flow and brittle fracture. In: Liebowitz, H. (ed.) Fracture: An Advanced Treatise, vol. II, pp. 313–496. Academic Press, New York (1968)
155. Pełczyński, T.: Wpływ stanu napięcia na przejście materiału w stan plastyczny. Przeglad Mechaniczny **7**, 204–208 (1951)
156. Pisarenko, G.S., Lebedev. A.A.: Deformation and Fracture of Materials under Combined Stress (in Russ.: Soprotivlenie materialov deformirovaniju i razrusheniju pri slozhnom napr-jazhennom sostojanii). Naukowa Dumka, Kiev (1969)
157. Pisarenko, G.S., Lebedev, A.A.: Deformation and Strength of Materials under Complex Stress State (in Russ.: Deformirovanie i prochnost' materialov pri slozhnom naprjazhennom sosto-janii). Naukowa Dumka, Kiev (1976)
158. Pęcherski, R.B., Szeptyński, P., Nowak, M.: An extension of Burzyński hypothesis of material effort accounting for the third invariant of stress tensor. Arch. Metall. Mater. **56**(2), 503–508 (2011)
159. Poncelet, J.V.: Mécanique Industrielle. Meline, Bruxelles (1839)

160. Poynting, J.H.: On pressure perpendicular to the shear planes in finite pure shears, and on the lengthening of loaded wires when twisted. Proc. R. Soc. Lond. Ser. A **82**(557), 546–559 (1909)
161. Poynting, J.H.: On the changes in the dimensions of a steel wire when twisted, and on the pressure of distortional waves in steel. Proc. R. Soc. Lond. Ser. A **86**(590), 534–561 (1912)
162. Poynting, J.H., Thomson, J.J.: A Text-Book of Physics, Properties of Matter. Charles Griffin & Company, London (1927)
163. Prager, W., Hodge, P.: Theorie ideal plastischer Körper. Springer, Wien (1954)
164. Raghava, R., Caddell, R., Yeh, G.: The macroscopic yield behaviour of polymers. J. Mater. Sci. **8**(2), 225–232 (1973)
165. Raniecki, B., Mróz, Z.: Yield or martensitic phase transformation conditions and dissipation functions for isotropic, pressure-insensitive alloys exhibiting sd effect. Acta Mech. **195**, 81–102 (2008)
166. Rankine, W.J.M.: Manual of Applied Mechanics. Griffin, London (1876)
167. Reckling, K.: Plastizitätstheorie und ihre Anwendung auf Festigkeitsprobleme. Springer, Berlin (1967)
168. Reiner, M.: Deformation Strain and Flow. An Elementary Introduction to Rheology. H. K. Lewis & Co. Ltd., London (1960)
169. Reiner, M., Abir, D.: Second-Order Effects in Elasticity, Plasticity and Fluid Dynamics. Jerusalem Academic Press, Jerusalem (1964)
170. Résal, J.: Résistance des matériaux. Librairie polytechnique, Baudry & cie., Paris (1898)
171. Reuss, A.: Vereinfachte Beschreibung der plastischen Formänderungsgeschwindigkeiten bei Voraussetzung der Schubspannungsfließbedingung. ZAMM **13**(5), 356–360 (1933)
172. Rivlin, R.S., Saunders, D.W.: Large elastic deformations of isotropic materials. vii. experiments on the deformation of rubber. Philos. Trans. R. Soc. Lond. Ser. A **243**(865), 251–288 (1951)
173. Roscoe, K.H., Burland, J.B.: On the generalized stress-strain behaviour of 'wet' clay. In: Heyman, J., Leckie, F.A. (eds.) Engineering Plasticity, Papers for Conference held in Cambridge, pp. 535–609. University Press, Cambridge March 1968
174. Sähn, S., Göldner, H., Fischer, K.F., Nickel, J.: Bruch- und Beurteilungskriterien in der Festigkeitslehre. Fachbuchverbuchverlag, Leipzig-Köln (1993)
175. de Saint-Venant, B.: Comptes rendus des séances de l'Académie des sciences (1871)
176. Sandel, G.D.: Über die Festigkeitsbedingungen: ein Beitrag zur Lösung der Frage der zulässigen Anstrengung der Konstruktionsmaterialen. Ph.D. thesis, TeH, Stuttgart (1919)
177. Sauter, J.: Neue und alte statische Festigkeitshypothesen. VDI, Reihe 1: Konstruktionstechnik / Maschinenelente Nr. 191, Düsseldorf (1990)
178. Sayir, M.: Zur Fließbedingung der Plastizitätstheorie. Ing. Arch. **39**, 414–432 (1970)
179. Schleicher, F.: Der Spannungszustand an der Fließgrenze (Plastizitätsbedingung). Ztschr f Math und Mech **6**(3), 199–216 (1926)
180. Schleicher, F.: Über die Sicherheit gegen Überschreiten der Fliessgrenze bei statischer Beanspruchung. Der Bauingenieur **9**(15), 253–261 (1928)
181. Schlimmer, M.: Zeitabhängiges mechanisches Werkstoffverhalten: Grundlagen, Experimente. Rechenverfahren für die Praxis, Springer, Berlin (1984)
182. Schmidt, R.: Über den Zusammenhang von Spannungen und Formänderungen im Verfestigungsgebiet. Ing. Arch. **3**(3), 215–235 (1932)
183. Schneider, W.: Mikromechanische Betrachtung von Bruchkriterien unidirektional verstärkten Schichten aus Glasfaser/Kunststoff. Fachbereich Maschinenbau D 17, Technische Hochschule Darmstadt, Darmstadt (1974)
184. Schneider, W.: Versagenskriterien für Kunststoffe unter mehrachsiger Kurzzeitbeanspruchung. In: Belastungsgrenzen von Kunststoff-Bauteilen, pp. 81–105. VDI-Verlag GmbH, Düsseldorf (1975)
185. Schneider, W., Bardenheier, R.: Versagenskriterien für Kunststoffe. Zeitschrift für Werkstofftechnik (J of Materials Technology) **6**(8), 269–280 (1975)
186. Schofield, A., Wroth, P.: Critical State Soil Mechanics. McGraw-Hill, London (1968)

187. Schreyer, H.: Smooth limit surfaces for metals, concrete, and geotechnical materials. J. Eng. Mech. **115**(9), 1960–1975 (1989)
188. Schur, I., Grunsky, H.: Vorlesungen über Invariantentheorie: Die Grundlehren der mathematischen Wissenschaften in Einzelndarstellungen. Springer, Berlin (1968)
189. Schwartz, R., Dugger, J.: Shear strength of plastic materials. Mod. Plast. **21**, 117–121, 164, 166 (1944)
190. Schwarzl, F., Stavermann, A.J.: Bruchspannung und festigkeit von hochpolymeren. In: Stuart, H. (ed.) Die Physik der Hochpolymeren: Theorie und Molekulare Deutung Technologischer Eigenschaften von Hochpolymeren Werkstoffen, vol. 4, pp. 165–176. Springer, Berlin (1964)
191. Sdobyrev, V.P.: Criterion for the long term strength of some heat-resistant alloys at a multiaxial loading (in Russ.: Kriterij dlitelnoj prochnosti dlja nekotorykh zharoprochnykh splavov pri slozhnom naprjazhennom sostojanii). Izvestija Akademii Nauk SSSR, Otdelenie tekhnicheskikh Nauk, Mechanika i Mashinostroenie **6**, 93–99 (1959)
192. Shanley, F.R.: Strength of Materials. McGraw-Hill, New York (1957)
193. Shesterikov, S.A.: On the theory of ideal plastic solid (in Russ.: K postroeniju teorii ideal'no plastichnogo tela). Prikladnaja matematika i mechanika, Rossijskaja Akademija Nauk **24**(3), 412–415 (1960)
194. Skrzypek, J.J.: Plasticity and Creep: Theory. CRC Press, Boca Raton, Examples and Problems (1993)
195. Sokolovsky, V.V.: Theory of Plasticity (in Russ.: Teorija plastichnosti). Gos. izd. tekhn.-teor. literatury, Moscow (1950)
196. Stadler, W.: Multicriteria Optimization in Engineering and the Sciences. Mathematical Concepts and Methods in Science and Engineering, vo. 37. Plenum Press, New York (1988)
197. Stepin, P.A.: Strength of Materials. Gordon & Breach, New York (1963)
198. Sternstein, S., Ongchin, L.: Yield criteria for plastic deformation of glassy high polymers in general stress fields. Div. Polym. Chem. **10**, 1117–1121 (1969)
199. Stockton, F.D., Drucker, D.C.: Fitting mathematical theory of plasticity to experimental results. J. Colloid Sci. **5**(3), 239–250 (1950)
200. Swift, H.W.: Plastic strain in an isotropic strain hardening material. Engineering: for innovators in technology, manufacturing and management **163**, 381–389 (1946)
201. Tarasenko, I.I.: On the criteria of brittle strength of materials (in Russ.: O kriterijach khrupkoj prochnosti materialov). Sbornik nauchnykh trudov, Leningradskij inzhenerno-stroitelnij institut **26**, 161–168 (1957)
202. Theocaris, P.: A general yield criterion for engineering materials, depending on void growth. Meccanica **21**(2), 97–105 (1986)
203. Thorkildsen, R.: Mechanical behaviour. In: Baer, E. (ed.) Engineering Design for Plastics, pp. 227–399. Reinhold, New York (1964)
204. Timoshenko, S.P.: History of Strength of Materials: With a Brief Account of the History of Theory of Elasticity and Theory of Structure. McGraw-Hill, New York (1953)
205. Timoshenko, S.P., Goodier, J.N.: Theory of Elasticity. McGraw-Hill, New York (1987)
206. Timoshenko, S.P., Young, D.H.: Elements of Strength of Materials. D. van Nostrand Company, Princeton (1962)
207. Torre, C.: Einfluss der mittleren Hauptnormalspannung auf die Fließ- und Bruchgrenze. Österreichisches Ingenieur-Archiv **I**(4/5), 316–342 (1947)
208. Torre, C.: Grenzbedingung für spröden Bruch und plastisches Verhalten bildsamer Metalle. Österreichisches, Ingenieur-Archiv **IV**(2):174–189 (1950)
209. Tresca, H.: Mémoire sur l'ecoulement des corps solides. Mémoires Pres par Div Savants **18**, 733–799 (1868)
210. Tschoegl, N.W.: Failure surfaces in principial stress space. J. Polym. Sci., Part C: Polym. Symp. **32**, 239–267 (1971)
211. Tsvelodub, I.Y.: Stability postulate and its applications in the creep theory for metallic materials (in Russ.: Postulat ustojchivosti i ego prilozhenija v teorii polzuchesti metallicheskich materialov). Institut Gidromechaniki, Novosibirsk (1991)

212. Tsvelodub, I.Y.: Multimodulus elasticity theory. J. Appl. Mech. Technical Phys. **49**(1), 129–135 (2008)
213. Vesik, S.: Handbook of Strength of Materials (in Russ.: Spravochnik po soprotivleniju materialov). Budivelnik, Kiev (1970)
214. Volkov, S.D.: Basics of the statistical theory of strength (in Russ.: Osnovy statisticheskoj teorii prochnosti). In: Ioffe, A.F., Kurdjymov, G.B., Zhurkov, S.N. (eds.) Nekotorye problemy prochnosti tverdogo tela, Izdatel'stvo Akademii Nauk SSSR, pp. 325–333. Leningrad, Moscow (1959)
215. Wack, B.: The torsion of a tube (or a rod): General cylindrical kinematics and some axial deformation and ratchet measurements. Acta Mech. **80**(1), 39–59 (1989)
216. Wang, D.A., Pan, J.: A non-quadratic yield function for polymeric foams. Int. J. Plast. **22**(3), 434–458 (2006)
217. Weigler, H., Becker, G.: Über das Bruch- und Verformungsverhalten von Beton bei mehrachsiger Beanspruchung. Der Bauingenieur **36**(10), 390–396 (1961)
218. Westergaard, H.M.: On the resitance of ductile materials to combined stress in two or three directions perpendicular to one another. J. Franklin Inst. **189**, 627–640 (1920)
219. Williams, J.G.: Stress Analysis of Polymers. Longman, London (1973)
220. Yagn, Y.I.: New methods of strength prediction (in Russ.: Novye metody pascheta na prochnost'). Vestnik inzhenerov i tekhnikov **6**, 237–244 (1931)
221. Yagn, Y.I.: Strength of Materials: Theory and Problems (in Russ.: Soprotivlenie materialov: teorja i zadachnik). Kubuch, Leningrad (1933)
222. Yu, M.H.: General behaviour of isotropic yield function (in Chinese). Scientific and Technological Research Paper of Xi'an Jiaotong University, pp 1–11 (1961)
223. Yu, M.H.: Brittle fracture and plastic yield criterion (in Chinese). Scientific and Technological Research Paper of Xi'an Jiaotong University, pp 1–25 (1962)
224. Yu, M.H.: Twin shear stress yield criterion. Int. J. Mech. Sci. **25**(1), 71–74 (1983a)
225. Yu, M.H.: Twin shear stress yield criterion. Int. J. Mech. Sci. **25**(11), 845–846 (1983b)
226. Yu, M.H.: Researches on the twin shear stress strength theory (in Chinese). Xi'an Jiaotong University Press, Xi'an (1988)
227. Yu, M.H.: Engineering Strength Theory (in Chinese). Higher Education Press, Beijing (1999)
228. Yu, M.H.: Advances in strength theories for materials under complex stress state in the 20th century. Appl. Mech. Rev. **55**(5), 169–218 (2002)
229. Yu, M.H.: Unified Strength Theory and its Applications. Springer, Berlin (2004)
230. Yu, M.H., He, L., Song, L.: Twin shear stress theory and its generalization. Scientia Sinica, Series A-Mathematical, Physical, Astronomical and Technical Sciences **28**(11), 1174–1183 (1985)
231. Zawadzki, J.: Ciśnienie zredukowane jako jeden z parametrów wytężenia (Przyrost właściwej energii swobodnej jako miara wytężenia). Rozprawy Inzynierskie **LXXIII**, 358–398 (1957)
232. Ziegler, H.: Zum plastischen Potential der Bodenmechanik. Z f angew Math und Phys **20**, 659–675 (1969)
233. Zienkiewicz, O., Pande, G.: Some useful forms of isotropic yield surfaces for soil and rock mechanics. In: Gudehus, G. (ed.) Finite Elements in Geomechanics, pp. 179–198. John Wiley, London, New York (1977)
234. Zyczkowski, M.: Combined Loadings in the Theory of Plasticity. PWN-Polish Scientific Publ, Warszawa (1981)
235. Zyczkowski, M.: Discontinuous bifurcations in the case of the Burzyński-Torre yield condition. Acta Mech. **132**(1), 19–35 (1999)

Plasticity of Cellular Metals (Foams)

Thomas Daxner

Abstract Cellular metals, e.g., made by solidification of molten metal foam, have interesting mechanical properties, among them high specific strength and stiffness coupled with inflammability and good damping properties. This makes them interesting for engineering applications which require the prediction of the onset of yielding under multi-axial stress states and the development of plastic strains over a strain range that may extend into the regime of full compaction of the foam micro-structure, as it is the case in applications for crash protection. This chapter investigates the micro-mechanical deformation mechanisms which govern the elasto-plastic behavior of cellular metals on the macro-mechanical level, where the cellular structure can be treated as a homogeneous material if the difference between the cell size and the component size is large enough. If this is the case suitable constitutive models can be applied for predicting the onset of macroscopic yielding, the evolution of plastic strains and the hardening behavior. Thus, a review of the most important material models proposed for simulating the effective elasto-plastic behavior of isotropic cellular metals is presented. This behavior is characterized by a distinct pressure sensitivity, which sets apart the behavior of cellular metals from the one of solid metals as described by classical (e.g., von Mises) theory of plasticity.

Keywords Cellular metal · Metal foam · Open-cell foam · Closed-cell foam · Yield surface · Flow rule · Hardening · Micromechanics · Multi-axial loading · Pressure dependent yielding

T. Daxner (✉)
CAE Simulation & Solutions, Pitkagasse 2/1/16, A-1210 Vienna, Austria
e-mail: daxner@cae-sim-sol.at

H. Altenbach and A. Öchsner (eds.), *Plasticity of Pressure-Sensitive Materials*,
Engineering Materials, DOI: 10.1007/978-3-642-40945-5_3,

1 Introduction

Cellular metals are a class of materials which is characterized by a foam or sponge-like structure on a length-scale that is typically much smaller than the one of the component. Since the individual cells are much smaller than the typical objects containing cellular metals the mechanical behavior of the cellular structure can generally be described in a homogenized manner, and the present chapter tries to give some insight into the available methods and challenges.

Most cellular materials fall into one of the two following categories:

- open-cell foams (sponges),
- closed-cell foams.

The cells, which are either closed or open according to this classification, are typically filled with air, which is a compressible medium. If the metallic structure is treated as a 'material' in the homogenized sense, this fact gives rise to the marked pressure-sensitivity of cellular metals in the plastic range.

The strength of the materials is primarily determined by the mechanical behavior of the metallic structure, especially in the quasi-static range, where the gas can escape from within the cells through missing or ruptured cell walls. In the case of closed-cell foams the metallic structure consists of a network of struts which meet in vertices and are connected by cell walls in a manner that is very similar to that of liquid foams. This is not surprising, because metallic foams are typically produced by cooling down and solidifying a liquid foam made from molten metal.

In the case of open-cell foams, the metallic skeleton does not form closed cells, because the cell walls are missing, either because they broke during the solidification process or because they where removed by subsequent manufacturing processes.

Interesting structures can be obtained by coating a polymer precursor foam with a metallic layer and subsequently removing the precursor structure. This leaves the struts hollow and, in the ideal case, separates two gas filled cavities, namely the one inside and the one between the struts.

No matter how the cellular metals are produced or which topology they exhibit, their effective mechanical response is rooted in the deformation of the cellular structure itself. To take this fact into account, we investigate deformation mechanisms in cellular metals in Sect. 3.

Transferring theoretical results from the structural level, which is characterized by cells with a size range of tenths of millimeters up to several millimeters, to the component level, which is typically much larger, is a process commonly called homogenization. It is one way of obtaining insight into the mechanical behavior of cellular metals under multi-axial loads. The preferred method, however, is to perform appropriate experiments, which go beyond classical uniaxial compression tests and take multiaxial loading conditions into account. These experiments require expensive equipment and special care owing to the fact that applying hydrostatic pressure by a fluid is difficult because the fluid has to be kept from penetrating into the structure. Nevertheless, corresponding experiments have been performed in the past and the interpretation of their results has lead to the formulation of constitutive laws for the effective mechanical behavior of cellular metals on the macro-mechanical level.

Section 4 reviews the most important contributions in the field of constitutive modeling of cellular metals. The focus is on works which were tailored to metallic foams, which reduces the number of candidate material models to a handful, and allows for a detailed look into their formulation and derivation. Preceding this main body of this study is a brief introduction to the basics of constitutive modeling for elasto-plastic materials, which will be given in the following Sect. 2.

2 Constitutive Modeling: Basics

2.1 Introduction

Before it is possible to deal with the particular phenomena characterizing the elasto-plastic behavior of cellular metals, it is necessary to acquaint oneself with the basics of continuum mechanics of solid materials and the foundations of the theory of plasticity. This section is intended to provide the reader with the knowledge and the mathematical tools necessary for understanding the formulation of the constitutive models for cellular metals which will be presented in later sections.

First, the mathematical notation used in this chapter will be introduced in Sect. 2.2. Next, the concept of stress will be recapitulated in Sect. 2.3. The description of deformation processes in terms of appropriate strain measures will be reviewed in Sect. 2.4. Finally, Sect. 2.5 describes the elements of the theory of plasticity, which provides the foundation for most of the constitutive laws for cellular metals.

2.2 Mathematical Notation

The most important mathematical terms and operations necessary for performing the derivations presented in this chapter will be briefly summarized in the following. The intention is to familiarize the reader with the mathematical notation used throughout this chapter.

The location of a point in space is described by a vector $\mathbf{x}$ from the origin of the reference system to the point. In order to do actual numerical calculations, the vector has to be related to a coordinate system 1-2-3 for obtaining its coordinates x_1, x_2, and x_3. Only Cartesian coordinate systems are considered here.

The scalar product of two vectors $\mathbf{a}$ and $\mathbf{b}$ gives a scalar c, which is the product of the length of vector $\mathbf{a}$ with the length of the projection of vector $\mathbf{b}$ onto vector $\mathbf{a}$, and vice versa:

$$c = \mathbf{a} \bullet \mathbf{b} \qquad c = a_i\, b_i \qquad \ldots = \sum_{i=1}^{3} a_i\, b_i, \tag{1}$$

where the use of Einstein's summation notation is indicated by the sum sign. The length l of a vector $\mathbf{v}$ can be calculated as $l = \sqrt{\mathbf{v} \bullet \mathbf{v}}$.

In the framework of tensor algebra a vector is equivalent to a first order tensor $\mathbf{x}$. Such a first order tensor $\mathbf{x}$ can be transformed into a new first order tensor $\mathbf{y}$ by applying a second order tensor $\mathbf{A}$ to it. By writing $\mathbf{y} = \mathbf{A}\,\mathbf{x}$ in tensorial notation, we imply the fact that $\mathbf{A}$ represents a linear transformation of $\mathbf{x}$ into $\mathbf{y}$. The coordinates of the second order tensor $\mathbf{A}$ with respect to an orthonormal coordinate system are indicated as $[\mathbf{A}]_{ij}$ or as A_{ij}, respectively. The coordinates y_i of the transformed tensor $\mathbf{y}$ are a linear combination of the coordinates x_j of the original tensor $\mathbf{x}$, where each original coordinate is weighted by the coordinate A_{ij} of the tensor $\mathbf{A}$. Using both tensor notation and index notation, we obtain

$$\mathbf{y} = \mathbf{A}\,\mathbf{x} \qquad y_i = A_{ij} x_j \qquad \ldots = \sum_{j=1}^{3} A_{ij} x_j. \tag{2}$$

A special second order tensor is the unit tensor $\mathbf{I}$ which has the coordinates $[\mathbf{I}]_{ij} = \delta_{ij}$, where δ_{ij} is the Kronecker Delta function which returns one for $i = j$ and zero for $i \neq j$. Applying the unit tensor to a second order tensor $\mathbf{A}$ leaves the latter tensor unchanged: $\mathbf{IA} = \mathbf{A}$.

A second order tensor $\mathbf{C}$ can be constructed from vectors $\mathbf{a}$ and $\mathbf{b}$ using the tensor, or dyadic product:

$$\mathbf{C} = \mathbf{a} \otimes \mathbf{b} \qquad C_{ij} = a_i\, b_j \tag{3}$$

The resulting tensor $\mathbf{C}$, applied to a vector $\mathbf{x}$, returns as the result the first vector $\mathbf{a}$, scaled by the product of the length of the second vector $\mathbf{b}$ times the length of the projection of the vector $\mathbf{x}$ onto the second vector $\mathbf{b}$, $(\mathbf{a} \otimes \mathbf{b})\mathbf{x} = \mathbf{a}(\mathbf{b} \bullet \mathbf{x})$.

A second-order tensor $\mathbf{C}$ which represents a transformation $\mathbf{B}$ followed by a transformation $\mathbf{A}$ can be constructed as follows:

$$\mathbf{C} = \mathbf{AB} \qquad C_{ij} = A_{ik} B_{kj} \qquad \ldots = \sum_{k=1}^{3} A_{ik} B_{kj} \tag{4}$$

A fourth order tensor $\mathbb{C}$ represents a linear transformation which is applied to a second order tensor $\mathbf{A}$ for obtaining a new second order tensor $\mathbf{B}$. We express this operation in tensor notation and index notation:

$$\mathbf{B} = \mathbb{C}\,\mathbf{A} \qquad B_{ij} = C_{ijkl} A_{kl} \qquad \ldots = \sum_{k=1}^{3} \sum_{l=1}^{3} C_{ijkl} A_{kl} \tag{5}$$

The coordinates of the fourth order tensor $\mathbb{C}$ are represented as either $[\mathbb{C}]_{ijkl}$ or C_{ijkl}.

Summing over two indices as in Eq. (5) is called double contraction. The double contraction of a second order tensor **A** and another second order tensor **B** gives a scalar c. This operation is written as follows:

$$c = \mathbf{A}:\mathbf{B} \qquad c = A_{ij}B_{ij} \qquad \ldots = \sum_{i=1}^{3}\sum_{j=1}^{3} A_{ij}B_{ij} \tag{6}$$

The *trace* tr (**A**) of a second order tensor **A** is defined as the sum of the coordinates A_{ii} $(i = 1 \ldots 3)$ on the main diagonal (in matrix notation), or the double contraction with the second order tensor of unity **I**:

$$\text{tr}\,(\mathbf{A}) = \sum_{i=1}^{3} A_{ii} = \mathbf{A}:\mathbf{I} \tag{7}$$

The derivative of a scalar-valued function

$$g(\boldsymbol{\sigma}) = g(\sigma_{11}, \sigma_{22}, \sigma_{33}, \sigma_{12}, \sigma_{21}, \sigma_{23}, \sigma_{32}, \sigma_{13}, \sigma_{31})$$

of a second-order tensor $\boldsymbol{\sigma}$ with respect to (the coordinates of) the tensor $\boldsymbol{\sigma}$ gives the gradient of g at $\boldsymbol{\sigma}$, which is itself a second order tensor:

$$\left[\frac{\partial g}{\partial \boldsymbol{\sigma}}\right]_{ij} = \frac{\partial g}{\partial \sigma_{ij}} \tag{8}$$

If the function $g(\boldsymbol{\sigma})$ is homogeneous of degree n in $\boldsymbol{\sigma}$, i.e.,

$$g(\alpha\boldsymbol{\sigma}) = \alpha^{n}\, g(\boldsymbol{\sigma}) \tag{9}$$

then Euler's theorem on homogeneous functions can be applied to obtain the relationship

$$\frac{1}{g}\boldsymbol{\sigma}:\frac{\partial g}{\partial \boldsymbol{\sigma}} = n \qquad \frac{\sigma_{ij}}{g}\frac{\partial g}{\partial \sigma_{ij}} = n \tag{10}$$

which can be useful for manipulating expressions related to plastic flow rules (as introduced in Sect. 2.5).

2.3 The Stress State

In this section, expressions and relationships from continuum mechanics, which are necessary for the mathematical description of the mechanical behavior of materials, are introduced.

The local stress conditions in a body are described by a symmetric second-order tensor $\boldsymbol{\sigma} = [\sigma_{ij}]$ called the stress tensor. The 3/D stress tensor has nine coordinates $\sigma_{ij}, i \in [1, 3], j \in [1, 3]$, six of which are independent owing to the symmetry of the tensor, i.e., $\sigma_{ji} = \sigma_{ij}$.

Coordinates with identical indices $i = j$ relate to normal loading with respect to the reference coordinate system, whereas coordinates with differing indices $i \neq j$ indicate shear loading.

The actual values of the coordinates of the stress tensor depend on the chosen reference coordinate system. However, three scalar properties I_1, I_2, and I_3, can be defined which are invariant with respect to a rotation of the reference system, These invariants are given by:

$$I_1 = \sigma_{kk} = \sigma_{11} + \sigma_{22} + \sigma_{33} \tag{11}$$

$$I_2 = \frac{1}{2}\left(\sigma_{ii}\sigma_{kk} - \sigma_{ij}\sigma_{ij}\right) \tag{12}$$

$$I_3 = \det \boldsymbol{\sigma} \tag{13}$$

It is possible to find a reference frame 1-2-3 for which all non-diagonal coordinates of the stress tensor, i.e., the shear stresses, vanish. The coordinates σ_{ii} along the diagonal of the stress tensor expressed in this specific reference system are then called the principal stresses, $\sigma_1 = \sigma_{11}$, $\sigma_2 = \sigma_{22}$, and $\sigma_3 = \sigma_{33}$. They are the eigenvalues of the stress tensor and can be found by solving the characteristic equation

$$\det(\boldsymbol{\sigma} - \sigma_i \mathbf{I}) = -\sigma_i^3 + I_1\,\sigma_i^2 - I_2\,\sigma_i + I_3 = 0. \tag{14}$$

For the investigation of the mechanical behavior of pressure sensitive materials, it is necessary to define the pressure p, which is a function of the normal stresses only:

$$p = -\frac{1}{3}\,(\sigma_{11} + \sigma_{22} + \sigma_{33}) = -\sigma_{\mathrm{m}} \tag{15}$$

Also defined in Eq. (15) is the mean stress σ_{m}, which has the negative value of the pressure, $\sigma_{\mathrm{m}} = -p$. Note the relationship $\sigma_{\mathrm{m}} = \frac{1}{3}\,I_1$ between the mean stress and the first invariant I_1 of the stress tensor, compare Eq. (11).

The contribution of the hydrostatic pressure to the stress tensor can be isolated as

$$\boldsymbol{\sigma}_{\mathrm{Hydro}} = -p\,\mathbf{I} = \sigma_{\mathrm{m}}\,\mathbf{I} \tag{16}$$

Correspondingly, the deviatoric part $\mathbf{s}$ of the stress tensor can be obtained by subtracting the hydrostatic part from the stress tensor,

$$\mathbf{s} = \boldsymbol{\sigma} - \boldsymbol{\sigma}_{\mathrm{Hydro}} \qquad \ldots = \boldsymbol{\sigma} - \sigma_{\mathrm{m}}\,\mathbf{I} = \boldsymbol{\sigma} + p\,\mathbf{I} \tag{17}$$

For the deviatoric stress tensor **s** three invariants J_1, J_2, and J_3 can be defined. The first invariant J_1 is zero by definition, because the deviatoric stress tensor contains no contribution from hydrostatic pressure. The second invariant J_2 is given by

$$J_2 = \frac{1}{2}\mathbf{s}:\mathbf{s} = \frac{1}{2}s_{ij}s_{ij}$$

It plays an important role in metal plasticity, since it can be used to express the von Mises equivalent stress σ_e in the following form:

$$\sigma_\mathrm{e} = \sqrt{3J_2} \quad \ldots = \sqrt{\frac{3}{2}\mathbf{s}:\mathbf{s}}. \tag{18}$$

For some materials, in particular metals, which are not sensitive to pressure with regard to yielding, the von Mises equivalent stress is an appropriate scalar measure for assessing the stress state in the material with respect to plastic yielding. For uniaxial tension and compression, the von Mises stress is equal to the applied stress. It is sometimes useful to re-write the von Mises stress in terms of principal stresses:

$$\sigma_\mathrm{e} = \sqrt{\frac{1}{2}\left[(\sigma_1-\sigma_2)^2+(\sigma_2-\sigma_3)^2+(\sigma_3-\sigma_1)^2\right]} \tag{19}$$

At this point, some partial derivatives of scalar-valued functions of the stress tensor with respect to the stress tensor coordinates, recall Eq. (8), are presented in Table 1.

Table 1 Some useful derivatives of quantities related to the description of yield surfaces for pressure-sensitive materials with respect to the coordinates σ_{ij} of the stress tensor $\boldsymbol{\sigma}$

Tensor notation of derivative	Selected terms in index notation
$\frac{\partial(\mathbf{s}:\mathbf{s})}{\partial\boldsymbol{\sigma}} = 2\mathbf{s}$	$\frac{\partial(s_{ij}s_{ij})}{\partial\sigma_{11}} = 2\left(\sigma_{11} - \frac{\sigma_{11}+\sigma_{22}+\sigma_{33}}{3}\right) = 2s_{11}$ $\frac{\partial(s_{ij}s_{ij})}{\partial\sigma_{12}} = 2\sigma_{12} = 2s_{12}$
$\frac{\partial\sigma_\mathrm{e}}{\partial\boldsymbol{\sigma}} = \frac{3}{2}\frac{\mathbf{s}}{\sigma_\mathrm{e}}$	$\frac{\partial\sigma_\mathrm{e}}{\partial\sigma_{11}} = \frac{1}{2\sigma_\mathrm{e}}\left[2\sigma_{11}-\sigma_{22}-\sigma_{33}\right] = \frac{3s_{11}}{2\sigma_\mathrm{e}}$ $\frac{\partial\sigma_\mathrm{e}}{\partial\sigma_{12}} = \frac{3\sigma_{12}}{2\sigma_\mathrm{e}}$
$\frac{\partial(\sigma_\mathrm{e}{}^2)}{\partial\boldsymbol{\sigma}} = 3\mathbf{s}$	$\frac{\partial(\sigma_\mathrm{e}{}^2)}{\partial\sigma_{11}} = 2\sigma_{11}-\sigma_{22}-\sigma_{33} = 3s_{11}$ $\frac{\partial(\sigma_\mathrm{e}{}^2)}{\partial\sigma_{12}} = 3\sigma_{12} = 3s_{12}$
$\frac{\partial\sigma_\mathrm{m}}{\partial\boldsymbol{\sigma}} = \frac{1}{3}\mathbf{I}$	$\frac{\partial\sigma_\mathrm{m}}{\partial\sigma_{11}} = \frac{1}{3}$
$\frac{\partial(\sigma_\mathrm{m}{}^2)}{\partial\boldsymbol{\sigma}} = \frac{2}{3}\sigma_\mathrm{m}\,\mathbf{I}$	$\frac{\partial(\sigma_\mathrm{m}{}^2)}{\partial\sigma_{11}} = \frac{2}{9}\left[\sigma_{11}+\sigma_{22}+\sigma_{33}\right]$
$\frac{\partial p}{\partial\boldsymbol{\sigma}} = -\frac{1}{3}\mathbf{I}$	$\frac{\partial p}{\partial\sigma_{11}} = -\frac{1}{3}$
$\frac{\partial(p^2)}{\partial\boldsymbol{\sigma}} = -\frac{2}{3}p\,\mathbf{I}$	$\frac{\partial(p^2)}{\partial\sigma_{11}} = \frac{2}{9}\left[\sigma_{11}+\sigma_{22}+\sigma_{33}\right]$

Left column: tensor notation, right column: selected terms in index notation

Materials, for which the mechanical response is independent of any material direction, are called isotropic materials. Since the choice of the reference coordinate system for the description of the constitutive behavior of these materials is not predetermined by any geometrical feature that introduces a direction dependency (such as fibers in a fiber-reinforced composite material) it is possible to describe the material behavior in terms of tensor invariants, and, in particular, in terms of the three principal stresses.

The yield surface, which plays a major role in any plasticity model (compare Sect. 4), is often depicted in the three-dimensional space that is described by interpreting the three principal stresses σ_1, σ_2, and σ_3 as coordinates in an orthonormal reference system. In this principal stress space, the following interesting entities can be identified:

- Hydrostatic axis: The space diagonal in the principal stress space, describing stress states that are purely hydrostatic and do not contain any deviatoric component, i.e., $\sigma_e = 0$.
- Deviatoric plane: Any plane perpendicular to the hydrostatic axis, i.e., all stress states with the same mean stress $\sigma_m = \text{const.}$
- π-plane: The deviatoric plane which contains the origin $\sigma_1 = \sigma_2 = \sigma_3 = 0$ of the principal stress coordinate system, and, therefore, all stress states with vanishing mean stress, $\sigma_m = 0$.
- Meridian: A plane which contains the hydrostatic axis.

Most of the constitutive laws for the description of the effective mechanical behavior of isotropic cellular metals can be formulated in terms of the mean stress σ_m and the von Mises equivalent stress σ_e. Surfaces in the principal stress space, which are defined implicitly by $F(\sigma_m, \sigma_e) = 0$ show rotational symmetry about the hydrostatic axis.

If this rotational symmetry cannot be assumed, a third parameter in addition to σ_m and σ_e has to be considered. This third parameter is often chosen to be an angle θ measured in the π-plane, namely between the projection of the σ_1-axis on the π-plane and a vector from the origin of the principal stress space to the projection of the stress state onto the π-plane along the direction of the hydrostatic axis. By this geometrical definition, the angle θ, which is often called the Lode angle, can assume values between 0 and 60°, provided that the principal stresses are sorted as $\sigma_1 \geq \sigma_2 \geq \sigma_3$. The Lode angle can be calculated from

$$\cos 3\theta = \frac{3\sqrt{3}}{2} \frac{J_3}{\sqrt{(J_2)^3}} \tag{20}$$

Note the dependency of this quantity on the third invariant J_3 of the stress deviator tensor. For purely hydrostatic stress states, θ becomes undefined, because inserting $J_2 = 0$ in Eq. (20) would cause a division by zero. The Lode angle describes the relationship between the principal stresses.

2.4 Deformation and Strain

Application of loads to solid bodies leads to a deformation of these bodies. The quantities and methods available for the description of this deformation are the topics of this section.

We investigate a process, which moves and deforms a body from an undeformed reference configuration to a deformed configuration. Points $\mathbf{X}$ in the undeformed configuration are mapped onto points in the deformed configuration by the function $\boldsymbol{\varphi}$, giving $\mathbf{x} = \boldsymbol{\varphi}(\mathbf{X}, t)$ at any given time t.

Correspondingly, a line element $d\mathbf{X}$ in the undeformed configuration is transformed into a line element in the actual configuration $d\mathbf{x}$ according to $d\mathbf{x} = \mathbf{F}\, d\mathbf{X}$, where $\mathbf{F}$ is a second order tensor called the deformation gradient, and is defined as

$$\mathbf{F} = \frac{\partial \boldsymbol{\varphi}}{\partial \mathbf{X}} \qquad F_{ij} = \frac{\partial \varphi_i}{\partial X_j} \tag{21}$$

The deformation gradient $\mathbf{F}$ contains all the necessary information for describing the deformation process at a given material point. In particular, it contains information about the actual deformation of the material and any super-imposed rigid body rotation. To separate these two contributions, a polar decomposition of $\mathbf{F}$ into an orthogonal rotation tensor $\mathbf{R}$ (with properties $\det(\mathbf{R}) = 1$, $\mathbf{R}^{\mathrm{T}} = \mathbf{R}^{-1}$) and a symmetric tensor $\mathbf{U}$, which is called the right (or material) stretch tensor, can be performed:

$$\mathbf{F} = \mathbf{R}\mathbf{U} \tag{22}$$

This decomposition means that the shape and/or volume of a volume element dV is first changed according to $\mathbf{U}$ and the volume element is then rotated by $\mathbf{R}$ into the final configuration dv. The three eigenvalues λ_i of $\mathbf{U}$ are called the principal stretches. The length of a line element dL_i in the undeformed configuration, which points into the direction of the i-th eigenvector $\mathbf{N}_i$ of $\mathbf{U}$ is stretched by the deformation process to a new length dl_i, which can be calculated using

$$\lambda_i = \frac{dl_i}{dL_i} \tag{23}$$

To actually calculate the right stretch tensor $\mathbf{U}$, we first introduce the right Cauchy-Green tensor $\mathbf{C}$:

$$\mathbf{C} = \mathbf{F}^{\mathrm{T}}\mathbf{F} = \mathbf{U}^{\mathrm{T}}\underbrace{\mathbf{R}^{\mathrm{T}}\mathbf{R}}_{\mathbf{I}}\mathbf{U} = \mathbf{U}^{\mathrm{T}}\mathbf{U} \tag{24}$$

This tensor has the eigenvalues Λ_i and the eigenvectors $\mathbf{N}_i$, and can be represented using the spectral decomposition theorem:

$$\mathbf{C} = \sum_{i=1}^{3} \Lambda_i \mathbf{N}_i \otimes \mathbf{N}_i \tag{25}$$

We can now build the right stretch tensor **U** in a similar manner noting that it has the same eigenvectors as **C** and its eigenvalues λ_i are the square root of the eigenvalues Λ_i of **C**, i.e., $\lambda_i = \sqrt{\Lambda_i}$:

$$\mathbf{U} = \sum_{i=1}^{3} \lambda_i \mathbf{N}_i \otimes \mathbf{N}_i \tag{26}$$

The principal stretches already give a lot of information about the deformation process, but they do not lend themselves naturally to the description of the strain in the material, because they assume a value of one for the undeformed state (for which, in the absence of residual stresses, the stresses are zero).

Therefore, appropriate strain measures have to be defined. For large deformations, logarithmic strain measures are often used. The logarithmic strain tensor in the reference configuration is defined as

$$\boldsymbol{\varepsilon} = \sum_{i=1}^{3} \ln(\lambda_i) \mathbf{N}_i \otimes \mathbf{N}_i \tag{27}$$

For homogenous uniaxial deformations, the axial logarithmic strain can be found as $\varepsilon_{\text{axial}}^{(\ln)} = \ln(l/l_0) = \ln(\lambda_{\text{axial}})$. It is zero for the undeformed state.

It is often useful to split the strain tensor into a part which describes the change of shape of a volume element and a part which describes the change of its volume.

Let us look at a cube-shaped volume element dV which is oriented along the eigenvectors $\mathbf{N}_i$. The undeformed edge length is given by dL, and the undeformed volume by $dV = dL^3$. If the volume element is transformed into the deformed configuration by the right stretch tensor **U**, its edges remain orthogonal (because they where parallel to the eigenvectors of **U**) but their length is now dl_1, dl_2, and dl_3, respectively. The volume in the deformed configuration is, thus, $dv = dl_1 dl_2 dl_3$. Relating the deformed volume dv to the undeformed volume dV, we can now define the logarithmic volumetric strain ε_{vol}:

$$\begin{aligned}\varepsilon_{\text{vol}} &= \ln \frac{dv}{dV} = \ln \frac{dl_1 dl_2 dl_3}{dL^3} = \ln(\lambda_1 \lambda_2 \lambda_3) = \ln(\lambda_1) + \ln(\lambda_2) + \ln(\lambda_3) \\ &= \varepsilon_1^{(\ln)} + \varepsilon_2^{(\ln)} + \varepsilon_3^{(\ln)}\end{aligned} \tag{28}$$

This shows the convenient property that the logarithmic volumetric strain is equal to the trace of the logarithmic strain tensor $\boldsymbol{\varepsilon}$.

The part of the strain tensor, which changes the volume of a volume element can now be isolated as: $\boldsymbol{\varepsilon}_{\text{vol}} = \frac{1}{3}\varepsilon_{\text{vol}}\mathbf{I}$. The remaining, deviatoric part **e** of strain tensor can be found by subtracting this tensor from the full strain tensor:

$$\mathbf{e} = \boldsymbol{\varepsilon} - \boldsymbol{\varepsilon}_{\text{vol}} = \boldsymbol{\varepsilon} - \frac{1}{3}\varepsilon_{\text{vol}}\mathbf{I} \tag{29}$$

Since the deviatoric strain tensor does not impose any volume change, its trace is equal to zero.

Similar to the von Mises equivalent stress σ_{e}, an equivalent or effective strain ε_{e} can be defined from the deviatoric part of the strain tensor:

$$\varepsilon_{\text{e}} = \sqrt{\frac{2}{3}\mathbf{e}:\mathbf{e}} \tag{30}$$

This definition implies two properties: (a) for a volume-preserving uniform extension process, the effective strain is equal to the axial strain, and (b) the effective strain is work conjugate to the von Mises equivalent stress σ_{e}.

In addition, the rate $\dot{\varepsilon}_{\text{vol}}$ of the logarithmic volumetric strain is work conjugate to the mean stress σ_{m}. Finally, an expression for the rate $\dot{W}$ of the internal mechanical work done by the stresses can be obtained exclusively in terms of effective properties:

$$\dot{W} = \sigma_{\text{e}}\dot{\varepsilon}_{\text{e}} + \sigma_{\text{m}}\dot{\varepsilon}_{\text{vol}} \tag{31}$$

2.5 Formal Introduction to Elasto-Plasticity

In this section, the common framework of the constitutive modeling of elastic-plastic materials will be reviewed.

A solid body under global loading experiences local stress fields in parts or all of its volume. These stresses lead to deformations, which can be expressed by strain tensors on the material point level. The local deformation state is described by a strain tensor $\boldsymbol{\varepsilon}$.

The strain state is related to the stress state by constitutive laws, which express the strain tensor $\boldsymbol{\varepsilon}$ in terms of the stress tensor $\boldsymbol{\sigma}$, the temperature T (if necessary), and internal variables S_i:

$$\boldsymbol{\varepsilon} = f(\boldsymbol{\sigma}, T, S_i) \tag{32}$$

The internal variables S_i describe the internal state of the material. Accumulated plastic strains, e.g., are important internal variables for plasticity. The description of the evolution of the material state throughout the deformation process requires evolution equations for the internal variables:

$$\frac{\mathrm{d}S_i}{\mathrm{d}t} = g_i(\boldsymbol{\sigma}, T, S_i) \tag{33}$$

The coupled system of Eqs. (32) and (33) has to be solved in order to obtain the deformation history of the material.

We now consider materials which deform inelastically, i.e., which are able to undergo irreversible, plastic deformations. These material often do not show yielding from the onset, but rather require a certain yield stress state to be exceeded before plastic deformation sets in. If the stress state is not critical with respect to plastic yielding, the material deforms elastically.

Classical plasticity theory (see e.g. [26]) describes this behavior by defining a surface in stress space the interior of which contains all stress states for which the deformation mechanism remains purely elastic. This surface is called yield surface. The yield surface is defined implicitly by the relationship

$$F(\boldsymbol{\sigma}, S_i, T) = 0 \tag{34}$$

where F is called the yield function. The interior of the yield surface contains all material states for which $F < 0$.

As soon as stress states reach the yield surface, and, consequently, the yield criterion $F(\boldsymbol{\sigma}, S_i) = 0$ is fulfilled, the material starts accumulating plastic strains $\boldsymbol{\varepsilon}_{\text{pl}}$ at a rate, which is defined by the plastic *flow rule* (given here in incremental form):

$$d\boldsymbol{\varepsilon}_{\text{pl}} = d\boldsymbol{\varepsilon}_{\text{pl}}(\boldsymbol{\sigma}, d\boldsymbol{\sigma}, S_i, T). \tag{35}$$

The flow rule is often defined using the gradient of a flow potential $G(\boldsymbol{\sigma}, S_i, T)$:

$$d\boldsymbol{\varepsilon}_{ij}^{(\text{pl})} = d\lambda \, \frac{\partial G}{\partial \sigma_{ij}} \tag{36}$$

Herein, $d\lambda$ is the increment of the plastic flow multiplier λ, which can be determined using the fact that the stress state always fulfills the yield condition (34) during plastic loading. By definition, the stress state always remains on the yield surface as long as the material point deforms plastically.

The size, the location and the shape of the yield surface, however, can change according to the applied constitutive theory. The evolution of the yield surface is described by the evolution laws for the relevant internal variables S_i, e.g., by appropriate *hardening laws*.

For many constitutive laws, the flow potential G is chosen to be identical to the yield function, $G \equiv F$. The corresponding flow rules are called *associated* flow rules.

In a general deformation process involving plastic deformation, contributions from elastic deformation mechanisms and plastic yielding are superimposed. If the elastic deformations remain small compared to the overall deformations, an additive split of the strain tensor $\boldsymbol{\varepsilon}$ into an elastic part $\boldsymbol{\varepsilon}_{\text{el}}$ and a plastic part $\boldsymbol{\varepsilon}_{\text{pl}}$ is appropriate:

$$\boldsymbol{\varepsilon} = \boldsymbol{\varepsilon}_{\text{el}} + \boldsymbol{\varepsilon}_{\text{pl}} \tag{37}$$

3 Deformation Mechanisms and Yielding in Cellular Metals

3.1 Onset of Failure

While cellular metals can be treated as homogeneous materials for most applications because of the size difference between their metallic structure and the size of corresponding components, their mechanical behavior is still governed by deformation mechanisms in the metallic structure itself.

For open-cell metallic foams, the bending and buckling of the struts is the dominating deformation mechanism. Usually, the struts of open-cell foams have two distinct geometrical properties:

- their cross-sections have the shapes of triangles with concave sides, which is a result of surface tension and drainage processes acting on the foam structure as long as it is in a molten state. The characteristic shape of these struts is called a Plateau border. Figure 1 shows the struts enclosing a single open cell.
- The cross-sectional area of the struts is smallest around their middle and increases towards the vertices. This means that the vertices are considerably stiffer than the struts and, consequently, rather rotate and move than deform.

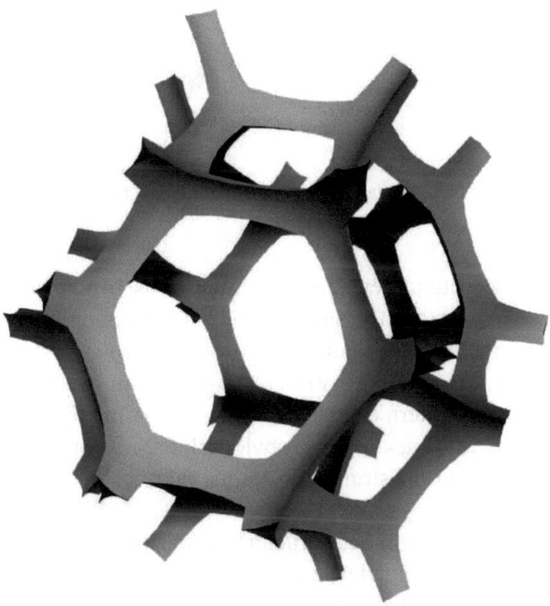

Fig. 1 Rendering of the Plateau border network forming the structure of an open-cell foam. Adapted from [2]

Fig. 2 Local bending deformation of a strut in an open-cell metal foam (image courtesy of Institute of Materials Science and Testing, Vienna University of Technology)

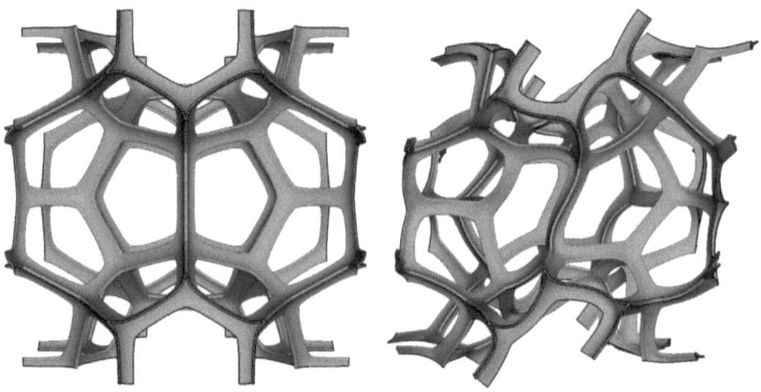

Fig. 3 Buckling of an open-cell structure with hollow struts under hydrostatic pressure

Struts in open-cell materials experience normal loads, bending moments, and torsional loads depending on the macromechanical loading conditions, their orientation and their connection to the surrounding framework of struts.

Figure 2 shows a typical deformation mode of a strut in an open-cell metallic foam. The deformation pattern can be the result of bending by transverse forces or moments, or elasto-plastic buckling. It is obvious that the deformation affects mainly the thin middle section of the struts whereas the thick vertices perform rigid body movements.

Subjecting regular cellular model structures to macroscopic hydrostatic pressure loading may cause struts to experience compressive stresses which can lead to buckling. This buckling can be global in nature, as is shown in Fig. 3 for a periodic unit cell model of an open-cell structure with hollow struts. In this simulation the periodicity of the buckling mode is tied to the geometrical periodicity of the unit cell. Simulation methods, which can capture buckling modes with wavelengths far exceeding the dimensions of the constituting unit cell model, deserve to be mentioned in this

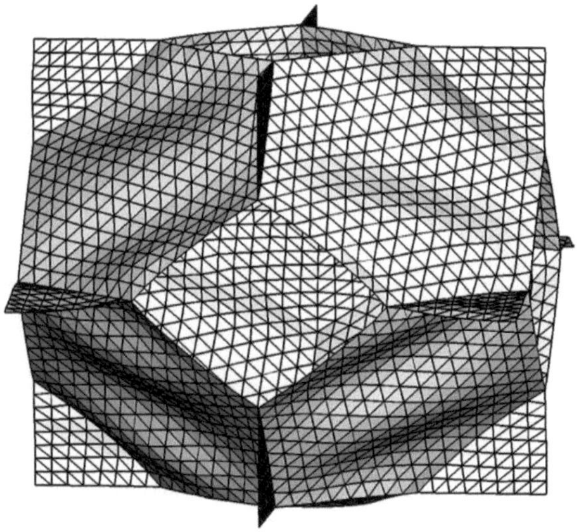

Fig. 4 Cell wall buckling in a unit cell model of a regular tetrakaidecahedral cell structure under macroscopic uniaxial tension

context [19, 20, 25, 32]. Even if the initial buckling mechanism was an elastic one, stress redistribution due to the excentric loading of the struts in the post-buckling regime can quickly lead to plastic yielding in the outer zones of the struts.

For closed-cell foams the presence of cell walls adds the additional mechanisms of bending and stretching of the cell walls. In the direction of compressive principal stresses, the cell walls may buckle, and they may rupture if the local stresses exceed the strength of the walls.

A tetrakaidecahedral unit cell model of a closed-cell foam can be used to illustrate an interesting phenomenon in connection with elastic buckling of cell walls, compare Fig. 4. Here, elastic buckling can be observed on the cell wall level even though the macroscopic loading state is one of uniaxial tension. The reason for this is the fact that the hexagonal faces, which are oblique to the loading direction, experience in-plane shear loading rather than pure tensile loading, causing shear buckling as soon as the critical stress is exceeded.

Micro-mechanical unit cell models can also be exploited to demonstrate the initial stages of deformation in a closed-cell foam under uniaxial compressive loading. Figure 5 shows a uniaxial compressive stress versus compressive strain diagram for a periodic tetradecahedral unit cell model. A straight line at the beginning of the stress-strain curve represents the regime of linear elastic deformation. It is not completely clear, if such a distinct linear regime exists in a real cellular metal owing to the fact that such a material usually contains many imperfections and inhomogeneities that can trigger plastic yielding under even the slightest macroscopic load.

As the macroscopic load is increased, the stresses inside the cellular structure rise and, inevitably, reach the yield stress of the solid material in the first critical spots.

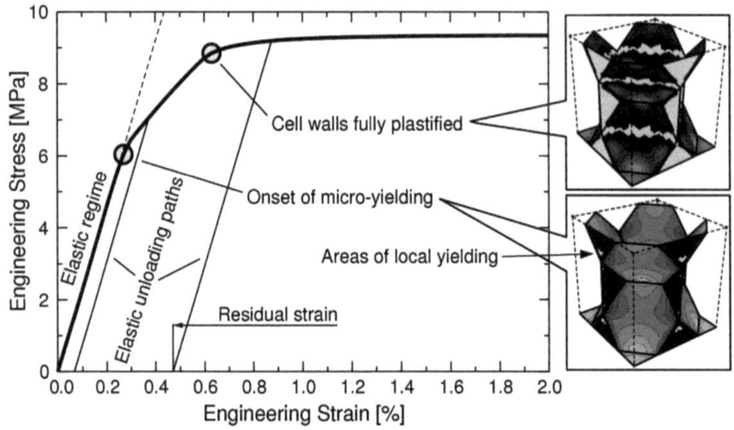

Fig. 5 Compressive stress versus strain diagram for the uniaxial compression of a regular, periodic tetrakaidecahedral cellular structure. The two small contour plots on the right side show the distributions of the von Mises equivalent stress at the onset of local yielding (*bottom*) and close to the limit stress (*top*) (from [9])

These spots are indicated by an arrow in Fig. 5, and they mark the influence regions of stress concentrations in the vicinity of the vertices of the model structure. After the initiation of yielding the stress-strain-curve starts to deviate visibly from the tangent in the origin of the stress versus strain curve. Elastic unloading beyond the onset of micro-yielding can be simulated for obtaining the residual plastic strains in the structure as is also shown in Fig. 5.

Because the detection of micro yielding as defined by the determination of the onset of yielding in any integration point or finite element node in the simulation model may be mesh-dependent or predict yielding at stresses much lower than the macroscopic yield stress, it can prove advantageous to define yielding based on the magnitude of the macroscopic plastic strain which remains after the structure has been unloaded, similar to the definition of the offset yield point for metals without a distinct elastic limit stress.

As the compressive load increases, larger and larger sections of the cell walls start to deform plastically. Finally, as the plastic deformation bands in the cell walls connect across the individual cells, the limit load of the unit cell model is nearly reached and the compressive stress-strain-curve shows a nearly horizontal plateau.

Even though a fairly simple finite element unit cell model was used for obtaining Fig. 5, it can nevertheless illustrate the sequence of events leading to plastic failure of cellular metals.

In solid metals the application of hydrostatic pressure does not cause plastic yielding, at least not within the confines of classic metal plasticity theory. In cellular metals, however, applying hydrostatic pressure on the macroscopic level will lead to local stress states in the cellular structure which are—depending on the symmetry and regularity of the structure—predominantly uniaxial and compressive. Thus, they

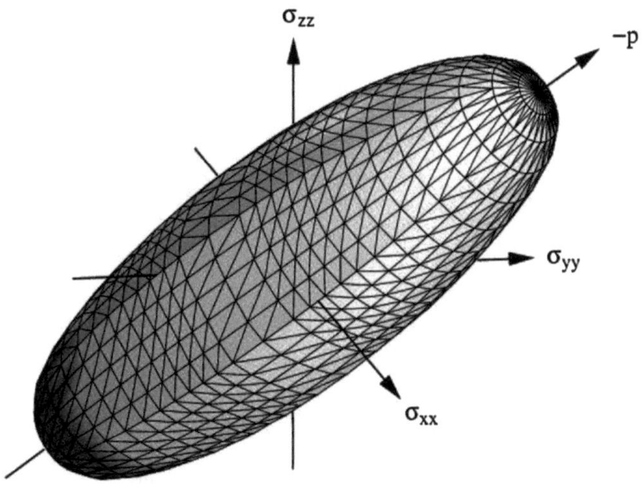

Fig. 6 Yield surface of a regular closed-cell foam predicted by a periodic Weaire-Phelan unit cell model. The dark cap at the end of the ellipsoidal in the regime of positive hydrostatic pressure values indicates stress states, for which yielding may be preceded by elastic buckling (from [10])

can and will cause plastic yielding, because the macroscopic hydrostatic load does not lead to hydrostatic stress states on the micromechanical level.

By loading a sample made of cellular metal along different paths in stress space, stress states for which the macromechanical behavior becomes inelastic can be determined and connected to form a yield surface in the stress space. This can be done experimentally or by means of numerical models.

Figure 6 shows a yield surface in principal stress space that was predicted for a periodic finite element unit cell model of a Weaire-Phelan structure, which is a good generic model for closed-cell foams. In accordance with the statements above, the yield surface intersects the hydrostatic axis at the points corresponding to the hydrostatic tensile and compressive yield stresses. The shape of the yield surface is the one of an ellipsoid which is elongated along the hydrostatic axis.

From a numerical point of view, Fig. 6 contains additional information in the form of the darkened cap at the end of the yield surface which corresponds to almost purely hydrostatic compression. In those dark areas on the yield surface the system matrix has negative eigenvalues, which means that elastic buckling precedes or prematurely initiates plastic yielding.

Figure 6 shows an ellipsoidal yield surface which is visibly elongated along the axis of hydrostatic stress states. This can be attributed to the fact that the stress states induced in the cell walls are mainly membrane-like for hydrostatic loads on the one hand and characterized by high bending stresses close to the vertices for uniaxial loads, on the other hand, which initiates yielding at deviatoric macroscopic stresses that are lower than the hydrostatic macroscopic yield stresses.

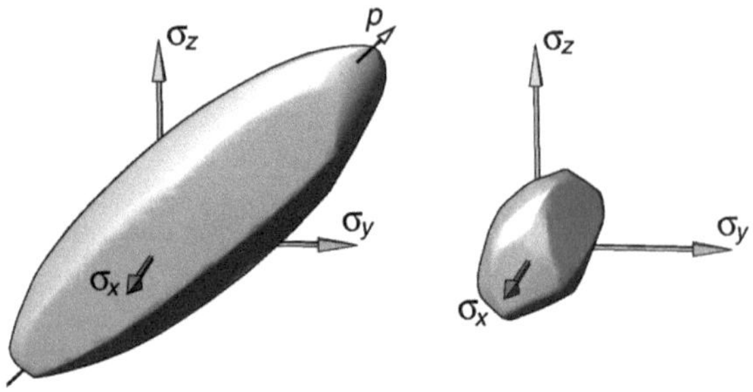

Fig. 7 Yield surfaces predicted for a regular (*left*) and an irregular (*right*) tetrakaidecahedral unit cell model of closed-cell foam, respectively [8]

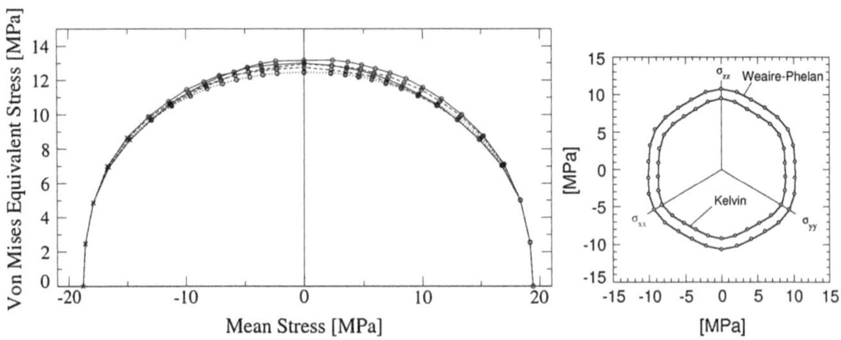

Fig. 8 Projection of the yield surfaces predicted for two periodic unit cell models of closed-cell foam into a diagram of von Mises equivalent stress σ_e versus mean stress σ_m (*left*). Cross-section of the yield surfaces by a deviatoric plane (*right*, from [10])

For real cellular metals, the difference between length and diameter of the ellipsoid is not expected to be so pronounced, as is illustrated in Fig. 7, which presents the yield surface of a regular tetrakaidecahedral finite element unit cell model (left) and the yield surface of a unit cell model with the same topology, but randomly perturbed vertex positions (right). The geometrically imperfect, and therefore more realistic model, shows a smaller yield surface with a much lower ratio between the length and the diameter of the ellipsoidal, which is closer to the experimental evidence. The reason behind this are the bending moments which are induced by excentrical loading of the microstuctural members in the case of the irregular unit cell model even for predominantly hydrostatic macroscopic pressure.

A common method of visualizing yield surfaces of cellular metals is the projection of points on the yield surface onto a diagram of von Mises equivalent stress σ_e versus mean stress σ_m. Figure 8 (left) displays the points on the yield surface in Fig. 6 in the corresponding form. In this diagram, the yield surface collapses into a point cloud which can be fitted by an elliptical curve in a first approximation.

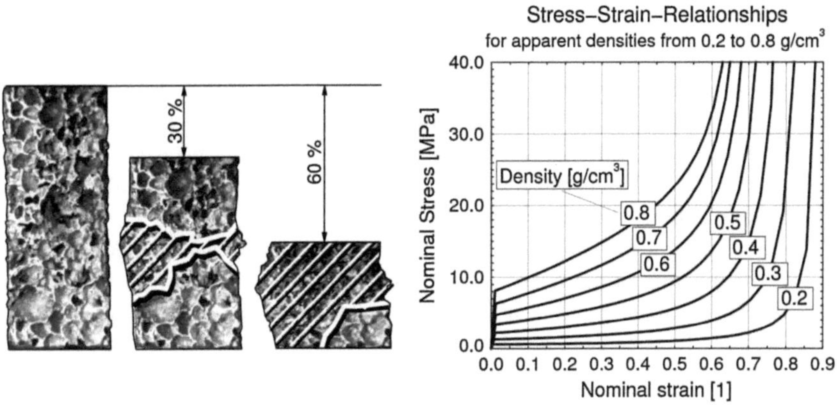

Fig. 9 A sample of a foam specimen in the unloaded state and at 30 and 60 % compressive nominal strain, respectively (*left*). Generic stress versus strain curves for different apparent aluminum foam densities are shown on the right (after [21])

The fact that the points do not sit on a single curve indicates that the yield surface does not have a circular cross-section in any deviatoric plane, $\sigma_m = \text{const}$. Figure 8 (right) shows such a cross section. Consequently, for an exact description of the yield surface, the von Mises equivalent stress σ_e and the mean stress σ_m are not sufficient and an additional measure such as the third invariant J_3 of the deviatoric stress tensor or the Lode angle (20) has to be supplied. Experimental evidence for an influence of the third invariant J_3 on the yield surface shape has been provided by [5, 6, 12].

3.2 Progressive Collapse and Densification

The plateau region of the uniaxial compressive stress-strain relationship is characterized by the successive collapse of layers of cells, starting from the weakest cell layer and spreading either into neighboring regions or other weak layers in different sections of the specimen. This process is indicated by the extent of the hatched regions in Fig. 9 (left) which mark the collapsed regions at two stages of compression of a particular sample of metallic foam. Generally, some amount of hardening can be expected, because the collapsed layers show high resistance to further compression and the remaining uncollapsed regions are stronger than the collapsed ones in their initial state. In addition, the metallic bulk material itself typically experiences strain hardening.

Increasing the load leaves only stiff and strong cells undamaged. In combination with the collapsed cells, which are nearly incompressible in relation to the undamaged ones, this means that the slope of the uniaxial compressive stress-strain relationship is getting steeper and steeper.

Finally, with the application of sufficiently high compressive stresses, all cells along the loading path are expected to collapse. This stage of uniaxial compression is called densification, because it is characterized by most of the void volume being squeezed out of the cellular structure and the mechanisms of the bending and stretching of cell struts and walls being replaced by the transfer of compressive forces along bridges of solid bulk material formed by the structural members of collapsed cells which are in contact with each other.

The following approximation of the nominal compressive densification strain ε_D as a linear function of the relative density $\rho_{rel} = \rho^*/\rho_S$, where ρ^* is the apparent density of the cellular material and ρ_S is the density of the bulk material, is suggested in [17]:

$$\varepsilon_D = 1 - 1.4\frac{\rho^*}{\rho_S} \tag{38}$$

The effective uniaxial compressive response of a cellular metal depends on

1. the relative density of the cellular metal,
2. the topology and the homogeneity of the cell structure,
3. the hardening behavior and the ductility of the bulk material.

The influence of the relative density on the effective uniaxial compressive stress-strain behavior was captured in the comprehensive study [21] including a large number of quasi-static compressive tests on Al99.5 foam specimens of varying density.

To express the dependency of the compressive stress-strain response of this particular material on its relative density, an analytical relationship between the nominal stress σ and the nominal strain ε proposed by Shim [31] was fitted to the experimentally obtained stress-strain curves. Below a collapse stress σ_0, and a corresponding collapse strain ε_0, respectively, the uniaxial compressive stress σ is assumed to be a linear function of the compressive strain ε. The plateau region and the densification regime are described by the superposition of a linear and an exponential function of the compressive strain, which are parameterized by shape parameters a, b, ε_0 and n according to the second line of the following equation:

$$\sigma(\varepsilon) = \sigma_0 \begin{cases} \dfrac{\varepsilon}{\varepsilon_0} & : \quad \varepsilon \le \varepsilon_0 \\ \exp\left(\dfrac{a(\varepsilon - \varepsilon_0)}{(a-\varepsilon)^n}\right) - b(\varepsilon - \varepsilon_0) & : \quad \varepsilon_0 < \varepsilon < a \end{cases} \tag{39}$$

Gradinger [21] derived the following relationships between the curve parameters and the apparent density ρ of the investigated material, which has to be inserted in $\left[\text{g/cm}^3\right]$:

$$\begin{aligned} a(\rho) &= -0.37004\rho + 1.0000 \\ b(\rho) &= 6.6964\rho^2 - 10.2790\rho + 2.3053 \\ \varepsilon_0(\rho) &= -0.0459\rho^2 + 0.0563\rho - 0.0055 \\ n(\rho) &= -1.3633\rho^2 + 1.2243\rho + 0.3321 \\ \sigma_0(\rho) &= 12.3430 \times \rho^{1.8807} \text{ [MPa]} \end{aligned} \tag{40}$$

The corresponding uniaxial compressive stress-strain curves are shown in Fig. 9, right. It is evident that the collapse of low density Al99.5 foam progressed along a long plateau regime of nearly constant compressive stress up to a nominal densification strain of nearly 90 %. In contrast, densification occurred between 60 and 70 % nominal compressive strains for aluminum foams of higher apparent density and the hardening modulus of these foams was considerably higher.

4 Constitutive Modeling of Cellular Metals

4.1 Introduction

Cellular materials have a complicated micro-structure which generally cannot be described in its entirety by testing and visualization methods. Nevertheless, the effective mechanical behavior of the cellular material is a consequence of the interaction of the deformation mechanisms of the countless structural members on the cellular level. Since it is not feasible to model the micro-structure, the mechanical response of the cellular metals has to be described in a macroscopic, averaged-out sense within the framework of the theory of plasticity. This approach requires that the overall dimensions of the structures made of foam are at least one or two orders of magnitude larger than the typical size of typical individual foam cells.

Constitutive laws for cellular metals have to primarily account for the fact that these materials can yield under purely hydrostatic stress states, and, consequently, can acquire volumetric plastic deformations when being loaded beyond the yield limit. This is clearly in contrast with the assumption of classical theory of plasticity for solid metals, that hydrostatic stresses — however high — will not cause plastic deformations, and that plastic flow does not result in changes of the volume.

Consequently, new constitutive laws had to be developed for cellular metals, and these constitutive theories will be the subject of this section. An early overview of constitutive laws for the simulation of metallic foams was compiled by Hanssen et al. [23, 24]. They compare constitutive formulations proposed by Schreyer et al. [28], Ehlers et al. [14, 15], Deshpande and Fleck [11], Miller [27], as well as one implemented in the finite element software Abaqus [7] and then proceed to validate constitutive formulations implemented in the finite element code `LS-DYNA` [22]. A general review of yield criteria for cellular materials was given in [1].

It contains a classification of several yield criteria according to their dependency (linear or quadratic) on the homogenized mean stress and on the homogenized von Mises equivalent stress. The subject of yielding of anisotropic cellular materials is also treated in considerable detail in this paper. Finally, a very comprehensive review of yield criteria and constitutive models for cellular metals is available in [30].

In the following sections, the most important constitutive models for cellular metals will be presented. This review is restricted to isotropic material behavior, because no complete constitutive model for anisotropic cellular metals was available. With the exception of the GAZT yield criterion (see Sect. 4.3), the derivation of which was found to be interesting with regard to transferring micro-mechanical considerations to the macro-mechanical level, only complete constitutive models including a plastic flow rule are considered.

The yield criteria $F_i(\sigma_e, p) = 0$ for the presented constitutive models all contain a term quadratic in the hydrostatic pressure p $(= -\sigma_m)$. They can be distinguished further into models for which the von Mises stress σ_e enters F in linear form [18, 27],

$$F_1(\sigma_e, p) = \frac{\sigma_e}{a_1} + \frac{p^2}{b_1} - 1 \tag{41}$$

or in quadratic form, with the yield surface being either symmetric about the origin when plotted in the (σ_e, p) plane [4, 11],

$$F_2(\sigma_e, p) = \frac{\sigma_e^2}{a_2} + \frac{p^2}{b_2} - 1 \tag{42}$$

or with a center that is offset along the p axis [7, 33],

$$F_3(\sigma_e, p) = \frac{\sigma_e^2}{a_3} + \frac{(p - p_0)^2}{b_3} - 1 \tag{43}$$

The yield functions for the constitutive models cited above will be presented in detail in Sects. 4.3 to 4.8. In addition, a more complex yield function which takes into account the third invariant J_3 of the stress deviator tensor [16] will be described in Sect. 4.9.

The constitutive models for cellular metals differ not only in terms of the formulation of their yield function F, but also in terms of their flow potential G. Several models assume associated plastic flow, i.e., $G = F$. Others define a non-associated flow potential, in most cases to allow for an independent calibration of the plastic Poisson's ratio. Lastly, the models differ in terms of the definition of the hardening variable(s) and the corresponding evolution laws. For more details the reader is referred to Sects. 4.4–4.9.

The treatment of the elastic part of the deformation of cellular metals is essentially the same across the constitutive theories presented here. Therefore, the following separate section is dedicated to this subject.

4.2 Linear Elastic Behavior

For the undeformed material, the strain tensor vanishes per definition, i.e., $\varepsilon_{ij} = 0$. Loading the material will induce stresses and strains. If the stress level is small, the deformation may remain purely elastic. Assuming small elastic deformations, the tensor of elastic strains $\boldsymbol{\varepsilon}_{\text{el}}$ can be related to the stress tensor $\boldsymbol{\sigma}$ by a linear relationship of the form

$$\boldsymbol{\sigma} = \mathbb{E}\boldsymbol{\varepsilon}_{\text{el}}, \qquad \text{or} \quad \sigma_{ij} = E_{ijkl}\varepsilon_{kl} \tag{44}$$

which is called Hooke's Law. It involves the fourth-order tensor of elasticity $\mathbb{E}$.

For the special case of an isotropic material, only two material parameters λ and μ called the Lamé parameters are necessary for the definition of the tensor of elasticity E_{ijkl}:

$$E_{ijkl} = \lambda\, \delta_{ij}\delta_{kl} + \mu\left(\delta_{ik}\delta_{jl} + \delta_{il}\delta_{jk}\right), \tag{45}$$

The Lamé coefficients λ and μ can be related to the Young's modulus E and the Poisson's ratio ν of the material:

$$\lambda = \frac{E\nu}{(1+\nu)(1-2\nu)}, \qquad \mu = \frac{E}{2(1+\nu)} \tag{46}$$

The Young's modulus E relates a uniaxial stress σ to the resulting strain ε in tension direction in a uniaxial tension test: $\sigma = E\varepsilon$.

While the theoretical description of elasto-plastic materials assumes the existence of an elastic deformation regime, it is very difficult to actually observe purely elastic behavior in experiments on cellular metals, especially under compression. The reason is that plastic deformations on the micro-level may appear at very low overall load levels, e.g., at the sample-test machine interface or around microstructural imperfections, even though these stresses are well below the limit or plateau stress of the materials. It is, therefore, often difficult to define a Young's modulus for cellular metals. In order to arrive at a well defined value for E, the unloading modulus is sometimes used, which is typically higher than the apparent modulus at the onset of loading.

The Poisson's ratio ν of an isotropic material is used in the relationship between the longitudinal stress σ_{11} and the transverse strains $\varepsilon_{22} = \varepsilon_{33} = -\nu\sigma_{11}/E$ in a uniaxial tension test. Because of the typically rough surfaces of cellular materials, it is very hard to measure the Poisson's ratio experimentally.

Another important elastic quantity is the shear modulus G, which relates the shear stress τ to the shear angle γ by $\tau = G\gamma$ in a shear test, and is defined as:

$$G = \frac{E}{2(1+\nu)} \tag{47}$$

Finally, the bulk modulus K couples the mean stress σ_{m} to the volumetric strain ε_{vol} according to $\sigma_{\text{m}} = K\varepsilon_{\text{vol}}$ and is given by

$$K = \frac{E}{3(1-2\nu)} \tag{48}$$

Hooke's law can be written in a very compact and useful form, when the deviatoric and the volumetric parts of the elastic strain tensor are inserted separately:

$$\boldsymbol{\sigma} = 2G\mathbf{e}^{(\mathrm{el})} + K\varepsilon_{\mathrm{vol}}^{(\mathrm{el})}\mathbf{I} \tag{49}$$

which implies for the deviatoric part $\mathbf{s}$ of the stress tensor that

$$\mathbf{s} = 2G\mathbf{e}^{(\mathrm{el})} \tag{50}$$

The elastic strain energy density W for an isotropic, linear elastic material can be specified in terms of the stress and strain tensors by:

$$W = \frac{1}{2}\boldsymbol{\sigma}:\boldsymbol{\varepsilon}^{(\mathrm{el})} = \frac{1}{2}\left[\mathbf{s} + \sigma_{\mathrm{m}}\mathbf{I}\right]:\left[\mathbf{e}^{(\mathrm{el})} + \frac{1}{3}\varepsilon_{\mathrm{vol}}^{(\mathrm{el})}\mathbf{I}\right] = \frac{1}{2}\left(\mathbf{s}:\mathbf{e}^{(\mathrm{el})} + \sigma_{\mathrm{m}}\varepsilon_{\mathrm{vol}}^{(\mathrm{el})}\right) \tag{51}$$

The last step in this equation uses the identity $\mathbf{I}:\mathbf{I} = 3$, as well as the fact that the double contraction of the deviatoric part of a second-order tensor and the second order tensor of unity gives zero, i.e., $\mathbf{s}:\mathbf{I} = \mathbf{I}:\mathbf{s} = 0$, and $\mathbf{e}^{(\mathrm{el})}:\mathbf{I} = \mathbf{I}:\mathbf{e}^{(\mathrm{el})} = 0$, because $\mathrm{tr}\,(\mathbf{s}) = 0$ and $\mathrm{tr}\left(\mathbf{e}^{(\mathrm{el})}\right) = 0$, respectively, compare Eq. (7). The last term in Eq. (51) indicates that σ_{m} and $\varepsilon_{\mathrm{vol}}$ are energetically conjugate. Using Eq. (50) it can be shown that

$$\frac{1}{2}\mathbf{s}:\mathbf{e}^{(\mathrm{el})} = \frac{1}{2}\sigma_{\mathrm{e}}\varepsilon_{\mathrm{e}}^{(\mathrm{el})} \tag{52}$$

which shows that σ_{e} and ε_{e} are also energetically conjugate.

Since the stress tensor $\boldsymbol{\sigma}$ and the strain tensor $\boldsymbol{\varepsilon}$ are directly related by Hooke's law, expressions for the strain energy density W can be derived which depend either only on the stress or only on the strain measures:

$$W = \frac{1}{2\bar{E}}\left(\sigma_{\mathrm{e}}^2 + \beta^2\sigma_{\mathrm{m}}^2\right) \tag{53}$$

and

$$W = \frac{\bar{E}}{2}\left(\left[\varepsilon_{\mathrm{e}}^{(\mathrm{el})}\right]^2 + \frac{1}{\beta^2}\left[\varepsilon_{\mathrm{vol}}^{(\mathrm{el})}\right]^2\right) \tag{54}$$

where two alternative elastic material parameters $\bar{E}$ and β are used:

$$\bar{E} = \frac{3E}{2(1+\nu)} \qquad \beta^2 = \frac{9(1-2\nu)}{2(1+\nu)} \tag{55}$$

Observing this, Chen and Lu [4] introduced a definition of a characteristic stress $\bar{\sigma}$,

$$\bar{\sigma}^2 = \sigma_e^2 + \beta^2 \sigma_m^2 \tag{56}$$

and a characteristic strain $\bar{\varepsilon}$,

$$\bar{\varepsilon}^2 = [\varepsilon_e]^2 + \frac{1}{\beta^2} [\varepsilon_{vol}]^2 \tag{57}$$

and used them in their definition of a constitutive model for metallic foam, see Sect. 4.6. Using (50) it can be shown that

$$W = \frac{1}{2} \bar{\sigma} \bar{\varepsilon},$$

i.e., that $\bar{\sigma}$ and $\bar{\varepsilon}$ are energetically conjugate, and that the relationship $\bar{\sigma} = \bar{E}\bar{\varepsilon}$ holds.

4.3 The Gibson-Ashby-Zhang-Triantafillou (GAZT) Model

An interesting early contribution to the description of the effective mechanical behavior of cellular materials has been made in [18]. Therein, a definition of a macroscopic limit surface is derived using exclusively generic micromechanical considerations and dimensional analysis.

First, a simple cubic unit cell is proposed, in which straight struts meet in vertices and corners at rectangular angles. The model struts have a length of l and a square cross-section with a side length of t. The relative density ρ_{rel}, which is the quotient $\rho_{rel} = \rho^*/\rho_S$ of the effective, homogenized density ρ^* and the density of the solid material ρ_S, can be expressed considering that the volume of a unit cell is $V^* = l^3$ and the volume of the solid phase is $V_S = C_1 l t^2$, with C_1 being a constant of proportionality which is related to the total length of all struts in the unit cell. With $\rho^* = \rho_S V_S / V^*$ the following relationship between the unit cell dimensions and the relative density is obtained:

$$\rho_{rel} = \frac{V_S}{V^*} = C_1 \left(\frac{t}{l}\right)^2 \tag{58}$$

Next, the plastic limit load under uniaxial macroscopic loading along the principal directions will be investigated. The macroscopic stress σ^* causes forces F in the cell struts, which are proportional to the application area, $F = C_2 \sigma^* l^2$, with C_2 being the corresponding constant of proportionality. The maximum bending moment in the struts is proportional to Fl. The plastic limit moment M_{pl} of a strut with a square cross-section is given by $M_{pl} = \sigma_{ys} t^3/4$, with σ_{ys} being the yield stress of the material, which is assumed to behave ideally plastic for this investigation. Once the bending moments M reach the plastic limit moment M_{pl}, plastic hinges start to form, and the cellular structure collapses. The corresponding uniaxial limit stress σ_{pl}^* can be found by solving $M(\sigma_{pl}^*)l = M_{pl}$ for σ_{pl}^*:

$$\sigma_{\text{pl}}^{*} = C_3 \,\frac{t^3}{l^3}\, \sigma_{\text{ys}} = C_3 \,{\rho_{\text{rel}}}^{3/2}\, \sigma_{\text{ys}} \tag{59}$$

Gibson et al. propose a value of $C_3 = 0.3$ for fitting this relationship to experimental data for the uniaxial plateau stress of foams.

Next, Gibson et al. recall that regular hexagonal 2D honeycomb structures under bi-axial loading ($\sigma_1 = \sigma_2$) do not experience bending moments in their cell walls. Instead, only normal section forces act in cell wall direction in these honeycombs. They investigate collapse of a 3D cellular material under hydrostatic loading by extending this finding to the tri-axial case, claiming that a hydrostatic macroscopic stress state will lead to purely axial compression or tension in the cell struts or walls. Ironically, this assumption is not true for their model microstructure, because it disregards the fact that any kind of normal loading on the unit cell produces bending moments in the struts perpendicular to the loading direction. These bending moments cannot be compensated by loads in the other principal directions.

Disregarding the bending stresses completely, axial stresses σ_{ax} can be obtained for the struts of the unit cell under a macroscopic mean stress σ_{m}. The total volume V_{S} of solid material in the unit cell is given by $V_{\text{S}} = l^3 \rho_{\text{rel}}$. This volume is now divided by three for obtaining an approximation of the volume of the group of struts that run into each of the three principal directions. The sum of the cross-sectional areas of these struts is given by $A = (V_{\text{S}}/3)/l$. For hydrostatic stress states, the macroscopic mean stress σ_{m} is acting on each face of the unit cube, resulting in a total normal traction force of $F_{\text{m}} = \sigma_{\text{m}} l^2$. Combining all of the above, the axial stress σ_{ax} for hydrostatic loading can be calculated as

$$\sigma_{\text{ax}} = \frac{F_{\text{m}}}{A} = \frac{\sigma_{\text{m}} l^2}{(l^3 \rho_{\text{rel}}/3)/l} = \frac{3\sigma_{\text{m}}}{\rho_{\text{rel}}} \tag{60}$$

Plastic collapse under hydrostatic loading occurs when $\sigma_{\text{ax}} = \sigma_{\text{ys}}$.

If the hydrostatic part of any macroscopic stress tensor is assumed to produce only normal stresses in the struts (or the cell walls) then the deviatoric stresses are expected to produce bending moments. Consequently, Gibson et al. describe the relationship between the von Mises stress σ_{e} and the average bending moment M in the struts as

$$M \propto l^3 \sigma_{\text{e}} = l^3 \sqrt{\frac{1}{2}\left[(\sigma_1 - \sigma_2)^2 + (\sigma_2 - \sigma_3)^2 + (\sigma_3 - \sigma_1)^2\right]} \tag{61}$$

The factor l^3 is based on the fact that traction forces $F \propto \sigma^* l^2$ and bending moments $M \propto Fl$.

For pure bending, the plastic limit moment of a strut is given by $M_{\text{pl}} = \sigma_{\text{ys}} t^3/4$. Superimposing an axial stress σ_{ax} on the strut reduces the limit moment, as can be seen from the following expression for M_{pl} which can be found after some re-arrangement:

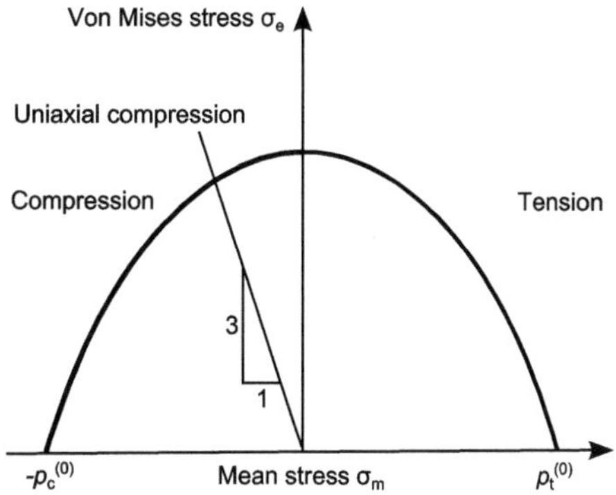

Fig. 10 Sketch of the GAZT yield surface as defined in Eq. (64) and [18]

$$M_{\mathrm{pl}} \propto \sigma_{\mathrm{ys}}\, t^3 \left[1 - \left(\frac{\sigma_{\mathrm{ax}}}{\sigma_{\mathrm{ys}}}\right)^2\right] \tag{62}$$

If the plastic limit moment, which is reduced by the axial stress due to the macroscopic hydrostatic stress according to Eq. (62), is equal to the bending moment induced by the macroscopic deviatoric stress, then the foam will collapse. The corresponding limit condition is found by inserting (60) into (62), considering that $t^3/l^3 = {\rho_{\mathrm{rel}}}^{3/2}$ and some rearranging:

$$\frac{\sigma_{\mathrm{e}}}{\sigma_{\mathrm{ys}}} = \pm\gamma\, {\rho_{\mathrm{rel}}}^{3/2} \left[1 - \left[\frac{3\sigma_{\mathrm{m}}^*}{\sigma_{\mathrm{ys}}\rho_{\mathrm{rel}}}\right]^2\right] \tag{63}$$

The constant γ is a new constant of proportionality, which can be approximated as $\gamma \approx 0.3$ for relevant relative densities ($\rho_{\mathrm{rel}} < 0.3$). The final form of the GAZT limit criterion for the plastic collapse of cellular materials under multiaxial loads is obtained by solving (59) for σ_{ys} using a factor of $C_3 = 0.3$ and inserting the result into (63) with $\gamma = 0.3$:

$$\pm \frac{\sigma_{\mathrm{e}}}{\sigma_{\mathrm{pl}}^*} + 0.81\rho_{\mathrm{rel}} \left(\frac{\sigma_{\mathrm{m}}^*}{\sigma_{\mathrm{pl}}^*}\right)^2 = 1 \tag{64}$$

The dependence on the von Mises stress σ_{e} is linear for this criterion, while the mean stress σ_{m} is squared. Figure 10 shows a generic plot of the yield surface defined by Eq. (64). The parameters $p_{\mathrm{c}}^{(0)}$ and $p_{\mathrm{t}}^{(0)}$ show the compressive and tensile hydrostatic pressures for initial yielding, respectively. Gibson et al. [18] provide similar derivations for failure criteria pertaining to brittle crushing in compression, fracture in tension and elastic buckling. They note, that the corresponding failure surfaces can intersect the failure surface (64) for plastic collapse, and limit the load-carrying capacity of the material further. They also suggest a possible extension of the theory to model failure in anisotropic foams.

4.4 The Miller Model

Miller [27] proposed a constitutive model for cellular materials which is specifically designed for being fitted to the following experimental test results: (a) the compressive and the tensile yield stresses under uniaxial loading conditions, (b) the uniaxial, compressive stress-strain response, and (c) the degree of lateral expansion in a uniaxial compression test.

The corresponding yield function F_{Miller} is an extension of the Drucker-Prager yield function (commonly used for modeling soil) by a term which is a multiple of the square of the hydrostatic pressure p. By adding this term, the plastic Poisson's ratio ν_{pl} can be varied independently of the uniaxial compressive and tensile yield stresses, something, which is not possible with the Drucker-Prager material model. The shape and the size of the yield surface are controlled by three parameters d, γ, and α, as can be seen from the definition of the yield function

$$F_{\text{Miller}} = \underbrace{\underbrace{\sigma_{\text{e}} - d}_{\text{v. Mises}} - \gamma p}_{\text{Drucker-Prager}} + \frac{\alpha}{d} p^2 \tag{65}$$

Braces in Eq. (65) indicate, which parts of Miller's yield function represent the simpler von Mises yield function for classical metal plasticity and the Drucker-Prager yield function, respectively. Figure 11 shows a sketch of the projection of the Miller yield surface $F_{\text{Miller}} = 0$ onto the von Mises stress versus mean stress plane.

Associated plastic flow is assumed and the increment $d\varepsilon_{ij}^{(\text{pl})}$ of the plastic strain tensor is, therefore, normal to the instantaneous yield surface during active yielding:

$$d\varepsilon_{ij}^{(\text{pl})} = d\lambda \, \frac{\partial F_{\text{Miller}}}{\partial \sigma_{ij}} \tag{66}$$

The definition of the yield function (65) implies, that the yield stresses under uniaxial tension and uniaxial compression differ. Denoting the compressive yield

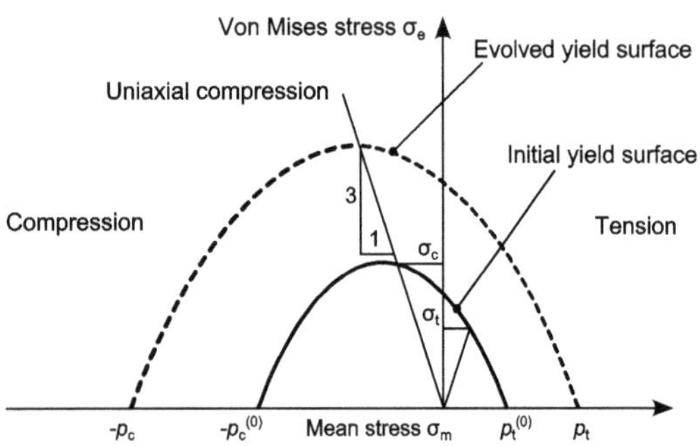

Fig. 11 Sketch of initial and hardened yield surfaces as predicted by the model proposed by Miller [27]

stress as σ_c leads to the definition of the invariants $\sigma_e = \sigma_c$ and $p = \sigma_c/3$, respectively, for uniaxial compression. Inserting these into (65) gives a quadratic equation for σ_c, with the solution[1]

$$\sigma_c = \frac{2d}{1 - \gamma/3 + \sqrt{(1-\gamma/3)^2 + 4\alpha/9}}, \tag{67}$$

which expresses the compressive uniaxial yield stress σ_c as a function of the yield surface shape parameters. For uniaxial tension, considering that $\sigma_e = \sigma_t$ and $p = -\sigma_t/3$ leads to a similar expression for the uniaxial tensile yield stress:

$$\sigma_t = \frac{2d}{1 + \gamma/3 + \sqrt{(1+\gamma/3)^2 + 4\alpha/9}} \tag{68}$$

The ratio β between the compressive uniaxial yield stress σ_c and the tensile uniaxial yield stress σ_t follows as

$$\beta = \frac{\sigma_c}{\sigma_t} = \frac{1 + \gamma/3 + \sqrt{(1+\gamma/3)^2 + 4\alpha/9}}{1 - \gamma/3 + \sqrt{(1-\gamma/3)^2 + 4\alpha/9}} \tag{69}$$

Since the ratio β can be determined from comparatively simple uniaxial compression and tension tests, it will later be useful for the calibration of the shape parameters of the yield function F_{Miller}.

The next constituent of Miller's material model is the description of the hardening behavior. Miller intends to separate the contribution of hardening of the cell wall material from the hardening which arises from the collapse of cells and the subsequent contact of cell walls. The latter effect is assumed to be a function of the logarithmic volumetric strain ε_{vol}, which relates the volume Δv of an infinitesimal volume element in the deformed configuration to the initial volume ΔV of the same volume element in the undeformed configuration:

$$\varepsilon_{\text{vol}} = \ln \frac{\Delta v}{\Delta V} \tag{70}$$

The stress response of the material in a uniaxial compression test is then described by the uniaxial compressive yield stress σ_c, which is defined as the product of a stress function $\bar{\sigma}_c(\bar{\varepsilon}_{\text{pl}})$, and the dimensionless function $\nu(\varepsilon_{\text{vol}})$, which depends only on the volumetric strain and is intended to describe the influence of densification:

$$\sigma_c = \bar{\sigma}_c(\bar{\varepsilon}_{\text{pl}})\, \nu(\varepsilon_{\text{vol}}) \tag{71}$$

[1] For obtaining this original form given in [27], this general relationship is helpful: $a - b = \frac{(a+b)(a-b)}{(a+b)} = \frac{a^2-b^2}{a+b}$.

In [27], $\nu(\varepsilon_{\text{vol}})$ assumes the value of unity for volumetric strains greater than a volumetric densification strain $\varepsilon_{\text{vol}}^{(\text{D})}(<0)$. If the volumetric strain falls below the corresponding densification strain, then $\nu(\varepsilon_{\text{vol}})$ grows rapidly to values much larger than one to model the increase of the stress levels caused by the loss of compliance due to contact of the cell struts or walls.

The function $\bar{\sigma}_{\text{c}}(\bar{\varepsilon}_{\text{pl}})$ depends on the equivalent plastic strain $\bar{\varepsilon}_{\text{pl}}$ and represents the initial mechanical response of the material and the plateau region of the stress-strain relationship. The equivalent plastic strain $\bar{\varepsilon}_{\text{pl}}$ controls the expansion of the yield surface, because it enters Eqs. (71) and (67), respectively, to give the yield function parameter d. The functions $\bar{\sigma}_{\text{c}}(\bar{\varepsilon}_{\text{pl}})$ and $\nu(\varepsilon_{\text{vol}})$ have to be chosen such that they fit the results of uniaxial compression tests.

To link the expansion of the yield surface under general stress states to the data obtained for the uniaxial case, the concept of accumulated plastic work is used. Herein, the increment of the plastic work given by the product of the increment of the equivalent plastic strain $d\bar{\varepsilon}_{\text{pl}}$ and the instantaneous yield stress σ_{c}, which would correspond to the accumulated plastic strain in a uniaxial compression test, is set equal to the actual increment of the plastic work in the volume element:

$$\sigma_{\text{c}}\, d\bar{\varepsilon}_{\text{pl}} = \sigma_{ij}\, d\varepsilon_{ij}^{(\text{pl})} \tag{72}$$

For the uniaxial compression test it follows that $d\bar{\varepsilon}_{\text{pl}} = d\varepsilon_{11}^{(\text{pl})}$, which is consistent with this definition.

The last item necessary for the calibration of Miller's constitutive theory is the plastic Poisson's ratio ν_{pl}, defined for a uniaxial compression test as the negative ratio of the plastic strains $\varepsilon_{22}^{(\text{pl})}$ in transverse direction and the plastic strains $\varepsilon_{11}^{(\text{pl})}$ in loading direction:

$$\nu_{\text{pl}} = \frac{-\varepsilon_{22}^{(\text{pl})}}{\varepsilon_{11}^{(\text{pl})}} = \frac{\varepsilon_{22}^{(\text{pl})}}{\bar{\varepsilon}_{\text{pl}}} \tag{73}$$

The right part of Eq. (73) follows from the fact that only $\sigma_{11} = -\sigma_{\text{c}} \neq 0$ in the compression test, and, therefore, the definition Eq. (72) of the equivalent plastic strain simplifies to $\sigma_{\text{c}}\bar{\varepsilon}_{\text{pl}} = \sigma_{11}\varepsilon_{11}^{(\text{pl})}$, leading to $\varepsilon_{11}^{(\text{pl})} = -\bar{\varepsilon}_{\text{pl}}$.

The increment of the plastic strain $d\varepsilon_{22}^{(\text{pl})}$ transverse to the loading direction 1 can be found by specializing the flow rule Eq. (66):

$$d\varepsilon_{22}^{(\text{pl})} = d\lambda\, \frac{\partial F_{\text{Miller}}}{\partial \sigma_{22}} \tag{74}$$

The increment of the plastic multiplier $d\lambda$ can be obtained by adapting the implicit definition (72) of the equivalent plastic strain increment $d\bar{\varepsilon}_{\text{pl}}$ to the conditions of uniaxial compression, where only $\sigma_{11} \neq 0$ and, therefore, only $d\varepsilon_{11}^{(\text{pl})}$ contributes to the increment of plastic work:

$$\sigma_c \, d\bar{\varepsilon}_{pl} = \sigma_{11} \, d\varepsilon_{11}^{(pl)} = \sigma_{11} d\lambda \, \frac{\partial F_{Miller}}{\partial \sigma_{11}} \tag{75}$$

Finally, the total plastic strain $\varepsilon_{22}^{(pl)}$ in the transverse direction can be derived from

$$\varepsilon_{22}^{(pl)} = \int_0^{\bar{\varepsilon}_{pl}} \frac{\sigma_c \, d\bar{\varepsilon}_{pl}}{\sigma_{11} \frac{\partial F_{Miller}}{\partial \sigma_{11}}} \frac{\partial F_{Miller}}{\partial \sigma_{22}} \tag{76}$$

For the derivation of the sub-expressions in Eq. (76), partial derivatives of the individual terms of the Miller yield function F_{Miller} are given below:

$$\frac{\partial \sigma_e}{\partial \sigma_{11}} = \frac{1}{2\sigma_e} [2\sigma_{11} - \sigma_{22} - \sigma_{33}] \tag{77}$$

$$\frac{\partial p}{\partial \sigma_{11}} = -\frac{1}{3} \tag{78}$$

$$\frac{\partial \left(p^2\right)}{\partial \sigma_{11}} = \frac{2}{9} (\sigma_{11} + \sigma_{22} + \sigma_{33}) \tag{79}$$

An auxiliary term

$$d_0 = \frac{1}{2} \left(1 - \gamma/3 + \sqrt{(1 - \gamma/3)^2 + 4\alpha/9}\right) \tag{80}$$

is now introduced to stay compatible with Miller. Note, that $\sigma_c = d/d_0$, compare Eq. (67). Inserting Eqs. (77) to (80) into the sub-expressions of Eq. (76) and considering, that for uniaxial compression $\sigma_{11} = -\sigma_c$, $\sigma_{22} = \sigma_{33} = 0$, and $\sigma_e = \sigma_c$ gives the intermediate results

$$\sigma_{11} \frac{\partial F_{Miller}}{\partial \sigma_{11}} = \sigma_c(\bar{\varepsilon}_{pl}, \varepsilon_{vol}) \left(1 - \frac{\gamma}{3} - \frac{2\alpha}{9d_0}\right) \tag{81}$$

$$\frac{\partial F_{Miller}}{\partial \sigma_{22}} = \frac{1}{2} + \frac{\gamma}{3} - \frac{2\alpha}{9d_0} \tag{82}$$

and, finally, an expression for the plastic Poisson's ratio ν_{pl} can be obtained as a function of the yield surface shape parameters:

$$\nu_{pl} = \frac{1/2 + \gamma/3 - 2\alpha/9d_0}{1 - \gamma/3 + 2\alpha/9d_0} \tag{83}$$

The purpose of this operation is to relate the yield surface shape parameters γ and α to the physically more meaningful ratio β between the compressive and the tensile uniaxial yield stresses, Eq. (69), and the plastic Poisson's ratio ν_{pl}:

$$\gamma = \frac{6\beta^2 - 12\beta + 6 + 9(\beta^2 - 1)/(1 + \nu_{pl})}{2(\beta + 1)^2} \tag{84}$$

$$\alpha = \frac{45 + 24\gamma - 4\gamma^2 + 4\nu_{pl}(2 + \nu_{pl})(-9 + 6\gamma - \gamma^2)}{16(1 + \nu_{pl})^2} \tag{85}$$

Miller implemented the constitutive law for metallic foams described above as a user-defined material subroutine (UMAT) in the commercial finite element code Abaqus. Details of the implementation are not given in [27]. It is important to note that β and ν_{pl}, and, consequently, α and γ are assumed to remain constant during plastic flow for the sake of simplicity. This assumption may not be valid for very large strains. Miller simulated compressive loading of a double notched specimen in plain strain, a Brinell hardness test, and an indentation of foam cores with aluminum face sheets investigating different model materials with varying yield surface parameters. In particular, he examined the influence of the plastic Poisson's ratio ν_{pl}.

A considerable part of the discussion in [27] is devoted to the comparison of the proposed material model to the GAZT model, which can be derived as a special case from the present model by choosing $\gamma = 0$, $\alpha = 0.81\rho_{rel}$, and $d = \sigma^*_{pl}$. Miller chose a relative density of $\rho_{rel} = 0.08$ and obtained a plastic Poisson's ratio $\nu_{pl} = 0.479$ by means of Eq. (83). Since this value is close to the limit value of $\nu_{pl} = 0.5$ for plastic incompressibiliy, he concludes that the GAZT material is nearly incompressible in uniaxial compression. This conclusion, however, is only valid under the assumption that the GAZT yield function is coupled with an associative flow rule. This assumption is not supported by the original paper [18] where no mention of the application of an associative flow rule is made.

4.5 The Deshpande-Fleck Foam Models

4.5.1 Introduction

Deshpande and Fleck [11] published data obtained by experiments, which subjected specimens of open and closed-cell metallic foams to axisymmetric compressive stress states, were presented. Based on the experimental findings, two constitutive models were developed, which will be presented in the following sections. The simpler of the two models, the so-called self-similar yield surface model, has influenced the simulation of components made from or containing metallic foams considerably, because it was implemented early in commercial finite element codes such as Abaqus.

The experimental work is equally impressive, because it comprises a method of probing the initial yield surfaces of metallic foams as well as determining the evolution of the yield surfaces under uniaxial and compressive loading. The hardening behavior under uniaxial compression and hydrostatic compression was also investigated. During the uniaxial compression tests, the diameter of the cylindrical specimens was measured in order to derive the plastic Poisson's ratio, altogether giving a reasonably complete picture of the behavior of the investigated metallic foams under multi-axial loading.

4.5.2 The Self-Similar Yield Surface Model

From their multi-axial compression experiments [11] Deshpande and Fleck concluded, that the investigated foams showed essentially isotropic mechanical behavior, and that the yield surfaces, which they probed with axisymmetric compressive stress states ranging from unixaxial compression to hydrostatic compression, could be approximated well by a yield function F_{DF},

$$F_{\mathrm{DF}} \equiv \hat{\sigma} - Y = 0, \tag{86}$$

where $\hat{\sigma}$ is a suitably defined equivalent stress and Y stands for the uniaxial yield strength. Since isotropic material behavior was considered, the equivalent stress $\hat{\sigma}$ was defined in terms of the von Mises stress σ_{e} and the mean stress σ_{m}:

$$\hat{\sigma} = \sqrt{\frac{1}{1+(\alpha/3)^2}\left(\sigma_{\mathrm{e}}^2 + \alpha^2\sigma_{\mathrm{m}}^2\right)} \tag{87}$$

This definition corresponds to an elliptical shape in a von Mises stress versus mean stress diagram, see Fig. 12. The aspect ratio of this ellipse is controlled by the shape parameter α. The limiting case of $\alpha = 0$ corresponds to the von Mises yield criterion, since $\hat{\sigma}$ becomes equal to σ_{e} for this case. Deshpande and Fleck report α values between 1.35 and 2.08 for the foams they investigated.

An associated flow rule was assumed, giving a direction of plastic flow $\dot{\boldsymbol{\varepsilon}}^{(\mathrm{pl})}$, which is normal to the yield surface:

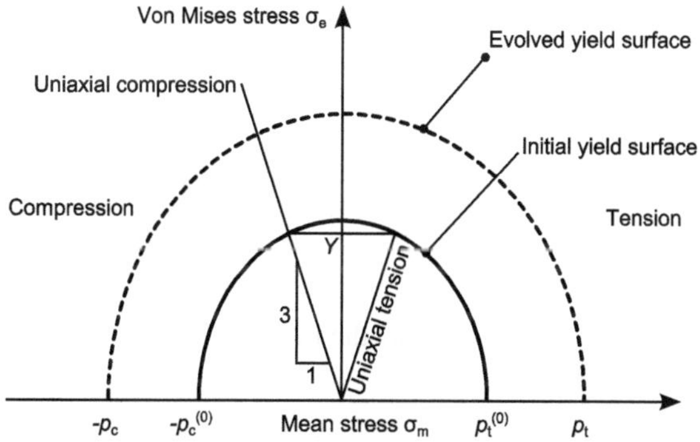

Fig. 12 Sketch of initial and hardened yield surfaces as predicted by the self-similar model proposed by Deshpande and Fleck in [11]

$$\dot{\boldsymbol{\varepsilon}}^{(\mathrm{pl})} = \dot{\lambda}\frac{\partial F_{\mathrm{DF}}}{\partial \boldsymbol{\sigma}} = \dot{\lambda}\frac{\partial \hat{\sigma}}{\partial \boldsymbol{\sigma}} \tag{88}$$

The plastic Poisson's ratio, which is defined as the negative ratio of the transverse logarithmic strain rate $\dot{\varepsilon}_{tt}^{(\mathrm{pl})}$ to the axial logarithmic strain rate $\dot{\varepsilon}_{xx}^{(\mathrm{pl})}$ in a uniaxial compression (or tension) test, can be calculated directly from the flow rule (88), with the help of Table 1 and noting that only $\sigma_{xx} \neq 0$:

$$\nu_{\mathrm{pl}} \equiv -\frac{\dot{\varepsilon}_{tt}^{(\mathrm{pl})}}{\dot{\varepsilon}_{xx}^{(\mathrm{pl})}} = -\frac{(\partial F_{\mathrm{DF}}/\partial \sigma_{tt})}{(\partial F_{\mathrm{DF}}/\partial \sigma_{xx})} = \frac{(1/2) - (\alpha/3)^2}{1 + (\alpha/3)^2} \tag{89}$$

Next, an equivalent strain rate $\dot{\hat{\varepsilon}}$ is introduced, which is the plastic work rate conjugate to the equivalent stress $\hat{\sigma}$:

$$\hat{\sigma}\dot{\hat{\varepsilon}} = \sigma_{ij}\dot{\varepsilon}_{ij}^{(\mathrm{pl})} \tag{90}$$

Inserting the flow rule (88) into (90) and solving for $\dot{\hat{\varepsilon}}$ reveals that the rate of this equivalent strain is, in fact, equivalent to the rate of the plastic multiplier $\dot{\lambda}$ in the flow rule (88):

$$\dot{\hat{\varepsilon}} = \underbrace{\frac{\sigma_{ij}}{\hat{\sigma}}\frac{\partial \hat{\sigma}}{\partial \sigma_{ij}}}_{=1}\dot{\lambda} = \dot{\lambda} \tag{91}$$

In (91), the application of Euler's theorem (10), which is justified by the fact that $\hat{\sigma}$ is homogeneous of degree one in σ_{ij}, is indicated. Inserting (91) into the plastic flow rule (88) gives

$$\dot{\boldsymbol{\varepsilon}}^{(\mathrm{pl})} = \dot{\hat{\varepsilon}}\frac{\partial F_{\mathrm{DF}}}{\partial \boldsymbol{\sigma}} = \dot{\hat{\varepsilon}}\frac{\partial \hat{\sigma}}{\partial \boldsymbol{\sigma}}. \tag{92}$$

In order to establish a connection between the equivalent plastic strain and the uniaxial yield stress Y, the consistency condition $\dot{F}_{\mathrm{DF}} = 0$ is written using the Jaumann stress rate $\breve{\sigma}_{ij}$:

$$\dot{F}_{\mathrm{DF}} = \frac{\partial F_{\mathrm{DF}}}{\partial \sigma_{ij}}\breve{\sigma}_{ij} + \frac{\partial F_{\mathrm{DF}}}{\partial Y}\dot{Y} = 0 \tag{93}$$

which can be simplified noting that $\partial F_{\mathrm{DF}}/\partial Y = -1$:

$$\dot{Y} = \frac{\partial F_{\mathrm{DF}}}{\partial \sigma_{ij}}\breve{\sigma}_{ij} \tag{94}$$

The hardening modulus H is defined as the ratio of the rate of the equivalent stress to the rate of the equivalent strain:

$$H \equiv \frac{\dot{\hat{\sigma}}}{\dot{\hat{\varepsilon}}} \tag{95}$$

By differentiating F_{DF} with respect to time, the identity $\dot{\hat{\sigma}} = \dot{Y}$ is obtained, which is inserted into Eq. (95) along with the expression for $\dot{Y}$ from (94) to obtain the relationship

$$\dot{\hat{\varepsilon}} = \frac{\dot{\hat{\sigma}}}{H} = \frac{\dot{Y}}{H} = \frac{1}{H}\frac{\partial F_{\mathrm{DF}}}{\partial \sigma_{ij}}\breve{\sigma}_{ij} \tag{96}$$

Finally, the flow rule (92) is rewritten using (96) to obtain the form of Eq. (4) in [11]:

$$\dot{\varepsilon}_{ij}^{(\mathrm{pl})} = \underbrace{\frac{1}{H}\frac{\partial F_{\mathrm{DF}}}{\partial \sigma_{kl}}\breve{\sigma}_{kl}}_{\dot{\lambda}=\dot{\hat{\varepsilon}}}\frac{\partial F_{\mathrm{DF}}}{\partial \sigma_{ij}} \tag{97}$$

Going back to the flow rule (92) and writing out the gradient of the equivalent stress $\partial\hat{\sigma}/\partial\boldsymbol{\sigma}$ gives:

$$\dot{\boldsymbol{\varepsilon}}^{(\mathrm{pl})} = \frac{1}{2\left[1+(\alpha/3)^2\right]}\frac{\dot{\hat{\varepsilon}}}{\hat{\sigma}}\left(3\mathbf{s} + \frac{2}{3}\alpha^2\sigma_{\mathrm{m}}\,\mathbf{I}\right) \tag{98}$$

The stress deviator $\mathbf{s}$ does not contribute to the volumetric plastic strain rate $\dot{\varepsilon}_{\mathrm{m}}$, because $\mathrm{tr}\,(\mathbf{s}) = 0$. Thus, the volumetric plastic strain rate $\dot{\varepsilon}_{\mathrm{m}}$ becomes:

$$\dot{\varepsilon}_{\mathrm{m}} = \mathrm{tr}\left(\dot{\boldsymbol{\varepsilon}}^{(\mathrm{pl})}\right) = \frac{\alpha^2}{1+(\alpha/3)^2}\,\dot{\hat{\varepsilon}}\,\frac{\sigma_{\mathrm{m}}}{\hat{\sigma}} \tag{99}$$

using $\mathrm{tr}\,(\mathbf{I}) = 3$. Inserting (98) into the definition (30) for the effective strain, the effective strain rate can be obtained in the following form:

$$\dot{\varepsilon}_{\mathrm{e}} = \frac{1}{1+(\alpha/3)^2}\,\dot{\hat{\varepsilon}}\,\frac{\sigma_{\mathrm{e}}}{\hat{\sigma}} \tag{100}$$

The equivalent plastic strain rate $\dot{\hat{\varepsilon}}$ can now be formulated in terms of the volumetric and the effective plastic strain rates by expressing σ_{m} in terms of $\dot{\varepsilon}_{\mathrm{m}}$ and $\dot{\hat{\varepsilon}}$ using Eq. (99) and by expressing σ_{e} in terms of $\dot{\varepsilon}_{\mathrm{e}}$ and $\dot{\hat{\varepsilon}}$ based on Eq. (100), followed by inserting the corresponding terms into the definition of the equivalent stress, Eq. (87). Solving the resulting equation for the equivalent strain rate $\dot{\hat{\varepsilon}}$ gives the following expression:

$$\dot{\hat{\varepsilon}}^2 = \left[1+\left(\frac{\alpha}{3}\right)^2\right]\left(\dot{\varepsilon}_{\mathrm{e}}^2 + \frac{1}{\alpha^2}\dot{\varepsilon}_{\mathrm{m}}^2\right), \tag{101}$$

which is useful for calculating the equivalent plastic strain rate $\dot{\hat{\varepsilon}}$ when $\dot{\varepsilon}_{\mathrm{e}}$ and $\dot{\varepsilon}_{\mathrm{m}}$ are known, e.g., from multiaxial experiments. In particular, the equivalent strain rate $\dot{\hat{\varepsilon}}$ enters the definition $H \equiv \dot{\hat{\sigma}}/\dot{\hat{\varepsilon}}$ of the hardening modulus H, as introduced in Eq. (95). The hardening modulus provides the connection between the rate form of the material law and the experimentally obtained stress-versus-strain data. Specifically, it is required for evaluating the flow rule (97).

Deshpande and Fleck proposed an expression for H which contained a maximum of information from their multi-axial tests. They chose H to depend on the current equivalent strain $\hat{\varepsilon}$ on the one hand and on the direction of the stress path on the other hand. The latter was described by $\eta \equiv |\sigma_m/\sigma_e|$ or the ratio $\sigma_e/\hat{\sigma}$.

The test setup for multi-axial testing consisted of a pressure cell filled with hydraulic fluid and holding a cylindrical foam specimen which was wrapped in insulating layers and further separated from the fluid by a rubber membrane. In this cell the specimen was subjected to a hydrostatic pressure p. In addition, a compressive force acting along the axis of the cylindrical specimen resulted in an additional axial compressive stress σ, bringing the total axial stress to $\sigma_{33} = -(p+\sigma)$. Under these axisymmetric loading conditions, the mean stress becomes $\sigma_m = -(p+\sigma/3)$, the von Mises stress is equal to $\sigma_e = |\sigma|$, and the equivalent stress $\hat{\sigma}$ can be calculated easily from Eq. (87).

During the multi-axial experiments, the axial plastic strain $\varepsilon_{33}^{(pl)}$ was measured. This strain can be inserted in Eq. (98) along with σ_m and σ_e, which gives an expression that can be solved for $\hat{\varepsilon}$:

$$\hat{\varepsilon} = \varepsilon_{33}^{(pl)} \frac{\sqrt{\left[1+\alpha^2\eta^2\right]\left[1+(\alpha/3)^2\right]}}{1+\alpha^2\eta/3} \tag{102}$$

The value for α can be found by fitting the yield surface to experimental data or by measuring the plastic Poisson's ratio. For several ratios η between the mean stress and the von Mises stress, the average tangent modulus $\bar{H}$ was then calculated as $\bar{H} = \Delta\hat{\sigma}/\Delta\hat{\varepsilon}$, where $\hat{\varepsilon}$ was a suitable initial increment of the equivalent plastic strain. This operation led to the result that the hardening modulus H can be approximated with reasonable accuracy as a linear function of the direction of the stress path, expressed by the ratio $\sigma_e/\hat{\sigma}$:

$$H = \left[\frac{\sigma_e}{\hat{\sigma}} h_\sigma + \left(1 - \frac{\sigma_e}{\hat{\sigma}}\right) h_p\right] \tag{103}$$

Since the hardening behavior of cellular metals is generally nonlinear, the coefficients h_σ and h_p depend on the instantaneous magnitude of the equivalent strain $\hat{\varepsilon}$. For uniaxial stress states, H becomes equal to h_σ, because $\hat{\sigma} = \sigma_e = |\sigma_{33}|$. The equivalent strain is equal to the absolute value of the axial plastic strain, $\hat{\varepsilon} = |\varepsilon_{33}^{(pl)}|$. Finally, the coefficient $h_\sigma = h_\sigma(\hat{\varepsilon})$ is equal to the slope of the curve of the Cauchy stress versus the logarithmic plastic strain in loading direction.

The second coefficient h_p can be found from evaluating a hydrostatic compression test in an analogous manner. The equivalent stress for pure hydrostatic loading by a pressure p follows from Eq. (87) by setting $\sigma_e = 0$:

$$\hat{\sigma}(p) = \frac{\alpha}{\sqrt{1+(\alpha/3)^2}}\, p \tag{104}$$

The rate of the equivalent strain can be found from Eq. (101) as a function of the rate of the logarithmic plastic volumetric strain $\dot{\varepsilon}_{\mathrm{m}}$ by setting $\dot{\varepsilon}_{\mathrm{e}} = 0$:

$$\dot{\hat{\varepsilon}} = -\frac{\sqrt{1 + (\alpha/3)^2}}{\alpha}\,\dot{\varepsilon}_{\mathrm{m}} \tag{105}$$

Finally, the coefficient h_p can be obtained:

$$h_p = \frac{-\alpha^2}{1 + (\alpha/3)^2}\,\frac{\dot{p}}{\dot{\varepsilon}_{\mathrm{m}}} \tag{106}$$

Deshpande and Fleck determined the coefficients h_σ and h_p from uniaxial compression and hydrostatic compression tests, respectively, for different materials. Then they were able to approximate the evolution of the hardening modulus H for intermediate stress ratios $\sigma_{\mathrm{m}}/\sigma_{\mathrm{e}}$, and to obtain the corresponding $\hat{\sigma}(\hat{\varepsilon})$ curves by integration. These were compared to the respective curves from axisymmetric compression tests with the same stress ratios. The agreement was very well for two different densities (8.4 and 16 % relative density, respectively) of Alporas foam, but the equivalent stresses were overestimated for equivalent strains above 0.3 for Duocel foam with 7 % relative density. In all cases, the response under purely hydrostatic compression and uniaxial compression could be fitted to the experimental results to the desired accuracy as part of the calibration process.

The self-similar yield surface model does not allow for a change of the shape of the yield surface and the yield surface remains centered in the $\sigma_{\mathrm{m}} - \sigma_{\mathrm{e}}$ plane. Differences in the hardening behavior under uniaxial loading and under hydrostatic loading, however, can be taken into account. A simplified version of this constitutive theory, which has strongly influenced modeling of cellular materials with commercial finite element codes, will be presented in the following sections.

4.5.3 The Simplified Self-Similar Yield Surface Model

For simplifying the calibration of material model input data, Deshpande and Fleck proposed a simplified version of the self-similar yield surface model, for which the hardening response does not depend any more on the behavior under hydrostatic pressure, but can be defined simply on the basis of the response under uniaxial compression and some assumptions regarding the shape of the yield surface.

Specifically, the function (103) for the hardening modulus H simplifies to

$$H(\hat{\varepsilon}) = h_\sigma(\hat{\varepsilon}) \tag{107}$$

This means that the hardening modulus corresponds to the slope of the uniaxial Cauchy stress versus logarithmic plastic strain curve. To fully define the material behavior, a suitable yield surface ellipticity α has to be chosen in addition to $H = H(\hat{\varepsilon})$.

In [11] it was pointed out that measured $\hat{\sigma}(\hat{\varepsilon})$ curves for different stress paths do not collapse into a single curve, as Eq. (107) would suggest. In particular, the hardening under hydrostatic compression loading is visibly more pronounced than the one under uniaxial compressive loading. Nevertheless, the simplified self-similar yield surface model has been implemented in many commercial finite element codes. One of these implementations will be described in the following.

4.5.4 The Abaqus Implementation of the Simplified Self-Similar Yield Surface Model

The simplified self-similar yield surface model developed by Deshpande and Fleck was originally implemented as a user material subroutine in the finite element code Abaqus [7] by Chen [3]. The definitions of the yield function and the equivalent stress were those of Eqs. (86) and (87) respectively.

Later, the material model became part of the standard material library of Abaqus, with the addition of allowing for an independent calibration of the evolution of the yield surface and the plastic Poisson's ratio. Recall, that for the original self-similar Deshpande/Fleck model, the plastic Poisson's ratio ν_{pl} and the yield surface shape parameter α are directly related by Eq. (89), owing to use of an associated flow rule.

In the Abaqus implementation, a non-associated plastic flow rule is introduced to allow for independent calibrations of the shape of the yield surface and the plastic Poisson's ratio. The corresponding flow potential introduces a new parameter β:

$$G_{\mathrm{CF}} = \sqrt{\sigma_{\mathrm{e}}^2 + \beta^2 p^2} \tag{108}$$

With the help of Table 1, the direction of plastic flow can be found:

$$\frac{\partial G_{\mathrm{CF}}}{\partial \boldsymbol{\sigma}} = \frac{1}{2G_{\mathrm{CF}}}\left(3\mathbf{s} - \frac{2}{3}\beta^2 p\mathbf{I}\right) \tag{109}$$

which is sufficient to calculate the plastic Poisson's ratio ν_{pl}:

$$\nu_{\mathrm{pl}} = \frac{1-(2/9)\beta^2}{2+(2/9)\beta^2} \tag{110}$$

and, in turns, the parameter β as a function of the plastic Poisson's ratio ν_{pl}:

$$\beta = \frac{3}{\sqrt{2}}\sqrt{\frac{1-2\nu_{\mathrm{pl}}}{1+\nu_{\mathrm{pl}}}} \tag{111}$$

For zero plastic expansion in the transverse direction during uniaxial compression, i.e., $\nu_{\mathrm{pl}} = 0$, a value of $\beta = 3/\sqrt{2} \approx 2.12$ follows immediately. For incompressible plastic flow, on the other hand, $\nu_{\mathrm{pl}} = 0.5$, $\beta = 0$, and $G_{\mathrm{CF}} = \sigma_{\mathrm{e}}$, which corresponds to the flow rule of classical (von Mises) plasticity.

In accordance with the simplified self-similar yield surface model proposed by Deshpande and Fleck, the hardening of the foam under multi-axial loading is described exclusively by relating the multi-axial strain state to the stress-strain relationship of a uniaxial compression test by means of an equivalent plastic strain $\bar{\varepsilon}^{(\mathrm{pl})}$.

The evolution of the equivalent plastic strain $\bar{\varepsilon}^{(\mathrm{pl})}$ is assumed to be governed by the principle of equivalent plastic work, leading to the relationship

$$\sigma_{\mathrm{c}}\dot{\bar{\varepsilon}}^{(\mathrm{pl})} = \boldsymbol{\sigma}:\dot{\boldsymbol{\varepsilon}}^{(\mathrm{pl})} \tag{112}$$

For uniaxial tension or compression, only the axial plastic strain contributes to the plastic work, and, consequently, the equivalent plastic strain is identical to the axial plastic strain for these loading cases.

Owing to its simplicity and its availability via the finite element code Abaqus this version of the simplified self-similar yield surface model has become quite popular.

4.5.5 The Differential Hardening Model

The experimental evidence collected by Deshpande and Fleck suggested that both the sizes and the shapes of the yield surfaces of isotropic metallic foams change during plastic loading depending on the direction of loading. In the context of their experiments, where a hydrostatic pressure load and an axial load were superimposed, the direction of loading was expressed by the quotient $|\sigma_{\mathrm{m}}/\sigma_{\mathrm{e}}|$.

Correspondingly, they suggested a quadratic yield function F_{DH} of the mean stress σ_{m} and the von Mises stress σ_{e} which separately takes into account the yield strength S under deviatoric loading and the yield strength P under hydrostatic loading:

$$F_{\mathrm{DH}} \equiv \left(\frac{\sigma_{\mathrm{e}}}{S}\right)^2 + \left(\frac{\sigma_{\mathrm{m}}}{P}\right)^2 - 1 \leq 0 \tag{113}$$

Under uniaxial loading, the yield surfaces of the metallic foams investigated in [11] changed predominantly in size, whereas their shapes remained unaffected. Under hydrostatic loading, on the other hand, the yield surfaces were found to become elongated along the hydrostatic axis. Hydrostatic loading affected the deviatoric yield strengths to varying degrees, depending on the material. For Duocel foam, the uniaxial compressive strength remained nearly constant during hydrostatic loading. For Alporas foam, on the other hand, hydrostatic loading consistently led to an increase in deviatoric strength. Uniaxial loading always increased both the deviatoric yield strength and the hydrostatic yield strength. Figure 13 schematically shows an initial yield surface and evolved yield surfaces for two different stress paths. The yield surfaces remain symmetric with respect to the σ_{e} axis, but their aspect ratio can change.

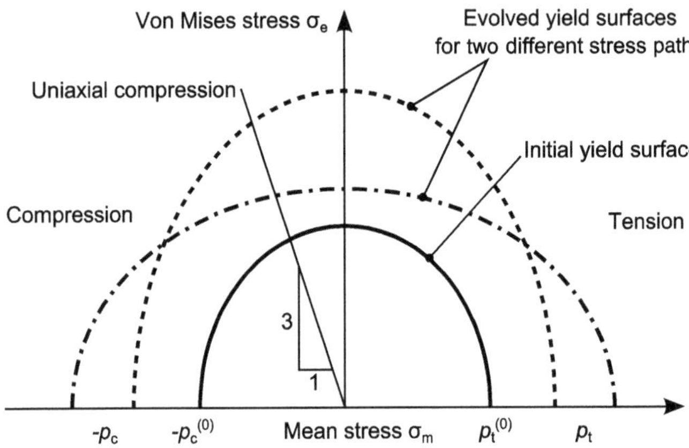

Fig. 13 Sketch of initial and hardened yield surfaces as predicted by the differential hardening model proposed by Deshpande and Fleck [11]

A general hardening rule was thus proposed, which relates the strength values P and S to kinematic quantities ε and γ according to the matrix equation

$$\begin{bmatrix} \dot{P} \\ \dot{S} \end{bmatrix} = \begin{bmatrix} h_{11} & h_{12} \\ h_{21} & h_{22} \end{bmatrix} \begin{bmatrix} \dot{\epsilon} \\ \dot{\gamma} \end{bmatrix} \tag{114}$$

The element h_{11} of the hardening matrix relates the rate $\dot{P}$ of the hydrostatic yield stress P to the rate $\dot{\epsilon}$ of a kinematic quantity ϵ, which is connected to the rate $\dot{\varepsilon}_{\mathrm{m}}$ of the mean strain. The element h_{22} connects the rate of the deviatoric yield strength $\dot{S}$ to the rate $\dot{\gamma}$ in a similar manner. The cross-hardening terms h_{12} and h_{21} connect the hydrostatic strength P and the deviatoric kinematic variable γ as well as the deviatoric strength S and the volumetric straining via ϵ, respectively.

Deshpande and Fleck found good agreement with their experimental data for a choice of the kinematic variables ϵ and γ which makes them plastic work conjugates of P and S, respectively:

$$P\dot{\epsilon} + S\dot{\gamma} = \sigma_{\mathrm{e}}\dot{\varepsilon}_{\mathrm{e}} + \sigma_{\mathrm{m}}\dot{\varepsilon}_{\mathrm{m}} \equiv \sigma_{ij}\dot{\varepsilon}_{ij}^{(\mathrm{pl})} \tag{115}$$

From this, the following relationships between ϵ and the mean plastic strain ε_{m} as well as between γ and the effective plastic strain ε_{e} can be established:

$$\dot{\epsilon} \equiv \frac{\sigma_{\mathrm{m}}}{P}\dot{\varepsilon}_{\mathrm{m}} \tag{116}$$

$$\dot{\gamma} \equiv \frac{\sigma_{\mathrm{e}}}{S}\dot{\varepsilon}_{\mathrm{e}} \tag{117}$$

In order to tie the coefficients $h_{\alpha\beta}$ of the hardening matrix to experimental results, the flow rule Eq. (97) has to be adapted in a suitable manner:

$$
\begin{aligned}
\boldsymbol{\varepsilon}^{(\mathrm{pl})} &= \frac{1}{H}\frac{\partial F_{\mathrm{DH}}}{\partial \boldsymbol{\sigma}}\left(\frac{\partial F_{\mathrm{DH}}}{\partial \boldsymbol{\sigma}}:\breve{\boldsymbol{\sigma}}\right) \\
&= \frac{1}{H}\left(\frac{3}{S^2}\mathbf{s}+\frac{2}{3}\frac{\sigma_{\mathrm{m}}}{P^2}\mathbf{I}\right)\left[\left(\frac{3}{S^2}\mathbf{s}+\frac{2}{3}\frac{\sigma_{\mathrm{m}}}{P^2}\mathbf{I}\right):(\dot{\mathbf{s}}+\dot{\sigma}_{\mathrm{m}}\mathbf{I})\right] \\
&= \frac{1}{H}\left(\frac{3}{S^2}\mathbf{s}+\frac{2}{3}\frac{\sigma_{\mathrm{m}}}{P^2}\mathbf{I}\right)\left[\frac{3}{S^2}\mathbf{s}:\dot{\mathbf{s}}+\frac{2}{P^2}\sigma_{\mathrm{m}}\,\dot{\sigma}_{\mathrm{m}}\right] \\
&= \frac{1}{H}\left(\frac{3}{S^2}\mathbf{s}+\frac{2}{3}\frac{\sigma_{\mathrm{m}}}{P^2}\mathbf{I}\right)\left[\frac{2\sigma_{\mathrm{e}}\dot{\sigma}_{\mathrm{e}}}{S^2}+\frac{2\sigma_{\mathrm{m}}\dot{\sigma}_{\mathrm{m}}}{P^2}\right]
\end{aligned} \tag{118}
$$

The expression in square brackets in Eq. (118) is a scalar multiplier, which can be derived using the relationships $\mathbf{s}:\mathbf{I}=0$ and $\mathbf{I}:\mathbf{I}=3$. For the reformulation of the last line of Eq. (118), the identity

$$
\dot{\sigma}_{\mathrm{e}} = \frac{3}{2}\frac{\mathbf{s}:\dot{\mathbf{s}}}{\sigma_{\mathrm{e}}} \tag{119}
$$

is used. Now, the plastic strain rate tensor $\dot{\boldsymbol{\epsilon}}^{(\mathrm{pl})}$ is split up into a volume-changing part $(\dot{\varepsilon}_{\mathrm{vol}}\mathbf{I})$ and a shape-changing part $\dot{\mathbf{e}}$:

$$
(\dot{\varepsilon}_{\mathrm{vol}}\mathbf{I}) = \frac{2}{3}\frac{1}{HP^2}\sigma_{\mathrm{m}}\mathbf{I}\left[\frac{2\sigma_{\mathrm{e}}\dot{\sigma}_{\mathrm{e}}}{S^2}+\frac{2\sigma_{\mathrm{m}}\dot{\sigma}_{\mathrm{m}}}{P^2}\right] \tag{120}
$$

$$
\dot{\mathbf{e}} = 3\frac{1}{HS^2}\mathbf{s}\left[\frac{2\sigma_{\mathrm{e}}\dot{\sigma}_{\mathrm{e}}}{S^2}+\frac{2\sigma_{\mathrm{m}}\dot{\sigma}_{\mathrm{m}}}{P^2}\right] \tag{121}
$$

For each part, the rate of the corresponding scalar quantity $\dot{\varepsilon}_{\mathrm{m}}$ and $\dot{\varepsilon}_{\mathrm{e}}$ (using $\sigma_{\mathrm{e}} = \sqrt{\frac{3}{2}\mathbf{s}:\mathbf{s}}$), respectively, is calculated:

$$
\dot{\varepsilon}_{\mathrm{m}} = \dot{\varepsilon}_{\mathrm{vol}}\,\mathrm{tr}(\mathbf{I}) = \frac{4\sigma_{\mathrm{m}}}{HP^2}\left[\frac{\sigma_{\mathrm{e}}\dot{\sigma}_{\mathrm{e}}}{S^2}+\frac{\sigma_{\mathrm{m}}\dot{\sigma}_{\mathrm{m}}}{P^2}\right] \tag{122}
$$

$$
\dot{\varepsilon}_{\mathrm{e}} = \sqrt{\frac{2}{3}\dot{\mathbf{e}}:\dot{\mathbf{e}}} = \frac{4\sigma_{\mathrm{e}}}{HS^2}\left[\frac{\sigma_{\mathrm{e}}\dot{\sigma}_{\mathrm{e}}}{S^2}+\frac{\sigma_{\mathrm{m}}\dot{\sigma}_{\mathrm{m}}}{P^2}\right] \tag{123}
$$

The hardening modulus H can be eliminated as an unknown by considering the consistency relation as an additional equation:

$$
\dot{F}_{\mathrm{DH}} = 0 = \frac{\partial F_{\mathrm{DH}}}{\partial \boldsymbol{\sigma}}:\breve{\boldsymbol{\sigma}} + \frac{\partial F_{\mathrm{DH}}}{\partial S}\dot{S} + \frac{\partial F_{\mathrm{DH}}}{\partial P}\dot{P} \tag{124}
$$

from which follows, after some rearranging, the hardening modulus H:

$$H \equiv \frac{4\sigma_m^2}{P^3}\left[h_{11}\frac{\sigma_m^2}{P^3} + h_{12}\frac{\sigma_e^2}{S^3}\right] + \frac{4\sigma_e^2}{S^3}\left[h_{21}\frac{\sigma_m^2}{P^3} + h_{22}\frac{\sigma_e^2}{S^3}\right] \tag{125}$$

Now, the hardening coefficients $h_{\alpha\beta}$ can be determined by specializing the expressions (122) and (123) for the mean and the effective strain rates as well as (125) for the hardening modulus H to conform to the loading and deformation conditions of the various experiments. For the hydrostatic compression test, $\sigma_m = -P$ and $\sigma_e = 0$ lead to $h_{11} = \dot{\sigma}_m/\dot{\varepsilon}_m$. At the same time, since $\dot{\gamma} = 0$ for hydrostatic compression of an isotropic material, h_{21} can be expressed as the slope of the deviatoric yield strength versus the plastic volumetric strain curve, $h_{21} = \dot{S}/|\dot{\varepsilon}_m|$, as follows from the second row of the hardening law (114) after setting $\dot{\gamma} = 0$ and applying Eq. (116).

For pure shear loading conditions, $\sigma_e = S$ follows from the yield condition (113) with $\sigma_m = 0$. If the shear loading is characterized by an applied shear stress τ, then $\sigma_e = |\tau|\sqrt{3}$. Thus, the deviatoric yield strength S is related to the shear yield stress τ_{yld} by $S = \tau_{yld}\sqrt{3}$. Specializing the expression (125) for the hardening modulus for conditions of pure shear and inserting into (123) gives the coefficient $h_{22} = \dot{\sigma}_e/\dot{\varepsilon}_e$. The rate $\dot{\varepsilon}_e$ of the effective strain follows from an engineering shear strain rate $\dot{\gamma}_{12}$ as $\dot{\varepsilon}_e = |\dot{\gamma}_{12}|/\sqrt{3}$. Finally, writing out the first row of the hardening law (114), setting $\dot{\epsilon} = 0$ for pure shear deformation, and applying Eq. (117) gives $h_{12} = \dot{P}/\dot{\varepsilon}_e$.

Deshpande and Fleck also propose a procedure for calibrating the coefficients $h_{\alpha\beta}$ from uniaxial compression tests instead of shear tests. Details can be found in the original paper [11]. Assuming that h_{11} depends only on ϵ, they determined the evolution of this coefficient from the hydrostatic compressive stress versus volumetric compressive strain curve. They found the ratio h_{21}/h_{11} to remain approximately constant and to assume values between 0.4 and 0.55 for the investigated materials. Furthermore, no significant cross-hardening between deviatoric straining $\dot{\gamma}$ and hydrostatic strength $\dot{P}$ was observed, i.e., $h_{12} = 0$. Finally, h_{22} was determined indirectly from the uniaxial and hydrostatic compressive test results.

The differential hardening model was calibrated from unixial compression and hydrostatic compression test results and subsequently used to predict the material behavior for intermediate proportional loading paths. The correlation to experimental results was slightly better than in the case of the self-similar model (see [11]), especially at higher strain levels, and in particular for a Duocel foam of 7.0% relative density. However, the higher accuracy can perhaps not compensate for the added complexity of this model, as Deshpande and Fleck remark themselves.

4.6 Chen and Lu Metallic Foam Material Model

Chen and Lu proposed a material model for metallic foams, which performed well when compared to the more complicated differential hardening model by Deshpande and Fleck (see Sect. 4.5.5). This model was a part of their phenomenological framework of constitutive modeling for elasto-plastic solids [4], which is based on the definitions of the characteristic stress $\bar{\sigma}$ in Eq. (56) and the characteristic strain $\bar{\varepsilon}$ in Eq. (57) as they were introduced in connection with isotropic elasticity in Sect. 4.2.

Then a 'stress potential' F_{CL} is introduced, which relates multi-axial stress states to the instantaneous total characteristic strain $\bar{\varepsilon}$:

$$F_{\mathrm{CL}} = \bar{\sigma}^2 + C(\bar{\varepsilon})\sigma_{\mathrm{m}}^2 - Y(\bar{\varepsilon}) = 0 \tag{126}$$

No distinction between elastic and plastic strain components is made by Chen and Lu. Therefore, all strain values mentioned in this section pertain to the total strain. Stresses and strains are connected by an associated flow rule $\dot{\epsilon}_{ij} = \dot{\lambda}\partial F_{\mathrm{CL}}/\partial\sigma_{ij}$ the factor $\dot{\lambda}$ being calculated from the consistency condition $\dot{F}_{\mathrm{CL}} = 0$.

The expressions for $\bar{\sigma}$ and $\bar{\varepsilon}$ simplify for some standard experimental settings. For a hydrostatic test with an applied hydrostatic stress of σ_{h} resulting in a volumetric strain of ε_{h}, one gets

$$\bar{\sigma} = \beta|\sigma_{\mathrm{h}}|, \qquad \bar{\varepsilon} = \frac{|\varepsilon_{\mathrm{h}}|}{\beta}, \tag{127}$$

with β as defined earlier in Eq. (55). For a strain ε_{u} caused by a uniaxial stress σ_{u}, the characteristic stress and strain measures become

$$\bar{\sigma} = \frac{\sqrt{9+\beta^2}}{3}|\sigma_{\mathrm{u}}|, \qquad \bar{\varepsilon} = \frac{3}{\sqrt{9+\beta^2}}|\varepsilon_{\mathrm{u}}|, \tag{128}$$

and, finally, for a shear stress τ inducing a engineering shear angle γ:

$$\bar{\sigma} = |\tau|\sqrt{3}, \qquad \bar{\varepsilon} = |\gamma|/\sqrt{3}. \tag{129}$$

The functions $C(\bar{\varepsilon})$ and $Y(\bar{\varepsilon})$ can be determined based on the available experimental results. For example, a characteristic stress-strain curve from a uniaxial compression test with a compressive applied stress of $\bar{\sigma}_{\mathrm{uc}}$ can be used in combination with a corresponding curve obtained in a hydrostatic compression test with an applied compressive hydrostatic stress of $\bar{\sigma}_{\mathrm{hc}}$. Noting that $|\sigma_{\mathrm{m}}| = \bar{\sigma}_{\mathrm{uc}}/\sqrt{9+\beta^2}$ in uniaxial compression and $|\sigma_{\mathrm{m}}| = \bar{\sigma}_{\mathrm{hc}}/\beta$ for hydrostatic compression, the following functions can be obtained after inserting the $\bar{\sigma}$ and σ_{m} values for the two experiments into the yield condition (126):

$$C(\bar{\varepsilon}) = \frac{\bar{\sigma}_{\mathrm{hc}}^2 - \bar{\sigma}_{\mathrm{uc}}^2}{\bar{\sigma}_{\mathrm{uc}}^2/(9+\beta^2) - \bar{\sigma}_{\mathrm{hc}}^2/\beta^2} \tag{130}$$

$$Y(\bar{\varepsilon}) = \bar{\sigma}_{\mathrm{hc}}^2\bar{\sigma}_{\mathrm{uc}}^2\frac{1/(9+\beta^2) - 1/\beta^2}{\bar{\sigma}_{\mathrm{uc}}^2/(9+\beta^2) - \bar{\sigma}_{\mathrm{hc}}^2/\beta^2} \tag{131}$$

For each level of equivalent strain $\bar{\varepsilon}$ the corresponding equivalent stress values $\bar{\sigma}_{uc}$ and $\bar{\sigma}_{hc}$ have to be entered in Eqs. (130) and (131) in order to get meaningful functions $C(\bar{\varepsilon})$ and $Y(\bar{\varepsilon})$.

Chen and Lu used these results to fit their functions C and Y to the experimental test results of Deshpande and Fleck [11]. They found that they could predict an intermediate stress versus strain path, which was not used in the calibration process, with the same accuracy as the one possible with the differential hardening model of Deshpande and Fleck [11], even though the calibration process is more simple for their model.

The constitutive equations of Chen and Lu are intended for predicting the multi-axial mechanical response of metallic foams in the compressive regime. Tension or elastic unloading are treated only rudimentarily. The missing separation of elastic and plastic strain contributions falls outside of the framework of classical plasticity theory. Nevertheless, the form of the stress potential F_{CL} and the definitions of the characteristic stress $\bar{\sigma}$ and the characteristic strain $\bar{\varepsilon}$ can inspire other constitutive approaches.

4.7 The Model by Zhang et al.

Even though it was originally developed for polymeric foam materials, the constitutive model proposed by Zhang et al. [33] deserves mention, because it combines a few interesting features which are similar to those already mentioned in previous chapters, and may have actually been providing the motivation for them.

The yield surface is defined in terms of squares of the hydrostatic pressure p and the von Mises equivalent stress σ_e and represents a half-ellipse when it is projected onto the p-σ_e plane (see Fig. 14):

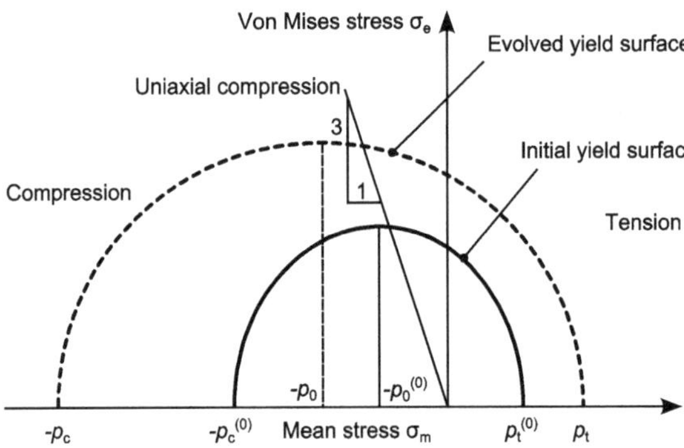

Fig. 14 Sketch of initial and hardened yield surfaces as predicted by the constitutive model proposed in [33]

$$F_{\text{Zhang}} = \frac{[p - p_0]^2}{a} + \frac{\sigma_e^2}{b} - 1 = 0 \tag{132}$$

The variable p_0 marks the center of the elliptical projection of the yield surface in the p-σ_e plane, compare Fig. 14. The variables a and b define the size and the shape of the yield surface, which are both allowed to change during the plastic deformation process. The terms $p_0(\varepsilon_{\text{vol}}^{(\text{pl})})$, $a(\varepsilon_{\text{vol}}^{(\text{pl})})$, and $b(\varepsilon_{\text{vol}}^{(\text{pl})})$ are all functions of the volumetric plastic strain $\varepsilon_{\text{vol}}^{(\text{pl})}$, i.e., volumetric hardening is assumed. This has the important consequence that the material's response during a shear deformation is perfectly plastic because no volumetric strain hardening occurs.

The present yield surface definition is quite similar to the one of the Abaqus crushable foam model, compare Eq. (135) in Sect. 4.8, but offers more degrees of freedom, because the tensile hydrostatic yield strength is unconstrained.

A non-associated flow rule equivalent to the one in the Abaqus implementation of the Deshpande and Fleck model, see Sect. 4.5.4, is part of the Zhang et al. model. The flow potential has the form

$$G_{\text{Zhang}} = \sqrt{\sigma_e^2 + \alpha p^2} \tag{133}$$

which is almost the same as the one in Eq. (108) with the exception that the constant α corresponds to β^2 in Eq. (108). Following the derivations in Sect. 4.5.4 the parameter α can be related to the plastic Poisson's ratio ν_{pl} via

$$\nu_{\text{pl}} = \frac{9 - 2\alpha}{2(9 + \alpha)} \quad \text{and} \quad \alpha = \frac{9(1 - 2\nu_{\text{pl}})}{2(1 + \nu_{\text{pl}})} \tag{134}$$

and the plastic Poisson's ratio can thus be set independently of other material parameters.

Zhang et al. [33] covers additional interesting aspects of constitutive modeling of polymeric foam materials, namely the numerical implementation of the material model in an explicit finite element code, as well as the mathematical description of the strain-rate sensitivity and and the temperature dependency of the material behavior.

4.8 The Abaqus Crushable Foam Model

The crushable foam model described in this section is a part of the standard material library of the commercial Finite Element Code Abaqus [7]. It became available for the modeling of cellular materials relatively early, at least compared to the other constitutive theories presented here. The crushable foam model was originally developed for PU foams, but an application and calibration of this model for metallic foams is possible, see, e.g., [29]. The model has the following characteristic features:

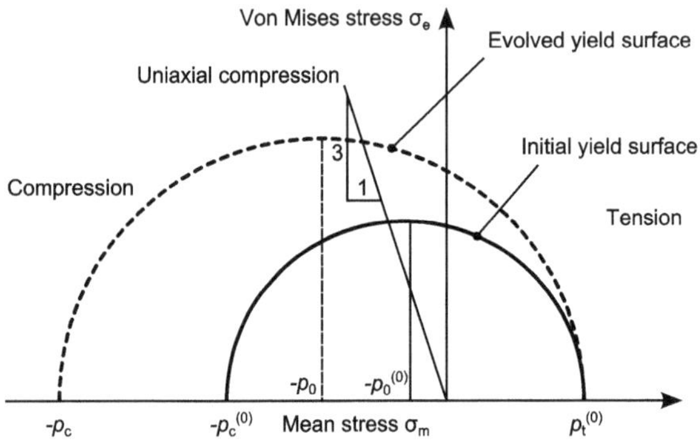

Fig. 15 Sketch of initial and hardened yield surfaces as predicted by the Abaqus crushable foam model [7]. The yield stress $p_\mathrm{t}^{(0)}$ for tensile hydrostatic loading and the aspect ratio of the ellipse remain constant

- a constant yield stress value $-p_\mathrm{t}^{(0)}$ for hydrostatic tension,
- a non-associated flow rule corresponding to a plastic Poisson's ratio of zero, and
- hardening, which is driven by the compressive volumetric plastic strain $-\varepsilon_\mathrm{vol}^{(\mathrm{pl})}$.

The yield surface of the crushable foam material law is defined implicitly by

$$F_\mathrm{CF}(\sigma_\mathrm{e}, p) = \sqrt{(\sigma_\mathrm{e})^2 + \alpha^2(p - p_0)^2} - B = 0 \tag{135}$$

which depends on two stress invariants, namely the von Mises equivalent stress σ_e and the hydrostatic pressure p $(= -\sigma_\mathrm{m})$, indicating that the material behavior is assumed to be isotropic. The yield surface can geometrically be described as a half-ellipse in the σ_e versus σ_m plane, see Fig. 15.

The vertex of the ellipse on the axis of positive mean stress is defined by the tensile hydrostatic yield stress $p_\mathrm{t}^{(0)}$. As a characteristic feature of the crushable foam model, the yield stress $p_\mathrm{t}^{(0)}$ is assumed to remain constant, even if the yield surface expands due to hardening. Consequently, this vertex stays fixed on the σ_m axis. The hydrostatic yield pressure p_c bounds the yield surface on the negative σ_m axis. It can evolve from an initial value $p_\mathrm{c}^{(0)}$ to greater values due to hardening.

The center of the yield surface corresponds to a pressure of

$$p_0 = \frac{p_\mathrm{c} - p_\mathrm{t}^{(0)}}{2} \tag{136}$$

The size of the yield surface is determined by the semi-axis length B in Eq. (135), whereas the parameter α controls the shape. While the yield surface is allowed to expand in the direction of positive hydrostatic pressure, the shape is defined to stay self-similar by keeping the parameter α constant. For fitting the shape parameter α to experimental results, two material parameters k and k_t can be determined:

$$k = \frac{\sigma_c^{(0)}}{p_c^{(0)}} \quad \text{and} \quad k_t = \frac{p_t^{(0)}}{p_c^{(0)}}. \tag{137}$$

The shape parameter (aspect ratio) α then follows as:

$$\alpha = \frac{3k}{\sqrt{(3k_t + k)(3 - k)}} \tag{138}$$

The radius B can be calculated by multiplying the half-length of the yield ellipse along the hydrostatic axis with the shape parameter α:

$$B = \alpha \frac{p_c + p_t^{(0)}}{2} \tag{139}$$

The fact that the tensile hydrostatic yield stress $-p_t^{(0)}$ is assumed to remain constant throughout the evolution of the yield surface and, therefore, during any plastic deformation process, takes into account that no hardening due to compaction of the cellular structure is possible under hydrostatic tension.

In order to fit the initial yield surface to experimental data, three points on the initial yield surface have to be determined experimentally. The following list gives candidates for these points:

- initial compressive yield stress in uniaxial compression $\sigma_c^{(0)}$,
- initial tensile yield stress in uniaxial tension $\sigma_t^{(0)}$,
- initial yield stress in simple shear $\tau_y^{(0)}$,
- initial yield stress under hydrostatic pressure $p_c^{(0)}$.

Not included in this list is the initial yield stress under hydrostatic tension $p_t^{(0)}$, because it is almost impossible to determine this value experimentally. It can, nevertheless, be calculated from other yield stresses, for example from $\sigma_c^{(0)}$, $\sigma_t^{(0)}$, and $p_c^{(0)}$, which are comparatively easy to find from experiments:

$$p_t^{(0)} = \frac{p_c^{(0)} \sigma_c^{(0)} \sigma_t^{(0)}}{3\, p_c^{(0)} (\sigma_c^{(0)} - \sigma_t^{(0)}) + \sigma_c^{(0)} \sigma_t^{(0)}} \tag{140}$$

According to [7] the yield strength in hydrostatic tension $p_t^{(0)}$ has to be expected to be considerably lower than the inital yield strength in hydrostatic compression $p_c^{(0)}$, resulting in ratios $p_t^{(0)}/p_c^{(0)} = 0.05$ to 0.10.

While the shape of the yield surface remains constant, its size increases with progressing plastic deformation. The hardening behavior is controlled by a plastic strain measure, which, for this model, is equal to the compressive volumetric strain $-\varepsilon_{vol}^{(pl)}$. The hardening behavior can be extracted from the stress-versus-strain-results of a uniaxial compression test assuming zero plastic Poisson's ratio. In this case, the uniaxial plastic strain is equal to the volumetric plastic strain.

The assumption of vanishing plastic deformation transverse to any given loading direction, expressed by $\nu_{\mathrm{pl}} = 0$, is, indeed, a feature of the crushable foam material model. It is based on experimental observations, which indicate that cellular materials do not tend to deform significantly in the lateral direction when they are tested in a uniaxial compressive test. Preventing plastic flow normal to the loading direction is achieved by choosing an appropriate non-associated flow rule of the form

$$d\boldsymbol{\varepsilon} = d\bar{\varepsilon}^{(\mathrm{pl})} \frac{\partial G_{\mathrm{CF}}}{\partial \boldsymbol{\sigma}} \tag{141}$$

containing the increment of the equivalent plastic strain $d\bar{\varepsilon}^{(\mathrm{pl})}$, which will be investigated in more detail below, and the flow potential function G_{CF}, which is defined in terms of the stress tensor invariants σ_{e} and p as:

$$G_{\mathrm{CF}} = \sqrt{(\sigma_{\mathrm{e}})^2 + \frac{9}{2} p^2} \tag{142}$$

It can be shown that this corresponds to the equivalent tensorial form

$$G_{\mathrm{CF}} = \sqrt{\frac{3}{2} \boldsymbol{\sigma} : \boldsymbol{\sigma}} \tag{143}$$

which can be differentiated with respect to the coordinates of the stress tensor to obtain the direction of plastic flow for the crushable foam model:

$$\frac{\partial G_{\mathrm{CF}}}{\partial \boldsymbol{\sigma}} = \frac{3\boldsymbol{\sigma}}{2G_{\mathrm{CF}}} \tag{144}$$

Equation (144) indicates a direction of plastic flow which is identical to the stress direction for radial paths in stress space. This means, that loading in any principal direction does, per definition, not cause any plastic deformation in the other principal directions.

Finally, the increment of the equivalent plastic strain $d\bar{\varepsilon}^{(\mathrm{pl})}$ follows from Eq. (141) if the scalar product of both sides with the stress tensor is formed:

$$\boldsymbol{\sigma} : d\boldsymbol{\varepsilon}^{(\mathrm{pl})} = \boldsymbol{\sigma} : \left(d\bar{\varepsilon}^{(\mathrm{pl})} \frac{\partial G_{\mathrm{CF}}}{\partial \boldsymbol{\sigma}} \right) = d\bar{\varepsilon}^{(\mathrm{pl})}\, \boldsymbol{\sigma} : \frac{3\boldsymbol{\sigma}}{2G_{\mathrm{CF}}} = d\bar{\varepsilon}^{(\mathrm{pl})}\, G_{\mathrm{CF}} \tag{145}$$

from which the increment of the equivalent plastic strain can be expressed as

$$d\bar{\varepsilon}^{(\mathrm{pl})} = \frac{\boldsymbol{\sigma} : d\boldsymbol{\epsilon}^{(\mathrm{pl})}}{G_{\mathrm{CF}}} \tag{146}$$

While the definition of the crushable foam yield surface, which relates the stress tensor invariants σ_{e} and p to each other in the form of the equation of an ellipse, is not unusual for cellular materials, the treatment of the hydrostatic tensile yield strength as being a constant, and the fact that radial loading does not cause any transverse plastic deformation, are unique characteristics of the crushable foam model.

4.9 The Ehlers Model for Cellular Metals

Ehlers and Müllerschön adapted a constitutive model, which was orgininally developed for porous and granular porous media, to represent plastic yielding of metal foam [14, 16]. The yield function

$$F_{\text{Ehlers}} = \sqrt{\left[J_2\left(1+\gamma\frac{J_3}{(J_2)^{3/2}}\right)^m + \frac{1}{2}\alpha I_1^2 + \delta^2 I_1^4\right]} + \beta I_1 + \varepsilon I_1^2 - \kappa = 0 \quad (147)$$

depends on the first invariant $I_1 = 3\sigma_\text{m} = -3p$ of the stress tensor, the second invariant $J_2 = \frac{1}{2}\mathbf{s}{:}\mathbf{s} = \sigma_\text{e}^2/3$ of the stress deviator tensor $\mathbf{s}$ and on the third invariant $J_3 = \det(\mathbf{s})$ of $\mathbf{s}$.

The characteristic which sets apart this material model from the others presented so far is its dependence on the third invariant J_3, which indicates a shape of the yield surface cross-section in the deviatoric plane, which is not circular, but rather triangular with rounded corners. The shape of the deviatoric cross-section is controlled by the parameters γ and m in (147). By setting $\gamma = 0$ the influence of the third deviatoric invariant is eliminated and the rounded triangular shape of the yield surface cross sections in the deviatoric plane changes into a circular shape.

For this material model, a non-associated flow rule is proposed with a plastic flow potential G_{Ehlers}, which is presented here in the form given in [13]:

$$G_{\text{Ehlers}} = \sqrt{\left[\psi_1 J_2 + \frac{1}{2}\alpha I_1^2 + \delta^2 I_1^4\right]} + \psi_2\beta I_1 + \varepsilon I_1^2 \quad (148)$$

In an earlier publication [14] the parameters ψ_i were set to unity, $\psi_1 = \psi_2 = 1$, and the plastic potential is fully defined by the parameters describing the yield surface.

The Ehlers model has the largest number of parameters required for defining the shape of the initial yield surface, namely seven. For hardening along arbitrary strain paths, these parameters may even be history-dependent. This makes material calibration somewhat complicated, as a large number of different tests (uniaxial compression/tension, axial-symmetric compression and biaxial loading in [16]) and a least-square fitting process are required. For an AlSi7Mg foam with an average apparent density of 0.3 g/cm^3 (produced by Hydro Aluminium) the following parameters are given for the initial yield surface in [16]: $\alpha = 0.0196$, $\beta = 0.07$, $\gamma = 1.4$, $\delta = 0.0176\,\text{MPa}^{-1}$, $\varepsilon = 0.00196\,\text{MPa}^{-1}$, $\kappa = 2.02\,\text{MPa}$, $m = 0.61$. Figure 16 shows a rendering of the yield surface in the principal stress space. By projecting the yield surface along the direction of the hydrostatic axis, the non-circular deviatoric

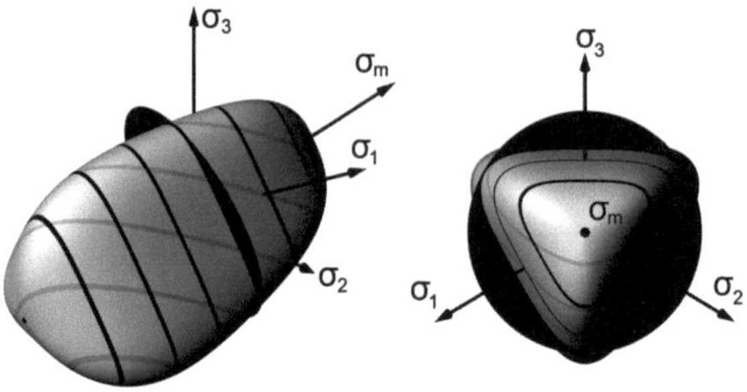

Fig. 16 Rendering of the Ehlers et al. yield surface in principal stress space using material parameters from [16]. The black contour lines indicate levels of constant mean stress in intervals of 1 MPa. The black disc represents the π plane and has a radius corresponding to $\sigma_e = 4.5$ MPa. The grey contour lines indicate cross-sections with constant σ_3

cross-sections can be made clearly visible, see Fig. 16 (right). The quality of the fit of the yield surface(s) to the experimental yield points is illustrated in [12] where both yield surface shapes which deviate from simple ellipsoidal shapes and triangular cross-sections in the deviatoric plane are well documented.

5 Discussion and Conclusions

Existing material laws for metallic foams are adequate for modeling simple deformation histories and predominantly radial stress paths. More complex mechanical processes such as successive perpendicular loading will require more sophisticated modeling techniques to account for the anisotropy caused by changing loading or deformation directions, see Deshpande and Fleck [11] and Hanssen [24].

Perhaps the most restricting feature about the available foam models is the absence of appropriate fracture models. Metal foams are very prone to softening and premature failure under tensile stress states, the overall behavior being governed by progressive failure of the cell walls in this regime. Since fracture is often observed in real-world applications of metal foams, the introduction of corresponding simulation methods is necessary.

A correct calibration of the material parameters based on experimental data is indispensable for the success of any finite element simulation. In most cases, a uniaxial compressive test is the minimum requirement for this calibration. For more sophisticated constitutive models additional data on the yield surface shape and the hardening behavior has to be provided by the user. This requires information about the mechanical behavior of the material under multi-axial loading conditions. Since multi-axial experimental data is scarce the user must rely on appropriate assumptions, which can be derived either from micro-mechanical finite element simulations or via parameter identification techniques (that is, by minimizing the discrepancy between simulations and experimental results, e.g., a sphere indentation test). The more parameters enter any given constitutive law, the more complicated the calibration process may become. As a consequence, comparatively simple constitutive laws such as the simplified self-similar yield surface model, compare Sect. 4.5.3, see more frequent application, and are often sufficient if the loading conditions are simple enough.

When cellular metals and metallic foams will see more widespread use, and safety-relevant applications such as crash-protection will require more accurate material models, then further advances in the constitutive modeling of cellular metals are to be expected. This contribution hopefully provides the reader with a good starting point for further developments.

Acknowledgments The author would like to thank Lorna J. Gibson, Ronald E. Miller, Vikram S. Deshpande, and Wolfgang Ehlers for helpful discussions regarding their respective publications. The author also wants to thank Maria Steininger for her help with obtaining some of the publications cited in this article.

References

1. Abrate, S.: Criteria for yielding or failure of cellular materials. J. Sandw. Struct. Mat. **10**(1), 5–51 (2008)
2. Bitsche, R.: Space-filling polyhedra as mechanical models for solidified dry foams. Vienna University of Technology, Vienna, Diploma thesis (2005)
3. Chen, C.: Manual for a UMAT user subroutine. Technical Report CUED/C-MICROMECH/TR.4, Department of Engineering, Cambridge University, Cambridge (1998)
4. Chen, C., Lu, T.J.: A phenomenological framework of constitutive modelling for incompressible and compressible elasto-plastic solids. Int. J. Sol. Struct. **37**, 7769–7786 (2000)
5. Combaz, E., Bacciarini, C., Charvet, R., Dufour, W., Dauphin, F., Mortensen, A.: Yield surface of polyurethane and aluminium replicated foam. Acta Mater **58**, 5168–5183 (2010)
6. Combaz, E., Bacciarini, C., Charvet, R., Dufour, W., Mortensen, A.: Multiaxial yield behaviour of Al replicated foam. J. Mech. Phys. Solids **59**, 1777–1793 (2011)
7. Dassault Systèmes (2012) Abaqus 6.12 Theory Manual. Dassault Systèmes Simulia Corp., Providence, Rhode Island.
8. Daxner, T., Denzer, R., Böhm, H.J., Rammerstorfer, F.G., Maier, M.: Simulation des elasto-plastischen Verhaltens von Metallschaum mit Hilfe von 2D und 3D Einheitszellen-Modellen. Mater-wiss u Werkst-techn **31**, 447–450 (2000)
9. Daxner, T., Böhm, H.J., Rammerstorfer, F.G.: Numerical investigation of local yielding in metallic foams. In: Banhart, J., Fleck, N.A., Mortensen, A. (eds.) Cellular Metals: Manufacture, pp. 413–418. Properties, Applications, Verlag MIT, Berlin (2003)
10. Daxner, T., Bitsche, R., Böhm, H.: Space-filling polyhedra as mechanical models for solidified dry foams. Mater. Trans. **47**(9), 2213–2218 (2006)
11. Deshpande, V.S., Fleck, N.A.: Isotropic constitutive models for metallic foams. J. Mech. Phys. Solids **48**, 1253–1283 (2000)
12. Droste, A.: Beschreibung und Anwendung eines elastisch-plastischen Materialmodells mit Schädigung für hochporöse Metallschäume. Ph.D. thesis, University of Stuttgart, Stuttgart (2002)
13. Ehlers, W., Avci, O.: Stress-dependent hardening and failure surfaces of dry sand. Int. J. Numer. Anal. Meth. Geomech. **37**, 787–809 (2013)

14. Ehlers, W., Droste, A.: A continuum model for highly porous aluminium foam. Techn. Mech. **19**(4):341–350 (1999a)
15. Ehlers, W., Droste, A.: FE simulation of metal foams based on the macroscopic approach of the theory of porous media. In: Banhart, J., Ashby, M.F., Fleck, N.A. (eds.) Metal Foams and Porous Metal Structures, pp. 299–302. Verlag MIT, Bremen (1999b)
16. Ehlers, W., Müllerschön, H., Klar, O.: On the behaviour of aluminium foams under uniaxial and multiaxial loading. In: Banhart, J., Ashby, M.F., Fleck, N.A. (eds.) Metal Foams and Porous Metal Structures, pp. 255–262. Verlag MIT, Bremen (1999)
17. Gibson, L.J., Ashby, M.F.: Cellular Solids: Structure and Properties, 2nd edn. Cambridge University Press, Cambridge, New York (1997)
18. Gibson, L.J., Ashby, M.F., Zhang, J., Triantafillou, T.C.: Failure surfaces for cellular materials under multiaxial loads—I. Modelling. Int J Mech Sci **31**(9), 635–663 (1989)
19. Gong, L., Kyriakides, S.: Compressive response of open cell foams. Part II: Initiation and evolution of crushing. Int. J. Sol. Struct. **42**(5–6):1381–1399 (2005)
20. Gong, L., Kyriakides, S., Triantafyllidis, N.: On the stability of Kelvin cell foams under compressive loads. J. Mech. Phys. Solids **53**, 771–794 (2005)
21. Gradinger, R.C.: Das mechanische Verhalten von Aluminiumschaum bei Druck—und Crushbelastung – Experimente und numerische Simulation. Vienna University of Technology, Vienna, Diploma thesis (1997)
22. Hallquist, J.O.: LS DYNA Theoretical Manual. Livermore Software Technology Corporation, Livermore, CA (1998)
23. Hanssen, A.G.: Structural crashworthiness of aluminium foam-based components. Ph.D. thesis, Norges Tekniske Høgskole, Trondheim, Norway (2000)
24. Hanssen, A.G., Hopperstad, O.S., Langseth, M., Ilstad, H.: Validation of constitutive models applicable to aluminium foams. Int. J. Mech. Sci. **44**(2), 359–406 (2002)
25. Laroussi, M., Sab, K., Alaoui, A.: Foam mechanics: nonlinear response of an elastic 3D-periodic microstructure. Int. J. Sol. Struct. **39**(13–14), 3599–3623 (2002)
26. Lubliner, J.: Plasticity Theory. Macmillan Publishing Company, New York (1990)
27. Miller, R.E.: A continuum plasticity model for the constitutive and indentation behaviour of foamed metals. Int. J. Mech. Sci. **42**, 729–754 (2000)
28. Schreyer, H.L., Zuo, Q.H., Maji, A.K.: An anisotropic plasticity model for foams and honeycomb. J. Eng. Mech. ASCE **120**(9), 1913–1930 (1994)
29. Seitzberger, M., Rammerstorfer, F.G., Degischer, H.P., Gradinger, R.: Crushing of axially compressed steel tubes filled with aluminium foam. Acta Mechanica **125**, 93–105 (1997)
30. Shahbeyk, S.: Yield/failure criteria, constitutive models, and crashworthiness applications of metal foams. In: Dukhan, N. (ed.) Metal Foams: Fundamentals and Applications, pp. 131–214. DEStech Publications, Lancaster, Pennsylvania (2013)
31. Shim, V.P.W., Tay, B.Y., Stronge, W.J.: Dynamic crushing of strain-softening cellular structures—a one-dimensional analysis. J. Eng. Mat. Tech. ASME **112**, 398–405 (1990)
32. Todt, M.: Long wave instabilities in periodic structures. Vienna University of Technology, Vienna, Diploma thesis (2008)
33. Zhang, J., Kikuchi, N., Li, V., Yee, A., Nusholtz, G.: Constitutive modeling of polymeric foam material subjected to dynamic crash loading. Int. J. Impact. Eng. **21**(5), 369–386 (1998)

Transmission Conditions for Thin Elasto-Plastic Pressure-Dependent Interphases

Gennady Mishuris, Wiktoria Miszuris, Andreas Öchsner and Andrea Piccolroaz

Abstract A thin soft elasto-plastic interphase between two different media is under consideration. The intermediate layer is assumed to be of infinitesimal thickness and is modeled by nonlinear transmission conditions which incorporate the elasto-plastic material behavior of the layer. The case of pressure-independent (von Mises) as well as pressure-dependent yield condition is theoretically treated. Finite element analysis of a bimaterial structure with such an imperfect elasto-plastic interface (von Mises) shows the efficiency of the approach and illustrates some restrictions of its application.

Keywords Interface · Inhomogeneous · Nonlinear · Deformation theory · von Mises material · Drucker-Prager material

G. Mishuris (✉) · W. Miszuris · A. Piccolroaz
Department of Mathematics and Physics, IMPACS, Penglais, Aberystwyth University, Aberystwyth, Ceredigion SY23 3BZ, UK
e-mail: ggm@aber.ac.uk

W. Miszuris
e-mail: wim@aber.ac.uk

A. Öchsner
Griffith University, Griffith School of Engineering, Southport 4222, Australia

A. Öchsner
The University of Newcastle, Callaghan, NSW 2308, Australia
e-mail: andreas.oechsner@gmail.com

A. Piccolroaz
University of Trento, via Mesiano 77, 38123 Trento, Italy
e-mail: andrea.piccolroaz@ing.unitn.it

H. Altenbach and A. Öchsner (eds.), *Plasticity of Pressure-Sensitive Materials*, Engineering Materials, DOI: 10.1007/978-3-642-40945-5_4,

1 Introduction

Thin interphases appearing in dissimilar bodies such as composite structures with adhesively bonded materials may influence significantly the whole spectrum of structural parameters: strength, dynamics, fracture, lifetime, and so on. Recently, significant efforts have been done to clarify various phenomena connected with the so-called elastic imperfect interface approach. It consists of replacing the real thin interphase between two different materials by an infinitesimal layer of zero thickness. This layer is then modeled by special transmission conditions which incorporate information about geometrical and mechanical properties of the thin interphase. At first, such proposed conditions were based on phenomenological approaches and have been sufficiently exploited (see [5, 8, 9] among others and the respective references). Later, various imperfect transmission conditions have been evaluated by asymptotic methods in [2, 4, 10, 18] for different types of interfaces and materials. The accurate asymptotic behavior of solutions of interface crack problems at the imperfect interface formulation have been investigated in [1, 23, 24] where it has been shown that the behavior may be very complicated and essentially depending on the material and geometrical properties of the imperfect interfaces. Possible error estimates and ranges of the edge zone effects connected with utilization of the imperfect models have been discussed in detail in [25, 26] by finite element analysis. This short review shows that *elastic* imperfect interfaces have been intensively investigated in different directions.

However, thin elasto-*plastic* interfaces play even a more important role in real applications [31] and results which are obtained up to now are absolutely insufficient and are mainly concentrated on problems of thin plastic interphases between stiff adherends [15, 17]. It is known in the context of thin structural adhesives that the mechanical response is nonlinear over a large range of strain [32] and that adhesives are sensitive to the hydrostatic pressure [22]. Nevertheless, many finite element simulations are restricted to pressure-independent yield conditions. This may result from the fact that these pressure-independent yield conditions (e.g. von Mises) are widely available in commercial codes and that only a very low number of input parameters are required. Sophisticated yield conditions which describe the adhesive behavior more realistically are not a priory available in commercial codes and must be implemented by the user [11]. Even having pressure-dependent yield conditions implemented or available in computational codes, there is still the remaining problem that in modern technology very thin adhesive layers (thin films) are used [33]. This fact makes it difficult to perform numerical calculations using the finite element method, since the need to build a complicated mesh of fundamentally different sized elements can lead to a loss in accuracy, unstable calculations and even loss of convergence [12]. In such a case, the so-called imperfect interface approach and the corresponding transmission conditions may result in new element formulations where the entire nonlinear behavior can be represented by a single element over the interphase thickness [30]. First attempts to incorporate plastic material behavior in the derivation of transmission conditions were done in [27] for the plane strain case and the von Mises yield condition. This chapter tries to progress this work and provides the theoretical foundation for transmission conditions for pressure-dependent interphases.

2 Transmission Conditions for Thin Soft Interphases

2.1 Problem Formulation

Here we will discuss how to evaluate transmission conditions for a thin soft elastic inhomogeneous interphase which is situated between two other elastic materials.

Consider three different elastic materials with Hooke's laws given by the equations:

$$\sigma_{ij}^{(l)} = 2\mu^{(l)}\varepsilon_{ij}^{(l)} + 3\lambda^{(l)}\varepsilon_{\rm m}^{(l)}\delta_{ij}, \quad \mathbf{x} \in \Omega_l, \quad l = 1, 2, 3. \tag{1}$$

Here $\sigma_{ij}^{(l)}$, $\varepsilon_{ij}^{(l)}$ are components of the stress and strain tensors, while $\mu^{(l)}$ and $\lambda^{(l)}$ are Lamé's parameters of the respective material ($l = 1, 2, 3$). Here and in what follows we will use the standard notation for the hydrostatic part of the tensors:

$$\sigma_m = \frac{1}{3}\sigma_{kk}, \quad \varepsilon_m = \frac{1}{3}\varepsilon_{kk}. \tag{2}$$

One can equivalently use any other pair out of the material parameters E, ν, G and K (Young's modulus, Poisson's ratio, shear modulus and bulk modulus, respectively):

$$E = \frac{\mu(3\lambda + 2\mu)}{\lambda + \mu}, \quad \nu = \frac{\lambda}{2(\lambda + \mu)}, \quad G = \lambda, \quad K = \lambda + \frac{2\mu}{3}. \tag{3}$$

We assume in our analysis that the material of the interphase ($l = 2$) is inhomogeneous, thus

$$\mu^{(2)} = \mu^{(2)}(\mathbf{x}), \quad \lambda^{(2)} = \lambda^{(2)}(\mathbf{x}), \tag{4}$$

and the material properties depend, whatever the reason is, on the position of the point in the interphase. For simplicity (in fact, this assumption will never be used in the further analysis) we assume that the materials of the adherends are homogeneous. All the elastic materials under considerations are assumed to be compressible. Thus, $0 \le \nu^{(l)} < 1/2 (l = 1, 3)$ and for any $\mathbf{x} \in \Omega_2$ there exists some value ν_0 such that

$$0 \le \nu^{(2)}(\mathbf{x}) \le \nu_0 < 1/2. \tag{5}$$

In absence of body forces, the balance equation reads

$$\operatorname{div} \boldsymbol{\sigma}^{(l)} = 0, \quad \mathbf{x} \in \Omega_l, \quad l = 1, 2, 3, \tag{6}$$

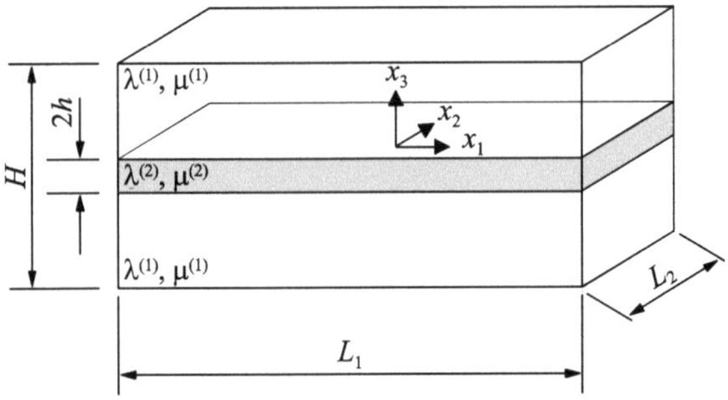

Fig. 1 General representation of the thin soft interphase (in *grey*)

where $\boldsymbol{\sigma}$ is the stress tensor. The strain tensor components are computed in the standard way:

$$\varepsilon_{ij} = \frac{1}{2}\left(\frac{\partial u_i}{\partial x_j} + \frac{\partial u_j}{\partial x_i}\right), \tag{7}$$

where $u_j (j = 1, 2, 3)$ are the components of the displacements vector, $\mathbf{u}$.

Substituting Eq. (7) into the Hooke's law (1) and then into the balance equation (6), one receives the Navier-Lamé equation:

$$\text{grad}\left((\lambda^{(l)} + \mu^{(l)})\text{div}\right)\mathbf{u}^{(l)} + (\nabla \cdot \mu^{(l)}\nabla)\mathbf{u}^{(l)} = 0, \quad \mathbf{x} \in \Omega_l, \tag{8}$$

where the material constants can be taken outside the operators in case of the adherends due to the assumptions (Fig. 1).

We assume that the interphase of constant thickness $2h$ is situated along the (x_1, x_2) plane. Then, the planes $x_3 = \pm h$ define the interphase boundaries. Along the boundaries, the following perfect (ideal) transmission conditions are satisfied:

$$\mathbf{u}^{(2)} = \mathbf{u}^{(1)}, \quad \sigma_{x_3}^{(2)} = \sigma_{x_3}^{(1)}, \quad x_3 = h, \tag{9}$$

$$\mathbf{u}^{(2)} = \mathbf{u}^{(3)}, \quad \sigma_{x_3}^{(2)} = \sigma_{x_3}^{(3)}, \quad x_3 = -h, \tag{10}$$

where $\sigma_{x_3}^{(l)}$ is the vector of stress acting on the interphase boundaries.

We will assume in the following that there exists a small parameter

$$\varepsilon \ll 1, \tag{11}$$

in the problem under consideration. This parameter controls on the one hand the thickness of the intermediate layer:

$$h = \varepsilon h_*, \tag{12}$$

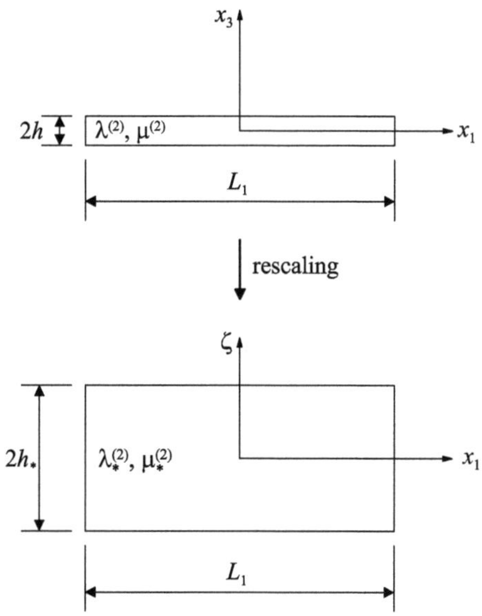

Fig. 2 Cross section of the interface according to the schematic specimen shown in Fig. 1

and, on the other hand, defines the softness of the interphase material:

$$E^{(2)}(\mathbf{x}) = \varepsilon E_*(\mathbf{x}). \tag{13}$$

Here we additionally assume that all material parameters $E_*(\mathbf{x})$ and $E^{(l)}$, $(l = 1, 3)$ are of the same order of magnitude regardless of the position of the point $\mathbf{x} \in \Omega_2$ inside the interphase. Note that under assumption that the Poisson's ratio is well separated from 1/2 ($\nu^{(2)} = \nu_* < 1/2$) and both Lamé's parameters are positive and exhibit the same asymptotic behavior:

$$0 < \mu^{(2)}(\mathbf{x}) = \varepsilon \mu_*(\mathbf{x}), \quad 0 < \lambda^{(2)}(\mathbf{x}) = \varepsilon \lambda_*(\mathbf{x}). \tag{14}$$

2.2 *Evaluation of the Transmission Conditions for Inhomogeneous Linear Interphases*

To derive the transmission conditions, let us rescale the x_3 independent variable inside the interphase in the following fashion (Fig. 2)

$$x_3 = \varepsilon \zeta, \quad \zeta \in (-h_*, h_*). \tag{15}$$

The Navier-Lamé equation (8) in the new independent variables $\bar{\mathbf{x}} = (x_1, x_2, \zeta)$ will take the following form:

$$\mathscr{L}_0\bar{\mathbf{u}} + \varepsilon\mathscr{L}_1\bar{\mathbf{u}} + \varepsilon^2\mathscr{L}_2\bar{\mathbf{u}} = 0, \quad \bar{\mathbf{x}} \in \Omega_2, \tag{16}$$

where $\bar{\mathbf{u}} = \bar{\mathbf{u}}(\bar{\mathbf{x}}) \equiv \bar{\mathbf{u}}^{(2)}(\mathbf{x})$, and the first two differential operators $\mathscr{L}_j$ are defined as follows

$$\mathscr{L}_0 = \frac{\partial}{\partial\zeta}\bar{\mathbb{A}}\frac{\partial}{\partial\zeta}, \quad \bar{\mathbb{A}} = \begin{pmatrix} \bar{\mu} & 0 & 0 \\ 0 & \bar{\mu} & 0 \\ 0 & 0 & \bar{\lambda}+2\bar{\mu} \end{pmatrix},$$

$$\mathscr{L}_1 = \begin{pmatrix} 0 & 0 & \frac{\partial}{\partial x_1}(\bar{\lambda}+\bar{\mu})\frac{\partial}{\partial\zeta} \\ 0 & 0 & \frac{\partial}{\partial x_2}(\bar{\lambda}+\bar{\mu})\frac{\partial}{\partial\zeta} \\ \frac{\partial}{\partial\zeta}(\bar{\lambda}+\bar{\mu})\frac{\partial}{\partial x_1} & \frac{\partial}{\partial\zeta}(\bar{\lambda}+\bar{\mu})\frac{\partial}{\partial x_2} & 0 \end{pmatrix},$$

while $\mathscr{L}_2$ is of dimension 2×2 analog to the original 3×3 operator with zeros completing its third row and third column. The Lamé's constants $\bar{\lambda}$ and $\bar{\mu}$ may vary essentially along the x_3-direction inside the interphase:

$$\bar{\mu}(x_1, x_2, \zeta) = \mu_*(x_1, x_2, \varepsilon\zeta), \quad \bar{\lambda}(x_1, x_2, \zeta) = \lambda_*(x_1, x_2, \varepsilon\zeta), \quad |\zeta| < h_*. \tag{17}$$

The traction operator can be written in the following manner:

$$\sigma_{x_3}^{(2)} = \mathscr{M}_0\bar{\mathbf{u}} + \varepsilon\mathscr{M}_1\bar{\mathbf{u}}, \quad \zeta = \pm h_*, \tag{18}$$

where

$$\mathscr{M}_0 = \bar{\mathbb{A}}\frac{\partial}{\partial\zeta}, \quad \mathscr{M}_1 = \begin{pmatrix} 0 & 0 & \bar{\mu}\frac{\partial}{\partial x_1} \\ 0 & 0 & \bar{\mu}\frac{\partial}{\partial x_2} \\ \bar{\lambda}\frac{\partial}{\partial x_1} & \bar{\lambda}\frac{\partial}{\partial x_2} & 0 \end{pmatrix}.$$

Finally, we can compute the components of the strain tensor $\bar{\varepsilon}(\bar{\mathbf{x}}) = \varepsilon^{(2)}(\mathbf{x})$ as

$$\bar{\varepsilon}_{ij} = \frac{1}{2}\left(\frac{\partial u_i}{\partial x_j} + \frac{\partial u_j}{\partial x_i}\right), \quad i, j = 1, 2,$$

$$\bar{\varepsilon}_{i3} = \frac{1}{2}\left(\frac{1}{\varepsilon}\frac{\partial u_i}{\partial\zeta} + \frac{\partial u_3}{\partial x_i}\right), \quad i = 1, 2, \quad \bar{\varepsilon}_{33} = \frac{1}{\varepsilon}\frac{\partial u_3}{\partial\zeta}. \tag{19}$$

Following the standard asymptotic procedure (see for example [28]), we will search for the solution inside the interphase in the form:

$$\bar{\mathbf{u}}(\bar{\mathbf{x}}) = \mathbf{v}_0(\bar{\mathbf{x}}) + \varepsilon\mathbf{v}_1(\bar{\mathbf{x}}) + \varepsilon^2\mathbf{v}_2(\bar{\mathbf{x}}) + \cdots. \tag{20}$$

We are only interested in the leading term of the solution $\mathbf{w} = \mathbf{v}_0$ which, as it follows from Eq. (16), should satisfy the equation

$$\mathscr{L}_0\mathbf{w} = 0, \quad |\zeta| < h_*. \tag{21}$$

Its solution can be computed by direct integration:

$$\bar{\mathbb{A}}(x_1, x_2, \zeta)\frac{\partial}{\partial \zeta}\mathbf{w}(x_1, x_2, \zeta) = \mathbf{a}(x_1, x_2), \tag{22}$$

and

$$\mathbf{w}(x_1, x_2, \zeta) = \int_{-h_*}^{\zeta} \bar{\mathbb{A}}^{-1}(x_1, x_2, t)dt \;\cdot \mathbf{a}(x_1, x_2) + \mathbf{b}(x_1, x_2), \quad |\zeta| < h_*, \tag{23}$$

where the unknown vector-functions $\mathbf{a}(x_1, x_2)$ and $\mathbf{b}(x_1, x_2)$ are immediately determined from the transmission conditions (10) under consideration of Eq. (18):

$$\mathbf{b}(x_1, x_2) = \mathbf{u}^{(3)}(x_1, x_2), \quad \mathbf{a}(x_1, x_2) = \sigma_{x3}^{(3)}(x_1, x_2).$$

As a result, the leading term of the solution inside the interphase ($-h_* < \zeta < h_*$) takes the following form:

$$\mathbf{w}(x_1, x_2, \zeta) = \int_{-h_*}^{\zeta} \bar{\mathbb{A}}^{-1}(x_1, x_2, t)dt \;\cdot \sigma_{x3}^{(3)}(x_1, x_2) + \mathbf{u}^{(3)}(x_1, x_2)\,. \tag{24}$$

Moreover, the stress vector

$$\sigma_{x3}^{(2)}(x_1, x_2, \zeta) \equiv \sigma_{x3}^{(2)}(x_1, x_2) = \sigma_{x3}^{(3)}(x_1, x_2) \tag{25}$$

does not depend on the independent variable ζ. Note that we still have not used the transmission conditions along the top boundary of the interphase ($\zeta = h_*$) which also has to be valid. This leads us to the following two imperfect transmission conditions which have to be satisfied:

$$\sigma_{x3}^{(1)}(x_1, x_2) = \sigma_{x3}^{(3)}(x_1, x_2), \tag{26}$$

and

$$\mathbf{u}^{(1)}(x_1, x_2) - \mathbf{u}^{(3)}(x_1, x_2) = \int_{-h_*}^{h_*} \bar{\mathbb{A}}^{-1}(x_1, x_2, t)dt \;\cdot \sigma_{x3}^{(3)}(x_1, x_2), \tag{27}$$

or in the original coordinates

$$\mathbf{u}^{(1)}(x_1, x_2) - \mathbf{u}^{(3)}(x_1, x_2) = \int_{-h}^{h} \mathbb{A}^{-1}(x_1, x_2, x_3)dx_3 \;\cdot \sigma_{x3}^{(3)}(x_1, x_2), \tag{28}$$

where:

$$\mathbb{A}^{-1} = \begin{pmatrix} \mu^{-1} & 0 & 0 \\ 0 & \mu^{-1} & 0 \\ 0 & 0 & (\lambda + 2\mu)^{-1} \end{pmatrix}.$$

The leading term of the solution inside the thin interphase takes the following form in the original variables:

$$\mathbf{u}^{(2)}(x_1, x_2, x_3) = \int_{-h}^{x_3} \mathbb{A}^{-1}(x_1, x_2, \xi)d\xi \cdot \sigma_{x_3}^{(3)}(x_1, x_2) + \mathbf{u}^{(3)}(x_1, x_2). \tag{29}$$

According to Eq. (19), we can also estimate the components of the strain tensor:

$$\varepsilon_{ij} = \mathcal{O}(1), \quad \varepsilon_{i3} = \frac{1}{2}\frac{\partial u_i^{(2)}}{\partial x_3} + \mathcal{O}(1), \quad i, j = 1, 2, \quad \varepsilon_{33} = \frac{\partial u_3^{(2)}}{\partial x_3}. \tag{30}$$

Thus the leading terms (of the order $\mathcal{O}(\varepsilon^{-1})$) are:

$$\varepsilon_{i3} = \frac{\sigma_{x_i}^{(1)}(x_1, x_2)}{2\mu(x_1, x_2, x_3)}, \quad i = 1, 2, \quad \varepsilon_{33} = \frac{\sigma_{x_3}^{(1)}(x_1, x_2)}{(\lambda + 2\mu)(x_1, x_2, x_3)}, \tag{31}$$

and, generally speaking, depend on the x_3-variable in contrast to the stress vector. In a particular case, when the material properties of the interphase do not depend on the x_3-variable, the leading asymptotic terms of the strain tensor are only functions of the two remaining variables.

Summarizing, Eqs. (26) and (28) constitute the sought for transmission conditions describing a soft imperfect *interface* in a dissimilar body. The conditions allow one to solve the outer problem (for materials denoted by the subscripts (1) and (3)) while the original thin soft inhomogeneous *interphase* is taken into account with the use of the asymptotic analysis.

This analysis, however, cannot be directly used in the case of a thin soft nonlinear interphase when the material properties depend on the solution within the interphase itself. The next subsection is therefore devoted to adjust the analysis for such a case.

2.3 Transmission Conditions for Nonlinear Interphases

We assume now that both Lamé's parameters inside the interphase depend on the strain tensor, or more precisely, are functions of two invariants of the strain tensor:

$$\mu^{(2)} = \mu^{(2)}(J_1^{\varepsilon}, J_2^{e}), \quad \lambda^{(2)} = \lambda^{(2)}(J_1^{\varepsilon}, J_2^{e}), \tag{32}$$

where the first invariant of the strain tensor is:

$$J_1^{\varepsilon} = \varepsilon_{kk}, \tag{33}$$

while the second invariant of the strain deviator tensor, J_2^e, is defined by

$$J_2^e = \frac{1}{2} e_{ij} e_{ij},$$

or using the components of the strain tensor:

$$J_2^e = \frac{1}{6}\Big[(\varepsilon_{11} - \varepsilon_{22})^2 + (\varepsilon_{22} - \varepsilon_{33})^2 + (\varepsilon_{11} - \varepsilon_{33})^2\Big] + \varepsilon_{12}^2 + \varepsilon_{13}^2 + \varepsilon_{23}^2 . \tag{34}$$

To simplify the analysis by using the result obtained in the previous subsection, we assume that the problem has been somehow solved and the complete distribution of the strain invariants are known a priory. As a result, we can write Eq. (32) and the representations (14) again with *unknown* values of the material parameters μ_* and λ_*. However, such assumption may be only justified if there exists two constants $\mu^{\top}$ and $\lambda^{\top}$ so that for any possible strain invariants the following estimate holds true:

$$0 < \mu^{(2)}(J_1^{\varepsilon}, J_2^e) \le \varepsilon \mu^{\top}, \quad 0 < \lambda^{(2)}(J_1^{\varepsilon}, J_2^e) \le \varepsilon \lambda^{\top}. \tag{35}$$

Thus, these assumptions play a crucial role in our analysis and have to be *always verified* when using the results of the analysis.

Assuming estimates (14) are valid, we can directly use the results of the previous section to justify the validity of the two transmission conditions (26) and (28) by taking into account Eq. (25) which is convenient to write in the following form:

$$\sigma_{x_3}^{(2)}(x_1, x_2) = \sigma_{x_3}^{(1)}(x_1, x_2) = \sigma_{x_3}^{(3)}(x_1, x_2), \tag{36}$$

$$[\mathbf{u}](x_1, x_2) = \int_{-h}^{h} \mathbb{A}^{-1}(x_1, x_2, x_3) dx_3 \cdot \sigma_{x_3}^{(2)}(x_1, x_2). \tag{37}$$

Here we use the standard notation for the jump of a function f across the interface $\Gamma : [f](x_1, x_2) = f(x_1, x_2, \Gamma_+) - f(x_1, x_2, \Gamma_-)$. We also can now compute the leading terms of the strain tensor invariants (by taking into account Eqs. (30) and (31)):

$$J_1^{\varepsilon} = \frac{\partial u_3^{(2)}}{\partial x_3} = \frac{\sigma_{33}(x_1, x_2)}{\lambda + 2\mu}, \tag{38}$$

$$J_2^e = \frac{1}{3}\varepsilon_{33}^2 + \varepsilon_{13}^2 + \varepsilon_{23}^2 = \frac{1}{3}\left(\frac{\partial u_3^{(2)}}{\partial x_3}\right)^2 + \frac{1}{4}\left(\frac{\partial u_1^{(2)}}{\partial x_3}\right)^2 + \frac{1}{4}\left(\frac{\partial u_2^{(2)}}{\partial x_3}\right)^2, \tag{39}$$

or taking into account the solution representation (29), the expression for the leading term of the strain components (31) is obtained as:

$$J_2^e = \frac{1}{3}\frac{\sigma_{33}^2(x_1,x_2)}{(\lambda+2\mu)^2} + \frac{1}{4}\frac{\sigma_{13}^2(x_1,x_2)}{\mu^2} + \frac{1}{4}\frac{\sigma_{23}^2(x_1,x_2)}{\mu^2}. \tag{40}$$

Note that due to our assumptions, the material parameters μ and λ are *unknown* functions of the coordinates (x_1, x_2, x_3). Simultaneously, according to Eq. (32), they are *known* functions of the strain invariants. Thus, Eqs. (38) and (40) allows us to uniquely determine the invariants as functions of the space variables for each particular case. However, all the stress components σ_{i3} of the stress vector σ_{x_3} are only functions of the space variables x_1 and x_2. This immediately implies that, in the case under consideration, the strain invariants are also functions of those two variables:

$$J_1^\varepsilon = J_1^\varepsilon(x_1,x_2), \quad J_2^e = J_2^e(x_1,x_2). \tag{41}$$

This crucial finding allows us to simplify the transmission conditions even further. Indeed, since the invariants (and therefore the material parameters μ and λ as well as the matrix A) are only functions of x_1- and x_2-variable, one can trivially compute the integral in the transmission condition (37) to obtain:

$$\frac{1}{2h}[\mathbf{u}](x_1,x_2) = \mathbb{A}^{-1}(x_1,x_2)\cdot\sigma_{x_3}^{(2)}(x_1,x_2). \tag{42}$$

Let us observe that the right-hand side of this transmission condition allows us to compute every term in the definition of the strain invariants in Eqs. (38) and (40). This leads to the following relationships:

$$J_1^\varepsilon = \frac{1}{2h}[u_3], \tag{43}$$

$$J_2^e = \frac{1}{12h^2}[u_3]^2 + \frac{1}{16h^2}[u_1]^2 + \frac{1}{16h^2}[u_2]^2. \tag{44}$$

This allows us to formulate the nonlinear transmission conditions for a thin soft nonlinear interphase in their final forms:

$$[\sigma_{x_3}] = 0, \quad \mathbb{A}([\mathbf{u}])\cdot[\mathbf{u}] = 2h\sigma_{x_3}, \tag{45}$$

where the matrix-function $\mathbb{A}$ is explicitly defined by $\mathbb{A}([\mathbf{u}]) \equiv \mathbb{A}(\mu(J_1^\varepsilon, J_2^e), \lambda(J_1^\varepsilon, J_2^e))$ and the relationships (43) and (44).

Remark 1. Let us remind that the assumptions (35) have to be valid to justify the transmission conditions (37). However, as it follows from the derivation, the estimate can be weakened a bit to the following formulation:

$$0 < \mu^{(2)}(J_1^\varepsilon, J_2^e) \leq \varepsilon\mu^\top, \quad 0 < (\lambda^{(2)} + 2\mu^{(2)})(J_1^\varepsilon, J_2^e) \leq \varepsilon\Lambda^\top. \qquad (46)$$

Remark 2. However, even if the estimate (46) cannot be proven *a priori*, the nonlinear transmission conditions (37) can be successfully used in modeling. To this end, *a posteriori* verification should confirm that the estimates (46) are valid for every point lying on the bimaterial interface.

Remark 3. The nonlinear elastic formulation (32) gives a chance to model thin soft plastic interphase situated between stiffer materials. For this reason, the deformation theory of plasticity can be adopted. Clearly, only a narrow range of monotonic loading can be analyzed with such an approach. On the other hand, bearing in mind the problem geometry (thin plastic layer) the range of possible strain-states in such a layer is sufficiently restricted. Thus, choosing the proper deformation theory, one can expect to avoid all numerical difficulties mentioned in the introduction and effectively use the evaluated transmission conditions. To verify what level of possible error could be introduced in the original model, we need to make efforts in two directions. First, to construct an appropriate deformation theory for a given yield criterion and a given plastic flow rule. Then, to compare the computations performed by the accurate plastic flow modeling with this based on the developed deformation theory to check the validity of the transmission conditions. Note that in the case of an incompressible plastic material obeying the von Mises yield criterion such an analysis has been done for the plane strain case in [27]. It shows for basic loading cases (tension, shear and combination of those two) an exceptional high accuracy comparable with that obtained for the classic elastic interphases. The only observable difference was that the size of the edge effect zone increases. This can be easy explained not by the failure of the asymptotic analysis in those zones but rather by the fact that the loading in the neighborhood of the edge cannot be considered as monotonic even if it is in the main part of the body.

Thus, in the next sections of the chapter we will discuss in detail how to construct an appropriate deformation theory for pressure dependent materials and discuss how the analysis can be validated and numerically verified.

3 Pressure-Dependent Deformation Theory

A deformation theory for the von Mises yield condition has been developed by many authors (see [14, 17, 21]). The theory is generally based on the unique stress-strain curve approach and the authors considered various types of curves (linear hardening, ideal plastic, power laws and many others (see for example the extensive review in the monograph by Jones [16])). In the general case, the development of a deformation theory of plasticity which will be valid in a wider range of monotonic load is not an easy task. To the best of the authors knowledge, this theory—in the general framework of large deformations—has been constructed by Lubarda [20] for the Drucker-Prager material. In Chen [6], one can find a suggestion on how to derive the theory based on a more generalized assumption. We believe that this has not been done yet because of the following two reasons:

(a) The plastic flow theory has been fully developed and represents the behavior of any plastic (visco-plastic) material much better than any deformation theory.
(b) The plastic flow theory has been effectively implemented in numerous numerical codes. Thus, the evaluation of a generalized deformation theory is cost ineffective or even makes no sense.

However, as we mentioned in the introduction, there are specific cases when such work can be justified. Namely, in the case of a thin plastic interphase, where a standard finite element approach needs a very fine mesh in the interphase to realistically approximate nonlinearities, the approach based on asymptotic methods and the notion of imperfect interface may bring some serious improvements in terms of time and accuracy of the computation. With this aim in mind, we are going to discuss a variant of the deformation theory which will be implemented in the asymptotic analysis performed in the previous section.

3.1 Main Assumptions

The simplest deformation theory is the J_2^s deformation theory proposed by Hencky [13, 17] and it is subjected to the following assumptions:

- The material is initially isotropic.
- The principal axes of the plastic strain tensor $\varepsilon_{ij}^{\mathrm{p}}$ are always coincident with those of the stress tensor σ_{ij}.
- The plastic deviatoric strain tensor e_{ij}^{p} is proportional to the deviatoric stress tensor s_{ij}.
- No plastic volumetric change occurs.

In this section, we generalize this theory to the case of pressure-sensitive materials, so that the last assumption is removed and the material is plastically compressible.

In the following, we make use of the volumetric and deviatoric decomposition of the stress and strain tensor (see Eq. (2)) as:

$$\sigma_{ij} = \sigma_{\mathrm{m}}\delta_{ij} + s_{ij} \,, \tag{47}$$

$$\varepsilon_{ij} = \varepsilon_{\mathrm{m}}\delta_{ij} + e_{ij} \,, \tag{48}$$

and the additive composition of the elastic and plastic strains

$$\varepsilon_{ij} = \varepsilon_{ij}^{\mathrm{e}} + \varepsilon_{ij}^{\mathrm{p}} = \left(\varepsilon_{\mathrm{m}}^{\mathrm{e}} + \varepsilon_{\mathrm{m}}^{\mathrm{p}}\right)\delta_{ij} + e_{ij}^{\mathrm{e}} + e_{ij}^{\mathrm{p}}, \tag{49}$$

in order to derive the closed-form relationship. According to assumptions (2) and (3), the plastic strain can be written as

$$\varepsilon_{ij}^{\mathrm{p}} = \phi_1 \sigma_{\mathrm{m}} \delta_{ij} + \phi_2 s_{ij} , \tag{50}$$

where ϕ_1 and ϕ_2 are functions which represent the hardening behavior of the material. In the case of loading it holds that $\phi_1 \neq 0$ and $\phi_2 > 0$ while in the case of unloading $\phi_1 = \phi_2 = 0$ prevails. Note that Eq. (50) can be derived following to Chen's suggestions [6].

The constitutive relation between stress and elastic strain decomposed in its volumetric and deviatoric parts is given by

$$e_{ij}^{\mathrm{e}} = \frac{s_{ij}}{2G}, \tag{51}$$

$$\varepsilon_{\mathrm{m}}^{\mathrm{e}} = \frac{\sigma_{\mathrm{m}}}{3K}. \tag{52}$$

The material is assumed to be elastic compressible, so that $0 \leq \nu < 1/2$ and $K < \infty$.

Substitution of Eqs. (50) (52) into (49) leads to the strain-stress relations for pressure-sensitive materials:

$$\varepsilon_{ij} = \left(\frac{1}{2G} + \phi_2 \right) s_{ij} + \left(\frac{1}{3K} + \phi_1 \right) \sigma_{\mathrm{m}} \delta_{ij} , \tag{53}$$

$$= \left(\frac{1+\nu}{E} + \phi_2 \right) s_{ij} + \left(\frac{1-2\nu}{E} + \phi_1 \right) \sigma_{\mathrm{m}} \delta_{ij} , \tag{54}$$

where the relations (3) between $G = \mu$, K, E and ν were used to convert the elastic constants. To derive the stress-strain relation, we replace in Eq. (53) s_{ij} by σ_{ij} using Eq. (47):

$$\varepsilon_{ij} = \left(\frac{1+\nu}{E} + \phi_2 \right) \sigma_{ij} + \left[- \left(\frac{3\nu}{E} + \phi_2 \right) + \phi_1 \right] \sigma_{\mathrm{m}} \delta_{ij} . \tag{55}$$

Now we use the facts from Eqs. (49), (50) and (52)

$$\sigma_{\mathrm{m}} = \frac{3K}{1 + 3K\phi_1} \varepsilon_{\mathrm{m}} = \frac{E}{1 - 2\nu + \phi_1 E} \varepsilon_{\mathrm{m}} , \tag{56}$$

to obtain

$$\varepsilon_{ij} = \left(\frac{1+\nu}{E} + \phi_2 \right) \sigma_{ij} - \left(\frac{3\nu}{E} + \phi_2 - \phi_1 \right) \frac{E}{1 - 2\nu + \phi_1 E} \varepsilon_{\mathrm{m}} \delta_{ij} , \tag{57}$$

and finally rearranged for the stress tensor and written it in form of Eq. (1) as:

$$\sigma_{ij} = \underbrace{\frac{E}{1+\nu+\phi_2 E}}_{2\tilde{\mu}} \varepsilon_{ij} + \underbrace{\frac{E}{1+\nu+\phi_2 E} \frac{3\nu+(\phi_2-\phi_1)E}{1-2\nu+\phi_1 E}}_{3\tilde{\lambda}} \varepsilon_{\mathrm{m}} \delta_{ij} , \tag{58}$$

where the generalized Lamé's coefficients have been introduced:

$$\tilde{\mu}(\phi_2) = \frac{E}{2(1+\nu+\phi_2 E)}, \tag{59}$$

$$\tilde{\lambda}(\phi_1,\phi_2) = \frac{3\nu+(\phi_2-\phi_1)E}{3(1+\nu+\phi_2 E)(1-2\nu+\phi_1 E)} E, \tag{60}$$

$$\tilde{\lambda}(\phi_1,\phi_2) + 2\tilde{\mu}(\phi_2) = \frac{3(1-\nu)+(\phi_2+2\phi_1)E}{3(1+\nu+\phi_2 E)(1-2\nu+\phi_1 E)} E . \tag{61}$$

It should be noted here that these coefficients coincide in the pure elastic regime ($\phi_1 = \phi_2 = 0$) with the elastic Lamé's coefficients (3). Further relationships for generalized elastic constants can be derived in the same manner:

$$\tilde{E}(\phi_1,\phi_2) = \frac{3E}{3+(2\phi_2+\phi_1)E}, \tag{62}$$

$$\tilde{\nu}(\phi_1,\phi_2) = \frac{3\nu+(\phi_2-\phi_1)E}{3+(2\phi_2+\phi_1)E}, \tag{63}$$

$$\tilde{K}(\phi_1) = \frac{E}{3(1-2\nu+\phi_1 E)}. \tag{64}$$

In order that the formulated deformation theory can be considered, it has to satisfy the ellipticity condition [3]

$$\tilde{\mu} > 0, \quad \tilde{K} > 0. \tag{65}$$

Note that these conditions are equivalent to

$$\tilde{E} > 0, \quad -1 < \tilde{\nu} < 0.5.$$

This is always true if $\phi_2 > 0$ and

$$\phi_1 > -\frac{1-2\nu}{E}. \tag{66}$$

However, the generalized Poisson's ratio $\tilde{\nu}$ may, generally speaking, become negative. Although, such a material behavior is possible and recently several artificial materials have been developed exhibiting such property (see for example [19]), we assume that it never happens in our analysis, thus:

$$\phi_2 - \phi_1 > -\frac{3\nu}{E}. \tag{67}$$

It is easy to show that

$$0 < \tilde{\mu}(\phi_2) \leq \mu, \quad 0 < \tilde{E}(\phi_1, \phi_2) \leq \frac{3E}{2(1+\nu)}. \tag{68}$$

However, it is not enough to prove the second estimate from Eq. (46). If we assume a bit stronger assumption than Eq. (66)

$$\phi_1 \geq -\gamma \frac{1-2\nu}{E}, \tag{69}$$

with some $\gamma < 1$, then

$$0 < \tilde{K}(\phi_1) \leq \frac{K}{1-\gamma}, \quad 0 < (\tilde{\lambda} + 2\tilde{\mu})(\phi_1, \phi_2) \leq \Lambda_{max}. \tag{70}$$

Here only the value of the constant Λ_{max} cannot be exactly predicted without knowing the information on the functions ϕ_1 and ϕ_2. Thus, the conditions (46) are valid under assumption (69) and the transmission conditions (37) can be used for such materials. Note that the case $\gamma = 0$ corresponds to the assumption $\phi_1 \geq 0$.

To define the functions ϕ_1 and ϕ_2, let us consider the relationships for the first and second invariants of the strain and stress tensor, i.e.

$$J_1^{\varepsilon^p} = \varepsilon_{kk}^p, \quad J_1^{\sigma} = \sigma_{kk}, \tag{71}$$

$$J_2^{e^p} = \frac{1}{2} e_{ij}^p e_{ij}^p, \quad J_2^s = \frac{1}{2} s_{ij} s_{ij}, \tag{72}$$

where the proportionality relation (50) can be used to obtain:

$$J_1^{\varepsilon^p} = 3\phi_1 \sigma_{\rm m} = J_1^{\sigma} \phi_1, \tag{73}$$

$$J_2^{e^p} = \frac{1}{2} \phi_2^2 s_{ij} s_{ij} = \phi_2^2 J_2^s. \tag{74}$$

Thus, the scalar functions ϕ_1 and ϕ_2 can be obtained in the case of a multiaxial stress state as

$$\phi_1 = \frac{J_1^{\varepsilon^p}}{J_1^{\sigma}}, \quad \phi_2 = \frac{\sqrt{J_2^{e^p}}}{\sqrt{J_2^s}}. \tag{75}$$

In order that these relations can be used in the analysis of the thin plastic layer presented in Sect. 2, one needs to operate the functions defined as total strain invariants. Thus, some additional work should be done.

First, one can directly evaluate from the nonlinear elasticity relationship (58) that the first invariants of the stress tensor and the total strain tensor are linearly dependent:

$$J_1^\sigma = (3\tilde{\lambda} + 2\tilde{\mu})\, J_1^\varepsilon = 3\tilde{K}\, J_1^\varepsilon. \tag{76}$$

Computing in a standard way the second invariants of the deviatoric tensors, one can conclude that

$$\sqrt{J_2^s} = 2\tilde{\mu}\sqrt{J_2^e}. \tag{77}$$

This, in turns, allows us to evaluate the generalized material parameters as functions of the invariants:

$$2\tilde{\mu} = \frac{\sqrt{J_2^s}}{\sqrt{J_2^e}}, \quad 3\tilde{K} = \frac{J_1^\sigma}{J_1^\varepsilon}. \tag{78}$$

Substituting these relations into Eqs. (59) and (64), one can define after some algebra:

$$\phi_1 = \frac{J_1^\varepsilon}{J_1^\sigma} - \frac{1-2\nu}{E}, \quad \phi_2 = \frac{\sqrt{J_2^e}}{\sqrt{J_2^s}} - \frac{1+\nu}{E}. \tag{79}$$

These relationships are more useful in the analysis than (75). This also gives the relationships between the deviators of the plastic strain and the total strain:

$$J_1^\varepsilon = J_1^{\varepsilon \mathrm{p}} + \frac{1-2\nu}{E} J_1^\sigma, \quad \sqrt{J_2^e} = \sqrt{J_2^{e\mathrm{p}}} + \frac{1+\nu}{E}\sqrt{J_2^s}. \tag{80}$$

3.2 Example: Uniaxial Stress Test

We choose this specific test as it is one of the most popular tests, widely accessible, and it may provide at least two sets of experimental data: $\sigma_x(\varepsilon_x)$ and $\varepsilon_y(\varepsilon_x)$.

The specific form of the scalar functions ϕ_1 and ϕ_2 for the case of a uniaxial stress state can be derived in the following way. Assuming that the direction of loading coincides with the x-direction ($\sigma_x \neq 0$; $\varepsilon_y^{\mathrm{p}} = \varepsilon_z^{\mathrm{p}}$), the scalar functions ϕ_1 and ϕ_2 become

$$\phi_1 = \frac{\varepsilon_x^{\mathrm{p}} + 2\varepsilon_y^{\mathrm{p}}}{\sigma_x}, \quad \phi_2 = \frac{|\varepsilon_x^{\mathrm{p}} - \varepsilon_y^{\mathrm{p}}|}{\sigma_x}. \tag{81}$$

Let us assume in the following an elastic-plastic material with linear hardening as shown in Fig. 3. The flow stress σ_x can be expressed as (cf. Fig. 3b)

$$\sigma_x(\varepsilon_x^{\mathrm{p}}) = E^{\mathrm{p}}\varepsilon_x^{\mathrm{p}} + \sigma_s\,, \tag{82}$$

whereas the transversal deformation can be expressed as (cf. Fig. 3d)

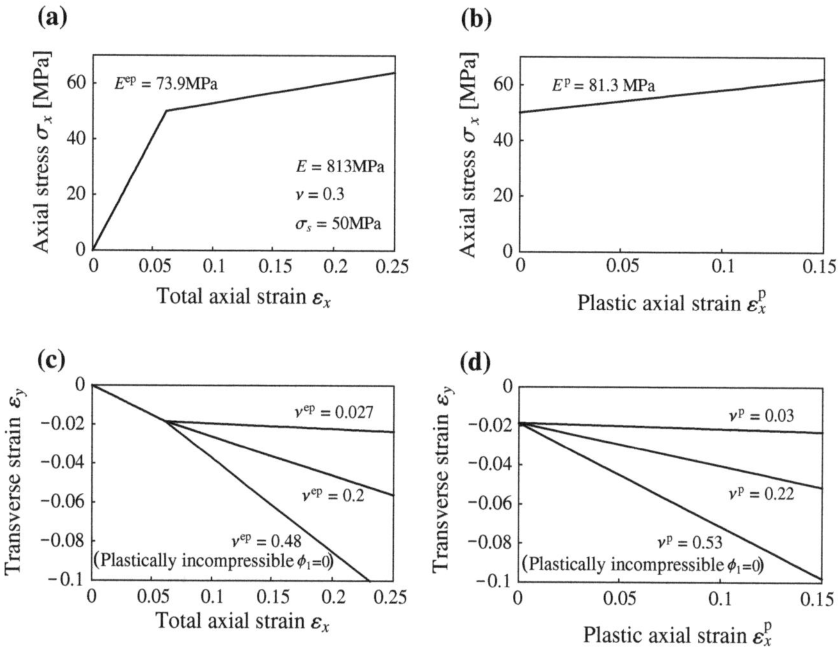

Fig. 3 Example of constitutive behaviour in a uniaxial stress test for a linear hardening material. **a** Axial stress σ_x as a function of the total axial strain ε_x; **b** axial stress σ_x as a function of the plastic axial strain $\varepsilon_x^{\mathrm{p}}$; **c** transverse contraction ε_y as a function of the total axial strain ε_x; **d** transverse contraction ε_y as a function of the plastic axial strain $\varepsilon_x^{\mathrm{p}}$. Three cases of transverse contraction are considered: plastically incompressible material $\varepsilon_{\mathrm{m}}^{\mathrm{p}} = 0$ ($\nu^{\mathrm{ep}} = 0.48$ and $\nu^{\mathrm{p}} = 0.53$), null plastic transverse contraction $\varepsilon_y^{\mathrm{p}} = 0$ ($\nu^{\mathrm{ep}} = 0.027$ and $\nu^{\mathrm{p}} = 0.03$) and an intermediate case ($\nu^{\mathrm{ep}} = 0.2$ and $\nu^{\mathrm{p}} = 0.22$)

$$\varepsilon_y(\varepsilon_x^{\mathrm{p}}) = -\nu^{\mathrm{p}}\varepsilon_x^{\mathrm{p}} + \varepsilon_y^{\mathrm{Y}} = -\nu^{\mathrm{p}}\varepsilon_x^{\mathrm{p}} - \frac{\nu}{E}\sigma_s. \tag{83}$$

The plastic part of ε_y can be derived as

$$\varepsilon_y^{\mathrm{p}} = \varepsilon_y - \varepsilon_y^{\mathrm{e}} = -\nu^{\mathrm{p}}\varepsilon_x^{\mathrm{p}} - \frac{\nu}{E}\sigma_s + \nu\frac{\sigma_x}{E} = -\left(\nu^{\mathrm{p}} - \nu\frac{E^{\mathrm{p}}}{E}\right)\varepsilon_x^{\mathrm{p}} \tag{84}$$

and the functions ϕ_1 and ϕ_2 can be obtained in dependence of the plastic strain in a uniaxial stress state as

$$\phi_1(\varepsilon_x^{\mathrm{p}}) = \frac{(E - 2\nu^{\mathrm{p}}E + 2\nu E^{\mathrm{p}})\varepsilon_x^{\mathrm{p}}}{E(E^{\mathrm{p}}\varepsilon_x^{\mathrm{p}} + \sigma_s)}, \tag{85}$$

$$\phi_2(\varepsilon_x^{\mathrm{p}}) = \frac{(E + \nu^{\mathrm{p}}E - \nu E^{\mathrm{p}})\varepsilon_x^{\mathrm{p}}}{E(E^{\mathrm{p}}\varepsilon_x^{\mathrm{p}} + \sigma_s)}, \tag{86}$$

Using the additive composition of the elastic and plastic strains, i.e. $\varepsilon_x = \varepsilon_x^{\mathrm{e}} + \varepsilon_x^{\mathrm{p}}$, one gets

$$\varepsilon_x^{\mathrm{p}} = \varepsilon_x - \varepsilon_x^{\mathrm{e}} = \varepsilon_x - \frac{\sigma_x}{E}. \tag{87}$$

The relationship between the stress σ_x and the total strain ε_x can be derived from Fig. 3 as

$$\sigma_x = E^{\mathrm{p}}\left(\varepsilon_x - \frac{\sigma_x}{E}\right) + \sigma_s, \tag{88}$$

so that

$$\sigma_x(\varepsilon_x) = E^{\mathrm{ep}}\varepsilon_x + \frac{E - E^{\mathrm{ep}}}{E}\sigma_s, \tag{89}$$

where

$$E^{\mathrm{ep}} = \frac{EE^{\mathrm{p}}}{E + E^{\mathrm{p}}}. \tag{90}$$

Replacing the stress σ_x in Eq. (87) by the relationship given in Eq. (82) results in

$$\varepsilon_x^{\mathrm{p}} = \frac{E\varepsilon_x - \sigma_s}{E + E^{\mathrm{p}}}, \tag{91}$$

and

$$\varepsilon_y(\varepsilon_x) = -\nu^{\mathrm{ep}}\varepsilon_x - \frac{\nu - \nu^{\mathrm{ep}}}{E}\sigma_s\,, \tag{92}$$

where

$$\nu^{\mathrm{ep}} = \frac{\nu^{\mathrm{p}}E}{E + E^{\mathrm{p}}}. \tag{93}$$

Finally, the dependence of ϕ_1 and ϕ_2 on the total strain ε_x as

$$\phi_1(\varepsilon_x) = \frac{1}{E}\left(1 - 2\nu^{\mathrm{p}} + 2\nu\frac{E^{\mathrm{p}}}{E}\right)\frac{E\varepsilon_x - \sigma_s}{E^{\mathrm{p}}\varepsilon_x + \sigma_s}, \tag{94}$$

$$\phi_2(\varepsilon_x) = \frac{1}{E}\left(1 + \nu^{\mathrm{p}} - \nu\frac{E^{\mathrm{p}}}{E}\right)\frac{E\varepsilon_x - \sigma_s}{E^{\mathrm{p}}\varepsilon_x + \sigma_s}. \tag{95}$$

Functions $\phi_1(\varepsilon_x)$ and $\phi_2(\varepsilon_x)$ corresponding to the materials of Fig. 3 are shown in Fig. 4. Interestingly, in case of the uniaxial stress state the ratio of the functions $\phi_1(\varepsilon_x)$ and $\phi_2(\varepsilon_x)$ is a constant value:

$$\frac{\phi_1(\varepsilon_x)}{\phi_2(\varepsilon_x)} = \frac{(1 - 2\nu^{\mathrm{p}})E + 2\nu E^{\mathrm{p}}}{(1 + \nu^{\mathrm{p}})E - \nu E^{\mathrm{p}}}. \tag{96}$$

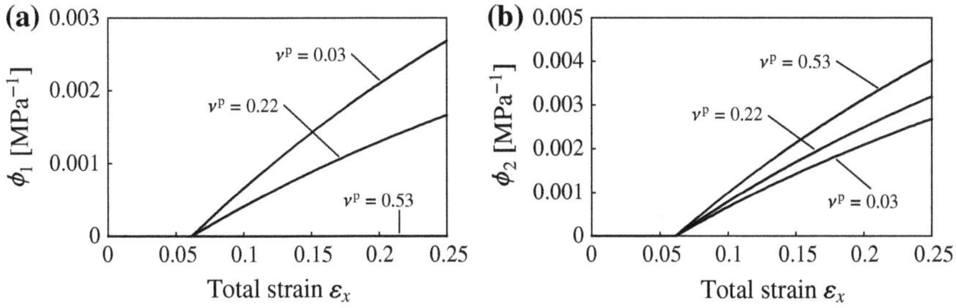

Fig. 4 Functions $\phi_1(\varepsilon_x)$ (**a**) and $\phi_2(\varepsilon_x)$ (**b**) corresponding to the materials of Fig. 3

Usually $E > E^{\mathrm{p}}$ and we assume that plastic contraction is always positive. This leads to the inequality $\nu^{\mathrm{p}}E > \nu E^{\mathrm{p}}$ (cf. (84)) and one can conclude that for the uniaxial stress state under consideration, the inequalities (66) and (67) are always valid as

$$\frac{\phi_1(\varepsilon_x)}{\phi_2(\varepsilon_x)} < \min\left\{1, \frac{1-2\nu^{\mathrm{p}}+2\nu}{1+\nu^{\mathrm{p}}-\nu}\right\}. \tag{97}$$

Finally, it should be noted that it is impossible to conclude a priori whether the ratio is positive, zero or negative. Only direct experimental data (or some additional phenomenological assumptions) allows to answer this question. One of the most popular and simplest assumption is that the plastic strain is incompressible or in accordance with the definition of the functions ϕ_j this ratio has to be always zero. This case is considered in next subsection in details.

3.2.1 von Mises Material

For a material obeying the von Mises yield criterion

$$F(J_2^s) = \sqrt{J_2^s} - k_s\,, \tag{98}$$

and the normality rule, the plastic strain turns out to be incompressible so that

$$\varepsilon_{ij}^{\mathrm{p}} = \phi_2 s_{ij}. \tag{99}$$

The theory developed in the previous section can be specialized to the case of a von Mises material by setting $\phi_1 = 0$ and assuming that

$$\phi_2 = \phi_2(J_2^e)\,, \tag{100}$$

where the function itself has to be defined from experiments. The generalized elastic constants (62) become in this case

$$\tilde{\mu}(\phi_2) = \tilde{G}(\phi_2) = \frac{E}{2(1+\nu+\phi_2 E)}, \tag{101}$$

$$\tilde{E}(\phi_2) = \frac{3E}{3+2\phi_2 E}, \tag{102}$$

$$\tilde{\nu}(\phi_2) = \frac{3\nu+\phi_2 E}{3+2\phi_2 E}, \tag{103}$$

$$\tilde{K} = \frac{E}{3(1-2\nu)}. \tag{104}$$

Note that the estimates given in Eq. (46) are valid here with $\gamma = 0$ and $\Lambda_{\max} = \lambda + 2\mu$ (see Eq. (70)). Thus, the evaluation of the transmission conditions (37) is therefore fully justified in this case. We can also evaluate the function k_s in the yield criterion. Indeed,

$$k_s = k_s(\varepsilon_x^{\mathrm{p}}) = \sqrt{J_2^s} = \frac{\sigma_x}{\sqrt{3}} = \frac{E^{\mathrm{p}}\varepsilon_x^{\mathrm{p}} + \sigma_s}{\sqrt{3}}, \tag{105}$$

and, in terms of the second invariant of the total deformation,

$$k_s = k_s(J_2^e) = \sqrt{J_2^s} = 2\tilde{\mu}\sqrt{J_2^e} = \frac{E\sqrt{J_2^e}}{1+\nu+\phi_2(J_2^e)E}. \tag{106}$$

Now we need to determine the unknown function $\phi_2 = \phi_2(J_2^e)$ from the uniaxial stress relationships (94)–(95). Note to this end that

$$\phi_1 = 0 \to \nu^{\mathrm{p}} = \frac{1}{2} + \nu\frac{E^{\mathrm{p}}}{E}, \tag{107}$$

$$\phi_2(\varepsilon_x) = \frac{3}{2}\frac{E\varepsilon_x - \sigma_s}{E(E^{\mathrm{p}}\varepsilon_x + \sigma_s)}. \tag{108}$$

This representation is valid however, only for a uniaxial stress state. A general form of the scalar function ϕ_2 for applications involving multiaxial stress states can be derived in the following way. In the plastic regime, the generalized Poisson's ratio $\tilde{\nu}$ is given by

$$\tilde{\nu} = -\frac{\varepsilon_y}{\varepsilon_x} = -\frac{\varepsilon_z}{\varepsilon_x}, \tag{109}$$

and the second invariant of the deviator of strain J_2^e can be expressed as

$$J_2^e = \frac{1}{6}[(\varepsilon_x-\varepsilon_y)^2 + (\varepsilon_y-\varepsilon_z)^2 + (\varepsilon_z-\varepsilon_x)^2] = \frac{(1+\tilde{\nu})^2}{3}\varepsilon_x^2, \tag{110}$$

which can be rearranged to obtain

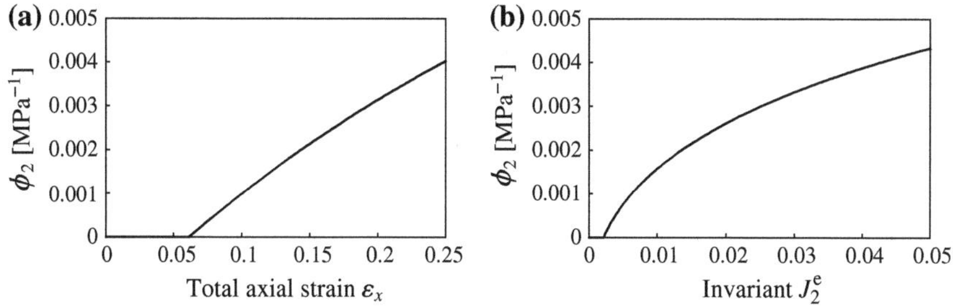

Fig. 5 Function ϕ_2 of Hencky's law as a function of (**a**) the total axial strain ε_x and (**b**) the second invariant of strain deviator tensor J_2^e. Material parameters are those of Fig. 3

$$\varepsilon_x = \frac{\sqrt{3J_2^e}}{1+\tilde{\nu}} = \frac{\sqrt{3J_2^e}}{1+\dfrac{3\nu+\phi_2 E}{3+2\phi_2 E}} = \frac{3+2\phi_2 E}{1+\nu+\phi_2 E}\sqrt{\frac{J_2^e}{3}}. \tag{111}$$

Introducing the last relationship into Eq. (108), the corresponding quadratic equation gives the dependency of the function ϕ_2 on the second invariant of the deviator of strain as:

$$\phi_2(J_2^e) = \frac{3E\sqrt{J_2^e} - \sqrt{3}(1+\nu)\sigma_s}{E\left(2E^{\mathrm{p}}\sqrt{J_2^e} + \sqrt{3}\sigma_s\right)}. \tag{112}$$

The evaluation of the function ϕ_2 for a uniaxial stress-strain curve given in Fig. 3 ($E = 813$ MPa; $\nu = 0.3$; $\sigma_s = 50$ MPa; $E^{\mathrm{p}} = 81.3$ MPa) is shown in Fig. 5.

3.2.2 Drucker-Prager Material

For a material obeying the Drucker-Prager yield criterion,

$$F(J_1^\sigma, J_2^s) = \alpha J_1^\sigma + \sqrt{J_2^s} - k_s\,, \tag{113}$$

where $\alpha > 0$ is usually assumed to be a constant while k_s is a positive function of the plastic strain:

$$k_s = k_s(\varepsilon^{\mathrm{p}}) > 0. \tag{114}$$

In the case of uniaxial stress test

$$k_s(\varepsilon_x^{\mathrm{p}}) = \alpha J_1^\sigma + \sqrt{J_2^s} = \left(\alpha + \frac{1}{\sqrt{3}}\right)\left(E^{\mathrm{p}}\varepsilon_x^{\mathrm{p}} + \sigma_s\right). \tag{115}$$

Assuming the normality rule (plastic strain increment proportional to the normal of the yield surface gives $n_{ij} = 2\alpha\sqrt{J_2^s}\delta_{ij} + s_{ij}$) and proportional loading we have (cf. [20])

$$\varepsilon_{ij}^{\mathrm{p}} = \phi_2 \left(2\alpha\sqrt{J_2^s}\delta_{ij} + s_{ij}\right). \tag{116}$$

The theory presented in Sect. 3 can be specialized to this type of material by setting

$$\phi_1 = \frac{6\alpha\sqrt{J_2^s}\phi_2}{J_1^\sigma}, \tag{117}$$

$$\phi_2 = \phi_2 . \tag{118}$$

It is important to underline here that both functions ϕ_1 and ϕ_2 equal to zero simultaneously in the elastic regime or are both different from zero in the plastic zone.

We need to represent now the functions ϕ_1 and ϕ_2 depending only on the invariants of the total strain tensor. Let us assume that we have been successful with the second function,

$$\phi_2 = \phi_2(J_1^\varepsilon, J_2^e), \tag{119}$$

then the remaining function ϕ_1 can be computed from Eq. (117) under taking into account Eqs. (78), (59) and (64)

$$\phi_1 = 6\alpha\frac{\sqrt{J_2^s}\phi_2}{J_1^\sigma} = 6\alpha\phi_2\frac{2\tilde{\mu}}{3\tilde{K}}\frac{\sqrt{J_2^e}}{J_1^\varepsilon} = 6\alpha\phi_2\frac{1-2\nu+\phi_1 E}{1+\nu+\phi_2 E}\frac{\sqrt{J_2^e}}{J_1^\varepsilon}, \tag{120}$$

or

$$\phi_1 = \phi_1(J_1^\varepsilon, J_2^e) = \frac{6\alpha(1-2\nu)\phi_2\sqrt{J_2^e}}{J_1^\varepsilon(1+\nu) + \phi_2 E\left(J_1^\varepsilon - 6\alpha\sqrt{J_2^e}\right)}. \tag{121}$$

Note that it is impossible to conclude something definitive on the sign of the ratio of the functions ϕ_1/ϕ_2. Indeed, in the elastic domain (for small values of J_2^e) $\phi_2 = 0$ and therefore $\phi_1 = 0$ too. On the other hand for $\phi_2 > 0$, one can always choose values of the first invariant J_1^ε to receive ϕ_1 positive or negative. However, we would like to underline here that those invariants are not independent as they are related via the yield stress condition and the corresponding strain-stress state achieved during the deformation. We will comment on this later in this subsection.

Using Eqs. (119) and (121), the generalized elastic constants defined in Eqs. (59)–(64) can be rewritten in terms of the only two invariants of the total strain tensor:

$$\tilde{\mu}(\phi_2) = \tilde{G}(\phi_2) = \frac{E}{2(1+\nu+\phi_2 E)}, \tag{122}$$

$$\tilde{E}(\phi_2) = \frac{3E}{1+\nu+\phi_2 E}\frac{J_1^\varepsilon(1+\nu+\phi_2 E) - 6\alpha\phi_2 E\sqrt{J_2^e}}{J_1^\varepsilon(3+2\phi_2 E) - 12\alpha\phi_2 E\sqrt{J_2^e}}, \tag{123}$$

$$\tilde{\nu}(\phi_2) = \frac{J_1^{\varepsilon}(3\nu + \phi_2 E) - 6\alpha\phi_2 E\sqrt{J_2^e}}{J_1^{\varepsilon}(3 + 2\phi_2 E) - 12\alpha\phi_2 E\sqrt{J_2^e}}, \tag{124}$$

$$\tilde{K}(\phi_2) = \frac{J_1^{\varepsilon}(1 + \nu + \phi_2 E) - 6\alpha\phi_2 E\sqrt{J_2^e}}{3J_1^{\varepsilon}(1 - 2\nu)(1 + \nu + \phi_2 E)} E\,. \tag{125}$$

The expressions reduce to those for the von Mises criterion when $\alpha = 0$. However, it is impossible to prove a priori the ellipticity conditions (65) as well as the upper estimates (46) necessary to justify the transmission conditions.

Finally, one can compute the function k_s from Eq. (114) taking into account that the yield criterion is represented by the equation $F(J_1^{\sigma}, J_2^s) = 0$ in the plastic regime where the function F is defined in Eq. (113). Using again the invariants representations (78) and the definition of the generalized elastic coefficients (59)–(64), we have

$$k_s = k_s(J_1^{\varepsilon}, J_2^e) = \alpha\frac{EJ_1^{\varepsilon}}{1 - 2\nu + \phi_1 E} + \frac{E\sqrt{J_2^e}}{1 + \nu + \phi_2 E}. \tag{126}$$

This can be rewritten as

$$k_s = \frac{E\sqrt{J_2^e}}{1 + \nu + \phi_2 E}\left(1 + 6\alpha^2\frac{\phi_2}{\phi_1}\right). \tag{127}$$

Finally, this can be equivalently represented in the following form:

$$k_s(J_1^{\varepsilon}, J_2^e) = \frac{E}{1 - 2\nu}\left(\alpha J_1^{\varepsilon} + \Psi\right), \tag{128}$$

where

$$\Psi = \frac{1 - 2\nu - 6\alpha^2\phi_2 E}{1 + \nu + \phi_2 E}\sqrt{J_2^e}\,. \tag{129}$$

Note that, generally speaking, $\Psi = \Psi(J_1^{\varepsilon}, J_2^e)$ as it follows from Eq. (119). Naturally, if one assumes $\alpha = 0$ formula (128) reduces to Eq. (106).

In case of the uniaxial stress test discussed in the previous subsection, Eqs. (96) and (117) allow us to determine the unknown parameter $\alpha > 0$ from the experiment

$$\alpha = \frac{J_1^{\sigma}}{6\sqrt{J_2^s}}\frac{(1 - 2\nu^{\mathrm{p}})E + 2\nu E^{\mathrm{p}}}{(1 + \nu^{\mathrm{p}})E - \nu E^{\mathrm{p}}} = \frac{1}{2\sqrt{3}}\frac{(1 - 2\nu^{\mathrm{p}})E + 2\nu E^{\mathrm{p}}}{(1 + \nu^{\mathrm{p}})E - \nu E^{\mathrm{p}}} < \frac{1}{2\sqrt{3}}. \tag{130}$$

Interestingly, while a priori we could not evaluate the sign of the ratio ϕ_1/ϕ_2 (see discussion after Eq. (121)), in case of the uniaxial stress test this value is definitely positive

$$0 < \frac{\phi_1(\varepsilon_x)}{\phi_2(\varepsilon_x)} = 2\alpha\sqrt{3} < 1\,. \tag{131}$$

This suggests that, at least for any stress-strain state close to the uniaxial stress, all the constants (122)–(125) are positive and suitable for justification of the transmission conditions. Moreover, this finding allows us to estimate the upper bound for the experimental value of the parameter ν^{p} for the uniaxial stress test

$$\nu^{\mathrm{p}} \leq \frac{1}{2} + \nu \frac{E^{p}}{E}, \tag{132}$$

where the maximal value corresponds to the von Mises criterion (incompressible plastic deformation). Thus, to complete the analysis, the only unknown function ϕ_2 should be defined. In general, this function depends on both strain invariants J_1^{ε} and J_2^{e}

$$\phi_2 = \phi_2(J_1^{\varepsilon}, J_2^{e}) \,. \tag{133}$$

Considering the uniaxial stress state, Eq. (94) has been satisfied due to Eq. (173), while (95) can be written in the form:

$$\phi_2(\varepsilon_x) = \frac{3}{2}\frac{\beta}{E}\frac{E\varepsilon_x - \sigma_s}{E^{\mathrm{p}}\varepsilon_x + \sigma_s}, \tag{134}$$

where

$$\beta = \frac{2}{3}\left(1 + \nu^{\mathrm{p}} - \nu\frac{E^{\mathrm{p}}}{E}\right). \tag{135}$$

Note that $\beta = 1$ holds for the von Mises yield criterion (cf. Eq. (108)). Repeating the same line of the reasoning as previously, we substitute

$$\varepsilon_x = \frac{\sqrt{3J_2^{e}}}{1+\tilde{\nu}}, \tag{136}$$

with

$$\tilde{\nu}(\phi_2) = \frac{J_1^{\varepsilon}(3\nu + \phi_2 E) - 6\alpha\phi_2 E\sqrt{J_2^{e}}}{J_1^{\varepsilon}(3 + 2\phi_2 E) - 12\alpha\phi_2 E\sqrt{J_2^{e}}}, \tag{137}$$

into the Eq. (134) and solve the corresponding quadratic equation to obtain

$$\phi_2(J_1^{\varepsilon}, J_2^{e}) = \frac{-b + \sqrt{b^2 - 4ac}}{2a}, \tag{138}$$

where

$$a = 2E^2(J_1^{\varepsilon} - 6\alpha\sqrt{J_2^{e}})\left(2\sqrt{3}E^{\mathrm{p}}\sqrt{J_2^{e}} + 3\sigma_s\right), \tag{139}$$

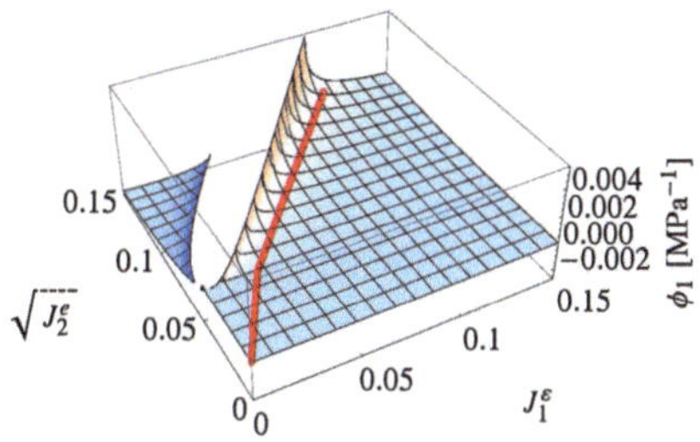

Fig. 6 Functions ϕ_1 and ϕ_2 of Drucker-Prager's law as functions of the two invariants of strain J_1^ε and J_2^e. Material parameters are those of Fig. 3, with $\nu^\mathrm{p} = 0.22$. The *red curve* is the loading path in the uniaxial stress test

$$b = 3E \left\{ 6\alpha\beta\sqrt{J_2^e}\left(2\sqrt{3}E\sqrt{J_2^e} - 3\sigma_s\right) + J_1^\varepsilon \left[(2 + 2\nu + 3\beta)\sigma_s + 2\sqrt{3}\left(E^\mathrm{p} - \beta E\right)\sqrt{J_2^e} \right] \right\}, \tag{140}$$

$$c = 9\beta J_1^\varepsilon \left((1+\nu)\sigma_s - \sqrt{3}\sqrt{J_2^e} E \right), \tag{141}$$

One can check that in case of $\alpha = 0$ and $\beta = 1$, Eq. (138) is transformed into (112).

The functions ϕ_1 and ϕ_2 are plotted in Fig. 6 against the invariants of the total strain, J_1^ε and J_2^e. The material parameters are those of Fig. 3, with intermediate $\nu^\mathrm{p} = 0.22$. In Fig. 6 it is also shown in red the loading path during the uniaxial stress test.

Remark 4. The functions ϕ_1 and ϕ_2 obtained in this section can be used only for a stress state close to the uniaxial stress state.

3.3 Example: Uniaxial Deformation Test

We present in this section an alternative derivation of the functions ϕ_1 and ϕ_2 from the uniaxial deformation test. Assuming that the direction of deformation coincides with the x-direction ($\varepsilon_x \neq 0$; $\sigma_y = \sigma_z$), the scalar functions ϕ_1 and ϕ_2 become

$$\phi_1 = \frac{\varepsilon_x^\mathrm{p} + 2\varepsilon_y^\mathrm{p}}{\sigma_x + 2\sigma_y}, \qquad \phi_2 = \frac{|\varepsilon_x^\mathrm{p} - \varepsilon_y^\mathrm{p}|}{\sigma_x - \sigma_y}. \tag{142}$$

Let us assume a linear hardening material, as shown in Fig. 7. In the elastic regime, the material behaviour is described by the following laws (Fig. 8)

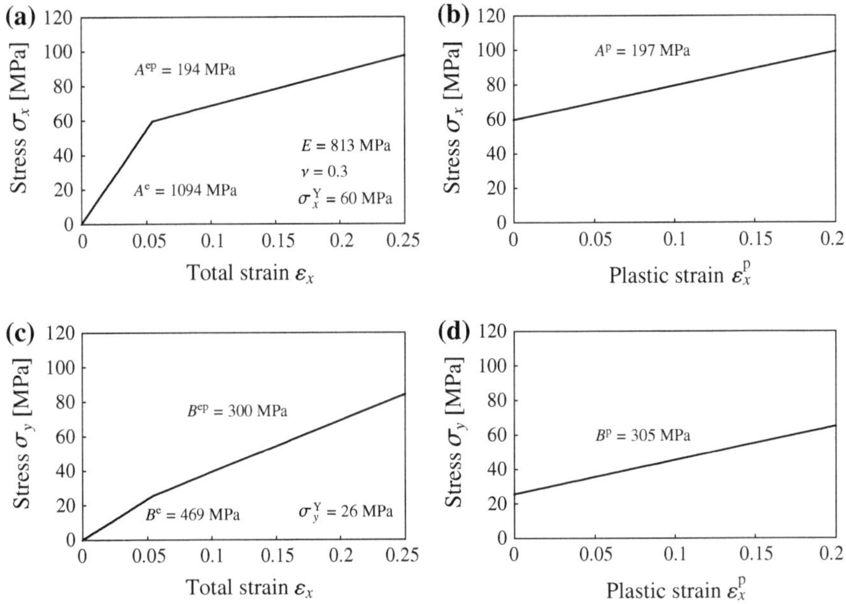

Fig. 7 Example of constitutive behaviour in a uniaxial deformation test for a linear hardening material. **a** Axial stress σ_x as a function of the total axial strain ε_x; **b** axial stress σ_x as a function of the plastic axial strain ε_x^p; **c** transverse stress σ_y as a function of the total axial strain ε_x; **d** transverse stress σ_y as a function of the plastic axial strain ε_x^p

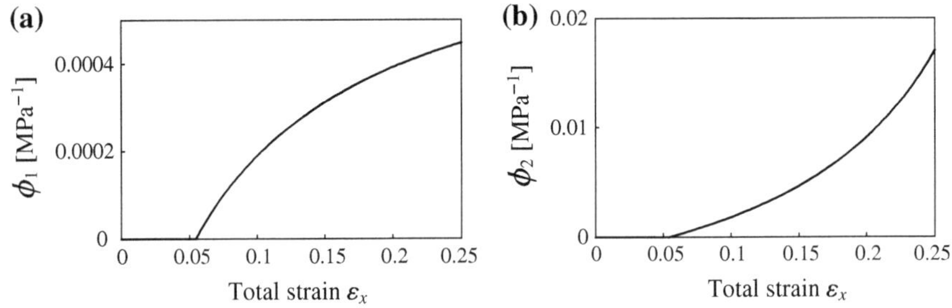

Fig. 8 Functions $\phi_1(\varepsilon_x)$ (**a**) and $\phi_2(\varepsilon_x)$ (**b**) corresponding to the material of Fig. 7

$$\sigma_x(\varepsilon_x) = A^e \varepsilon_x , \tag{143}$$

$$\sigma_y(\varepsilon_x) = B^e \varepsilon_x , \tag{144}$$

where

$$A^e = \frac{1-\nu}{(1+\nu)(1-2\nu)} E, \tag{145}$$

$$B^{\mathrm{e}} = \frac{\nu}{(1+\nu)(1-2\nu)} E. \tag{146}$$

In the plastic regime, the stress is given by

$$\sigma_x(\varepsilon_x) = A^{\mathrm{ep}} \varepsilon_x + \sigma_{x0}\,, \tag{147}$$

$$\sigma_y(\varepsilon_x) = B^{\mathrm{ep}} \varepsilon_x + \sigma_{y0}\,, \tag{148}$$

where

$$\sigma_{x0} = \frac{(1-\nu)E - (1+\nu)(1-2\nu)A^{\mathrm{ep}}}{(1-\nu)E} \sigma_x^{\mathrm{Y}}, \tag{149}$$

$$\sigma_{y0} = \frac{\nu E - (1+\nu)(1-2\nu)B^{\mathrm{ep}}}{(1-\nu)E} \sigma_x^{\mathrm{Y}}. \tag{150}$$

The axial strain ε_x can be written as

$$\varepsilon_x = \varepsilon_x^{\mathrm{p}} + \varepsilon_x^{\mathrm{e}} = \varepsilon_x^{\mathrm{p}} + \frac{1}{E}(\sigma_x - 2\nu\sigma_y). \tag{151}$$

A substitution of (151) into (149) and (150) gives

$$\sigma_x(\varepsilon_x) = A^{\mathrm{p}} \varepsilon_x^{\mathrm{p}} + \sigma_x^{\mathrm{Y}}\,, \tag{152}$$

$$\sigma_y(\varepsilon_x) = B^{\mathrm{p}} \varepsilon_x^{\mathrm{p}} + \frac{\nu}{1-\nu} \sigma_x^{\mathrm{Y}}\,, \tag{153}$$

where

$$A^{\mathrm{p}} = \frac{E A^{\mathrm{ep}}}{E - A^{\mathrm{ep}} + 2\nu B^{\mathrm{ep}}}, \tag{154}$$

$$B^{\mathrm{p}} = \frac{E B^{\mathrm{ep}}}{E - A^{\mathrm{ep}} + 2\nu B^{\mathrm{ep}}}. \tag{155}$$

Inversion of (154) and (155) gives

$$A^{\mathrm{ep}} = \frac{A^{\mathrm{p}} E}{A^{\mathrm{p}} + E - 2\nu B^{\mathrm{p}}}\,, \tag{156}$$

$$B^{\mathrm{ep}} = \frac{B^{\mathrm{p}} E}{A^{\mathrm{p}} + E - 2\nu B^{\mathrm{p}}}\,. \tag{157}$$

Finally, the dependence of ϕ_1 and ϕ_2 on the total strain ε_x is obtained as

$$\phi_1(\varepsilon_x) = \frac{\varepsilon_x}{(A^{\mathrm{ep}} + 2B^{\mathrm{ep}})\varepsilon_x + \sigma_{x0} + 2\sigma_{y0}} - \frac{1-2\nu}{E}, \tag{158}$$

$$\phi_2(\varepsilon_x) = \frac{\varepsilon_x}{(A^{\mathrm{ep}} - B^{\mathrm{ep}})\varepsilon_x + \sigma_{x0} - \sigma_{y0}} - \frac{1+\nu}{E}. \tag{159}$$

3.3.1 von Mises Material

For a material obeying the von Mises criterion, the yield stresses in uniaxial deformation test σ_x^{Y}, σ_y^{Y} are related to the yield stress in uniaxial stress test σ_s through the relation

$$\sigma_x^{\mathrm{Y}} = \frac{1-\nu}{1-2\nu}\sigma_s\,, \quad \sigma_y^{\mathrm{Y}} = \frac{\nu}{1-2\nu}\sigma_s\,. \tag{160}$$

Moreover, the condition $\phi_1 = 0$ must be satisfied for the incompressibility of plastic strain. This condition, together with Eq. (158), gives a relationship between A^{ep} and B^{ep},

$$B^{\mathrm{ep}} = \frac{E - (1-2\nu)A^{\mathrm{ep}}}{2(1-2\nu)}, \tag{161}$$

or, equivalently, using Eqs. (154) and (155), between A^{p} and B^{p},

$$B^{\mathrm{p}} = \frac{E + 2\nu A^{\mathrm{p}}}{2(1-\nu)}. \tag{162}$$

The fulfilment of the yield condition (98) during plastic loading implies additionally the following relationship between A^{p} and B^{p}

$$A^{\mathrm{p}} - B^{\mathrm{p}} = E^{\mathrm{p}}. \tag{163}$$

Consequently, by solving Eqs. (162) and (163), it is possible to obtain relationships between the parameters of the uniaxial deformation test, A^{p}, B^{p}, A^{ep}, B^{ep}, with the parameter of the uniaxial stress test E^{p}. These relationships read as follows

$$A^{\mathrm{p}} = \frac{E + 2(1-\nu)E^{\mathrm{p}}}{2(1-2\nu)}, \tag{164}$$

$$B^{\mathrm{p}} = \frac{E + 2\nu E^{\mathrm{p}}}{2(1-2\nu)}, \tag{165}$$

$$A^{\mathrm{ep}} = \frac{E(E + 2(1-\nu)E^{\mathrm{p}})}{(1-2\nu)(3E + 2(1+\nu)E^{\mathrm{p}})}, \tag{166}$$

$$B^{\mathrm{ep}} = \frac{E(E + 2\nu E^{\mathrm{p}})}{(1-2\nu)(3E + 2(1+\nu)E^{\mathrm{p}})}, \tag{167}$$

A substitution of (166) and (167) into (159) allows us to obtain ϕ_2 as a function of the total axial strain ε_x

$$\phi_2(\varepsilon_x) = 3\,\frac{E\varepsilon_x - (1+\nu)\sigma_s}{E(2E^{\mathrm{p}}\varepsilon_x + 3\sigma_s)}\,. \tag{168}$$

Finally, in the uniaxial deformation test the second invariant of the deviator of strain is given by

$$J_2^e = \frac{\varepsilon_x^2}{3}, \tag{169}$$

and consequently the general relation (112) of the von Mises material is recovered.

3.3.2 Drucker-Prager Material

For a material obeying the Drucker-Prager criterion, the yield stresses in the uniaxial deformation test σ_x^{Y}, σ_y^{Y} are related to the yield stress in the uniaxial stress test σ_s through the relation

$$\sigma_x^{\mathrm{Y}} = \frac{(1+\sqrt{3}\alpha)(1-\nu)\sigma_s}{(1-2\nu)+\sqrt{3}\alpha(1+\nu)}\,, \quad \sigma_y^{\mathrm{Y}} = \frac{(1+\sqrt{3}\alpha)\nu\sigma_s}{(1-2\nu)+\sqrt{3}\alpha(1+\nu)}\,. \tag{170}$$

We take now the trace and devitoric part of Eq. (116),

$$J_1^{\varepsilon \mathrm{p}} = 6\phi_2\alpha\sqrt{J_2^s}\,, \tag{171}$$

$$\sqrt{J_2^{e\mathrm{p}}} = \phi_2\sqrt{J_2^s}\,, \tag{172}$$

and the ratio between the last two equations gives

$$\alpha = \frac{J_1^{\varepsilon \mathrm{p}}}{\sqrt{J_2^{e\mathrm{p}}}} = \frac{\varepsilon_x^{\mathrm{p}} + 2\varepsilon_y^{\mathrm{p}}}{2\sqrt{3}(\varepsilon_x^{\mathrm{p}} - \varepsilon_y^{\mathrm{p}})}\,. \tag{173}$$

Due to the lateral constraint in the uniaxial deformation test, we have

$$\varepsilon_y^{\mathrm{p}} = -\varepsilon_y^{\mathrm{e}} = -\frac{1}{E}\left(-\nu\sigma_x + (1-\nu)\sigma_y\right) = \frac{\nu A^{\mathrm{p}} - (1-\nu)B^{\mathrm{p}}}{E}\varepsilon_x^{\mathrm{p}}\,. \tag{174}$$

Substituting (174) into (173), we get

$$\alpha = \frac{1}{2\sqrt{3}}\frac{E + 2\nu A^{\mathrm{p}} - 2(1-\nu)B^{\mathrm{p}}}{E - \nu A^{\mathrm{p}} + (1-\nu)B^{\mathrm{p}}}\,. \tag{175}$$

Moreover, the fulfilment of the yield criterion (113) implies

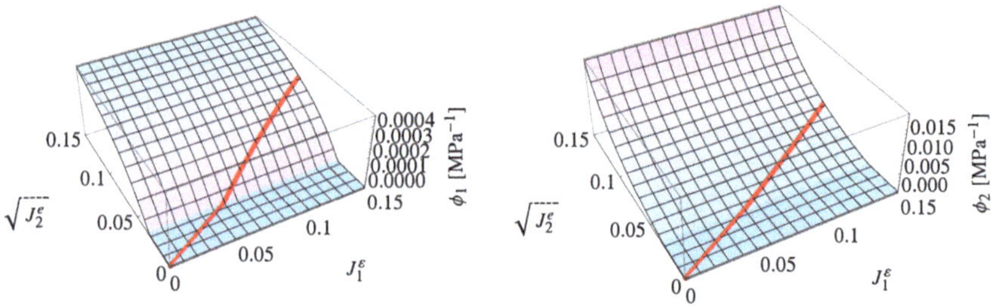

Fig. 9 Functions ϕ_1 and ϕ_2 of Drucker-Prager's law as functions of the two invariants of strain J_1^ε and J_2^e. Material parameters are those of Fig. 3, with $\nu^\mathrm{p} = 0.22$. The *red curve* is the loading path in the uniaxial deformation test

$$k_s(\varepsilon_x^\mathrm{p}) = \left[\left(\alpha + \frac{1}{\sqrt{3}}\right) A^\mathrm{p} + \left(2\alpha - \frac{1}{\sqrt{3}}\right) B^\mathrm{p}\right] \varepsilon_x^\mathrm{p} + \left(\alpha \frac{1+\nu}{1-\nu} + \frac{1}{\sqrt{3}} \frac{1-2\nu}{1-\nu}\right) \sigma_x^\mathrm{Y}. \tag{176}$$

A comparison of (176) and (115) allows us to write

$$\alpha = \frac{E^\mathrm{p} + B^\mathrm{p} - A^\mathrm{p}}{\sqrt{3}(A^\mathrm{p} + 2B^\mathrm{p} - E^\mathrm{p})}. \tag{177}$$

Finally, by solving (175) and (177), we obtain the characteristic parameters of the uniaxial deformation test as functions of the characteristic parameters of the Drucker-Prager material with linear hardening

$$A^\mathrm{p} = \frac{\left(12\alpha^2 - 4\sqrt{3}\alpha + 1\right) E + 2\left(3\alpha^2 + 2\sqrt{3}\alpha + 1\right)(1-\nu) E^\mathrm{p}}{6\alpha^2(1+\nu) + 2\sqrt{3}\alpha(2-\nu) + 2 - 4\nu}, \tag{178}$$

$$B^\mathrm{p} = \frac{\left(36\alpha^3 - 9\alpha + \sqrt{3}\right) E - 2\left(18\alpha^3 + 9\sqrt{3}\alpha^2 - \sqrt{3}\right) \nu E^\mathrm{p}}{2\left(\sqrt{3} - 6\alpha\right)\left(3\alpha^2(1+\nu) + \sqrt{3}\alpha(2-\nu) + 1 - 2\nu\right)}, \tag{179}$$

$$A^\mathrm{ep} = \frac{\left(\sqrt{3} - 6\alpha\right) E \left[2\left(3\alpha^2 + 2\sqrt{3}\alpha + 1\right)(1-\nu) E^\mathrm{p} - \left(4\left(\sqrt{3} - 3\alpha\right)\alpha - 1\right) E\right]}{3\left(\sqrt{3} - 6\alpha\right) E \left[6\alpha^2(1+\nu) + 1 - 2\nu\right] - 2\left(9\alpha^2\left(2\alpha + \sqrt{3}\right) - \sqrt{3}\right) E^\mathrm{p}(1+\nu)(1-2\nu)}, \tag{180}$$

$$B^\mathrm{ep} = \frac{E\left[\left(36\alpha^3 - 9\alpha + \sqrt{3}\right) E + 2\left(\sqrt{3} - 9\alpha^2\left(2\alpha + \sqrt{3}\right)\right) \nu E^\mathrm{p}\right]}{3\left(\sqrt{3} - 6\alpha\right) E \left[6\alpha^2(1+\nu) + 1 - 2\nu\right] - 2\left(9\alpha^2\left(2\alpha + \sqrt{3}\right) - \sqrt{3}\right) E^\mathrm{p}(1+\nu)(1-2\nu)} \tag{181}$$

A substitution of (180), (181) and (169) into (158) and (159) provides the functions ϕ_1 and ϕ_2. An example is shown in Fig. 9.

Remark 5. The functions ϕ_1 and ϕ_2 obtained in this section can be used only for a deformation state close to the uniaxial deformation state. Thus, in the case of the

Druker-Prager material, the two classic tests give different functions $\phi_j = \phi_j(J_1^\varepsilon, J_2^e)\ (j = 1, 2)$ which ideally should be used only for deformation states close to those where the functions have been evaluated. Thus, the question remains open whether it is possible to evaluate the functions $\phi_j = \phi_j(J_1^\varepsilon, J_2^e)\ (j = 1, 2)$ applicable for an arbitrary strain-stress state as this was in the case for the von Mises criterion.

3.4 Plane Stress State

The two-dimensional plane stress case shown in Fig. 10 is commonly used for the analysis of thin, flat plates loaded in the plane of the plate (x-y plane).

Under the plane stress conditions, i.e. $\sigma_z = \sigma_{xz} = \sigma_{yz} = 0$ and $\varepsilon_{xz} = \varepsilon_{yz} = 0$, the following component equations can be extracted from Eq. (58):

$$\sigma_x = 2\widetilde{\mu}\,\varepsilon_x + \widetilde{\lambda}(\varepsilon_x + \varepsilon_y + \varepsilon_z)\,, \tag{182}$$

$$\sigma_y = 2\widetilde{\mu}\,\varepsilon_y + \widetilde{\lambda}(\varepsilon_x + \varepsilon_y + \varepsilon_z)\,, \tag{183}$$

$$0 = 2\widetilde{\mu}\,\varepsilon_z + \widetilde{\lambda}(\varepsilon_x + \varepsilon_y + \varepsilon_z)\,, \tag{184}$$

$$\sigma_{xy} = 2\widetilde{\mu}\,\varepsilon_{xy}\,. \tag{185}$$

From Eq. (184) we get immediately that

$$\varepsilon_z = -\frac{\widetilde{\lambda}}{2\widetilde{\mu} + \widetilde{\lambda}} \cdot \left(\varepsilon_x + \varepsilon_y\right)\,. \tag{186}$$

The last equation can be rearranged to the following form:

$$\varepsilon_z = -m \cdot \left(\varepsilon_x + \varepsilon_y\right) \tag{187}$$

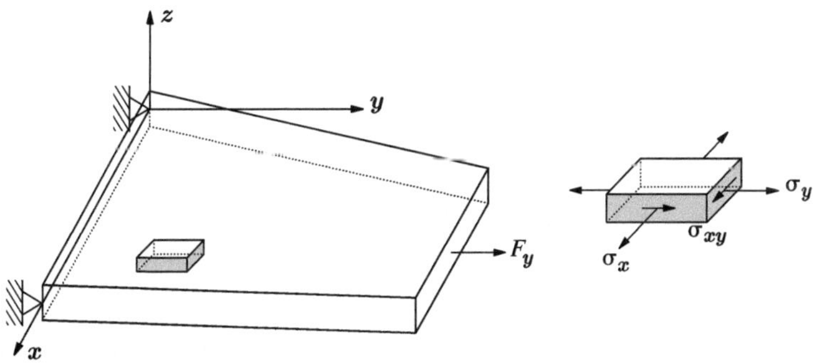

Fig. 10 Two-dimensional problem: plane stress state

with

$$m = \frac{\widetilde{\lambda}}{2\widetilde{\mu} + \widetilde{\lambda}} = \frac{3\nu + (\phi_2 - \phi_1)E}{3(1-\nu) + (2\phi_1 + \phi_2)E} . \tag{188}$$

Substituting Eq. (186) into Eqs. (182), (183) and (182) to eliminate ε_z, yields that the generalized nonlinear equation (58) holds also for the plane stress state (index '$\mathrm{p}\sigma$') in the following notation:

$$(\sigma_{ij})_{\mathrm{p}\sigma} = 2\tilde{\mu}_{\mathrm{p}\sigma}(\varepsilon_{ij})_{\mathrm{p}\sigma} + 3\tilde{\lambda}_{\mathrm{p}\sigma}\varepsilon_{\mathrm{m}}\delta_{ij} , \tag{189}$$

where the generalized elasto-plastic constants for the plane stress case take the form:

$$\tilde{\mu}_{\mathrm{p}\sigma}(\phi_2) = \tilde{\mu}(\phi_2) , \quad \tilde{\lambda}_{\mathrm{p}\sigma}(\phi_1, \phi_2) = \frac{2\widetilde{\mu}\widetilde{\lambda}}{2\widetilde{\mu} + \widetilde{\lambda}} , \quad \widetilde{\nu}_{\mathrm{p}\sigma}(\phi_1, \phi_2) = \frac{\widetilde{\lambda}}{2\widetilde{\mu} + 3\widetilde{\lambda}} . \tag{190}$$

The parameter m defined in Eq. (188) takes in the new notation the form:

$$m = m(\phi_1, \phi_2) = \frac{1}{2} \cdot \frac{\tilde{\lambda}_{\mathrm{p}\sigma}(\phi_1, \phi_2)}{\tilde{\mu}_{\mathrm{p}\sigma}(\phi_2)} . \tag{191}$$

Using arguments similar to those in Sect. 2, one can obtain imperfect transmission conditions in the form of Eq. (45) with the vectors of displacements and tractions consisting of two respective components only.

3.5 Plane Strain State

The two-dimensional plane strain case shown in Fig. 11 is commonly used for the analysis of elongated prismatic bodies of uniform cross section subjected to uniform loading along their longitudinal axis (z-axis) but without any component in direction of the z-axis (e.g. pressure p_1 and p_2 in Fig. 11), such as in the case of tunnels, soil slopes, and retaining walls.

Under the conditions of plane strain, the stress components σ_{yz} and σ_{xz} are zero, and Eq. (58) can be directly reduced to the form (index '$\mathrm{p}\varepsilon$' for plane strain)

$$(\sigma_{ij})_{\mathrm{p}\varepsilon} = 2\tilde{\mu}_{\mathrm{p}\varepsilon}(\varepsilon_{ij})_{\mathrm{p}\varepsilon} + 3\tilde{\lambda}_{\mathrm{p}\varepsilon}\varepsilon_{\mathrm{m}}\delta_{ij} , \tag{192}$$

where $\varepsilon_{\mathrm{m}} = \frac{1}{3}(\varepsilon_x + \varepsilon_y)$. By imposing the plane strain condition $\varepsilon_z = 0$, we get from Eq. (58)

$$\sigma_z = \widetilde{\lambda}(\phi_1, \phi_2) \cdot (\varepsilon_x + \varepsilon_y) . \tag{193}$$

Then using arguments similar (identical) to those in Sect. 2, one obtains (45) with vectors of displacements and tractions consisting of two components [25]:

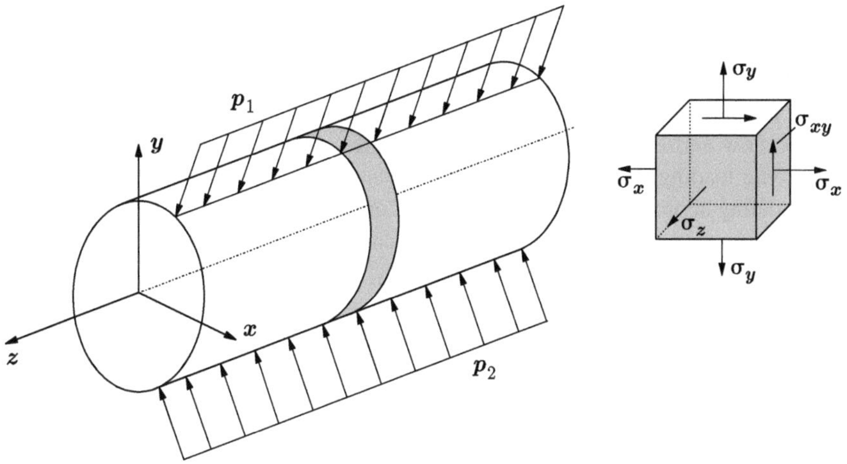

Fig. 11 Two-dimensional problem: plane strain state

$$[\sigma_{x_j x_3}] = 0, \quad F_{x_j}([u_{x_1}], [u_{x_3}]) = \sigma_{x_j x_3}, \quad j = 1, 3. \tag{194}$$

4 Validation of the Transmission Conditions: Plane Strain Case

4.1 Description of the FEM Model

In order to exemplarily investigate and validate the transmission conditions given in Sect. 2.3, a plane strain case is chosen as a benchmark problem. For the finite element approach, the structure shown in Fig. 1 is represented as a two-dimensional mesh in the x_1-x_3 plane. The specific geometrical dimensions are $L_1 = 10$ mm, $H = 1$ mm and $2h = 0.01$ mm. The thickness L_2 is not directly modeled in the case of a plane mesh and only assigned as a geometric property to all elements ($L_2 = 1$ mm). As a result of the assigned numerical values to the geometric dimensions, the value $\varepsilon = 2h/H = 0.01$ can be considered as the small parameter.

Let us furthermore assume that the outer material layers ($+h \leq x_3 \leq +H/2 \vee -H/2 \leq x_3 \leq -h$, see Fig. 1) are made of the same common aluminum alloy AlCuMg1 [29] where Young's modulus is $E^{(1)} = 72700$ MPa and Poisson's ratio is $\nu^{(1)} = 0.34$ and that these outer layers remain in the pure linear-elastic range. Two different elasto-plastic interphases are considered: a linear hardening material model and an elastic-perfectly plastic material. The following properties were taken in the simulations. Elastic parameters of the interphases are the same: $E^{(2)} = 813$ MPa, $\nu^{(2)} = 0.3$. In the plastic region which is appearing after reaching the initial yield stress of value $k_{\mathrm{t}}^{\mathrm{init},(2)} = 50$ MPa, the constant hardening modulus $E_{\mathrm{p}}^{(2)} = 81.3$ MPa is prescribed for the hardening material and $k_{\mathrm{t}} = k_{\mathrm{t}}^{\mathrm{init},(2)}$ for the ideal plasticity case. Furthermore, it is assumed in both hardening cases that the material of the interphase is obeying the von Mises yield condition and the associated flow rule. Let us underline that all commercial FEM codes are based on the more general theory of plastic flow [7, 14, 17]. As it has been mentioned above, the results with these models (plastic flow and deformation theories) coincide only under monotonic or nearly monotonic loading. Because of this, only *monotonic* external loading is applied in the modeling approach (Dirichlet boundary condition at the top of the sample). For the top surface, i.e. $x_3 = +H/2$, two loading cases are distinguished in the following simulations: A simple tensile load case with $u_{x_1}(x_1, H/2) = 0$, $u_{x_3}(x_1, H/2) = v_{x_3}$ and a combined load case with $u_{x_1}(x_1, H/2) = v_{x_1}$, $u_{x_3}(x_1, H/2) = v_{x_3}$, where the v_i are prescribed displacements in the range from 0 to 0.6 % of v_i/H (hardening interphase) or to 0.4 % of v_i/H (ideal plasticity interphase), applied in 100 increments (hardening interphase) or 200 increments (ideal plasticity interphase). The reduction of the load steps is in the case the perfect plastic law in the plastic region advisable in order to increase the accuracy of the calculation. The bottom surface is fixed in all directions, i.e. $u_{x_1}(x_1, -H/2) = 0$ and $u_{x_3}(x_1, -H/2) = 0$.

A commercial finite element code (MSC.Marc, MSC Software Corporation, Santa Ana, CA, USA) is used for the simulation of the mechanical behavior of the bimaterial interphase whereat the two-dimensional FE mesh is built up of four-node, isoparametric elements with bilinear interpolation functions. This element for plane strain applications (element type 11 in MSC.Marc) has only two degrees of freedom at each node, i.e. a displacement in the horizontal and vertical direction. In order to cover all possible edge effects [26] (cf. Fig. 1, left ($x_1 \to -L_1/2$) and right hand side ($x_1 \to +L_1/2$) of the interphase), a strong mesh refinement is performed in these regions, Fig. 12.

The density of the elements along the interphase varies strongly from approx. 1×10^5 elements per mm near the free edges ($x_1 \to \pm L_1/2$) to 200 elements per mm in the middle of the specimen ($x_1 \to 0$). Furthermore, the mesh is generated in such a way that it is possible to evaluate the displacements and stresses along the axes of symmetry ($x_1 = 0 \vee x_3 = 0$), and along all the interfaces between the interphase elasto-plastic material and the outer material layers ($x_3 = +h \vee x_3 = -h$), as well as along the lines parallel ($x_3 =$ const. $\wedge -h < x_3 < +h$) and perpendicular ($x_1 =$ const. $\wedge -h < x_3 < +h$) to the interfaces and lying within the interphase layer.

4.2 *Evaluation of the Mechanical Values: Displacement, Stress and Strain*

For the case of the elasto-plastic interphase with hardening under the conditions of simple tensile loading, distributions of all displacements and stress components in direction perpendicular to the interface through the whole sample in its middle part (along the line $x_1 = 0$, cf. Fig. 2) are shown in Figs. 13 and 14. Results presented in

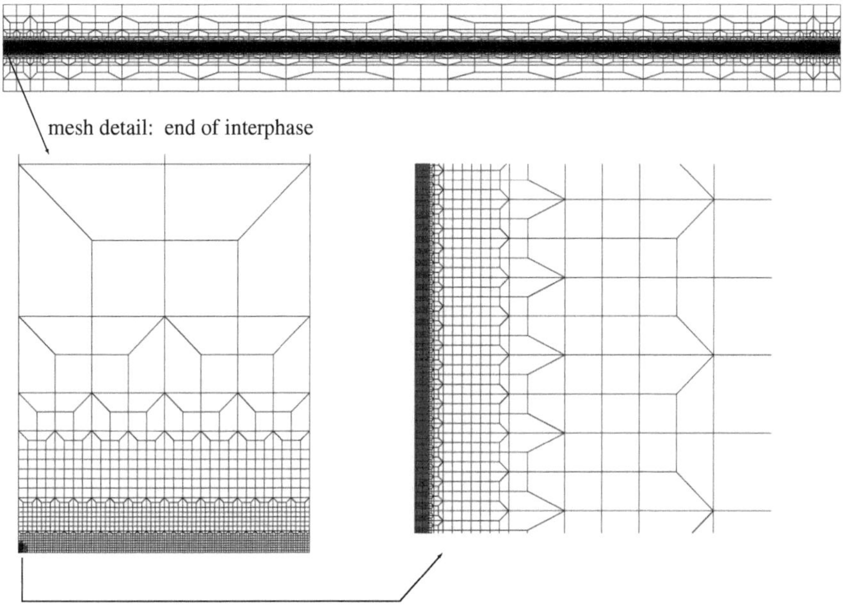

Fig. 12 Two-dimensional FE mesh: strong mesh refinement in the investigated area

Figs. 13a and 14a correspond to the elastic regime while Figs. 13b and 14b are valid for the plastic deformation. As one can see stresses within the interface are constant whereas the displacements are linear functions which completely coincides with the theoretical predictions.

As a result, equivalent von Mises stress and the equivalent plastic strain do not change within the interphase in direction perpendicular to its boundaries (for a fixed x_1). Its variation along the middle line of the elasto-plastic interphase is presented for several increments in Fig. 14.

In the case of the hardening interphase under combined loading, the same particularities can be observed with respect to distributions of the displacements and stresses inside the thin interphase and outside the interphase within the surrounding materials. In Fig. 16, the results concerning the von Mises stress and equivalent plastic strain are presented in the same way as it has been done in Fig. 15. A slightly different behavior can be observed which shows now the influence of the additional secondary loading in the x_1-direction.

In the case of the ideal plasticity interphase under simple loading, the results concerning the behavior of the solution within the interphase in direction perpendicular to the symmetry line $x_3 = 0$ are similar to those shown in Figs. 13 and 14 at point $x_1 = 0$ and hold without any conceptual change, i.e. constant stresses and linear displacements at each increment. Distributions of the equivalent von Mises stress and the equivalent plastic strain along the middle line of the elasto-plastic interphase

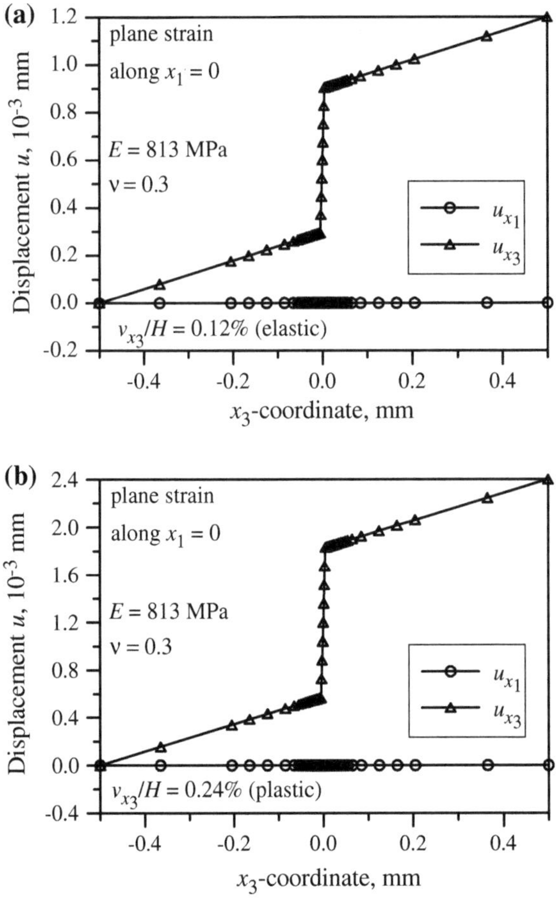

Fig. 13 Displacement distribution along $x_1 = 0$ (cf. Fig. 2) for an elastic and plastic stage inside the hardening interphase (hardening interphase case; simple tensile loading)

($x_3 = 0$) are presented for several increments in Fig. 17. One can clearly observe the ideal plasticity plateau starting from a total deformation of $v_{x_3}/H = 0.14\%$.

In the case of the ideal plasticity interphase under combined loading, we restrict ourselves to show the same results as for the simple loading case. Respective equivalent von Mises stress and equivalent plastic strain curves are presented in Fig. 18.

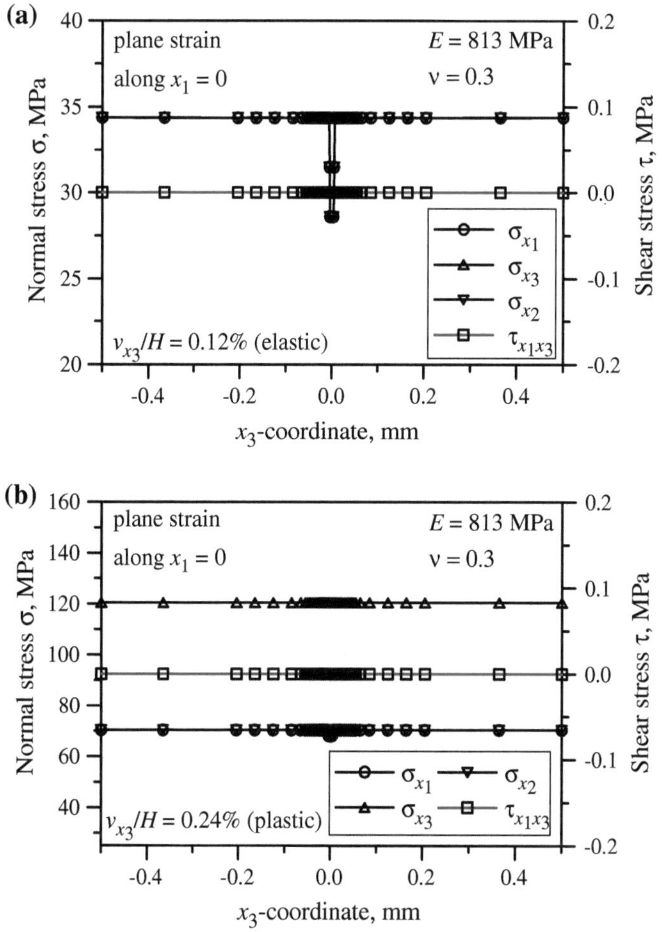

Fig. 14 Stress distribution along $x_1 = 0$ (cf. Fig. 2) for an elastic and plastic stage inside the hardening interphase (hardening interphase case; simple tensile loading)

4.3 Validation of the First and Second Transmission Condition

4.3.1 Simple Tensile Loading and Hardening Law

Due to the symmetry of the loading and the sample geometry, two of the transmission conditions (194), i.e. $[\sigma_{x_1x_3}] = 0$ and $F_{x_1}([u_{x_1}], [u_{x_3}]) = \sigma_{x_1x_3}$, are satisfied identically because of $[u_{x_1}] = 0$ and $\sigma_{x_1x_3} = 0$ holds in this case. The remaining two conditions $[\sigma_{x_1x_3}] = 0$ and $F_{x_3}(0, [u_{x_3}]) = \sigma_{x_3}$ have to be verified. The first one is the same as in the case of the pure elastic imperfect interface [25], has the same order of accuracy as discovered in [25] and is because of this of less interest in comparison with the second one.

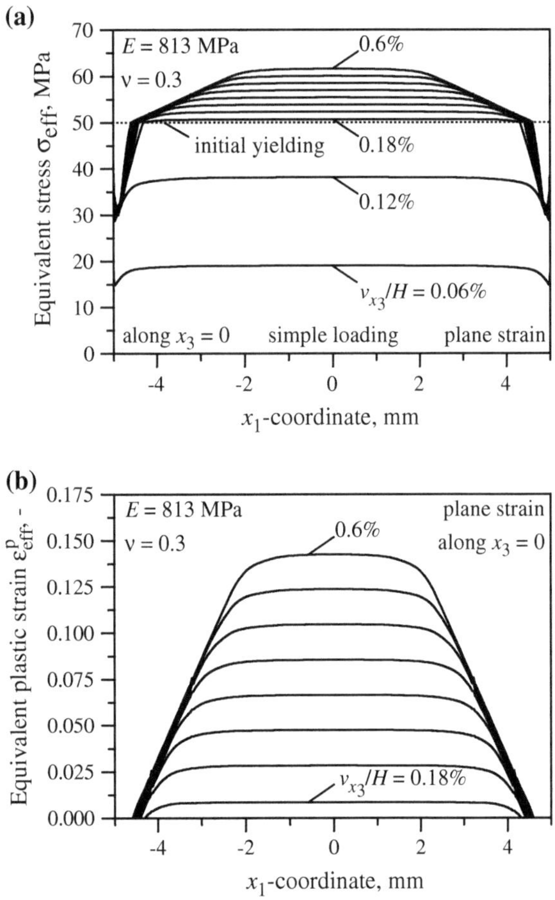

Fig. 15 Distribution of equivalent stress and strain along $x_3 = 0$ (cf. Fig. 2) for different levels of deformation (hardening interphase case; simple tensile loading)

In Fig. 19a, comparisons of the left- and right-hand sides of the condition $F_{x_3}(0, [u_{x_3}]) = \sigma_{x_3}$ are presented. The traction is drawn by the solid line while the values of the left-hand side function in (194) is depicted by circles in several points. The visible plastic zone appears in the middle of the interface at the 30th increment with a deformation ratio of $v_{x_3}/H = 0.18\%$. The accuracy of the evaluated transmission condition (194) is in the same range as it has been checked for the pure elastic interface [25]. Moreover, the region where the transmission conditions are valid does not change practically regardless the interphase material is in the elastic or plastic region, Fig. 19b. To highlight this fact, a magnification of the same functions as in Fig. 19a is presented in Fig. 19b. The 1% accuracy criterion has been chosen to indicate the validity regions for the transmission conditions. The regions are of 2–3 thickness of the interphase. It is also important to note that the plastic zones appearing near the

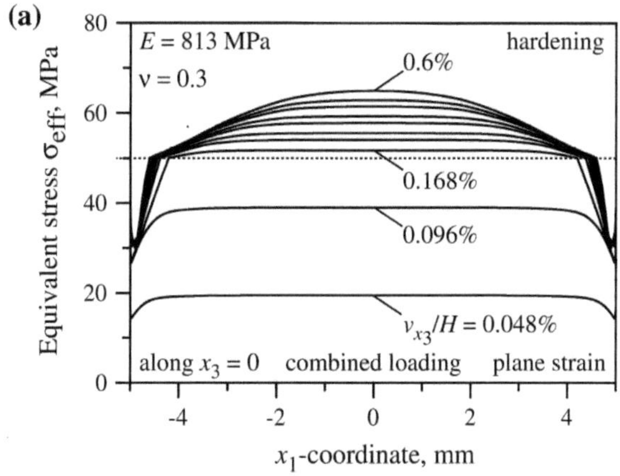

Fig. 16 Distribution of equivalent stress and strain along $x_3 = 0$ (cf. Fig. 2) for different levels of deformation (hardening interphase case; combined loading)

free edges are very small and are invisible in the scale of Fig. 19a. The zone where the transmission conditions are not longer valid coincides more or less with the range of the singularity dominated domains for the elastic interface [26] and becomes to be smaller with accumulated plastic deformation.

4.3.2 Combined Loading and Hardening Law

The question of the validity of the transmission conditions is in the case o the combined hardening much more interesting than in the case of the simple loading. Now both of them are not trivial. Moreover, a second non-zero jump $[u_{x_1}]$ is presented in the functions F_{x_1}, F_{x_3} appearing in the transmission conditions (194). It is interesting

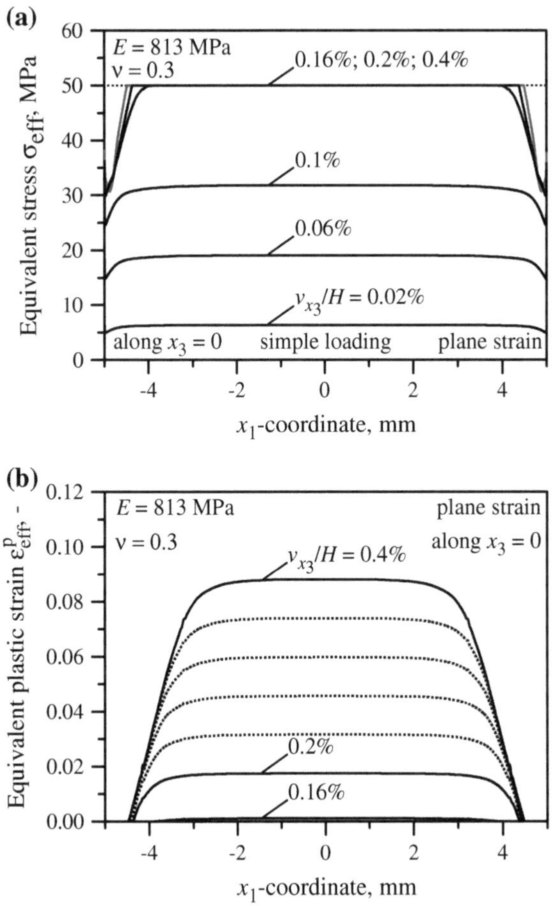

Fig. 17 Distribution of equivalent stress and strain along $x_3 = 0$ (cf. Fig. 2) for different levels of deformation (ideal plasticity interphase case; simple loading)

to note that the validity region is at least not smaller than in the case of the simple tensile loading. To manifest this, we present Fig. 20 where the same values are depicted as in Fig. 19. The same accuracy for the evaluated transmission conditions arises for the second transmission condition dealing with the jump $[u_{x_1}]$. We skip this picture only because it cannot be compared with the case of the simple tensile loading. One of the crucial points to underline is the fact that the stress-strain state of the 2D bimaterial structure under consideration is not pure monotonic due to the definition in [17]. Thus, it would be natural to expect a more essential difference between the numerical model based on the plastic flow theory and the analytically predicted interfacial conditions based on the deformation theory in comparison with the accuracy observed for the pure elastic interface. However, as it follows from the results presented in Figs. 19 and 20, the accuracy of the transmission conditions is much better

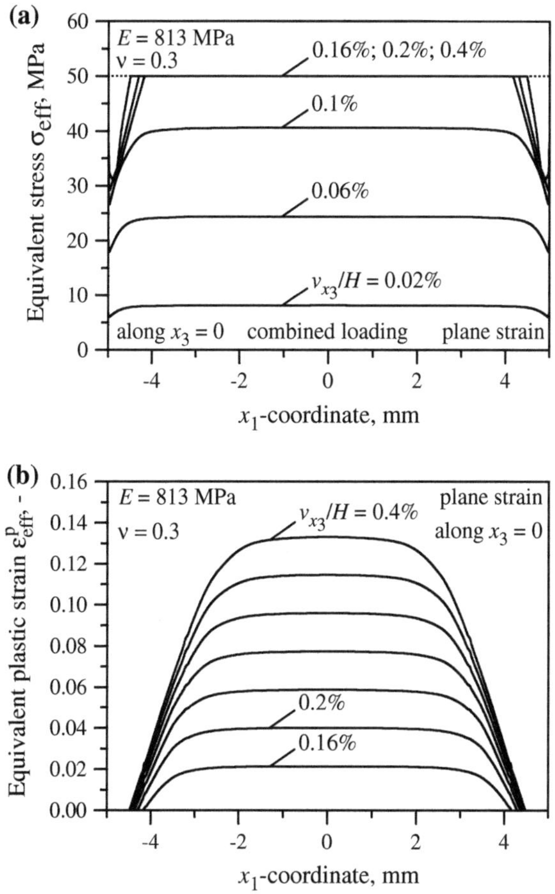

Fig. 18 Distribution of equivalent stress and strain along $x_3 = 0$ (cf. Fig. 2) for different levels of deformation (ideal plasticity interphase case; combined loading)

than one can even expect due to the limitation of the deformation theory. However, this is only true for a hardening interphase law. It will be shown in the following that the results are slightly worse in the case of perfect plasticity. It should be noted here that the outer material layers remained in the pure elastic regime at any stage of the applied deformation.

4.3.3 Perfect Plasticity

The verification of the transmission condition (194) in this case is presented in Fig. 12. Still very good agreement with the theoretical results can be observed over the whole range of the interface.

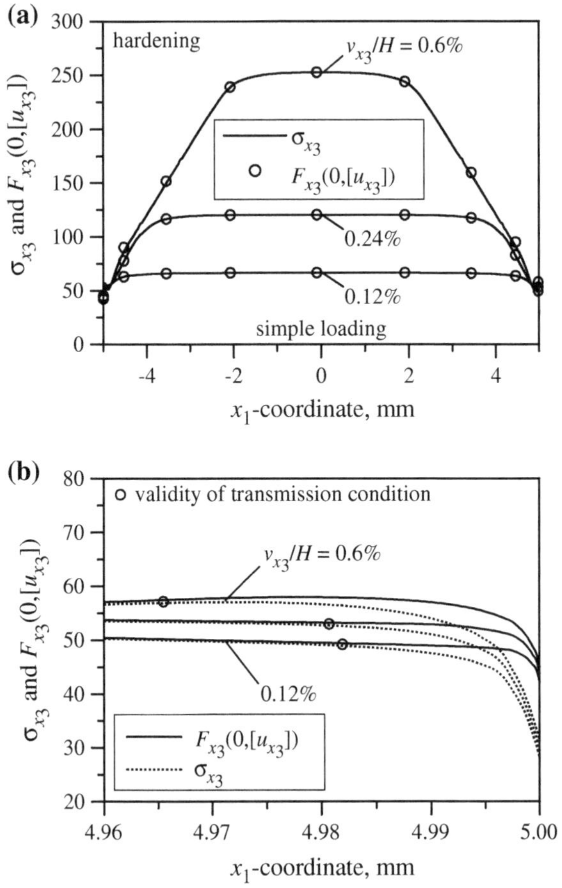

Fig. 19 Determination of the validity of the transmission condition for an elasto-plastic interphase (hardening interphase case; simple tensile loading)

The verification of the validity of the transmission conditions for the combined loading can be done based in Fig. 22. A very important difference in comparison with the hardening law can be observed in the case of the ideal plasticity law. Namely, the region where the transmission conditions are valid is smaller than that in the case of the hardening plastic law (compare Figs. 19b and 20b) and this region essentially depends on the level of plastic deformation (compare Figs. 21b and 22b). To clarify the difference, some estimates of the zone ends have been presented in Table 1 for the hardening and the ideal plasticity law for different levels of the deformation. However, in both cases application of combined loading provided slightly better results for the applicability of the transmission conditions. This is an important result. First of all because a combined external loading is more frequent in technical

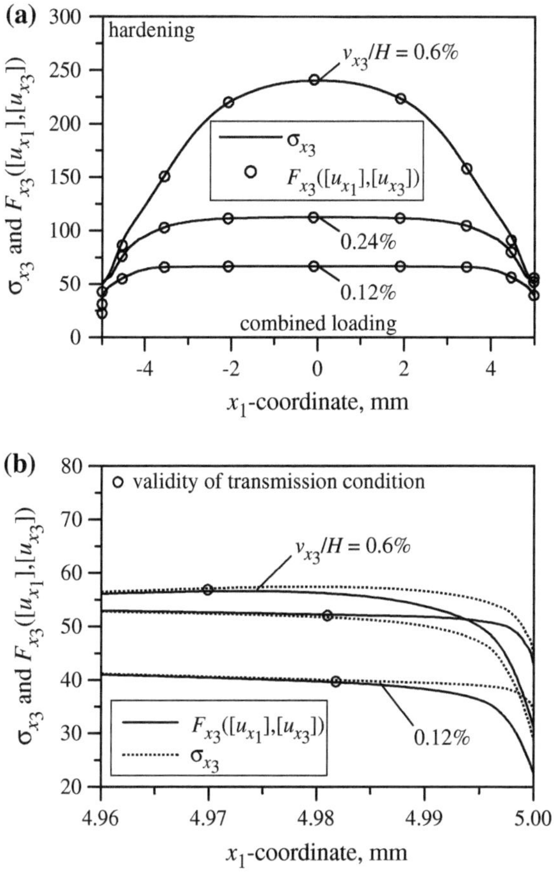

Fig. 20 Determination of the validity of the transmission condition for an elasto-plastic interphase (hardening interphase case; combined loading)

applications. On the other hand, it shows that the worse accuracy appears in simple loading cases which researchers usually apply for testing.

5 Discussions and Conclusions

The theoretical foundation for pressure-independent and pressure dependent transmission conditions for thin elasto-plastic interphases has been provided within this chapter. The high accuracy of the approach could be confirmed exemplarily on the example of a plane strain case under consideration of the von Mises yield condition. The future research work will concentrate on the corresponding derivations for the plane stress case as well as the general three-dimensional case. Finally, the numerical

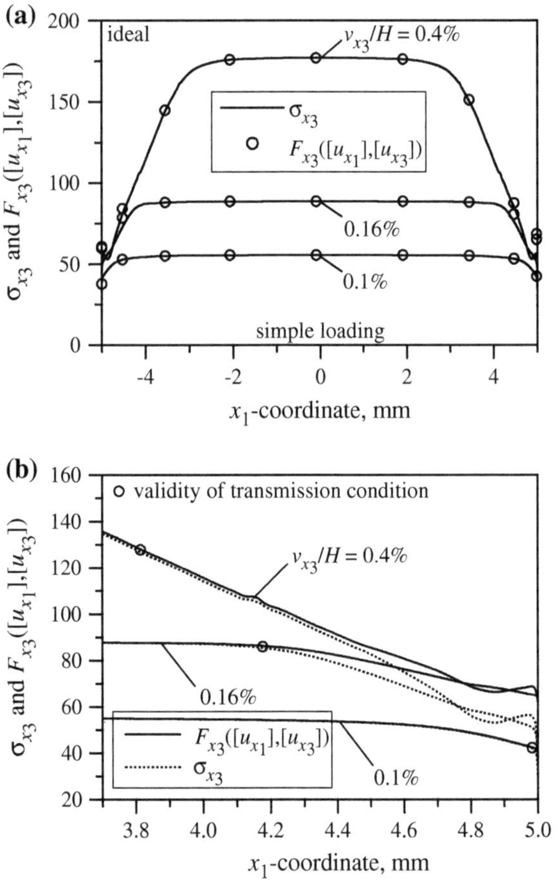

Fig. 21 Determination of the validity of the transmission condition for an elasto-plastic interphase (ideal plasticity interphase; simple tensile loading)

investigation of transmission conditions for pressure-sensitive yield conditions must be performed to obtain a general understanding of the involved errors. This approach allows to significantly simplify finite element computations for composite materials with complex geometry with thin elasto-plastic interfaces (glues, adhesives) where multi-scaled meshing and re-meshing make the computations too costly.

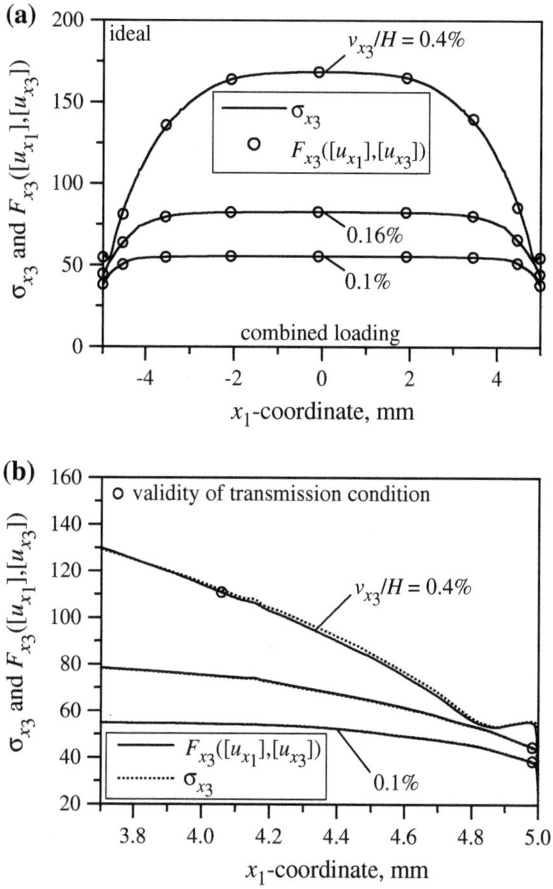

Fig. 22 Determination of the validity of the transmission condition for an elasto-plastic interphase (ideal plasticity interphase; combined loading)

Table 1 Validity of the transmission condition in terms of $\delta/(2h)$

Hardening	Deformation	0.12 %	0.22 %	0.6 %
	Simple tensile	1.81	1.93	3.45
	Combined loading	1.82	1.90	3.01
Ideal	Deformation	0.1 %	0.16 %	0.4 %
	Simple tensile	1.78	82.35	118.6
	Combined loading	1.82	1.80	94.2

Acknowledgments GM and WM thank to FP7 IAPP project PARM-2 (PIAPP-GA-2011-284544) for support of this research. WM gratefully acknowledges facilities and hospitality of the industrial partner, EUROTECH, during her secondment there in the framework of the project.

References

1. Antipov, Y.A., Avila-Pozos, O., Kolaczkowski, S.T., Movchan, A.B.: Mathematical model of delamination cracks on imperfect interfaces. Int. J. Solids Struct. **38**, 6665–6697 (2001)
2. Avila-Pozos, O., Klabring, A., Movchan, A.B.: Asymptotic model of ortotropic highly inhomogeneous layered structure. Mech. Mater. **31**, 101–115 (1999)
3. Bigoni, D.: Nonlinear Solid Mechanics Bifurcation Theory and Material Instability. Cambridge University Press, Cambridge (2012)
4. Benveniste, Y., Miloh, T.: Imperfect soft and stiff interfacses in two-dimentional elasticity. Mech. Mater. **33**, 309–323 (2001)
5. Benveniste, Y.: The effective mechanical behaviour of composite materials with imperfect contact between the constituens. Mech. Mater. **4**, 197–208 (1985)
6. Chen, W.F., Han, D.J.: Plasticity for Structural Engineers. Springer, New York (1988)
7. Chen, W.F.: Constitutive Equations for Engineering Materials. Elsevier, Amsterdam (1994)
8. Erdogan, F.: Fracture mechanics of interfaces. In: Balkoma, A.A. (ed.) Proceedings of the First International Conference on Damage and Failure of Interfaces DFI-I/Vienna/22-24 September 1997, Rotterdam-Brookfield, pp 3–36 (1997)
9. Hashin, Z.: Thermoeleastic properties of fiber composites with imperfect inyterface. Mech. Mater. **8**, 3333–3348 (1990)
10. Hashin, Z.: Thin interphase/imperfect interface in elasticity with application to coated fiber composites. J. Mech. Phys. Solids **50**, 2509–2537 (2002)
11. Hassanipour, M., Öchsner, A.: Implementation of a pressure sensitive yield criterion for adhesives into a commercial finite element code. J. Adhesion **87**, 1125–1147 (2011)
12. Hatheway, A.E.: Evaluating stresses in adhesive bond lines. In: MSC/NASTRAN Users' Conference March 13–17, 1989, Universal City (1989)
13. Hencky, H.: Zur Theorie plastischer Deformationen und der hierdurch im Material hervorgerufenen Nachspannungen. Z. Angew Math. Mech. **4**, 323–334 (1924)
14. Hill, R.: The Mathematical Theory of Plasticity. Oxford University Press, Oxford (1950)
15. Ikeda, T., Yamashita, A., Lee, D., Miyazaki, N.: Failure of a ductile adhesive layer constrained by hard adherends. Trans. ASME J. Eng. Mater. Technol. **122**, 80–85 (2000)
16. Jones, R.M.: Deformation Theory of Plasticity. Bull Ridge Publishing, Blacksburg (2009)
17. Kachanov, L.M.: Foundations of the Theory of Plasticity. North-Holland, Amsterdam (1971)
18. Klabring, A., Movchan, A.B.: Asymptotic modelling of adheasive joints. Mech. Mater. **28**, 137–145 (1998)
19. Lakes, R.S.: Negative Poisson's ratio materials. Science **238**, 551 (1987)
20. Lubarda, V.A.: Deformation theory of plasticity revisited. Proc. Mont. Acad. Sci. Arts **13**, 117–143 (2000)
21. Lubliner, J.: Plasticity Theory. Macmillan, New York (1990)
22. Mahnken, R., Schlimmer, M.: Simulation of strength difference in elasto-plasticity for adhesive materials. Int. J. Numer. Methods Eng. **63**, 1461–1477 (2005)
23. Mishuris, G.: Interface crack and nonideal interface approach. Mode III. Int. J. Fract. **107**(3), 279–296 (2001)
24. Mishuris, G., Kuhn, G.: Asymptotic behaviour of the elastic solution near the tip of a crack situated at a nonideal interface. Z. Angew Math. Mech. **81**(12), 811–826 (2001)
25. Mishuris, G., Öchsner, A., Kuhn, G.: Imperfect interfaces in dissimilar elastic body: FEM-analysis. In: Ren, Z., Kuhn, G., Skerget, L., Hribersek, M. (eds.) Advanced Computational Engineering Mechanics. Proc. of the First Workshop, Maribor Slovenia, October 9–11, 2003, University of Maribor Publishers, Maribor (2003)
26. Mishuris, G., Öchsner, A.: Edge effects connected with thin interphases in composite materials. Compos. Struct. **68**, 409–417 (2005)
27. Mishuris, G., Öchsner, A.: 2D modelling of a thin elasto-plastic interphase between two different materials: plane strain case. Compos. Struct. **80**, 361–372 (2007)
28. Movchan, A.B., Movhan, N.V.: Mathematical Modelling of Solids with Nonregular Boundaries. CRC Press, London (1995)

29. Öchsner, A., Winter, W., Kuhn, G.: Damage and fracture of perforated aluminum alloys. Adv. Eng. Mater. **2**, 423–426 (2000)
30. Öchsner, A., Mishuris, G.: A new finite element formulation for thin non-homogeneous heat-conducting adhesive layers. Int. J. Adhes. Adhes. **22**, 1365–1378 (2008)
31. Rosselli, F., Carbutt, P.: Structural bonding applications for the transportation industry. SAMPE J. **37**(6), 7–13 (2001)
32. Sancaktar, E.: Complex constitutive adhesive models. In: Silva, L.F.M., Öchsner, A. (eds.) Modeling of Adhesively Bonded Joints, pp. 95–130. Springer, Berlin (2008)
33. Yu, H.H., He, M.Y., Hutchinson, J.W.: Edge effects in thin film delamination. Acta Mater. **49**, 93–107 (2001)

Effect of Pressure-Dependency of the Yield Criterion on the Strain Rate Intensity Factor

Sergei Alexandrov, Elena Lyamina and Yeau-Ren Jeng

Abstract In the case of several rigid plastic models, the equivalent strain rate (quadratic invariant of the strain rate tensor) approaches infinity in the vicinity of maximum friction surfaces. The strain rate intensity factor is the coefficient of the leading singular term in a series expansion of the equivalent strain rate in the vicinity of such surfaces. This coefficient controls the magnitude of the equivalent strain rate in a narrow material layer near maximum friction surfaces. On the other hand, the equivalent strain rate is involved in many conventional equations describing the evolution of parameters characterizing material properties. Experimental data show that a narrow layer in which material properties are quite different from those in the bulk often appears in the vicinity of surfaces with high friction in metal forming processes. This experimental fact is in qualitative agreement with the aforementioned evolution equations involving the equivalent strain rate. However, when the maximum friction law is adopted, direct use of such equations is impossible since the equivalent strain rate in singular. A possible way to overcome this difficulty is to develop a new type of evolution equations involving the strain rate intensity factor instead of the equivalent strain rate. This approach is somewhat similar to the conventional approach in the mechanics of cracks when fracture criteria from the strength of materials are replaced with criteria based on the stress intensity factor in the vicinity of crack tips. The development of the new approach requires a special experimental program to establish relations between the magnitude of the strain rate intensity factor and the

S. Alexandrov and E. Lyamina
A.Yu. Ishlinsky Institute for Problems in Mechanics, Russian Academy of Sciences, Moscow, Russia
e-mail: sergei_alexandrov@spartak.ru

E. Lyamina
e-mail: lyamina@inbox.ru

S. Alexandrov and Y.-R. Jeng
Department of Mechanical Engineering and Advanced Institute of Manufacturing with High-tech Innovations, National Chung Cheng University, Chia-Yi, Taiwan
e-mail: imeyrj@ccu.edu.tw

H. Altenbach and A. Öchsner (eds.), *Plasticity of Pressure-Sensitive Materials*, Engineering Materials, DOI: 10.1007/978-3-642-40945-5_5,

evolution of material properties in a narrow material layer near surfaces with high friction as well as a theoretical method to deal with singular solutions for rigid plastic solids. Since no numerical method has been yet developed to determine the strain rate intensity factor, the present chapter focuses on analytical and semi-analytical solutions from which the dependence of the strain rate intensity factor on process and material parameters are found. In particular, the effect of pressure-dependency of the yield criterion on the strain rate intensity factor is emphasized using the double shearing model.

1 Introduction

In the case of rigid perfectly plastic solids, the maximum friction surfaces are defined by the condition that the friction stress at sliding is equal to the shear yield stress of the material [1]. In the case of planar flow, this definition is equivalent to the statement that the maximum friction surface coincides with a characteristic or an envelope of characteristics. The latter definition is naturally generalized on the double shearing model [2]. A distinguished feature of maximum friction surfaces is that the second invariant of the strain rate tensor (this quantity is also called the equivalent strain rate) approaches infinity in the vicinity of such surfaces for several rigid plastic models [1–5]. This theoretical feature of the solutions can be related to the formation of a layer of intensive plastic deformation in the vicinity of frictional interfaces in real metal forming processes [6–8], though no specific theory is available. Therefore, it is of interest to understand the effect of material and process parameters on the magnitude of the strain rate intensity factor. The traditional finite element method cannot be used to find the strain rate intensity factor because it is the coefficient of a singular term. The extended finite element method [9] is in general applicable in this case but no specific code has been developed yet. Therefore, semi-analytical solutions available in the literature are used in the present chapter to reveal the effect of pressure-dependency of the yield criterion on the strain rate intensity factor. The solutions for pressure-independent material are based on Tresca's yield criterion and its associated flow rule. The solutions for pressure-dependent material are based on the double-shearing model [10].

2 Strain Rate Intensity Factor

The strain rate intensity factor has been previously introduced for several rigid plastic models. Most of solutions in which the strain rate intensity factor appears are available for the classical phenomenological theory of plasticity and the double shearing model. The former is a model of pressure-independent plasticity. A great account on this model is given in [11]. The double shearing model is a model of pressure-dependent plasticity based on the Mohr-Coulomb yield criterion. This model is described in detail in [10].

The constitutive equations of the rigid perfectly plastic pressure-independent model are a yield criterion and its associated flow rule. By assumption, the yield criterion is independent of the first invariant of the stress tensor (or the hydrostatic stress). Therefore, it can be represented by a locus in two-dimensional space, where the second and third invariants of the stress tensor are taken as Cartesian coordinates. The second invariant of the stress tensor is also called the equivalent stress, σ_{eq}. A number of yield criteria independent of the hydrostatic stress (pressure-independent yield criteria) have been proposed in the literature, though it is commonly accepted that the criteria due to von Mises and Tresca are most representative of initial yielding in isotropic, metallic materials [12]. The monograph [11] mainly deals with the von Mises criterion. It is worthy of note that the formulation of plane strain problems in dimensionless form is independent of the yield criterion chosen. In other words, any plane strain solution for the von Mises criterion is the solution for any other pressure-independent criterion. The model based on Tresca's yield criterion under conditions of axial symmetry is described in detail in [13, 14].

The constitutive equations of the double shearing model are the Mohr-Coulomb yield criterion, the equation of incompressibility and the equation that connects stresses and velocities. Extensions of this theory to include plastic volume change are also available in the literature (see, for example, [15]) but they are not considered in the present chapter. At a specific set of parameters, the double shearing model reduces to the model of classical pressure-independent plasticity based on Tresca's yield criterion. Since the objective of the present chapter is to demonstrate the effect of pressure-dependency of the yield criterion on the strain rate intensity factor, the Tresca's yield criterion will be used to determine the strain rate intensity factor in axisymmetric problems for pressure-independent materials.

The strain rate intensity factor has been defined in [1] as the coefficient of the leading singular term in a series expansion of the equivalent strain rate in the vicinity of maximum friction surfaces. This work has been restricted to the classical pressure-independent model. The term maximum friction surface is used to indicate that the maximum friction law is adopted on that surface. The original formulation of the maximum friction law for pressure-independent material is

$$\tau_f = \tau_s \tag{1}$$

at sliding. Here τ_f is the friction stress and τ_s is the shear yield stress. It is worthy to note that τ_s is constant for perfectly plastic materials. It is known from the general theory (see, for example, [14]) that the equations of plane strain and axisymmetric deformation are hyperbolic (in the latter case, Tresca's yield criterion should be adopted). Moreover, the characteristics for the stresses and the velocities coincide and, therefore, there are only two distinct characteristic directions at a point. The shear stress along the characteristic is equal to τ_s. Thus the boundary condition (1) is equivalent to the statement that the friction surface coincides with a characteristic or an envelope of characteristics. Let ϕ be the angle between the major principal stress σ_1 and the tangent to the friction surface, measured from the tangent anti-clockwise. In the case of the model of pressure-independent plasticity the characteristics are

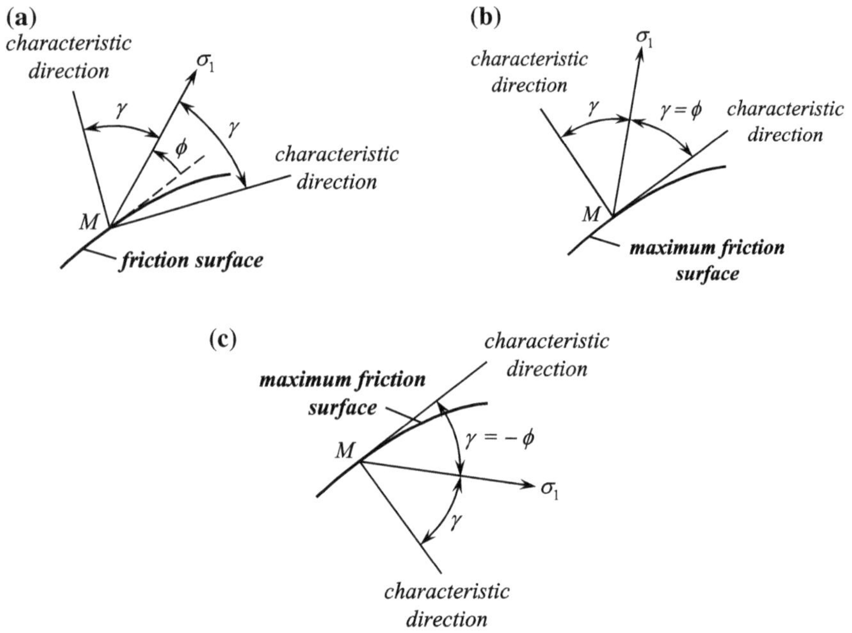

Fig. 1 Orientation of the characteristic directions relative to the tangent to friction surfaces. **a** arbitrary friction surface. **b** maximum friction surface with $\phi > 0$. **c** maximum friction surface with $\phi < 0$

inclined to the direction of σ_1 at $\pm\pi/4$. However, for later convenience it is assumed here that this angle is $\pm\gamma$. It is evident that the original case is obtained at $\gamma = \pi/4$. The characteristic directions at a generic point, M, on an arbitrary friction surface are illustrated in Fig. 1a. If the maximum friction law is valid on this surface then the orientation of the characteristics becomes such as shown in Fig. 1b, c. It is evident that $\phi = \gamma$ in Fig. 1b and $\phi = -\gamma$ in Fig. 1c. Therefore, since $\gamma = \pi/4$ in the case under consideration, the boundary condition (1) is equivalent to

$$\phi = \pm\frac{\pi}{4} \tag{2}$$

at sliding.

The equations of the double sharing model are also hyperbolic [10]. Therefore, the maximum friction law in the form of Eq. (2) can be extended to this model with no difficulty. In particular, it has been shown in [10] that in the case of plane strain and axisymmetric deformation $\gamma = \pi/4 + \varphi/2$ (Fig. 1). Here φ is the angle of internal friction, a material constant. Eq. (2) transforms to

$$\phi = \pm\left(\frac{\pi}{4} + \frac{\varphi}{2}\right) \tag{3}$$

at sliding. The strain rate intensity factor for this model has been introduced in [2].

When the maximum friction law is adopted, two qualitatively different options are possible. A natural way to distinguish these options is to use the maximum friction law in the form of Eqs. (2) and (3). Either of these equations is valid if the maximum friction surface coincides with (i) a characteristic or (ii) an envelope of characteristics. Option (i) imposes severe restrictions on the velocity field. In particular, the rate of extension along the friction surface is zero in the case of plane strain deformation of pressure-independent material. Therefore, in most cases Option (ii) occurs and it will be assumed throughout this chapter.

In order to precisely define the strain rate intensity factor, it is necessary to introduce the equivalent strain rate. The standard definition for this quantity involves the equivalent stress and the rate of plastic work [11]. In this chapter, however, the equivalent strain rate is understood as a pure kinematic quantity (the quadratic invariant of the strain rate tensor) defined by

$$\xi_{eq} = \sqrt{\frac{2}{3}\xi_{ij}\xi_{ij}}, \tag{4}$$

where ξ_{ij} are the components of the strain rate tensor. This definition coincides with the standard definition for the von Mises yield criterion combined with its associated flow rule. The equivalent strain rate approaches infinity in the vicinity of maximum friction surfaces. In particular,

$$\xi_{eq} = \frac{D}{\sqrt{s}} + o\left(\frac{1}{\sqrt{s}}\right) \quad \text{as} \quad s \to 0, \tag{5}$$

where s is the normal distance to the friction surface and D is the strain rate intensity factor. Under various assumptions concerning the pressure-independent yield criterion and modes of deformation, this result has been obtained in [1, 16–20]. For materials obeying the double shearing model the asymptotic expansion (5) for plane strain and axisymmetric flow has been found in [2, 21], respectively. Particular solutions show that (5) is also satisfied for other rigid plastic models [3–5, 22–25]. Reviews of solutions for the strain rate intensity factor are given in [26, 27].

It is always possible to choose such a coordinate system that the normal strain rates are bounded and one of the shear strain rates approaches infinity in the vicinity of maximum friction surfaces. Denote this shear strain rate by ξ_τ. Then, it follows from (4) and (5) that

$$|\xi_\tau| = \frac{\sqrt{3}}{2}\frac{D}{\sqrt{s}} + o\left(\frac{1}{\sqrt{s}}\right) \quad \text{as} \quad s \to 0. \tag{6}$$

In order to provide some insights into distinguished features of the maximum friction law and material models leading to (5), this asymptotic expansion is below derived for plane strain deformation of pressure-independent materials.

It is convenient to introduce a Cartesian coordinate system (x, y, z) whose z-axis is orthogonal to planes of flow. Let σ_{xx}, σ_{yy} and σ_{xy} be the components of the stress tensor in this coordinate system. Then, any pressure-independent yield criterion can be written in the form

$$\left(\sigma_{xx}-\sigma_{yy}\right)^2+4\sigma_{xy}^2=4\tau_s^2. \tag{7}$$

The flow rule associated with this yield criterion is

$$\xi_{xx}=\lambda\left(\sigma_{xx}-\sigma_{yy}\right),\quad \xi_{yy}=\lambda\left(\sigma_{yy}-\sigma_{xx}\right),\quad \xi_{xy}=2\lambda\sigma_{xy} \tag{8}$$

where ξ_{xx}, ξ_{yy} and ξ_{xy} are the components of the strain rate tensor in the Cartesian coordinates and $\lambda \geq 0$. Eliminating λ in (8) gives

$$\xi_{xx}+\xi_{yy}=0,\quad \frac{\xi_{xy}}{\xi_{xx}-\xi_{yy}}=\frac{\sigma_{xy}}{\sigma_{xx}-\sigma_{yy}}. \tag{9}$$

The first equation of this system is the equation of incompressibility. The strain rate components are expressed through the velocity components, u_x and u_y, as

$$\xi_{xx}=\frac{\partial u_x}{\partial x},\quad \xi_{yy}=\frac{\partial u_y}{\partial y},\quad \xi_{xy}=\frac{1}{2}\left(\frac{\partial u_y}{\partial x}+\frac{\partial u_x}{\partial y}\right). \tag{10}$$

Substituting (10) into (9) yields

$$\frac{\partial u_x}{\partial x}+\frac{\partial u_y}{\partial y}=0,\quad \left(\frac{\partial u_x}{\partial y}+\frac{\partial u_y}{\partial x}\right)\left(\sigma_{xx}-\sigma_{yy}\right)=2\left(\frac{\partial u_x}{\partial x}-\frac{\partial u_y}{\partial y}\right)\sigma_{xy}. \tag{11}$$

The constitutive equations should be supplemented with the equilibrium equations

$$\frac{\partial \sigma_{xx}}{\partial x}+\frac{\partial \sigma_{xy}}{\partial y}=0,\quad \frac{\partial \sigma_{xy}}{\partial x}+\frac{\partial \sigma_{yy}}{\partial y}=0. \tag{12}$$

The system of five Eqs. (7), (11) and (12) in the five unknowns σ_{xx}, σ_{yy}, σ_{xy}, u_x, u_y, is the basis for the calculation of the distribution of stress and velocity in the plastic region. It is known that this system is hyperbolic [11]. The characteristic directions make an angle of $\pi/4$ with the major principal stress direction. Let ψ be the angle between the major principal stress σ_1 and the x-axis, measured from the axis anticlockwise. Then, the orientation of the characteristic curves relative to the x-axis is (Fig. 2)

$$\phi_1=\psi-\frac{\pi}{4},\quad \phi_2=\psi+\frac{\pi}{4}. \tag{13}$$

Let ω be a tool surface (curve in planes of flow) where the condition (2) is satisfied, and consider Eqs. (7), (11) and (12) at an arbitrary point, M, on that surface. The tool is regarded as fixed. The Cartesian coordinate system is taken to be situated at M with the y-axis directed along the normal to ω, away from the rigid tool and

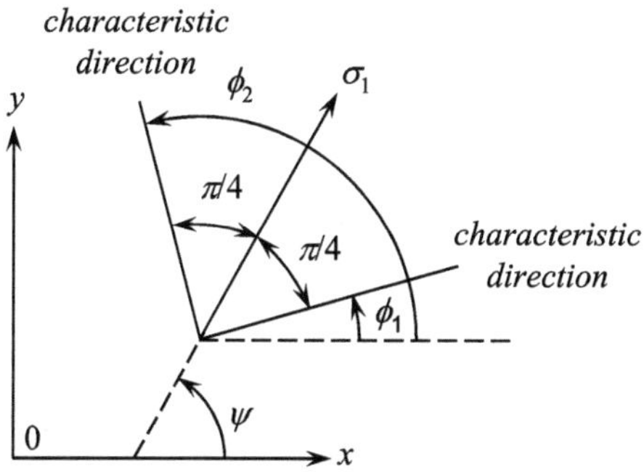

Fig. 2 Orientation of the major principal stress and characteristic directions relative to the x-axis

towards the plastic material (Fig. 3). Since the condition (2) is equivalent to (1), in this coordinate system

$$\left|\sigma_{xy}\right| = \tau_s \tag{14}$$

at M. Therefore, it follows from (7) that

$$\sigma_{xx} = \sigma_{yy} \tag{15}$$

at M. Using $(11)_1$ Eq. $(11)_2$ can be rewritten as

$$\left(\frac{\partial u_x}{\partial y} + \frac{\partial u_y}{\partial x}\right)\left(\sigma_{xx} - \sigma_{yy}\right) = 4\frac{\partial u_x}{\partial x}\sigma_{xy}. \tag{16}$$

Substituting (14) and (15) into (16) shows that it is necessary to examine the cases

$$\frac{\partial u_x}{\partial x} = 0 \quad \text{at} \quad y = 0 \tag{17}$$

and

$$\left|\frac{\partial u_x}{\partial y} + \frac{\partial u_y}{\partial x}\right| \to \infty \quad \text{as} \quad y \to 0. \tag{18}$$

It follows from (17) and the orientation of the x-axis (Fig. 3) that the rate of extension along the friction surface is zero. Since M is an arbitrary point on the friction surface, this means that a characteristic curve coincides with the friction surface. Such solutions have been excluded from consideration. Therefore, it is necessary to assume that Eq. (18) is valid. In order to determine the asymptotic behaviour of the velocity field in the vicinity of the friction surface, some additional assumptions are necessary. In particular,

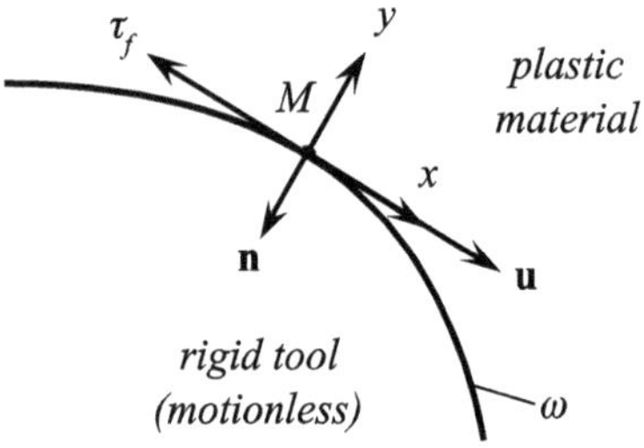

Fig. 3 Orientation of the local coordinate system at M

(i) Velocity components are bounded at M,
(ii) In-surface derivatives of the velocity components are bounded at M,
(iii) Solution can be represented by a power series in the vicinity of M.

With no loss of generality, it is possible to choose the direction of the x-axis such that $\xi_{xy} > 0$ at M. In particular, its direction should coincide with the direction of the velocity vector, **u**, of a material particle located at M at a given instant (Fig. 3). In Fig. 3, **n** is the outer normal to the plastic material and τ_f is the friction stress applied to this material. Then, using assumptions (ii) and (iii) Eq. (18) transforms to

$$\frac{\partial u_x}{\partial y} = u_1 y^{-\alpha} + o\left(y^{-\alpha}\right) \quad \text{as} \quad y \to 0, \tag{19}$$

where u_1 may depend on x and $u_1 > 0$. Integrating (19) yields

$$u_x = u_0 + \frac{u_1}{(1-\alpha)} y^{1-\alpha} + o\left(y^{1-\alpha}\right) \quad \text{as} \quad y \to 0, \tag{20}$$

where u_0 may depend on x and $u_0 > 0$. It follows from assumption (i) and (20) that

$$\alpha < 1. \tag{21}$$

On the other hand, (18) is satisfied if and only if $\alpha > 0$ in (19). Combining this inequality and (21) leads to

$$0 < \alpha < 1. \tag{22}$$

Using (20) Eq. $(11)_1$ can be transformed to

$$\frac{du_0}{dx} + \frac{y^{1-\alpha}}{(1-\alpha)} \frac{du_1}{dx} + \frac{\partial u_y}{\partial y} + o\left(y^{1-\alpha}\right) = 0 \quad \text{as} \quad y \to 0. \tag{23}$$

Integrating with the boundary condition $u_y = 0$ for $y = 0$ (at point M) and taking into account (22) gives

$$u_y = -\frac{du_0}{dx} y + o(y) \quad \text{as} \quad y \to 0. \tag{24}$$

Using (10), (20) and (24) the non-zero strain rate components can be represented as

$$\xi_{xx} = -\xi_{yy} = \frac{du_0}{dx} + o(1), \quad \xi_{xy} = \frac{u_1}{2} y^{-\alpha} + o\left(y^{-\alpha}\right) \quad \text{as} \quad y \to 0. \tag{25}$$

It is worthy to note here that, by assumption, $\xi_{xx} \neq 0$ (or $du_0/dx \neq 0$) and $u_1 \neq 0$ at M. Since $\xi_{xy} > 0$ by the choice of the coordinate system, it follows from $(8)_3$ that $\sigma_{xy} > 0$ as well. Therefore, using assumption (iii) the distribution of stresses in the vicinity of the friction surface can be represented as

$$\begin{aligned} \sigma_{xx} &= \sigma_{xx}^{(0)} + \sigma_{xx}^{(1)} y^{\gamma_{11}} + o\left(y^{\gamma_{11}}\right), \\ \sigma_{yy} &= \sigma_{yy}^{(0)} + \sigma_{yy}^{(1)} y^{\gamma_{22}} + o\left(y^{\gamma_{22}}\right), \\ \sigma_{xy} &= \sigma_{xy}^{(0)} + \sigma_{xy}^{(1)} y^{\gamma_{12}} + o\left(y^{\gamma_{12}}\right) \end{aligned} \tag{26}$$

as $y \to 0$. Here $\sigma_{xx}^{(0)}$, $\sigma_{xx}^{(1)}$, $\sigma_{yy}^{(0)}$, $\sigma_{yy}^{(1)}$, $\sigma_{xy}^{(0)}$, and $\sigma_{xy}^{(1)}$ can depend on x. Equations (14) and (15) demand

$$\sigma_{xx}^{(0)} = \sigma_{yy}^{(0)} \quad \text{and} \quad \sigma_{xy}^{(0)} = \tau_s \tag{27}$$

at $x = 0$ (at point M). Substituting (19), (24) and (26) into (16) and using (27) yields

$$u_1 y^{-\alpha} \left(\sigma_{xx}^{(1)} y^{\gamma_{11}} - \sigma_{yy}^{(1)} y^{\gamma_{22}}\right) = 4 \frac{du_0}{dx} \tau_s \tag{28}$$

to leading order. Since the right hand side of this equation is $O(1)$ as $y \to 0$, it is necessary to examine the cases

$$\gamma_{11} - \alpha = 0 \tag{29}$$

and

$$\gamma_{22} - \alpha = 0. \tag{30}$$

Substituting (26) into (12) gives

$$\begin{aligned} \frac{d\sigma_{xx}^{(0)}}{dx} + \frac{d\sigma_{xx}^{(1)}}{dx} y^{\gamma_{11}} + \sigma_{xy}^{(1)} \gamma_{12} y^{\gamma_{12}-1} &= 0, \\ \frac{d\sigma_{xy}^{(0)}}{dx} + \frac{d\sigma_{xy}^{(1)}}{dx} y^{\gamma_{12}} + \sigma_{yy}^{(1)} \gamma_{22} y^{\gamma_{22}-1} &= 0 \end{aligned} \tag{31}$$

to leading order. Combining (30) and $(31)_2$ yields $\alpha = 1$ or $\gamma_{12} = \alpha - 1$. The former contradicts (22). The latter combined with (22) leads to $\gamma_{12} < 0$. According to (26) this inequality results in $\left|\sigma_{xy}\right| \to \infty$ as $y \to 0$. Therefore, it is necessary to assume that (29) is valid. Then, substituting (26) into (7) and using (27) yields

$$y^{2\alpha} \left[\sigma_{xx}^{(1)} - \sigma_{yy}^{(1)} y^{(\gamma_{22}-1)}\right]^2 + 4\left[\tau_s + \sigma_{xy}^{(1)} y^{\gamma_{12}}\right]^2 = 4\tau_s^2 \tag{32}$$

as $y \to 0$. Using straight multiplication, Eq. (32) is simplified to

$$y^{2\alpha}\left[\sigma_{xx}^{(1)}\right]^2 + 8\tau_s\sigma_{xy}^{(1)}y^{\gamma_{12}} = 0 \quad \text{as} \quad y \to 0. \tag{33}$$

It follows from this equation that

$$2\alpha = \gamma_{12}. \tag{34}$$

Substituting (29) and (34) into $(31)_1$ shows that $2\alpha - 1 = \alpha$ or $2\alpha - 1 = 0$. The former gives $\alpha = 1$ and therefore contradicts (22). The latter results in

$$\alpha = \frac{1}{2}. \tag{35}$$

The asymptotic expansion (5) immediately follows from (4), (25) and (35).

3 Plane Strain Solutions for Pressure-Independent Material

3.1 Basic Equations

Section 3 is concerned with plane strain solutions for pressure-independent materials. In this section, two coordinate systems will be used, namely a Cartesian coordinate system $(x,\ y,\ z)$ and a cylindrical coordinate system $(r,\ \theta,\ z)$. All the solutions considered are independent of z. The constitutive equations in the Cartesian coordinate system are (7) and (9). The orientation of the characteristic curves relative to the x-axis is given by (13). The strain rate components ξ_{zz}, ξ_{xz} and ξ_{yz} as well as the stress components σ_{xz} and σ_{yz} vanish. The non-zero strain rate components are expressed through the velocity components according to (10). The equilibrium equations are given by (12). The transformation equations for stress components in xy-planes are (Fig. 2)

$$\sigma_{xx} = \frac{\sigma_1 + \sigma_2}{2} + \tau_s \cos 2\psi,\ \sigma_{yy} = \frac{\sigma_1 + \sigma_2}{2} - \tau_s \cos 2\psi,\ \sigma_{xy} = \tau_s \sin 2\psi. \tag{36}$$

It has been taken into account here that Eq. (7) in terms of the principal stresses σ_1 and σ_2 becomes

$$\sigma_1 - \sigma_2 = 2\tau_s. \tag{37}$$

Let σ_{rr}, $\sigma_{\theta\theta}$, σ_{zz}, σ_{rz}, $\sigma_{z\theta}$ and $\sigma_{r\theta}$ be the components of the stress tensor in the cylindrical coordinate system. The components σ_{rz} and $\sigma_{z\theta}$ vanish. The yield criterion (7) transforms to

$$(\sigma_{rr} - \sigma_{\theta\theta})^2 + 4\sigma_{r\theta}^2 = 4\tau_s^2. \tag{38}$$

Let ξ_{rr}, $\xi_{\theta\theta}$, ξ_{zz}, ξ_{rz}, $\xi_{z\theta}$ and $\xi_{r\theta}$ be the components of the strain rate tensor in the cylindrical coordinate system. The components ξ_{zz}, ξ_{rz} and $\xi_{z\theta}$ vanish. Equation (9) become

$$\xi_{rr} + \xi_{\theta\theta} = 0, \quad \frac{\xi_{r\theta}}{\xi_{rr} - \xi_{\theta\theta}} = \frac{\sigma_{r\theta}}{\sigma_{rr} - \sigma_{\theta\theta}}. \tag{39}$$

Equation (13) is valid but ψ is to be understood as the angle between the major principal stress σ_1 and the r-axis, measured from the axis anti-clockwise (Fig. 4). Thus the orientation of the characteristic directions relative to the r-axis is

$$\phi_1 = \psi - \frac{\pi}{4}, \quad \phi_2 = \psi + \frac{\pi}{4}. \tag{40}$$

The non-zero components of the strain rate tensor are

$$\xi_{rr} = \frac{\partial u_r}{\partial r}, \quad \xi_{\theta\theta} = \frac{1}{r}\left(\frac{\partial u_\theta}{\partial \theta} + u_r\right), \quad \xi_{r\theta} = \frac{1}{2}\left(\frac{\partial u_\theta}{\partial r} - \frac{u_\theta}{r} + \frac{1}{r}\frac{\partial u_r}{\partial \theta}\right), \tag{41}$$

where u_r and u_θ are the radial and circumferential velocities, respectively. The equilibrium equations are

$$\frac{\partial \sigma_{rr}}{\partial r} + \frac{1}{r}\frac{\partial \sigma_{r\theta}}{\partial r} + \frac{\sigma_{rr} - \sigma_{\theta\theta}}{r} = 0, \quad \frac{\partial \sigma_{r\theta}}{\partial r} + \frac{1}{r}\frac{\partial \sigma_{\theta\theta}}{\partial \theta} + \frac{2\sigma_{r\theta}}{r} = 0. \tag{42}$$

The transformation equations for stress components in $r\theta$-planes are (Fig. 4)

$$\sigma_{rr} = \frac{\sigma_1 + \sigma_2}{2} + \tau_s \cos 2\psi, \; \sigma_{\theta\theta} = \frac{\sigma_1 + \sigma_2}{2} - \tau_s \cos 2\psi, \; \sigma_{r\theta} = \tau_s \sin 2\psi. \tag{43}$$

Here Eq. (37) has been taken into account.

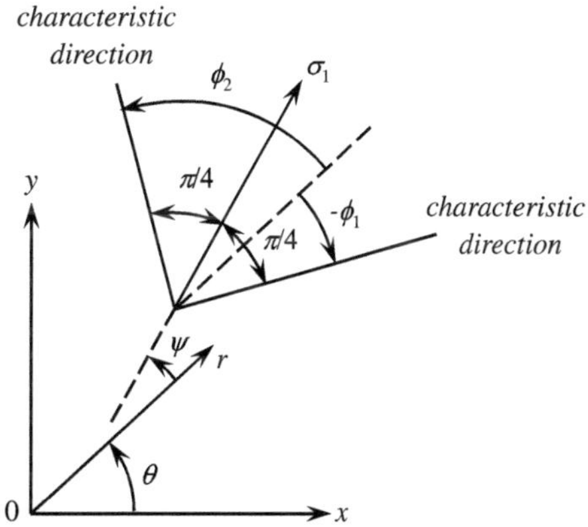

Fig. 4 Orientation of the major principal stress and characteristic directions relative to the r-axis

3.2 Compression of a Plastic Layer Between Parallel Plates

Consider compression of a wide plastic layer between two parallel plates. An approximate solution of this problem, known as Prandtl's problem, can be found in any monograph on plasticity theory (see, for example, [11]). The thickness of the layer is $2H$ and its width is $2L$. By assumption, $H/L \ll 1$. It is possible to choose the Cartesian coordinate system such that its axes x and y coincide with the axes of symmetry of the layer (Fig. 5). Therefore, it is sufficient to find the solution in the domain $0 \le x \le L$ and $0 \le y \le H$. The maximum friction law is valid at $y = H$. The velocity boundary conditions are

$$u_y = 0 \tag{44}$$

at $y = 0$,

$$u_y = -V \tag{45}$$

at $y = H$ and

$$u_x = 0 \tag{46}$$

at $x = 0$. Here V is the speed of the plate. The stress boundary conditions, in addition to the friction law, are

$$\sigma_{xy} = 0 \tag{47}$$

at $y = 0$, $x = 0$ and $x = L$ and

$$\sigma_{xx} = 0 \tag{48}$$

at $x = L$. The solution given in [11] ignores the boundary condition (47) at $x = 0$ and $x = L$. The boundary conditions (46) and (48) are replaced with the following integral conditions

$$\int_0^H u_x|_{x=0} dy = 0 \tag{49}$$

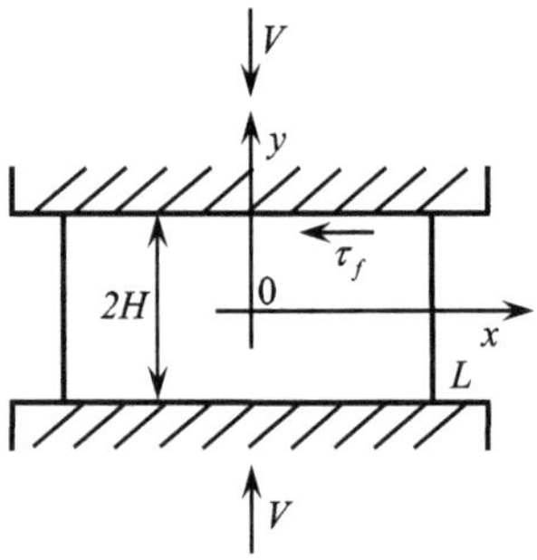

Fig. 5 Illustration of the boundary value problem

and

$$\int_0^H \sigma_{xx}|_{x=L} dy = 0, \tag{50}$$

respectively. Under the assumptions formulated, the velocity field satisfying Eq. (9) as well as the boundary conditions (44) and (45) is given by [11]

$$u_x = V\frac{x}{H} + 2V\left[1 - \left(\frac{y}{H}\right)^2\right]^{1/2} + U, \quad u_y = -V\frac{y}{H}, \tag{51}$$

where U is a constant of integration. Its value can be determined from the boundary condition (49). However, it has no effect on the strain rate intensity factor and, therefore, is not found here. It has been shown in [11] that the stress field used to determine the value of the right hand side of $(9)_2$ satisfies Eqs. (7) and (12) as well as the boundary conditions (1) at $y = H$, (47) at $y = 0$ and (50). Using (10) the components of the strain rate tensor are determined from (51). Then, the equivalent strain rate is found from (4) as

$$\xi_{\text{eq}} = \frac{2}{\sqrt{3}} \frac{V}{\sqrt{H^2 - y^2}}. \tag{52}$$

In the case under consideration $s = H - y$. Therefore, Eq. (52) can be represented in the form

$$\xi_{\text{eq}} = \sqrt{\frac{2}{3}} \frac{V}{\sqrt{s}\sqrt{H}} + o\left(\frac{1}{\sqrt{s}}\right) \quad \text{as} \quad s \to 0. \tag{53}$$

Comparing (5) and (53) gives

$$D = \sqrt{\frac{2}{3}} \frac{V}{\sqrt{H}}. \tag{54}$$

3.3 *Flow of Plastic Material Through an Infinite Wedge-Shaped Channel*

This is also one of the classical problems of plasticity. Its solution used in this section has been given in [11]. The process is illustrated in Fig. 6. Material flows to the line of intersection of two plates. The plates are inclined to each other at an angle 2α. The axis $\theta = 0$ of the cylindrical coordinate system coincides with the axis of symmetry of the flow. Therefore, it is sufficient to find the solution in the domain $0 \le \theta \le \alpha$. The maximum friction law is supposed at $\theta = \alpha$. The velocity boundary conditions are

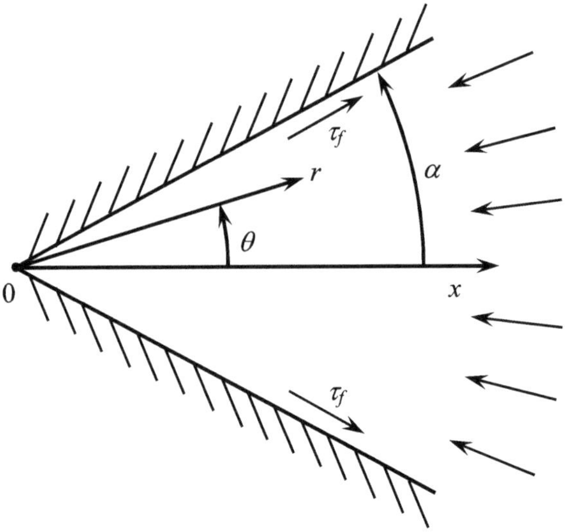

Fig. 6 Illustration of the boundary value problem

$$u_\theta = 0 \tag{55}$$

at $\theta = 0$ and $\theta = \alpha$. The stress boundary condition, in addition to the friction law, is

$$\sigma_{r\theta} = 0 \tag{56}$$

at $\theta = 0$. The velocity field satisfying Eq. (39) as well as the boundary conditions (55) are

$$u_r = -\frac{B}{r\,(c - 2\cos 2\psi)}, \quad u_\theta = 0, \tag{57}$$

where B is proportional to the material flux, c can be determined numerically using the maximum friction law and the boundary condition (56), and ψ is related to θ by the following equation

$$\cos 2\psi \left(\frac{d\psi}{d\theta} + 1\right) = \frac{c}{2} \tag{58}$$

whose solution is

$$\theta = -\psi + c \arctan\left[\left(\frac{c+2}{c-2}\right)^{1/2} \tan\psi\right]\left(c^2 - 4\right)^{-1/2}. \tag{59}$$

It has been shown in [11] that the stress field used to determine the value of the right hand side of $(39)_2$ satisfies Eqs. (38) and (42).

The maximum friction surface is determined by the equation $\theta = \alpha$. Therefore, $\phi_1 = 0$ or $\phi_2 = 0$ in (40). In order to choose between these two options, it is necessary to take into account that $\sigma_{r\theta} > 0$ at $\theta = \alpha$ (Fig. 6). Then, it is evident from (43) that $0 < \psi < \pi/2$ at $\theta = \alpha$. The equation $\phi_2 = 0$ contradicts this inequality. Therefore, $\phi_1 = 0$ and $\psi = \pi/4$ at $\theta = \alpha$. The equation for c is obtained from (59) at $\theta = \alpha$ and $\psi = \pi/4$. The resulting equation has been solved numerically and its solution is illustrated in Fig. 7. Using (41) the components of the strain rate tensor are determined from (57). Then, the equivalent strain rate is found from (4) as

$$\xi_{eq} = \frac{2}{\sqrt{3}} \frac{B}{r^2 (c - 2\cos 2\psi) \cos 2\psi}. \tag{60}$$

Expanding the right hand side of this expression in a series near $\psi = \pi/4$ and using (59) give

$$\xi_{eq} = \frac{\sqrt{2}B}{\sqrt{3} r^2 c \sqrt{c} (\alpha - \theta)^{1/2}} + o\left[(\alpha - \theta)^{-1/2}\right] \quad \text{as} \quad \theta \to \alpha. \tag{61}$$

Comparing (5) and (61) it is possible to conclude that the strain rate intensity factor is

$$D = \sqrt{\frac{2}{3}} \frac{B}{(rc)^{3/2}}. \tag{62}$$

Let Q be the material flux per unit length. Then,

$$Q = -2 \int_0^\alpha u_r r d\theta. \tag{63}$$

Substituting (57) into (63) and replacing integration with respect to θ with integration with respect to ψ by means of (58) result in

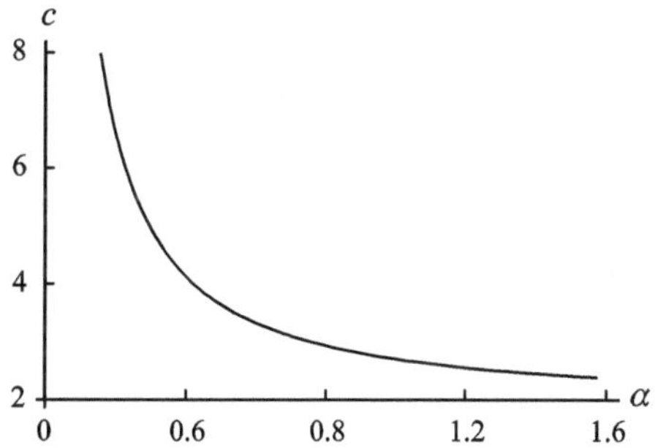

Fig. 7 Variation of c with α

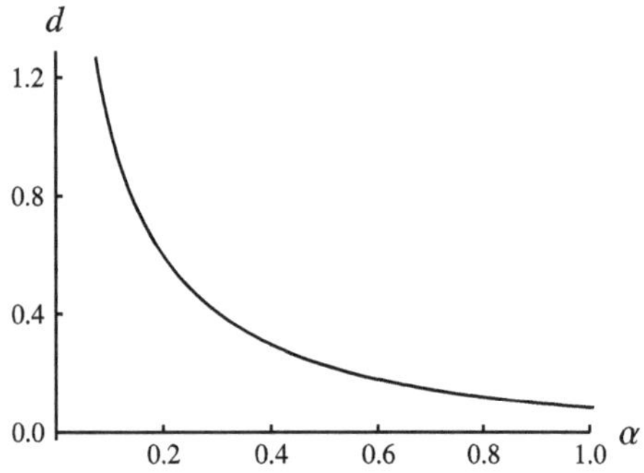

Fig. 8 Variation of the dimensionless strain rate intensity factor with α

$$B = \frac{Q}{4}\left[\int_0^{\pi/4} \frac{\cos 2\psi}{(c - 2\cos 2\psi)^2} d\psi\right]^{-1}. \tag{64}$$

Eliminating B in (62) by means of (64) shows that the strain rate intensity factor is a linear function of Q. In order to reveal the effect of α on the strain rate intensity factor, it is convenient to introduce its dimensionless representation by

$$d = \frac{Dr^{3/2}}{Q}. \tag{65}$$

It follows from (62), (64) and (65) that

$$d = \frac{1}{2\sqrt{6}c^{3/2}}\left[\int_0^{\pi/4} \frac{\cos 2\psi}{(c - 2\cos 2\psi)^2} d\psi\right]^{-1}. \tag{66}$$

Since the dependence of c on α has been found (Fig. 7), the variation of d with α is immediately determined from (66). This dependence is illustrated in Fig. 8.

3.4 Compression of a Plastic Layer Between Cylindrical Surfaces

The boundary value problem is illustrated in Fig. 9. Its solution has been given in [28]. The outer surface of radius R_2 is rigid and motionless whereas the inner surface of current radius R_1 expands. The rate of this expansion is $\dot{R}_1$. The solution is restricted to instantaneous flow. It is natural to use the cylindrical coordinate system shown in Fig. 9. The flow is symmetric with respect to the axis $\theta = 0$. It is therefore sufficient to obtain the solution in the region $0 \le \theta \le \theta_0$ where θ_0 is the orientation of the edge of the layer. The velocity boundary conditions are

Fig. 9 Illustration of the boundary value problem

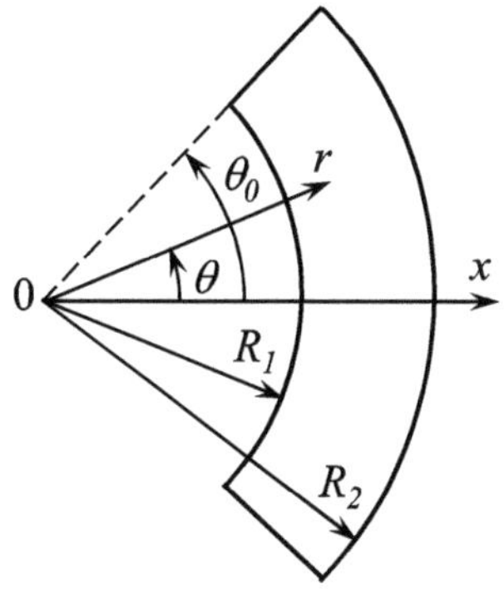

$$u_r = 0 \quad \text{at} \quad r = R_2, \tag{67}$$

$$u_r = \dot{R}_1 \quad \text{at} \quad r = R_1 \tag{68}$$

and

$$u_\theta = 0 \quad \text{at} \quad \theta = 0. \tag{69}$$

The stress boundary conditions, in addition to the maximum friction law, are

$$\sigma_{r\theta} = 0 \quad \text{at} \quad \theta = 0 \quad \text{and} \quad \theta = \theta_0, \tag{70}$$

and

$$\sigma_{\theta\theta} = 0 \quad \text{at} \quad \theta = \theta_0. \tag{71}$$

It is evident that the problem under consideration can be viewed as a generalization of the Prandtl's problem (Sect. 3.2). Therefore, the same assumptions are made. In particular, end effects are neglected such that the solution is not valid in the vicinity of $\theta = 0$ and $\theta = \theta_0$. Accordingly, the boundary conditions (70) are ignored, and the boundary conditions (69) and (71) should be replaced with integral conditions. On the other hand, the boundary conditions at $r = R_1$ and $r = R_2$ are exactly satisfied. These boundary conditions include (67), (68) and the maximum friction law at $r = R_1$ and $r = R_2$. Since the maximum friction law acts on both contact surfaces, two strain rate intensity factors are obtained. The velocity field satisfying Eqs. (39) as well as the boundary conditions (67) and (68) are [28]

$$\frac{u_r}{\dot{R}_1} = U_r(r), \quad \frac{u_\theta}{\dot{R}_1} = -\left(r\frac{dU_r}{dr} + U_r\right)\theta + U_\theta(r), \tag{72}$$

where $U_r(r)$ and $U_\theta(r)$ are functions of r given by

$$U_r = \frac{R_1\left(r^2 - R_2^2\right)}{r\left(R_1^2 - R_2^2\right)},$$

$$U_\theta = \frac{\sqrt{2}R_2}{\left[R_2^2 + R_1^2 + \left(R_2^2 - R_1^2\right)\sin 2\psi\right]^{1/2}} \times$$

$$\times \left\{u_0 + \int_{\pi/4}^{\psi} \frac{\left[R_2^2 + 3R_1^2 + \left(R_2^2 - R_1^2\right)\sin 2\gamma\right]\sin 2\gamma}{\left[R_2^2 + R_1^2 + \left(R_2^2 - R_1^2\right)\sin 2\gamma\right]} d\gamma\right\},$$

where γ is a dummy variable of integration and ψ is related to r by the following equation

$$\sin 2\psi = \frac{2R_1^2R_2^2}{\left(R_2^2 - R_1^2\right)r^2} - \frac{\left(R_2^2 + R_1^2\right)}{\left(R_2^2 - R_1^2\right)}. \tag{73}$$

It has been shown in [28] that the stress field used to determine the value of the right hand side of $(39)_2$ satisfies Eqs. (38) and (42) as well as the boundary condition (1) at $r = R_1$ and $r = R_2$. Using (41) the components of the strain rate tensor are determined from (72) and (73) as

$$\xi_{rr} = -\xi_{\theta\theta} = \frac{\dot{R}_1R_1\left(r^2 + R_2^2\right)}{\left(R_1^2 - R_2^2\right)r^2}, \quad \xi_{r\theta} = \frac{\dot{R}_1R_1\left(R_2^2 + r^2\right)}{\left(R_1^2 - R_2^2\right)r^2}\tan 2\psi.$$

Then, the equivalent strain rate is found from (4) as

$$\xi_{\text{eq}} = \frac{2\dot{R}_1R_1\left(R_2^2 + r^2\right)}{\sqrt{3}\left(R_2^2 - R_1^2\right)r^2\cos 2\psi}. \tag{74}$$

It has been assumed here that $-\pi/4 \le \psi \le \pi/4$. It follows from (73) that $\psi = \pi/4$ at $r = R_1$ and $\psi = -\pi/4$ at $r = R_2$. Eliminating in (74) the value of r by means of (73) and expanding the right hand side of (74) in a series in the vicinity of $\psi = \pi/4$ (or $r = R_1$) and $\psi = -\pi/4$ (or $r = R_2$) gives

$$\begin{aligned}
\xi_{\text{eq}} &= \frac{\dot{R}_1\left(R_2^2 + R_1^2\right)}{\sqrt{3}R_1\left(R_2^2 - R_1^2\right)(\pi/4 - \psi)} + o\left[\left(\frac{\pi}{4} - \psi\right)^{-1}\right] \quad \text{as} \quad \psi \to \frac{\pi}{4} \\
\xi_{\text{eq}} &= \frac{2\dot{R}_1R_1}{\sqrt{3}\left(R_2^2 - R_1^2\right)(\psi + \pi/4)} + o\left[\left(\psi + \frac{\pi}{4}\right)^{-1}\right] \quad \text{as} \quad \psi \to -\frac{\pi}{4}.
\end{aligned} \tag{75}$$

On the other hand, Eq. (73) in the vicinity of the maximum friction surfaces is represented in the form

$$\frac{\pi}{4} - \psi = \frac{\sqrt{2}R_2\sqrt{r-R_1}}{\sqrt{R_1\left(R_2^2-R_1^2\right)}} + o\left(\sqrt{r-R_1}\right) \quad \text{as} \quad r \to R_1,$$
$$\frac{\pi}{4} + \psi = \frac{\sqrt{2}R_1\sqrt{R_2-r}}{\sqrt{R_2\left(R_2^2-R_1^2\right)}} + o\left(\sqrt{R_2-r}\right) \quad \text{as} \quad r \to R_2. \tag{76}$$

Substituting (76) into (75) leads to

$$\xi_{eq} = \frac{\dot{R}_1\left(R_2^2+R_1^2\right)}{\sqrt{6}R_2\sqrt{R_1\left(R_2^2-R_1^2\right)}\sqrt{r-R_1}} + o\left(\frac{1}{\sqrt{r-R_1}}\right), \quad \text{as} \quad r \to R_1$$
$$\xi_{eq} = \sqrt{\frac{2}{3}}\frac{\dot{R}_1\sqrt{R_2}}{\sqrt{R_2^2-R_1^2}\sqrt{R_2-r}} + o\left[\frac{1}{\sqrt{R_2-r}}\right] \quad \text{as} \quad r \to R_2. \tag{77}$$

Comparing (5) and (77) it is possible to find that the strain rate intensity factors are

$$D_1 = \frac{\dot{R}_1\left(1+R_1^2/R_2^2\right)}{\sqrt{6}\sqrt{R_1}\sqrt{1-R_1^2/R_2^2}}, \qquad D_2 = \frac{2\dot{R}_1}{\sqrt{6}\sqrt{R_2}\sqrt{1-R_1^2/R_2^2}}. \tag{78}$$

Here D_1 corresponds to the maximum friction surface $r = R_1$ and D_2 to the maximum friction surface $r = R_2$. It is convenient to represent the final result in the form of the ratio D_1/D_2 because it is non-dimensional. It follows from (78) that

$$\Delta = \frac{D_1}{D_2} = \frac{1+R_1^2/R_2^2}{2\left(R_1/R_2\right)^{1/2}}.$$

The variation of Δ with R_1/R_2 is depicted in Fig. 10. It is seen from this figure that $\Delta > 1$ in the region $R_1/R_2 < r_{cr}$ and $\Delta < 1$ in the region $R_1/R_2 > r_{cr}$ where $r_{cr} \approx 0.3$. Using the hypothesis that the strain rate intensity factor controls the formation of the layer of intensive plastic deformation in the vicinity of friction

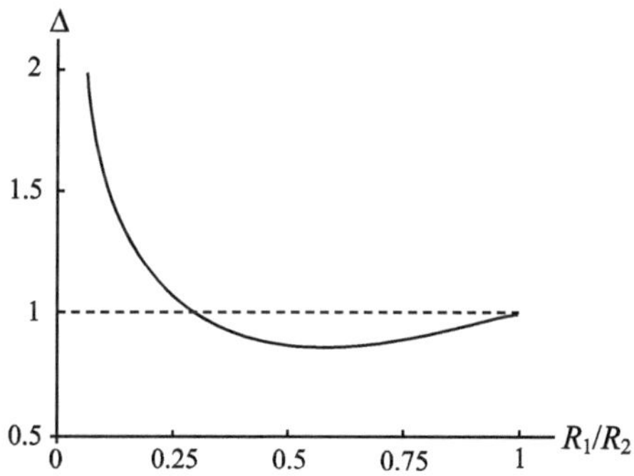

Fig. 10 Variation of Δ with the ratio of the radii

surfaces [8], it is possible to conclude from these inequalities that the thickness of this layer is larger near the friction surface $r = R_1$ if $R_1/R_2 < r_{cr}$, and it is larger near the friction surface $r = R_2$ if $R_1/R_2 > r_{cr}$.

3.5 Compression of a Plastic Layer Between Rotating Plates I

The boundary value problem is illustrated in Fig. 11. Two semi-infinite rough plates hinged together at their ends and inclined to each other at an angle 2α rotate towards each other with angular velocity of magnitude ω about an axis O through the hinge. The cylindrical coordinate system is taken with $\theta = 0$ taken as the perpendicular bisector of the angle 2α. Since $\theta = 0$ is an axis of symmetry for the flow, it is sufficient to find the solution in the region $\theta \geq 0$. By definition, ω is taken to be positive for the clockwise rotation of the upper plate. The maximum friction law is assumed at $\theta = \alpha$. The solution of this problem has been proposed in [29]. In fact, that solution was given for tension of the layer and, moreover, a velocity discontinuity surface appeared instead of the maximum friction surface in the problem under consideration. However, since the model is pressure-independent, these distinctions have no effect on the velocity field other than the sense of the velocity components. Qualitative behavior of the solution depends on the value of α. In particular, the solution exhibits sticking at the plates together with a rigid zone in the region adjacent to the plates for $\alpha > \pi/4$. In this case the velocity field is not singular and, therefore, the solution is not of interest for the purpose of the present chapter. The sliding regime of friction accompanied by a singular velocity field occurs for $\alpha < \pi/4$. The special case $\alpha = \pi/4$ will be treated separately.

The velocity boundary conditions are

$$u_\theta = 0 \quad \text{at} \quad \theta = 0 \tag{79}$$

and

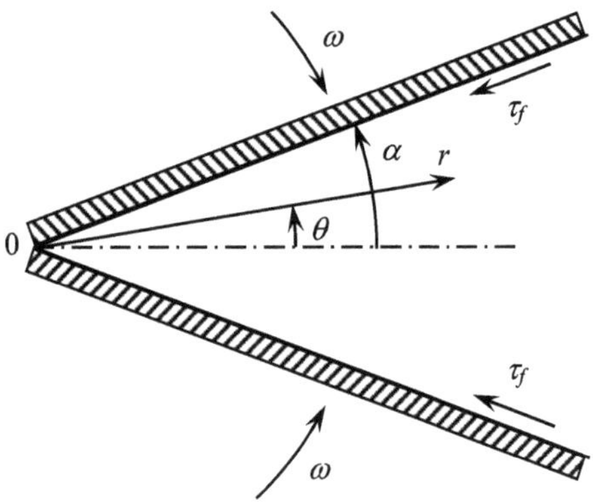

Fig. 11 Illustration of the boundary value problem

$$u_\theta = -\omega r \quad \text{at} \quad \theta = \alpha. \tag{80}$$

It is also assumed that there is no material flux through O. The stress boundary condition, in addition to the maximum friction law, is

$$\sigma_{r\theta} = 0 \quad \text{at} \quad \theta = 0. \tag{81}$$

The velocity field found in [29] is

$$u_r = -\frac{\omega r}{2}\left(c - 2\cos 2\psi\right), \quad u_\theta = \omega r \sin 2\psi, \tag{82}$$

where c is constant and ψ is related to θ by the following equation

$$\frac{d\psi}{d\theta} = \frac{(c - 2\cos 2\psi)}{2\cos 2\psi}. \tag{83}$$

The constraints imposed on the flow demand $u_r > 0$ and $\sigma_{rr} \geq \sigma_{\theta\theta}$. A consequence of the former inequality is $\sigma_{r\theta} < 0$ at $\theta = \alpha$. It follows from this inequality, the inequality $\sigma_{rr} \geq \sigma_{\theta\theta}$ and Eq. (43) that

$$-\frac{\pi}{4} \leq \psi \leq 0. \tag{84}$$

Since the friction surface is determined by the equation $\theta = \alpha$, the maximum friction law demands $\phi_1 = 0$ or $\phi_2 = 0$ in (40). Using (84) it is possible to conclude that $\phi_2 = 0$ and, therefore, $\psi = -\pi/4$ for $\theta = \alpha$. The solution of Eq. (83) satisfying this condition is

$$\theta = \alpha + 2\int_{-\pi/4}^{\psi} \frac{\cos 2\gamma d\gamma}{(c - 2\cos 2\gamma)}, \tag{85}$$

where γ is a dummy variable of integration. Substituting the condition $\psi = -\pi/4$ at $\theta = \alpha$ into (82) also shows that the boundary condition (80) is satisfied. It follows from (43), (81) and (84) that $\psi = 0$ at $\theta = 0$. Therefore, the equation for c is obtained from (85) in the form

$$\alpha = -2\int_{-\pi/4}^{0} \frac{\cos 2\gamma d\gamma}{(c - 2\cos 2\gamma)}. \tag{86}$$

Substituting the condition $\psi = 0$ at $\theta = 0$ into (82) also shows that the boundary condition (79) is satisfied. The variation of c with α found from (86) is depicted in Fig. 12. Expanding the right hand site of (83) in a series in the vicinity of $\psi = -\pi/4$ gives

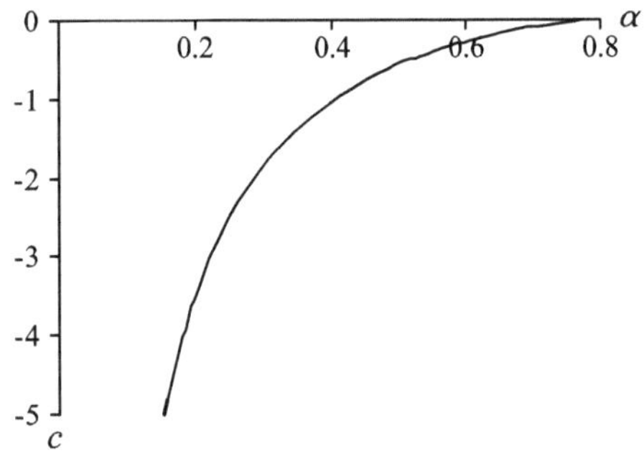

Fig. 12 Variation of c with α

$$\left[\frac{4}{c}\left(\psi+\frac{\pi}{4}\right)+o\left(\psi+\frac{\pi}{4}\right)\right]d\psi=d\theta \quad \text{as} \quad \psi\to-\frac{\pi}{4}.$$

Integrating with the use of the boundary condition $\psi=-\pi/4$ for $\theta=\alpha$ results in

$$\psi+\frac{\pi}{4}=\frac{\sqrt{-c\left(\alpha-\theta\right)}}{\sqrt{2}}+o\left(\sqrt{\alpha-\theta}\right) \quad \text{as} \quad \theta\to\alpha. \tag{87}$$

It is worthy to note here that $c\leq 0$ (Fig. 12). The strain rate components are found from (41), (82) and (83) as

$$\xi_{rr}=-\xi_{\theta\theta}=-\frac{\omega}{2}\left(c-2\cos 2\psi\right), \quad \xi_{r\theta}=\frac{\omega}{2}\tan 2\psi\left(c-2\cos 2\psi\right).$$

Then, it follows from (4) that

$$\xi_{\text{eq}}=\frac{\omega}{\sqrt{3}}\frac{\left(2\cos 2\psi-c\right)}{\cos 2\psi}. \tag{88}$$

It has been taken into account there that $c\leq 0$ (Fig. 12) and $\cos 2\psi\geq 0$ according to (84). Using (87), Eq. (88) in the vicinity of the maximum friction surface is represented as

$$\xi_{\text{eq}}=\frac{\omega\sqrt{-c}}{\sqrt{6}\sqrt{\alpha-\theta}}+o\left[\frac{1}{\sqrt{\alpha-\theta}}\right] \quad \text{as} \quad \theta\to\alpha. \tag{89}$$

Comparing (5) and (89) leads to the strain rate intensity factor in the form

$$D=\omega\sqrt{\frac{-cr}{6}}. \tag{90}$$

It follows from the solution that c vanishes at $\alpha=\pi/4$. Then, it is evident from (90) that the strain rate intensity factor vanishes at $\alpha=\pi/4$. Therefore, the solution

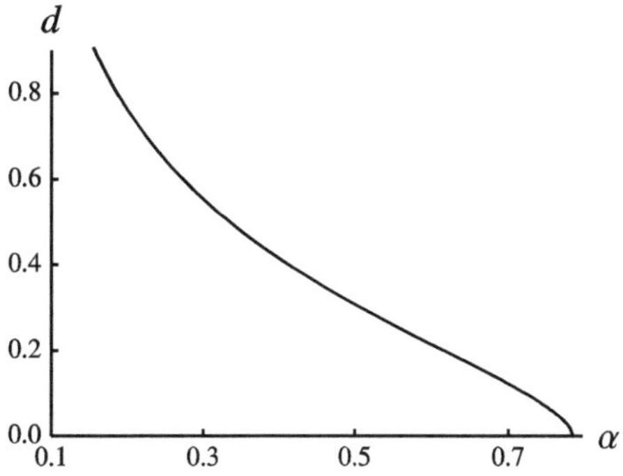

Fig. 13 Variation of the dimensionless strain rate intensity factor with α

is singular for $0 < \alpha < \pi/4$ and is not singular at $\alpha = \pi/4$. The variation of the dimensionless strain rate intensity factor defined by $d = D/\left(\omega\sqrt{r}\right)$ with α is illustrated in Fig. 13.

3.6 Compression of a Plastic Layer Between Rotating Plates II

The statement of the boundary value problem solved in the previous section is now slightly modified assuming that $u_r < 0$. This change in the direction of the radial velocity leads to the corresponding change in the direction of the friction stress (Fig. 14). The boundary conditions (79), (80) and (81) are valid. The maximum friction law is supposed at $\theta = \alpha$. As before, it is sufficient to find the solution in the region $\theta \geq 0$. The stress boundary conditions coincide with those in Sect. 3.3. Since ψ is solely determined from the stress equations, Eq. (58) is valid in the case under consideration. The velocity field is sought in the form

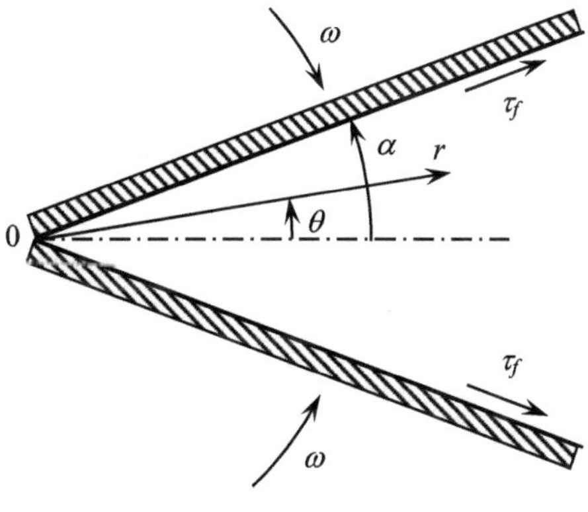

Fig. 14 Illustration of the boundary value problem

$$u_r = \frac{\omega r}{2}\frac{dg(\psi)}{d\theta} + \omega\frac{G(\psi)}{r}, \quad u_\theta = -\omega r g(\psi), \tag{91}$$

where $g(\psi)$ and $G(\psi)$ are arbitrary functions of ψ. It is possible to verify by inspection that the velocity field (91) automatically satisfies the incompressibility Eq. $(39)_1$. The inequality $u_r < 0$ demands $\sigma_{r\theta} > 0$ at $\theta = \alpha$. Therefore, it follows from (43) that $\pi/2 \geq \psi \geq 0$. The constraints imposed on the flow demand $\sigma_{rr} > \sigma_{\theta\theta}$. A consequence of this inequality and (43) is $\pi/4 \geq \psi \geq -\pi/4$. Finally,

$$\frac{\pi}{4} \geq \psi \geq 0. \tag{92}$$

Since the friction surface is determined by the equation $\theta = \alpha$, the maximum friction law demands $\phi_1 = 0$ or $\phi_2 = 0$ in (40). Using (92) it is possible to conclude that $\phi_1 = 0$ and, therefore, $\psi = \pi/4$ for $\theta = \alpha$. The same boundary condition was used in Sect. 3.3. Therefore, the solution (59) of Eq. (58) is valid in the case under consideration. The variation of c with α is depicted in Fig. 7.

Substituting (91) into (41) yields

$$\xi_{rr} = -\xi_{\theta\theta} = \frac{\omega}{2}\frac{dg}{d\psi}\frac{d\psi}{d\theta} - \frac{\omega G}{r^2},$$
$$\xi_{r\theta} = \frac{\omega}{4}\frac{d^2 g}{d\theta^2} + \frac{\omega}{2r^2}\frac{dG}{d\psi}\frac{d\psi}{d\theta} = \frac{\omega}{4}\left[\frac{d^2 g}{d\psi^2}\left(\frac{d\psi}{d\theta}\right)^2 + \frac{dg}{d\psi}\frac{d^2\psi}{d\theta^2}\right] + \frac{\omega}{2r^2}\frac{dG}{d\psi}\frac{d\psi}{d\theta}.$$

Eliminating here $d\psi/d\theta$ and $d^2\psi/d\theta^2$ by means of (58) gives

$$\xi_{rr} = -\xi_{\theta\theta} = \frac{\omega}{4}\frac{(c - 2\cos 2\psi)}{\cos 2\psi}\frac{dg}{d\psi} - \frac{\omega G}{r^2},$$
$$\xi_{r\theta} = \frac{\omega(c - 2\cos 2\psi)}{16\cos^2 2\psi}\left[(c - 2\cos 2\psi)\frac{d^2 g}{d\psi^2} + 2c\tan 2\psi\frac{dg}{d\psi}\right] + \frac{\omega}{4r^2}\frac{(c - 2\cos 2\psi)}{\cos 2\psi}\frac{dG}{d\psi}. \tag{93}$$

Substituting (93) into $(39)_2$ and eliminating stress components by means of (43) result in

$$\frac{(c - 2\cos 2\psi)^2}{16\cos^2 2\psi}\left(\frac{d^2 g}{d\psi^2} + 2\tan 2\psi\frac{dg}{d\psi}\right) + \left[\frac{(c - 2\cos 2\psi)}{4\cos 2\psi}\frac{dG}{d\psi} + G\tan 2\psi\right]\frac{1}{r^2} = 0.$$

This equation may have a solution if and only if

$$\cos 2\psi\frac{d^2 g}{d\psi^2} + 2\sin 2\psi\frac{dg}{d\psi} = 0, \quad (c - 2\cos 2\psi)\frac{dG}{d\psi} + 4G\sin 2\psi = 0. \tag{94}$$

Using (91) and taking into account that $\psi = 0$ at $\theta = 0$ and $\psi = \pi/4$ at $\theta = \alpha$ the boundary conditions (79) and (80) are transformed to

$$g = 0 \tag{95}$$

at $\psi = 0$ and

$$g = 1 \tag{96}$$

at $\psi = \pi/4$, respectively. The solution to $(94)_1$ satisfying these boundary conditions is

$$g = \sin 2\psi. \tag{97}$$

The general solution to $(94)_2$ is

$$G = \frac{G_0}{c - 2\cos 2\psi}, \tag{98}$$

where G_0 is a constant of integration. In order to find its value, a boundary condition in integral form similar to (49) should be prescribed. For example,

$$\int_0^\alpha u_r|_{r=R}\, d\theta = 0, \tag{99}$$

where the value of R should be prescribed. Using (91) and taking into account (95) and (96) the condition (99) is written in the form

$$1 + \frac{2}{R^2}\int_0^\alpha G d\theta = 0.$$

Replacing here integration with respect to θ with integration with respect to ψ by means of (58) and using (98) yields

$$G_0 = -\frac{R^2}{4}\left[\int_0^{\pi/4} \frac{\cos 2\psi}{(c - 2\cos 2\psi)^2} d\psi\right]^{-1}. \tag{100}$$

It follows from (93), (97) and (98) that

$$\begin{aligned} \xi_{rr} = -\xi_{\theta\theta} &= \frac{\omega}{2}(c - 2\cos 2\psi) - \frac{\omega G_0}{r^2 (c - 2\cos 2\psi)}, \\ \xi_{r\theta} &= \frac{\omega}{2\cos 2\psi}\left[(c - 2\cos 2\psi)\sin 2\psi - \frac{2G_0}{r^2}\frac{\sin 2\psi}{(c - 2\cos 2\psi)}\right]. \end{aligned} \tag{101}$$

It is evident that the normal strain rates are bounded and $|\xi_{r\theta}| \to \infty$ as $\psi \to \pi/4$. Therefore, Eq. (6) in which ξ_τ is replaced with $\xi_{r\theta}$ is valid. Using (101) the shear strain rate in the vicinity of the friction surface is represented as

$$\xi_{r\theta} = \frac{\omega c}{4\,(\pi/4 - \psi)}\left[1 - \frac{2G_0}{r^2c^2}\right] + o\left(\frac{1}{\pi/4 - \psi}\right) \quad \text{as} \quad \psi \to \frac{\pi}{4}. \tag{102}$$

Equation (58) transforms to

$$\frac{d\theta}{d\psi} = \frac{4}{c}\left(\frac{\pi}{4} - \psi\right) + o\left(\frac{\pi}{4} - \psi\right) \quad \text{as} \quad \psi \to \frac{\pi}{4}.$$

Integrating with the boundary condition $\theta = \alpha$ at $\psi = \pi/4$ gives

$$\alpha - \theta = \frac{2}{c}\left(\frac{\pi}{4} - \psi\right)^2 + o\left[\left(\frac{\pi}{4} - \psi\right)^2\right] \quad \text{as} \quad \psi \to \frac{\pi}{4}. \tag{103}$$

Replacing $\pi/4 - \psi$ in (102) with $\alpha - \theta$ by means of (103) results in

$$\xi_{r\theta} = \frac{\omega\sqrt{2c}}{4\sqrt{\alpha - \theta}}\left(1 - \frac{2G_0}{r^2c^2}\right) + o\left(\frac{1}{\sqrt{\alpha - \theta}}\right) \quad \text{as} \quad \theta \to \alpha. \tag{104}$$

It is seen from Fig. 7 that $c > 2$ and from Eq. (100) that $G_0 < 0$. Therefore, $\xi_{r\theta} > 0$. Then, combining (6) and (104) gives

$$D = \frac{\omega\sqrt{c}\sqrt{r}}{\sqrt{6}}\left(1 - \frac{2G_0}{r^2c^2}\right). \tag{105}$$

Since the dependence of c on α is known (Fig. 7), it follows from (100) that the dimensionless strain rate intensity factor defined by $d = D/\left(\omega\sqrt{r}\right)$ depends on α and the ratio r/R. Its variation with r/R at several values of α is depicted in Fig. 15 (curve 1 corresponds to $\alpha = \pi/36$, curve 2 to $\alpha = \pi/18$, curve 3 to $\alpha = \pi/12$, curve 4 to $\alpha = \pi/9$, and curve 5 to $\alpha = \pi/6$). In the solution for compression of a plastic layer between parallel plates (see Sect. 3.2) a rigid zone appears at the center of the layer [11]. The length of this zone at the friction surface is of order of the thickness of the layer. By analogy to this solution it is reasonable to assume that there is a rigid zone near the cross-section $r = R$ and that its length at the friction surface is equal to $R\alpha$. The solution found is not valid in the rigid zone. Therefore, the right ends of the curves in Fig. 15 are determined by the equation $r/R = 1 - \alpha$. The dependence of d on α at several values of r/R is illustrated in Fig. 16 (curve 1 corresponds to $r/R = 0.4$, curve 2 to $r/R = 0.3$, curve 3 to $r/R = 0.25$, and curve 4 to $r/R = 0.2$).

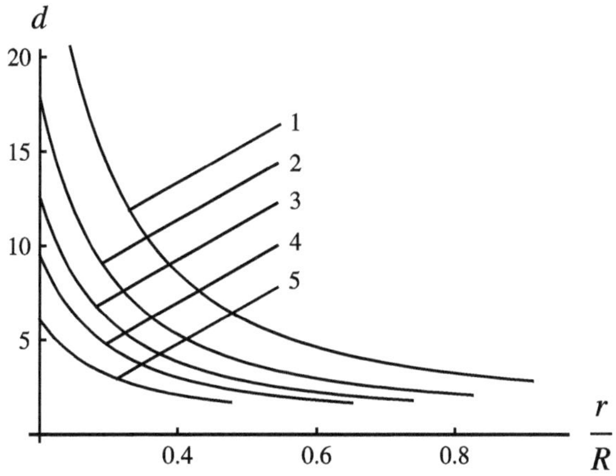

Fig. 15 Variation of the dimensionless strain rate intensity factor with r/R at several α- values

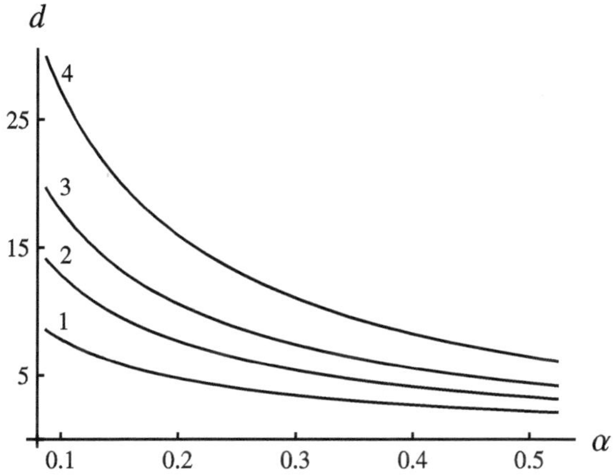

Fig. 16 Variation of the dimensionless strain rate intensity factor with α at several r/R- values

3.7 Simultaneous Shearing and Expansion of a Hollow Cylinder

The boundary value problem considered in this section consists of a planar deformation comprising the simultaneous shearing and expansion of a hollow cylinder (Fig. 17). The internal and external radii of the cylinder are denoted by a and b, respectively. It is convenient to use the cylindrical coordinates whose z-axis coincides the axis of the cylinder. In this coordinate system, the internal surface of the cylinder is determined by the equation $r = a$ and the external surface by the equation $r = b$. The deformation of the cylinder is caused by an expanding and rotating rod inserted into its hole. The radius of the rod is a and the rate of its expansion is $\dot{a}$. The angular velocity of the rod is ω and its direction is shown in Fig. 17. The external

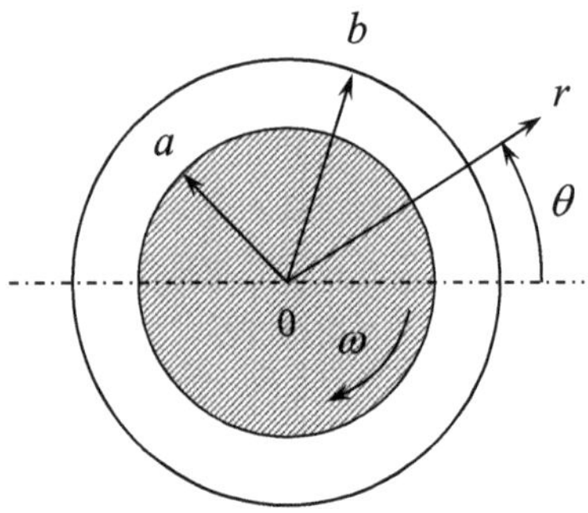

Fig. 17 Illustration of the boundary value problem

radius of the cylinder is fixed against rotation. Therefore, the velocity boundary conditions are

$$u_r = \dot{a} \tag{106}$$

for $r = a$ and

$$u_\theta = 0 \tag{107}$$

for $r = b$. One of the stress boundary conditions is $\sigma_{rr} = -p_a < 0$ for $r = a$. p_a is given, but its value has no effect of the strain rate intensity factor. The final boundary condition is the maximum friction law at $r = a$. In general, two different regimes may be identified at this boundary, sticking and sliding. However, the velocity field is singular if and only if the regime of sliding occurs. Therefore, the solution considered in this section is restricted to this regime. The general solution is provided in [30]. The direction of ω requires $\sigma_{r\theta} > 0$. On the other hand, it is evident that $\sigma_{rr} < \sigma_{\theta\theta}$. Therefore, it follows from (43) that

$$\frac{\pi}{4} \leq \psi < \frac{\pi}{2}. \tag{108}$$

The maximum friction surface is orthogonal to the r-axis. Therefore, the maximum friction law demands $\phi_1 = \pi/2$ or $\phi_2 = \pi/2$ at $r = a$. Comparing (40) and (108) gives

$$\psi = \frac{\pi}{4} \tag{109}$$

for $r = a$.

The solution satisfying Eqs. (38), (39) and (42) as well as the boundary conditions formulated has been given in [30]. The regime of sliding when (109) is valid requires

$$\omega a > \dot{a}\sqrt{1 - \frac{a^4}{b^4}}.$$

The velocity field in the case of sliding is

$$u_r = \frac{\dot{a}a}{r}, \quad u_\theta = -\frac{\dot{a}r}{a}\left(\sqrt{1-\frac{a^4}{b^4}} - \sqrt{1-\frac{a^4}{r^4}}\right). \tag{110}$$

Substituting (110) into (41) leads to

$$\xi_{rr} = -\xi_{\theta\theta} = -\frac{\dot{a}a}{r^2}, \quad \xi_{r\theta} = \frac{\dot{a}a^3}{r^4}\left(1-\frac{a^4}{r^4}\right)^{-1/2}.$$

Then, the equivalent strain rate follows from (4) as

$$\xi_{\mathrm{eq}} = \frac{2}{\sqrt{3}}\frac{\dot{a}a}{\sqrt{r^4-a^4}}.$$

Expanding the right hand side of this equation is a series in the vicinity of the surface $r = a$ gives

$$\xi_{\mathrm{eq}} = \frac{\dot{a}}{\sqrt{3}\sqrt{a}\sqrt{r-a}} + o\left(\frac{1}{\sqrt{r-a}}\right) \quad \text{as} \quad r \to a. \tag{111}$$

It follows from (5) and (111) that the strain rate intensity factor is

$$D = \frac{\dot{a}}{\sqrt{3}\sqrt{a}}. \tag{112}$$

4 Axisymmetric Solutions for Pressure-Independent Material

4.1 Basic Equations

Section 4 is concerned with axisymmetric solutions for pressure-independent material. In this section, two coordinate systems will be used, namely a cylindrical coordinate system (r, θ, z) and a spherical coordinate system (r, θ, ϑ). The solutions in the cylindrical coordinate system are independent of θ and the solutions in the spherical coordinate system are independent of ϑ. Let σ_1, σ_2 and σ_3 be the principal stresses. Then, Tresca's yield criterion adopted in this section can be written in the form

$$|\sigma_1 - \sigma_2| \le 2\tau_s, \quad |\sigma_2 - \sigma_3| \le 2\tau_s, \quad |\sigma_3 - \sigma_1| \le 2\tau_s. \tag{113}$$

This yield criterion is represented by a regular hexagonal prism in a three-dimensional space where the principal stresses are taken as Cartesian coordinates. The cross-section of this prism by the plane $\sigma_1 + \sigma_2 + \sigma_3 = 0$ is shown in Fig. 18. It is seen

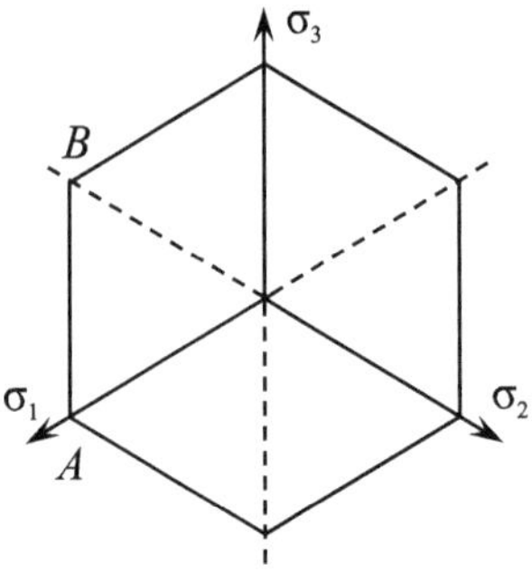

Fig. 18 Tresca's yield locus

from this figure that the yield surface is singular. Therefore, various plastic regimes can in general arise. However, of particular interest are those corresponding to edges of the yield surface (corners in Fig. 18). The circumferential stress is one of the principal stresses. It is possible to assume that σ_3 is the circumferential stress. It is also possible to assume, with no loss of generality, that $\sigma_1 \geq \sigma_2$. The case $\sigma_1 = \sigma_2$ is not considered here. Therefore, $\sigma_1 > \sigma_2$ and Eqs. (113) reduce to

$$\sigma_1 - \sigma_2 = 2\tau_s, \quad \sigma_1 - \sigma_3 = 2\tau_s \tag{114}$$

or

$$\sigma_1 - \sigma_2 = 2\tau_s, \quad \sigma_3 - \sigma_2 = 2\tau_s. \tag{115}$$

Equations (114) correspond to point A and Eq. (115) to point B (Fig. 18).

Let σ_{rr}, $\sigma_{\theta\theta}$, σ_{zz}, σ_{rz}, $\sigma_{z\theta}$ and $\sigma_{r\theta}$ be the components of the stress tensor, and ξ_{rr}, $\xi_{\theta\theta}$, ξ_{zz}, ξ_{rz}, $\xi_{z\theta}$ and $\xi_{r\theta}$ be the components of the strain rate tensor in the cylindrical coordinate system. In the case under consideration, $\xi_{r\theta} = \xi_{z\theta} = 0$, $\sigma_{r\theta} = \sigma_{z\theta} = 0$ and the circumferential velocity $u_\theta = 0$. The non-zero strain rate components are expressed through the velocity components, u_r and u_z, as

$$\xi_{rr} = \frac{\partial u_r}{\partial r}, \quad \xi_{\theta\theta} = \frac{u_r}{r}, \quad \xi_{zz} = \frac{\partial u_z}{\partial z}, \quad \xi_{rz} = \frac{1}{2}\left(\frac{\partial u_r}{\partial z} + \frac{\partial u_z}{\partial r}\right). \tag{116}$$

The equilibrium equations are

$$\frac{\partial \sigma_{rr}}{\partial r} + \frac{\partial \sigma_{rz}}{\partial z} + \frac{\sigma_{rr} - \sigma_{\theta\theta}}{r} = 0, \quad \frac{\partial \sigma_{rz}}{\partial r} + \frac{\partial \sigma_{zz}}{\partial z} + \frac{\sigma_{rz}}{r} = 0. \tag{117}$$

Taking into account (114) or (115) the transformation equations for stress components in rz-planes are represented as

$$\sigma_{rr} = \frac{1}{2}(\sigma_1 + \sigma_2) + \tau_s \cos 2\psi, \quad \sigma_{zz} = \frac{1}{2}(\sigma_1 + \sigma_2) - \tau_s \cos 2\psi, \tag{118}$$
$$\sigma_{rz} = \tau_s \sin 2\psi,$$

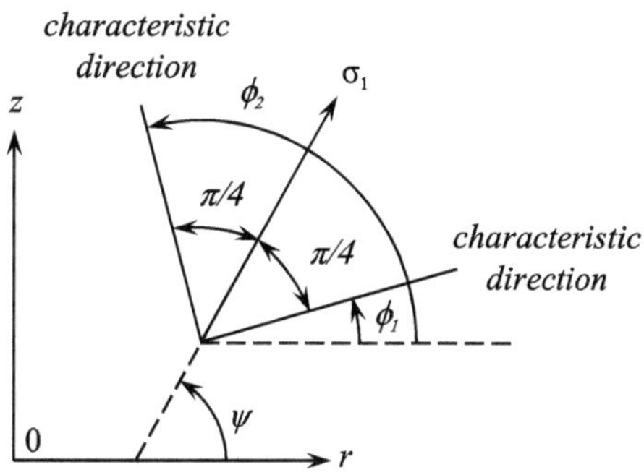

Fig. 19 Orientation of the major principal stress and characteristic directions relative to the r-axis of cylindrical coordinates

where ψ is the orientation of the principal stress σ_1 relative to the r-axis (Fig. 19). These equations show that $\sigma_{rr} + \sigma_{zz} = \sigma_1 + \sigma_2$. Then, it follows from Eqs. (114) and (115) that

$$\sigma_3 = \sigma_{\theta\theta} = \frac{1}{2}(\sigma_{rr} + \sigma_{zz}) + \varepsilon\tau_s, \tag{119}$$

where ε can take the values ± 1. It is evident that $\varepsilon = -1$ corresponds to point A and $\varepsilon = +1$ to point B (Fig. 18).

The flow rule associated with the yield criterion (114) or (115) leads to the equation of incompressibility and the isotropy condition. Using (116) the equation of incompressibility $\xi_{rr} + \xi_{\theta\theta} + \xi_{zz} = 0$ is transformed to

$$\frac{\partial u_r}{\partial r} + \frac{u_r}{r} + \frac{\partial u_z}{\partial z} = 0. \tag{120}$$

The isotropy condition is

$$\frac{\sigma_{rz}}{(\sigma_{rr} - \sigma_{zz})} = \frac{\xi_{rz}}{(\xi_{rr} - \xi_{zz})}.$$

Substituting (118) into this equation gives

$$\frac{2\xi_{rz}}{(\xi_{rr} - \xi_{zz})} = \tan 2\psi. \tag{121}$$

The conditions that an element of the material shall be in a state of stress corresponding to points A and B (Fig. 18) are that the associated velocity field satisfies the inequalities

$$\varepsilon u_r \geq 0, \quad (\xi_{rr} - \xi_{zz})^2 + 4\xi_{rz}^2 \geq \frac{u_r^2}{r^2}. \tag{122}$$

It is known (see, for example, [14]) that the characteristic directions make angles $\pm\pi/4$ with the direction of the principal stress σ_1. Therefore (Fig. 19),

$$\phi_1 = \psi - \frac{\pi}{4}, \quad \phi_2 = \psi + \frac{\pi}{4} \tag{123}$$

where ϕ_1 and ϕ_2 are the angles between the characteristic directions and the r-axis.

Let σ_{rr}, $\sigma_{\theta\theta}$, $\sigma_{\vartheta\vartheta}$, $\sigma_{r\vartheta}$, $\sigma_{\vartheta\theta}$ and $\sigma_{r\theta}$ be the components of the stress tensor, and ξ_{rr}, $\xi_{\theta\theta}$, $\xi_{\vartheta\vartheta}$, $\xi_{r\vartheta}$, $\xi_{\vartheta\theta}$ and $\xi_{r\theta}$ be the components of the strain rate tensor in the spherical coordinate system. In the case under consideration, $\xi_{r\vartheta} = \xi_{\vartheta\theta} = 0$, $\sigma_{r\vartheta} = \sigma_{\vartheta\theta} = 0$ and the circumferential velocity $u_\vartheta = 0$. The non-zero strain rate components are expressed through the velocity components, u_r and u_θ, as

$$\begin{aligned} \xi_{rr} &= \frac{\partial u_r}{\partial r}, \quad \xi_{\theta\theta} = \frac{1}{r}\left(\frac{\partial u_\theta}{\partial \theta} + u_r\right), \\ \xi_{\vartheta\vartheta} &= \frac{1}{r\sin\theta}\left(u_r \sin\theta + u_\theta \cos\theta\right), \quad \xi_{r\theta} = \frac{1}{2}\left(\frac{\partial u_\theta}{\partial r} - \frac{u_\theta}{r} + \frac{1}{r}\frac{\partial u_r}{\partial \theta}\right). \end{aligned} \tag{124}$$

The equilibrium equations are

$$\begin{aligned} \frac{\partial \sigma_{rr}}{\partial r} + \frac{1}{r}\frac{\partial \sigma_{r\theta}}{\partial \theta} + \frac{(2\sigma_{rr} - \sigma_{\theta\theta} - \sigma_{\vartheta\vartheta} + \sigma_{r\theta}\cot\theta)}{r} &= 0, \\ \frac{\partial \sigma_{r\theta}}{\partial r} + \frac{1}{r}\frac{\partial \sigma_{\theta\theta}}{\partial \theta} + \frac{(\sigma_{\theta\theta} - \sigma_{\vartheta\vartheta})\cot\theta + 3\sigma_{r\theta}}{r} &= 0. \end{aligned} \tag{125}$$

Taking into account (114) or (115) the transformation equations for stress components in $r\theta$-planes are represented as

$$\begin{aligned} \sigma_{rr} &= \frac{1}{2}(\sigma_1 + \sigma_2) + \tau_s \cos 2\psi, \quad \sigma_{\theta\theta} = \frac{1}{2}(\sigma_1 + \sigma_2) - \tau_s \cos 2\psi, \\ \sigma_{r\theta} &= \tau_s \sin 2\psi, \end{aligned} \tag{126}$$

where ψ is the orientation of the principal stress σ_1 relative to the r-axis (Fig. 20). These equations show that $\sigma_{rr} + \sigma_{\theta\theta} = \sigma_1 + \sigma_2$. Then, it follows from Eqs. (114) and (115) that

$$\sigma_3 = \sigma_{\vartheta\vartheta} = \frac{1}{2}(\sigma_{rr} + \sigma_{\theta\theta}) + \varepsilon\tau_s. \tag{127}$$

As before, $\varepsilon = -1$ corresponds to point A and $\varepsilon = +1$ to point B (Fig. 18).

Using (124) the equation of incompressibility, $\xi_{rr} + \xi_{\theta\theta} + \xi_{\vartheta\vartheta} = 0$, is transformed to

$$\frac{\partial u_r}{\partial r} + \frac{1}{r}\left(\frac{\partial u_\theta}{\partial \theta} + u_r\right) + \frac{1}{r\sin\theta}\left(u_r \sin\theta + u_\theta\cos\theta\right) = 0. \tag{128}$$

Using (126) the isotropy condition, $\sigma_{r\theta}(\xi_{rr} - \xi_{\theta\theta}) = (\sigma_{rr} - \sigma_{\theta\theta})\xi_{r\theta}$, is written as

$$\frac{2\xi_{r\theta}}{(\xi_{rr} - \xi_{\theta\theta})} = \tan 2\psi. \tag{129}$$

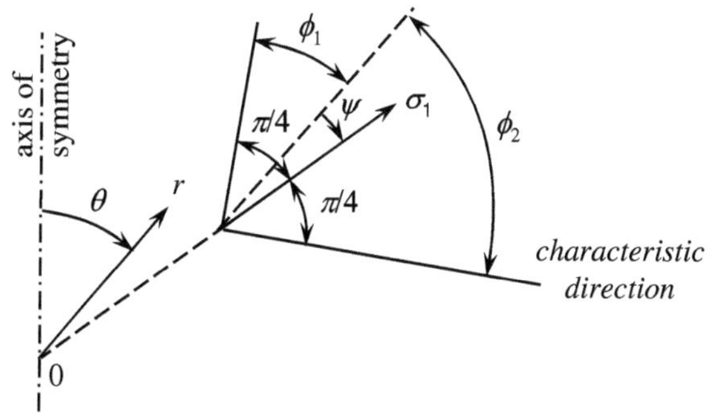

Fig. 20 Orientation of the major stress and characteristic directions relative to the r-axis of spherical coordinates

The inequalities (122) in the spherical coordinate system become

$$\varepsilon\left(u_r \sin\theta + u_\theta \cos\theta\right) \geq 0, \quad \left(\xi_{rr} - \xi_{\theta\theta}\right)^2 + 4\xi_{r\theta}^2 \geq \frac{\left(u_r \sin\theta + u_\theta \cos\theta\right)^2}{r^2 \sin^2\theta}. \tag{130}$$

Since the characteristic directions make angles $\pm\pi/4$ with the direction of the principal stress σ_1, it is seen from Fig. 20 that

$$\phi_1 = \psi - \frac{\pi}{4}, \quad \phi_2 = \psi + \frac{\pi}{4}, \tag{131}$$

where ϕ_1 and ϕ_2 are the angles between the characteristic directions and the r-axis.

4.2 Compression of a Hollow Cylinder on a Rigid Fibre

The boundary value problem is illustrated in Fig. 21. An axisymmetric hollow cylinder of internal radius a and external radius b is subject to compression by pressure p uniformly distributed over its outer surface. The length of the cylinder is $2L$. A rigid fibre of radius a is inserted into the hole of the cylinder. The cylindrical coordinate system is chosen such that the plane $z = 0$ coincides with the plane of symmetry of the cylinder. Then, it is sufficient to find the solution in the domain $0 \leq z \leq L$. Symmetry demands

$$u_z = 0 \tag{132}$$

and

$$\sigma_{rz} = 0 \tag{133}$$

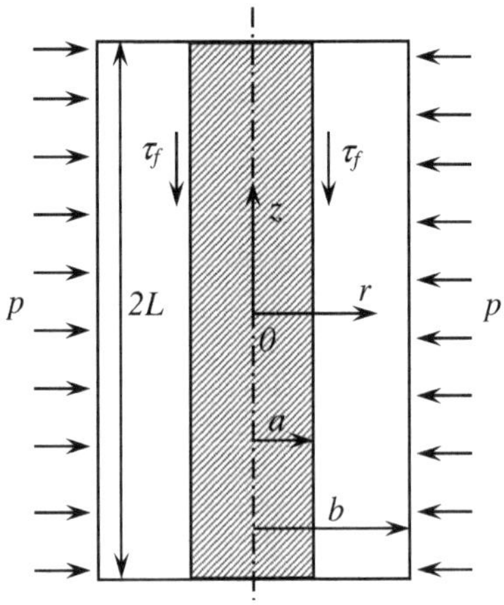

Fig. 21 Illustration of the boundary value problem

for $z = 0$. Since the fibre is rigid,

$$u_r = 0 \tag{134}$$

for $r = a$. The rate of contraction of the external radius of the cylinder is denoted by U. Then,

$$u_r = -U \tag{135}$$

for $r = b$. The surface $z = L$ is traction free. Therefore,

$$\sigma_{rz} = 0 \tag{136}$$

and

$$\sigma_{zz} = 0 \tag{137}$$

for $z = L$. By assumption, the surface $r = b$ is free of shear stress. Then,

$$\sigma_{rz} = 0 \tag{138}$$

for $r = b$. The final boundary condition is the maximum friction law at $r = a$. The direction of the friction stress is shown in Fig. 21. Therefore, the maximum friction law becomes

$$\sigma_{rz} = \tau_s \tag{139}$$

for $r = a$. The boundary value problem defined is an axisymmetric analogue to the problem considered in Sect. 3.2. Its approximate solution proposed in [31] ignores the boundary conditions (133) and (136). The boundary conditions (132) and (137) are replaced with the following integral conditions

$$\int_a^b u_z|_{z=0}\, r dr = 0 \tag{140}$$

and

$$\int_a^b \sigma_{zz}|_{z=L}\, r dr = 0, \tag{141}$$

respectively. Thus the solution is not valid in the vicinity of $z = 0$ and $z = L$.

The velocity field proposed in [31] is

$$\frac{u_r}{U} = -\frac{b\left(r^2 - a^2\right)}{r\left(b^2 - a^2\right)}, \tag{142}$$

$$\frac{u_z}{U} = \frac{b}{\left(b^2 - a^2\right)}\left[2z + \int \frac{\left(3r^2 + a^2\right) f(r)}{r^2\sqrt{1 - f(r)^2}} dr\right] + K, \quad f(r) = \frac{a\left(b^2 - r^2\right)}{r\left(b^2 - a^2\right)}.$$

Here K is the constant of integration. Its value can be found by means of the boundary condition (140). However, it is not necessary for determining the strain rate intensity factor. It is possible to verify by inspection that the velocity field (142) satisfies the equation of incompressibility (120). It has been shown in [31] that the velocity field also satisfies the inequalities (122) if $\varepsilon = -1$. Then, the associated state of stress must correspond to point A (Fig. 18). Such a stress field satisfying Eqs. (114), (117), (119) and (121) as well as the boundary conditions (138), (139) and (141) has been found in [31]. Substituting (142) into (116) yields

$$\xi_{rr} = -\frac{Ub\left(r^2 + a^2\right)}{r^2\left(b^2 - a^2\right)}, \quad \xi_{\theta\theta} = -\frac{Ub\left(r^2 - a^2\right)}{r^2\left(b^2 - a^2\right)}, \quad \xi_{zz} = \frac{2Ub}{\left(b^2 - a^2\right)}, \tag{143}$$
$$\xi_{rz} = \frac{Ub\left(3r^2 + a^2\right) f(r)}{2\left(b^2 - a^2\right) r^2\sqrt{1 - f(r)^2}}.$$

Since $f(a) = 1$, it is evident that the normal strain rates are bounded and $\xi_{rz} \to \infty$ as $r \to a$. Therefore, Eq. (6) in which ξ_τ is replaced with ξ_{rz} is valid. Eliminating the function $f(r)$ in (143) by means of (142) the shear strain rate in the vicinity of the maximum friction surface $r = a$ is represented as

$$\xi_{rz} = \frac{\sqrt{2}Ub\sqrt{a}}{\sqrt{b^4 - a^4}\sqrt{r - a}} + o\left(\frac{1}{\sqrt{r - a}}\right) \quad \text{as} \quad r \to a. \tag{144}$$

Combining (6) and (144) gives

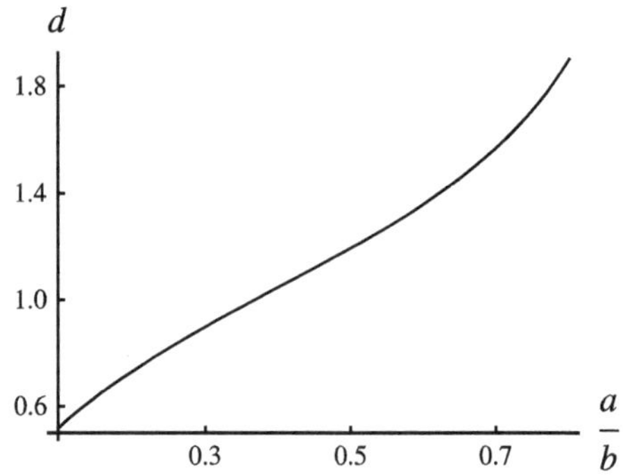

Fig. 22 Variation of the dimensionless strain rate intensity factor with the ratio a/b

$$D = \frac{2\sqrt{2}Ub\sqrt{a}}{\sqrt{3}\sqrt{b^4 - a^4}}. \tag{145}$$

It is convenient to introduce the dimensionless strain rate intensity factor by

$$d = D\sqrt{b}/U.$$

Then, it follows from (145) that

$$d = \frac{2\sqrt{2a}}{\sqrt{3b}}\left(1 - \frac{b^4}{a^4}\right)^{-1/2}. \tag{146}$$

The variation of d with a/b is shown in Fig. 22.

4.3 Flow of Plastic Material Through an Converging Conical Channel

Consider a converging conical channel (total angle 2α) through which plastic material is being forced (Fig. 23). The material flows to the virtual apex O. The origin of the spherical coordinate system is taken at O and the surface of the channel is determined by the equation $\theta = \alpha$. The maximum friction law is supposed at $\theta = \alpha$. The direction of friction stresses τ_f is dictated by the direction of flow (Fig. 23). The solution used in this section has been proposed in [32]. The velocity boundary conditions are

$$u_\theta = 0 \tag{147}$$

at $\theta = 0$ and $\theta = \alpha$. The stress boundary condition, in addition to the maximum friction law, is

$$\sigma_{r\theta} = 0 \tag{148}$$

Fig. 23 Illustration of the boundary value problem

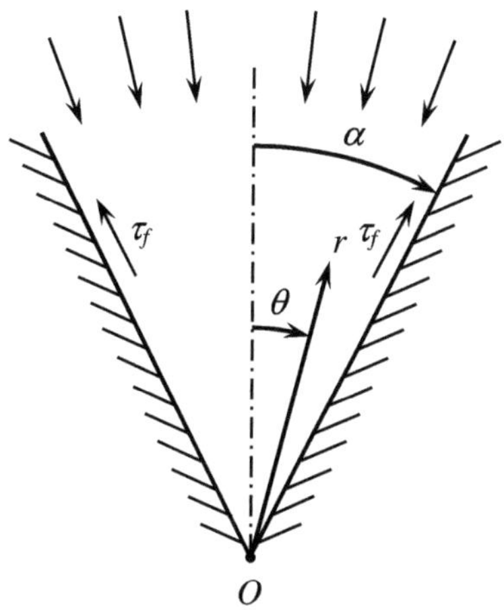

at $\theta = 0$. It is assumed in [32] that $u_\theta = 0$ everywhere. Then, the boundary conditions (147) are automatically satisfied. In the case of Tresca's yield criterion the radial velocity is given by [32]

$$u_r = -\frac{B}{r^2}\exp\left[-3\int\limits_{\alpha}^{\theta}\frac{t}{\sqrt{1-t^2}}d\gamma\right], \tag{149}$$

where B is proportional to the material flux, t is a function of θ and γ is a dummy variable of integration. The function $t\,(\theta)$ is determined by the following equation

$$\frac{dt}{d\theta} + t\cot\theta + 3\sqrt{1-t^2} = c, \tag{150}$$

where c is a constant of integration. The physical meaning of the function $t\,(\theta)$ is that $\sigma_{r\theta} = \tau_s t\,(\theta)$. Therefore, it follows from (148) and the maximum friction law (1) that

$$t = 0 \tag{151}$$

at $\theta = 0$ and

$$t = 1 \tag{152}$$

at $\theta = \alpha$. It is convenient to introduce the following substitution

$$t = \cos\mu. \tag{153}$$

Then, Eq. (150) transforms to

$$-\sin\mu\frac{d\mu}{d\theta}+\cos\mu\cot\theta+3\sin\mu=c \tag{154}$$

and the boundary conditions (151) and (152) to

$$\mu=\frac{\pi}{2}\quad \text{at}\quad \theta=0 \tag{155}$$

and

$$\mu=0\quad \text{at}\quad \theta=\alpha, \tag{156}$$

respectively. It is seen from (155) that the second term in (154) reduces to the expression $0\cdot\infty$ at $\theta=0$. Assume that

$$\mu=\frac{\pi}{2}+a_1\theta+o(\theta)\quad \text{as}\quad \theta\to 0. \tag{157}$$

Substituting (157) into (154) it is possible to express a_1 in terms of c. Then, (157) becomes

$$\mu=\frac{\pi}{2}+\frac{(3-c)}{2}\theta+o(\theta)\quad \text{as}\quad \theta\to 0. \tag{158}$$

Since the coefficient of the derivative in (154) vanishes at $\theta=\alpha$, it is convenient to rewrite this equation as

$$\frac{d\theta}{d\mu}=\frac{\sin\mu}{3\sin\mu+\cos\mu\cot\theta-c}. \tag{159}$$

Using (158) the function $\theta(\mu)$ is represented as

$$\theta=\frac{2}{(c-3)}\left(\frac{\pi}{2}-\mu\right)+o\left(\frac{\pi}{2}-\mu\right)\quad \text{as}\quad \mu\to\frac{\pi}{2}. \tag{160}$$

It follows from this equation that

$$\theta_a=\frac{2\delta_a}{(c-3)} \tag{161}$$

to leading order. Here $\delta_a=\pi/2-\mu_a<<1$. Using the boundary condition $\theta=\theta_a$ for $\mu=\mu_a$ at some value of c Eq. (159) can be solved numerically in the range $0\le\mu\le\mu_a$. Then, an iterative procedure should be adopted to find the value of c satisfying the boundary condition (156). The variation of c with α is depicted in Fig. 24. It has been shown in [32] that there exists a distribution of the normal stresses associated with the velocity field considered and the solution of Eq. (159) found. The state of stresses corresponds to point A of the yield surface (Fig. 18) and satisfies the equilibrium Eq. (125).

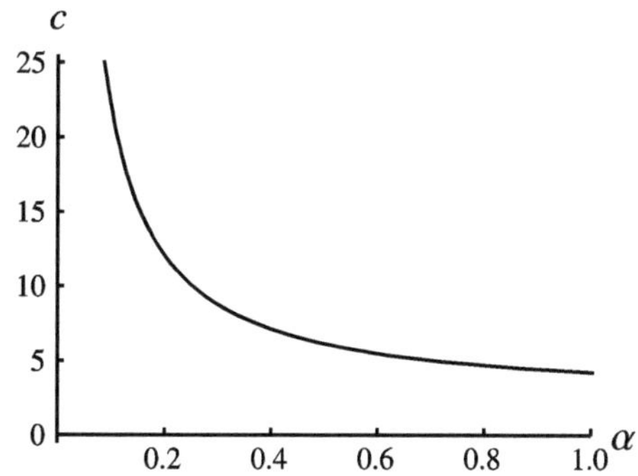

Fig. 24 Variation of c with α

Using (153) Eq. (149) transforms to

$$u_r = -\frac{B}{r^2}\exp\left[-3\int_{\alpha}^{\theta}\cot\mu d\gamma\right]. \qquad (162)$$

The components of the strain rate tensor are determined from (124). It is evident that the normal strain rates are bounded and $\xi_{r\theta} \to \infty$ as $\theta \to \alpha$. Therefore, Eq. (6) in which ξ_τ is replaced with $\xi_{r\theta}$ is valid. Using (124) and (162) the shear strain rate in the vicinity of the maximum friction surface $\theta = \alpha$ is represented as

$$\xi_{r\theta} = \frac{U}{r^3}\mu^{-1} + o\left(\mu^{-1}\right) \quad \text{as} \quad \mu \to 0, \qquad (163)$$

where

$$U = \frac{3B}{2}\exp\left(3\int_{0}^{\alpha}\cot\mu d\theta\right). \qquad (164)$$

Equation (159) in the vicinity of the friction surface $\theta = \alpha$ can be written as

$$\frac{d\theta}{d\mu} = \frac{\mu}{\cot\alpha - c} \quad \text{as} \quad \theta \to \alpha.$$

Integrating with the boundary condition (156) yields

$$\mu = \sqrt{c - \cot\alpha}\sqrt{\alpha - \theta} + o\left(\sqrt{\alpha - \theta}\right) \quad \text{as} \quad \theta \to \alpha. \qquad (165)$$

Substituting (165) into (163) gives

$$\xi_{r\theta} = \frac{U}{r^3\sqrt{c - \cot\alpha}\sqrt{\alpha - \theta}} + o\left(\frac{1}{\sqrt{\alpha - \theta}}\right) \quad \text{as} \quad \theta \to \alpha. \qquad (166)$$

Comparing (6) and (166) shows that

$$D = \frac{2}{\sqrt{3}} \frac{U}{\sqrt{c - \cot \alpha}} r^{-5/2}. \tag{167}$$

The material flux is defined by

$$Q = -2\pi \int_0^\alpha u_r r^2 \sin \theta d\theta. \tag{168}$$

Substituting (162) into (168) gives

$$Q = 2\pi B \int_0^\alpha \exp\left[-3 \int_\alpha^\theta \cot \mu d\gamma\right] \sin \theta d\theta. \tag{169}$$

It is convenient to introduce the dimensionless strain rate intensity factor as

$$d = \frac{D r^{5/2}}{Q}. \tag{170}$$

Then, it follows from (164), (167), (169), and (170) that

$$d = \frac{\sqrt{3}}{2\pi\sqrt{c - \cot \alpha}} \exp\left(3 \int_0^\alpha \cot \mu d\theta\right) \left\{\int_0^\alpha \exp\left[3 \int_\theta^\alpha \cot \mu d\gamma\right] \sin \theta d\theta\right\}^{-1}. \tag{171}$$

Using the solution of Eq. (159) along with the value of c found (Fig. 24) the integrals involved in this equation can be evaluated. As a result, the variation of d with α is obtained. This variation is illustrated in Fig. 25.

4.4 Radial Flow Between Two Conical Surfaces

Consider radial flow between two conical surfaces shown in Fig. 26. The material flows to the virtual apex O. The origin of the spherical coordinate system is taken at O and the surfaces of the channel are determined by the equations $\theta = \theta_0$ and $\theta = \theta_1$. The maximum friction law is supposed at both $\theta = \theta_0$ and $\theta = \theta_1$. The direction of friction stresses τ_f is dictated by the direction of flow (Fig. 26). A solution to this problem for the von Mises yield criterion has been proposed in [33]. It was based on the general solution given in [32] where the solution for the Tresca yield criterion considered in the previous section was also proposed. The latter will be adopted in the present section to describe radial flow between two conical surfaces.

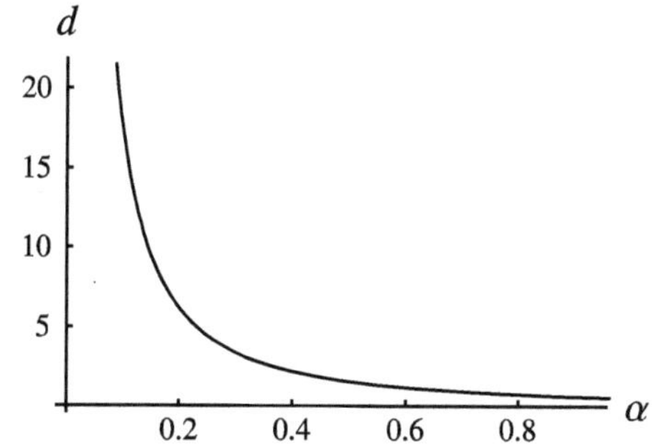

Fig. 25 Variation of the dimensionless strain rate intensity factor with α

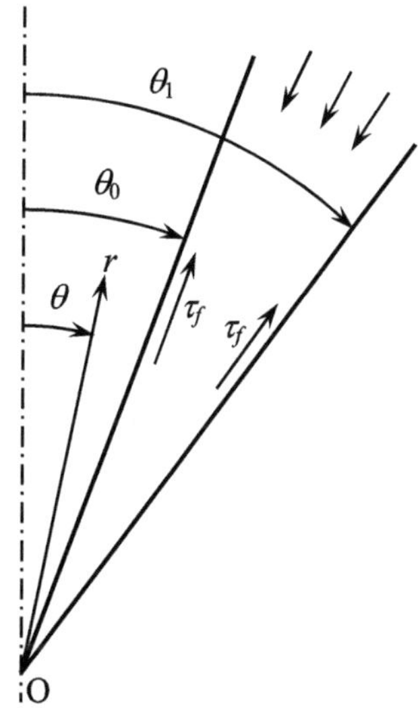

Fig. 26 Illustration of the boundary value problem

The velocity boundary conditions are

$$u_\theta = 0 \quad \text{at} \quad \theta = \theta_0 \quad \text{and} \quad \theta = \theta_1. \tag{172}$$

Therefore, the assumption that $u_\theta = 0$ everywhere adopted in the previous section is applicable in the case under consideration. Then, the radial velocity is given by (149). Using (153) this equation is transformed to (162). Equations (154) and (159) are valid. However, the boundary conditions (155) and (156) are replaced with

$$\mu = \pi \quad \text{at} \quad \theta = \theta_0 \tag{173}$$

and

$$\mu = 0 \quad \text{at} \quad \theta = \theta_1. \tag{174}$$

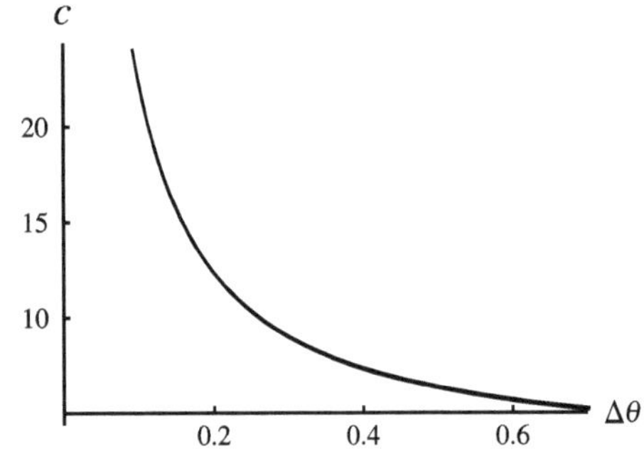

Fig. 27 Variation of c with $\Delta\theta$

Solving Eq. (159) along with the boundary conditions (173) and (174) numerically gives the dependence of c on θ_0 and θ_1. The variation of c with $\Delta\theta = \theta_1 - \theta_0$ at several values of θ_0 in the range $\pi/36 \leq \theta_0 \leq \pi/4$ is depicted in Fig. 27. It is seen from this figure that the value of c is practically independent of θ_0 and θ_1 separately (no difference between the curves is visible in the figure). Equation (162) is replaced with

$$u_r = -\frac{B}{r^2}\exp\left[-3\int\limits_{\theta_0}^{\theta}\cot\mu d\gamma\right]. \tag{175}$$

The components of the strain rate tensor are determined from (124). It is evident that the normal strain rates are bounded and $|\xi_{r\theta}| \to \infty$ as $\theta \to \theta_0$ and $\theta \to \theta_1$. Therefore, Eq. (6) in which ξ_τ is replaced with $\xi_{r\theta}$ is valid. Using (124) and (175) the shear strain rate in the vicinity of the maximum friction surfaces $\theta = \theta_1$ and $\theta = \theta_0$ is represented as

$$\xi_{r\theta} = \frac{U}{r^3}\mu^{-1} + o\left(\mu^{-1}\right) \quad \text{as} \quad \mu \to 0 \tag{176}$$

and

$$\xi_{r\theta} = -\frac{3B}{2r^3}(\pi-\mu)^{-1} + o\left[(\pi-\mu)^{-1}\right] \quad \text{as} \quad \mu \to \pi, \tag{177}$$

respectively. Here

$$U = \frac{3B}{2}\exp\left(-3\int\limits_{\theta_0}^{\theta_1}\cot\mu d\theta\right). \tag{178}$$

Using (173) and (174) Eq. (154) in the vicinity of points $\mu = 0$ and $\mu = \pi$ can be represented as

$$\frac{d\mu}{d\theta} = \frac{\cot\theta_0 + c}{\mu - \pi} + o\left[(\mu-\pi)^{-1}\right] \quad \text{as} \quad \mu \to \pi \tag{179}$$

and

$$\frac{d\mu}{d\theta} = \frac{\cot\theta_1 - c}{\mu} + o\left(\frac{1}{\mu}\right) \quad \text{as} \quad \mu \to 0. \tag{180}$$

Integrating (179) and (180) with the use of the boundary conditions (173) and (174) yields

$$\pi - \mu = \sqrt{2(\cot\theta_0 + c)}\sqrt{\theta - \theta_0} + o\left(\sqrt{\theta - \theta_0}\right) \quad \text{as} \quad \theta \to \theta_0 \tag{181}$$

and

$$\mu = \sqrt{2(c - \cot\theta_1)}\sqrt{\theta_1 - \theta} + o\left(\sqrt{\theta_1 - \theta}\right) \quad \text{as} \quad \theta \to \theta_1. \tag{182}$$

Substituting (181) into (177) and (182) into (176) gives

$$\xi_{r\theta} = -\frac{3B}{2r^3\sqrt{2(\cot\theta_0 + c)}\sqrt{\theta - \theta_0}} + o\left(\frac{1}{\sqrt{\theta - \theta_0}}\right) \quad \text{as} \quad \theta \to \theta_0 \tag{183}$$

and

$$\xi_{r\theta} = \frac{U}{r^3\sqrt{2(c - \cot\theta_1)}\sqrt{\theta_1 - \theta}} + o\left(\frac{1}{\sqrt{\theta_1 - \theta}}\right) \quad \text{as} \quad \theta \to \theta_1, \tag{184}$$

respectively. Comparing (6) with (183) and (184) gives

$$D_{in} = \frac{\sqrt{3}}{\sqrt{2}}\frac{B}{\sqrt{\cot\theta_0 + c}}r^{-5/2} \tag{185}$$

and

$$D_{\text{ex}} = \frac{\sqrt{2}}{\sqrt{3}}\frac{U}{\sqrt{c - \cot\theta_1}}r^{-5/2}. \tag{186}$$

Here D_{in} is the strain rate intensity factor related to the inner friction surface $\theta = \theta_0$ and D_{ex} is the strain rate intensity factor related to the outer friction surface $\theta = \theta_1$.

The material flux is defined by

$$Q = -2\pi\int_{\theta_0}^{\theta_1} u_r r^2 \sin\theta d\theta. \tag{187}$$

Substituting (175) into (187) gives

$$Q = 2\pi B \int_{\theta_0}^{\theta_1} \exp\left[-3\int_{\theta_0}^{\theta} \cot\mu d\gamma\right] \sin\theta d\theta. \tag{188}$$

As before, it is convenient to introduce the dimensionless strain rate intensity factor by Eq. (170). Then, it follows from (170), (178), (185), (186), and (188) that

$$d_{in} = \frac{\sqrt{6}}{4\pi\sqrt{\cot\theta_0 + c}} \left\{ \int_{\theta_0}^{\theta_1} \exp\left[-3\int_{\theta_0}^{\theta} \cot\mu d\gamma\right] \sin\theta d\theta \right\}^{-1} \tag{189}$$

and

$$d_{\mathrm{ex}} = \frac{\sqrt{6}}{4\pi\sqrt{c - \cot\theta_1}} \exp\left(-3\int_{\theta_0}^{\theta_1} \cot\mu d\theta\right) \left\{ \int_{\theta_0}^{\theta_1} \exp\left[-3\int_{\theta_0}^{\theta} \cot\mu d\gamma\right] \sin\theta d\theta \right\}^{-1}. \tag{190}$$

Using (159) the integrals involved in these equations are represented as

$$\int_{\theta_0}^{\theta_1} \cot\mu d\theta = -\int_{0}^{\pi} \frac{\cos\mu}{(3\sin\mu + \cos\mu\cot\theta - c)} d\mu,$$

$$\int_{\theta_0}^{\theta_1} \exp\left[-3\int_{\theta_0}^{\theta} \cot\mu d\gamma\right] \sin\theta d\theta$$

$$= -\int_{0}^{\pi} \exp\left[3\int_{\mu}^{\pi} \frac{\cos\gamma}{(3\sin\gamma + \cos\gamma\cot\theta - c)} d\gamma\right] \frac{\sin\theta\sin\mu}{(3\sin\mu + \cos\mu\cot\theta - c)} d\mu.$$

Using the solution of Eq. (159) along with the value of c found (Fig. 27) these integrals can be evaluated. As a result, the variation of d_{in} and d_{ex} with θ_0 and θ_1 is obtained. As before, it is convenient to use the parameters θ_0 and $\Delta\theta$ instead of θ_0 and θ_1. Figures 28 and 29 illustrate the dependence of d_{in} and d_{ex} on $\Delta\theta$ at several values of θ_0. It is also of great interest to understand the variation of the ratio $d_r = d_{\mathrm{ex}}/d_{in} = D_{\mathrm{ex}}/D_{in}$ with process parameters. Its dependence on $\Delta\theta$ at several values of θ_0 is illustrated in Fig. 30. In Figs. 28, 29 and 30, curve 1 corresponds to $\theta_0 = \pi/36$, curve 2 to $\theta_0 = \pi/18$, curve 3 to $\theta_0 = \pi/9$, curve 4 to $\theta_0 = \pi/6$, and curve 5 to $\theta_0 = \pi/4$.

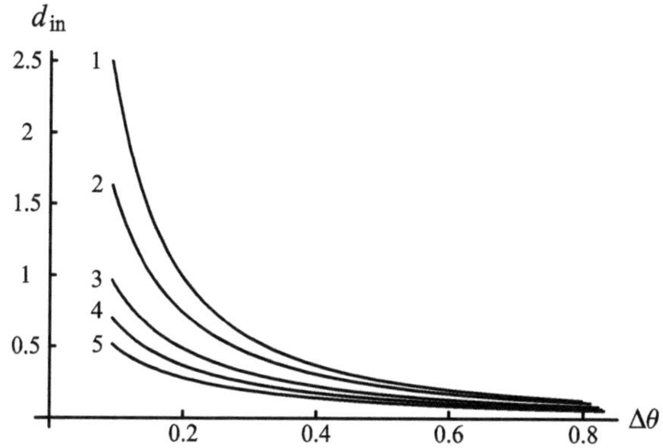

Fig. 28 Variation of the dimensionless strain rate intensity factor d_{in} with $\Delta\theta$ at several θ_0-values

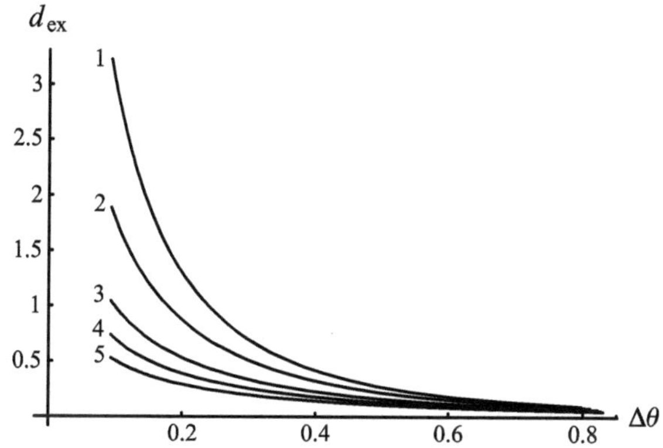

Fig. 29 Variation of the dimensionless strain rate intensity factor d_{ex} with $\Delta\theta$ at several θ_0-values

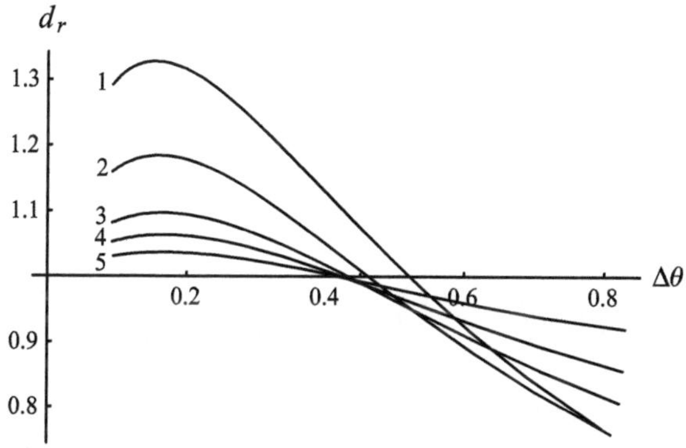

Fig. 30 Variation of the ratio of the strain rate intensity factors with $\Delta\theta$ at several θ_0-values

5 Plane Strain Solutions for the Double-Shearing Model

5.1 Basic Equations

Section 5 is concerned with plane strain solutions for the double-shearing model. In this section, two coordinate systems will be used, namely a Cartesian coordinate system (x, y, z) and a cylindrical coordinate system (r, θ, z). All the solutions considered are independent of z. The principal stress coinciding with the stress σ_{zz} is denoted by σ_3. The constitutive equations in terms of stress and strain rate have been given in [10, 34]. In particular, the Coulomb-Mohr yield criterion is

$$q - p \sin\varphi = k \cos\varphi \tag{191}$$

where k is the cohesion and

$$p = -\frac{\sigma_1 + \sigma_2}{2}, \quad q = \frac{\sigma_1 - \sigma_2}{2} > 0. \tag{192}$$

Using the transformation equations for stress components in xy-planes and (192) the stress components in the Cartesian coordinates are expressed as (Fig. 2)

$$\sigma_{xx} = -p + q\cos 2\psi, \quad \sigma_{yy} = -p - q\cos 2\psi, \quad \sigma_{xy} = q\sin 2\psi. \tag{193}$$

It follows from these equations that

$$p = -\frac{\sigma_{xx} + \sigma_{yy}}{2}, \quad q = \frac{1}{2}\sqrt{\left(\sigma_{xx} - \sigma_{yy}\right)^2 + 4\sigma_{xy}^2}. \tag{194}$$

It is seen from this equation that the yield criterion (191) reduces to the pressure-independent yield criterion (7) at $\varphi = 0$ assuming that $\tau_s = k$. The velocity equations are

$$\xi_{xx} + \xi_{yy} = 0, \quad 2\xi_{xy}\cos 2\psi - \left(\xi_{xx} - \xi_{yy}\right)\sin 2\psi + 2\sin\varphi\left(\omega_{xy} + \dot{\psi}\right) = 0, \tag{195}$$

where ω_{xy} is the component of spin relative to the Cartesian coordinates. By definition,

$$\omega_{xy} = \frac{1}{2}\left(\frac{\partial u_x}{\partial y} - \frac{\partial u_y}{\partial x}\right), \quad \dot{\psi} = \frac{\partial\psi}{\partial t} + u_x\frac{\partial\psi}{\partial x} + u_y\frac{\partial\psi}{\partial y}, \tag{196}$$

where $\partial\psi/\partial t$ is the derivative of ψ at a point which is fixed relative to the Cartesian coordinates. Equation $(195)_1$ coincides with $(9)_1$. Using (193) to eliminate ψ Eq. $(195)_2$ is reduced to $(9)_2$ at $\varphi = 0$. Substituting (10) and $(196)_1$ into (195) yields

$$\frac{\partial u_x}{\partial x} + \frac{\partial u_y}{\partial y} = 0,$$
$$(\cos 2\psi + \sin\varphi)\frac{\partial u_x}{\partial y} + (\cos 2\psi - \sin\varphi)\frac{\partial u_y}{\partial x} \tag{197}$$
$$-\left(\frac{\partial u_x}{\partial x} - \frac{\partial u_y}{\partial y}\right)\sin 2\psi + 2\sin\varphi\dot{\psi} = 0.$$

The constitutive equations should be supplemented with the equilibrium Eq. (12). It is known that the resulting system is hyperbolic [34]. The stress and velocity characteristics coincide. The orientation of the characteristic curves relative to the x-axis is

$$\phi_1 = \psi - \frac{\pi}{4} - \frac{\varphi}{2}, \quad \phi_2 = \psi + \frac{\pi}{4} + \frac{\varphi}{2}. \tag{198}$$

Figure 2 can serve as a geometric interpretation of these relations if $\pi/4$ is replaced with $\pi/4 + \varphi/2$.

Similar equations are valid in the cylindrical coordinates. In particular,

$$\sigma_{rr} = -p + q\cos 2\psi, \quad \sigma_{\theta\theta} = -p - q\cos 2\psi, \quad \sigma_{r\theta} = q\sin 2\psi, \tag{199}$$

where

$$p = -\frac{\sigma_{rr} + \sigma_{\theta\theta}}{2}, \quad q = \frac{1}{2}\sqrt{(\sigma_{rr} - \sigma_{\theta\theta})^2 + 4\sigma_{r\theta}^2}. \tag{200}$$

As before, ψ is now understood as the orientation of the stress σ_1 relative to the r-axis (Fig. 4). The velocity equations are

$$\xi_{rr} + \xi_{\theta\theta} = 0, \quad 2\xi_{r\theta}\cos 2\psi - (\xi_{rr} - \xi_{\theta\theta})\sin 2\psi + 2\sin\varphi\left(\omega_{r\theta} + \dot{\psi}\right) = 0, \tag{201}$$

where

$$\omega_{r\theta} = \frac{1}{2}\left(\frac{\partial u_r}{r\partial\theta} - \frac{\partial u_\theta}{\partial r} - \frac{u_\theta}{r}\right), \quad \dot{\psi} = \frac{\partial\psi}{\partial t} + u_r\frac{\partial\psi}{\partial r} + \frac{u_\theta}{r}\frac{\partial\psi}{\partial\theta}. \tag{202}$$

Substituting (41) and $(202)_1$ into (201) yields

$$r\frac{\partial u_r}{\partial r} + \frac{\partial u_\theta}{\partial\theta} + u_r = 0,$$
$$(\cos 2\psi - \sin\varphi)\frac{\partial u_\theta}{\partial r} + \left(\frac{\partial u_r}{r\partial\theta} - \frac{u_\theta}{r}\right)(\cos 2\psi + \sin\varphi) \tag{203}$$
$$-\left(\frac{\partial u_r}{\partial r} - \frac{\partial u_\theta}{r\partial\theta} - \frac{u_r}{r}\right)\sin 2\psi + 2\sin\varphi\dot{\psi} = 0.$$

The constitutive equations should be supplemented with the equilibrium Eq. (42). The orientation of the characteristic curves relative to the r-axis is given by the following equations similar to (198)

$$\phi_1 = \psi - \frac{\pi}{4} - \frac{\varphi}{2}, \quad \phi_2 = \psi + \frac{\pi}{4} + \frac{\varphi}{2}. \tag{204}$$

Figure 4 can serve as a geometric interpretation of these relations if $\pi/4$ is replaced with $\pi/4 + \varphi/2$.

5.2 Compression of a Plastic Layer Between Parallel Plates

This boundary value problem has been formulated and solved for pressure-independent material in Sect. 3.2 (see Fig. 5). An extension of this solution to the double shearing model has been proposed in [35]. In particular, the velocity field is given by

$$\frac{u_x}{V} = \frac{x}{H} - \frac{\cos 2\psi}{A} + U, \quad \frac{u_y}{V} = -\frac{y}{H}, \tag{205}$$

where U and A are constants of integration and ψ is related to y by the following equation

$$(\sin\phi + \cos 2\psi)\frac{d\psi}{dy} = \frac{A}{H}. \tag{206}$$

The value of U is not essential for determining the strain rate intensity factor. It is evident that the distribution of u_y in (205) satisfies the boundary conditions (44) and (45). The maximum friction surface is determined by the equation $y = H$. Therefore, $\phi_1 = 0$ or $\phi_2 = 0$ at $y = H$ in (198). The direction of the friction stress (Fig. 5) dictates that $\sigma_{xy} < 0$ at $y = H$. Therefore, it follows from (193) that $-\pi/2 < \psi < 0$ at $y = H$. Equation $\phi_1 = 0$ contradicts this inequality. Therefore, $\phi_2 = 0$ and

$$\psi = \psi_w = -\frac{\pi}{4} - \frac{\varphi}{2} \tag{207}$$

at $y = H$. On the other hand, it is reasonable to assume that $\sigma_{xx} > \sigma_{yy}$ at $y = 0$. Then, taking into account the boundary condition (47) at $y = 0$ and (193) it is possible to conclude that

$$\psi = 0 \tag{208}$$

at $y = 0$. Integrating (206) with the use of the boundary condition (207) results in

$$2A\left(\frac{y}{H} - 1\right) = 2\sin\varphi\left(\psi + \frac{\pi}{4} + \frac{\varphi}{2}\right) + \sin 2\psi + \cos\varphi. \tag{209}$$

Substituting the boundary condition (208) into this solution determines A as

$$A = -\sin\varphi\left(\frac{\pi}{4} + \frac{\varphi}{2}\right) - \frac{\cos\varphi}{2}. \tag{210}$$

Eliminating A in (206) by means of (210) it is possible to represent the resulting equation in the vicinity of the friction surface as

$$\frac{d\psi}{dy} = -\frac{\tan\varphi\,(\pi + 2\varphi) + 2}{8H\,(\psi - \psi_w)} + o\left[(\psi - \psi_w)^{-1}\right] \quad \text{as} \quad \psi \to \psi_w. \tag{211}$$

Integrating with the use of the boundary condition $\psi = \psi_w$ at $y = H$ yields

$$(\psi - \psi_w)^2 = \frac{[\tan\varphi\,(\pi + 2\varphi) + 2]}{4H}(H - y) + o\,(H - y) \quad \text{as} \quad y \to H. \tag{212}$$

Substituting (205) into (10) and using (206) give

$$\xi_{xx} = \frac{V}{H}, \quad \xi_{yy} = -\frac{V}{H}, \quad \xi_{xy} = \frac{V}{H}\frac{\sin 2\psi}{(\sin\varphi + \cos 2\psi)}. \tag{213}$$

Substituting (213) into (4) and expanding in a series in the vicinity of $\psi = \psi_w$ yield

$$\begin{aligned} \xi_{\text{eq}} &= \frac{2}{\sqrt{3}}\frac{V}{H}\sqrt{1 + \frac{\sin^2 2\psi}{(\sin\varphi + \cos 2\psi)^2}} \\ &= \frac{V}{\sqrt{3}H\,(\psi - \psi_w)} + o\left[(\psi - \psi_w)^{-1}\right] \quad \text{as} \quad \psi \to \psi_w. \end{aligned} \tag{214}$$

Combining (212) and (214) gives

$$\xi_{\text{eq}} = \frac{2V}{\sqrt{3H}\,[\tan\varphi\,(\pi + 2\varphi) + 2]\sqrt{H - y}} + o\left(\frac{1}{\sqrt{H - y}}\right) \text{ as } y \to H. \tag{215}$$

It follows from (5) and (215) that

$$D = \frac{2V}{\sqrt{3H\,[\tan\varphi\,(\pi + 2\varphi) + 2]}}. \tag{216}$$

The strain rate intensity factor given by (54) is recovered from (216) at $\varphi = 0$. In order to demonstrate the effect of pressure-dependency of the yield criterion on the strain rate intensity factor, it is convenient to introduce the dimensionless strain rate intensity factor, d, as the ratio of the strain rate intensity factor given by (216) to the strain rate intensity factor given by (54). As a result,

$$d = \frac{\sqrt{2}}{\sqrt{\tan\varphi\,(\pi + 2\varphi) + 2}}. \tag{217}$$

The variation of d with φ is depicted in Fig. 31. It is seen from this figure that the strain rate intensity factor decreases as the value of φ increases.

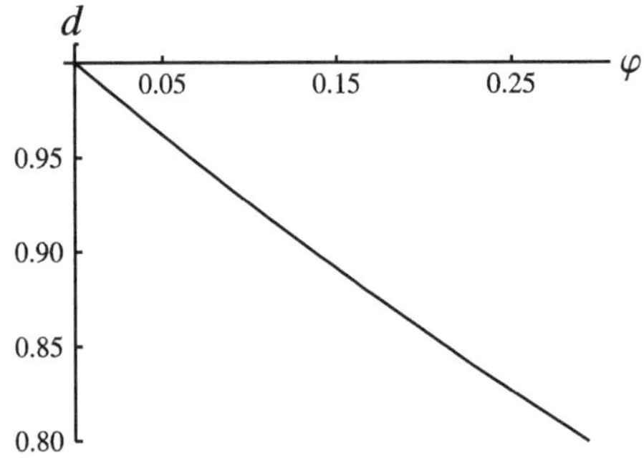

Fig. 31 Variation of the dimensionless strain rate intensity factor with φ

5.3 Flow of Plastic Material Through an Infinite Wedge-Shaped Channel

This boundary value problem has been formulated and solved for pressure-independent material in Sect. 3.3 (see Fig. 6). An extension of this solution to the double shearing model has been proposed in [36]. In particular, the velocity field is given by

$$u_r = -\frac{B}{r\left(\cos 2\psi + \sin\varphi - c\right)}, \qquad u_\theta = 0, \tag{218}$$

where B and c are constants of integration and ψ is related to θ by the following equation

$$(\sin\varphi + \cos 2\psi)\left(\frac{d\psi}{d\theta} + 1\right) = c. \tag{219}$$

It is evident that the distribution of u_θ in (218) satisfies the boundary condition (55) at $\theta = 0$ and $\theta = \alpha$. The maximum friction surface is determined by the equation $\theta = \alpha$. Therefore, $\phi_1 = 0$ or $\phi_2 = 0$ in (204). The direction of the friction stress (Fig. 6) dictates that $\sigma_{r\theta} > 0$ at $\theta = \alpha$. The equation $\phi_2 = 0$ contradicts this inequality. Therefore, $\phi_1 = 0$ and

$$\psi = \psi_w = \frac{\pi}{4} + \frac{\varphi}{2} \tag{220}$$

at $\theta = \alpha$. On the other hand, it is reasonable to assume that $\sigma_{rr} > \sigma_{\theta\theta}$ at $\theta = 0$. Then, taking into account the boundary condition (56) and (199) it is possible to conclude that

$$\psi = 0 \tag{221}$$

at $\theta = 0$. The solution to Eq. (219) satisfying the boundary condition (220) is

$$\theta = \alpha - \int_{\psi}^{\psi_w} \frac{(\sin\varphi + \cos 2\gamma)}{(c - \sin\varphi - \cos 2\gamma)} d\gamma. \tag{222}$$

Here γ is a dummy variable of integration. Substituting the boundary condition (221) into the solution (222) gives the following equation for c

$$\alpha = \int_{0}^{\psi_w} \frac{(\sin\varphi + \cos 2\gamma)}{(c - \sin\varphi - \cos 2\gamma)} d\gamma. \tag{223}$$

This equation has been solved numerically. The variation of c with α for several φ-values is shown in Fig. 32. The broken line corresponds to $\varphi = 0$ (pressure-independent material), curve 1 to $\varphi = 0.1$, curve 2 to $\varphi = 0.2$, and curve 3 to $\varphi = 0.3$. Equation (219) in the vicinity of the maximum friction surface is represented as

$$\frac{d\psi}{d\theta} = \frac{c}{2\cos\varphi\,(\psi_w - \psi)} + o\left[(\psi_w - \psi)^{-1}\right] \quad \text{as} \quad \psi \to \psi_w. \tag{224}$$

Integrating with the use of the boundary condition (220) yields

$$\psi_w - \psi = \left(\frac{c}{\cos\varphi}\right)^{1/2} (\alpha - \theta)^{1/2} + o\left[(\alpha - \theta)^{1/2}\right] \quad \text{as} \quad \theta \to \alpha. \tag{225}$$

It has been taken into account here that $c > 0$ (Fig. 32). The shear strain rate is determined from (41), (218) and (219) as

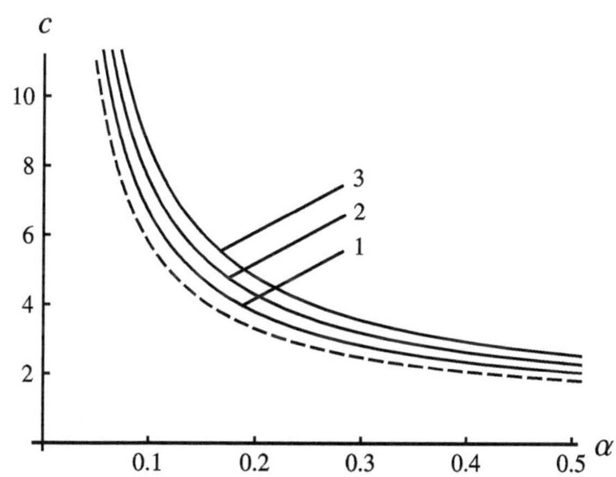

Fig. 32 Variation of c with α at several φ-values

$$\xi_{r\theta} = \frac{B \sin 2\psi}{r^2 (\sin\varphi + \cos 2\psi)(\cos 2\psi + \sin\varphi - c)}. \tag{226}$$

It is evident that the normal strain rates are bounded and $\xi_{r\theta} \to \infty$ as $\psi \to \psi_w$. Therefore, Eq. (6) in which ξ_τ is replaced with $\xi_{r\theta}$ is valid. Expanding the right hand side of (225) in a series in the vicinity of $\psi = \psi_w$ gives

$$\xi_{r\theta} = -\frac{B}{2cr^2(\psi_w - \psi)} + o\left[(\psi_w - \psi)^{-1}\right] \quad \text{as} \quad \psi \to \psi_w. \tag{227}$$

Combining (225) and (227) leads to

$$\xi_{r\theta} = -\frac{B\sqrt{\cos\varphi}}{2c^{3/2}r^2(\alpha - \theta)^{1/2}} + o\left[(\alpha - \theta)^{-1/2}\right] \quad \text{as} \quad \theta \to \alpha. \tag{228}$$

Comparing (6) and (228) shows that

$$D = -\frac{B\sqrt{\cos\varphi}}{\sqrt{3}\,(\mathrm{cr})^{3/2}}. \tag{229}$$

Substituting (218) into (63) and using (219) yields

$$B = -\frac{Q}{2}\left[\int_0^{\psi_w} \frac{(\sin\varphi + \cos 2\psi)}{(\cos 2\psi + \sin\varphi - c)^2} d\psi\right]^{-1}. \tag{230}$$

Since c has been found (Fig. 32), the integral here can be evaluated. Eliminating B in (229) by means of (230) shows that the strain rate intensity factor is a linear function of Q. In order to reveal the effect of α and φ on the strain rate intensity factor, it is convenient to introduce the dimensionless strain rate intensity factor, d, as the ratio of the strain rate intensity factor given by (229) to the strain rate intensity factor given by (62). The variation of d with α at several φ-values is depicted in Fig. 33. In this figure, curve 1 corresponds to $\varphi = 0.1$, curve 2 to $\varphi = 0.15$, curve 3 to $\varphi = 0.2$, curve 4 to $\varphi = 0.25$, and curve 5 to $\varphi = 0.3$. It is seen from this figure that the strain rate intensity factor at $\varphi \neq 0$ is smaller than the strain rate intensity factor for pressure-independent material in the entire range of parameters used in this study.

5.4 Compression of a Plastic Layer Between Cylindrical Surfaces

This boundary value problem has been formulated and solved for pressure-independent material in Sect. 3.4 (see Fig. 10). An extension of this solution to the double shearing model has been proposed in [37]. Let ψ_w be the value of ψ at $r = R_1$ and ψ_f be the value of ψ at $r = R_2$. The direction of flow in the region $\theta > 0$ demands

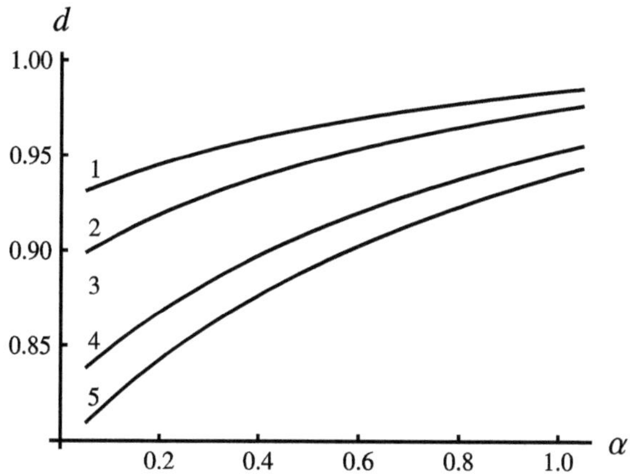

Fig. 33 Variation of the dimensionless strain rate intensity factor with α at several φ-values

$\sigma_{r\theta} > 0$ in the vicinity of the surface $r = R_1$ and $\sigma_{r\theta} < 0$ in the vicinity of the surface $r = R_2$. Then, it follows from (199) that

$$0 < \psi_w < \frac{\pi}{2} \quad \text{and} \quad \frac{\pi}{2} < \psi_f < \pi. \tag{231}$$

The maximum friction surfaces are orthogonal to the r-axis. Therefore, $\phi_1 = \pi/2$ or $\phi_2 = \pi/2$ in (198). The condition $\phi_1 = \pi/2$ contradicts (231) for ψ_w, and the condition $\phi_2 = \pi/2$ contradicts (231) for ψ_f. Therefore,

$$\psi = \psi_w = \frac{\pi}{4} - \frac{\varphi}{2} \quad \text{for} \quad r = R_1 \tag{232}$$

and

$$\psi = \psi_f = \frac{3\pi}{4} + \frac{\varphi}{2} \quad \text{for} \quad r = R_2. \tag{233}$$

It is seen from (232) and (233) that ψ is independent of θ at the maximum friction surfaces. Therefore, a natural assumption to find an approximate solution is that ψ is independent of θ. In this case substituting (199) into (42) and eliminating p by means of (191) result in

$$\frac{(\cos 2\psi \sin\varphi - 1)}{\sin\varphi}\frac{\partial \ln q}{\partial r} + \sin 2\psi \frac{\partial \ln q}{r\partial\theta} - 2\sin 2\psi \frac{d\psi}{dr} + \frac{2\cos 2\psi}{r} = 0, \tag{234}$$

$$-\frac{(\cos 2\psi \sin\varphi + 1)}{\sin\varphi}\frac{\partial \ln q}{r\partial\theta} + \sin 2\psi \frac{\partial \ln q}{\partial r} + 2\cos 2\psi \frac{d\psi}{dr} + \frac{2\sin 2\psi}{r} = 0.$$

Eliminating the derivative $\partial \ln q/\partial r$ between these two equations gives

$$-\cos\varphi\cot\varphi\frac{\partial\ln q}{r\partial\theta}+2\left(\cos 2\psi-\sin\varphi\right)\frac{d\psi}{dr}+\frac{2\sin 2\psi}{r}=0. \tag{235}$$

Since ψ is independent of θ, the first term of this equation must be independent of θ as well. It is possible if and only if

$$\ln\left(\frac{q}{k}\right)=\frac{2c\tan\varphi}{\cos\varphi}\theta+Q\left(\psi\right) \tag{236}$$

and

$$r\left(\cos 2\psi-\sin\phi\right)\frac{d\psi}{dr}=c-\sin 2\psi. \tag{237}$$

Here c is constant and $Q\left(\psi\right)$ is an arbitrary function of ψ. The function $Q\left(\psi\right)$ has no effect on the strain rate intensity factor. Therefore, Eq. (236) is not considered in the present chapter. The solution for the function $Q\left(\psi\right)$ is given in [37]. Equation (237) can be integrated in elementary functions. It is however more convenient to represent its solution satisfying the boundary condition (232) in the form

$$\ln\frac{r}{R_1}=\int\limits_{\psi_w}^{\psi}\frac{\left(\cos 2\gamma-\sin\varphi\right)}{\left(c-\sin 2\gamma\right)}d\gamma. \tag{238}$$

Here γ is a dummy variable of integration. Substituting the boundary condition (233) into (238) leads to the following equation for c

$$\ln\frac{R_2}{R_1}=\int\limits_{\psi_w}^{\psi_f}\frac{\left(\cos 2\gamma-\sin\varphi\right)}{\left(c-\sin 2\gamma\right)}d\gamma. \tag{239}$$

Numerical solution to this equation is illustrated in Fig. 34. The broken line corresponds to $\varphi=0$ (pressure-independent material), curve 1 to $\varphi=0.1$, curve 2 to $\varphi=0.2$, and curve 3 to $\varphi=0.3$.

The main assumption concerning the velocity field is that the radial velocity is independent of θ. Since ψ is also independent of θ, the radial velocity can be represented as

$$u_r=\dot{R}_1 v\left(\psi\right), \tag{240}$$

where $v(\psi)$ is an arbitrary function of ψ. It follows from (67), (68), (232) and (233) that the function $v(\psi)$ must satisfy the conditions

$$v=0 \quad \text{at} \quad \psi=\psi_f \quad (\text{or} \quad r=R_2) \tag{241}$$

and

$$v=1 \quad \text{at} \quad \psi=\psi_w \quad (\text{or} \quad r=R_1). \tag{242}$$

Substituting (240) into $(203)_1$ and taking into account that $v(\psi)$ is independent of θ yields

$$\frac{\partial u_\theta}{\dot{R}_1 \partial \theta} = -\frac{d(rv)}{dr}. \tag{243}$$

The right hand side of this equation is independent of θ. Therefore, integrating (243) gives

$$\frac{u_\theta}{\dot{R}_1} = -\theta \frac{d(rv)}{dr} + V(\psi), \tag{244}$$

where $V(\psi)$ is an arbitrary function of ψ.

A distinguished feature of the present solution, as compared to the solutions given in Sects. 5.2 and 5.3, is that $\dot{\psi} \neq 0$. In particular, substituting (237) and (240) into $(202)_2$ and taking into account that $\partial\psi/\partial\theta = 0$ and $\partial\psi/\partial t = \dot{R}_1 \partial\psi/\partial R_1$ gives

$$\frac{\dot{\psi}}{\dot{R}_1} = \frac{\partial \psi}{\partial R_1} + \frac{v(c - \sin 2\psi)}{r(\cos 2\psi - \sin\varphi)}. \tag{245}$$

In order to find the derivative $\partial\psi/\partial R_1$, it is necessary to differentiate (238). Since ψ_w is constant, one gets

$$d\psi = \frac{(c - \sin 2\psi)}{r(\cos 2\psi - \sin\varphi)} dr + \frac{(c - \sin 2\psi)}{(\cos 2\psi - \sin\varphi)} \left[\frac{dc}{dR_1} \int_{\psi_w}^{\psi} \frac{(\cos 2\gamma - \sin\varphi)}{(c - \sin 2\gamma)^2} d\gamma - \frac{1}{R_1} \right] dR_1.$$

It follows from this equation that

$$\frac{\partial \psi}{\partial R_1} = \frac{(c - \sin 2\psi)}{(\cos 2\psi - \sin\varphi)} \left[\frac{dc}{dR_1} \int_{\psi_w}^{\psi} \frac{(\cos 2\gamma - \sin\varphi)}{(c - \sin 2\gamma)^2} d\gamma - \frac{1}{R_1} \right]. \tag{246}$$

In order to find the derivative dc/dR_1, it is necessary to differentiate (239). Since ψ_w, ψ_f and R_2 are constant, one gets

$$\frac{dc}{dR_1} = \frac{1}{R_1} \left[\int_{\psi_w}^{\psi_f} \frac{(\cos 2\gamma - \sin\varphi)}{(c - \sin 2\gamma)^2} d\gamma \right]^{-1}. \tag{247}$$

Substituting (247) into (246) gives

$$\frac{\partial \psi}{\partial R_1} = \frac{(c - \sin 2\psi)}{R_1(\cos 2\psi - \sin\varphi)} \left[\int_{\psi_w}^{\psi_f} \frac{(\cos 2\gamma - \sin\varphi)}{(c - \sin 2\gamma)^2} d\gamma \right]^{-1} \left[\int_{\psi_w}^{\psi} \frac{(\cos 2\gamma - \sin\varphi)}{(c - \sin 2\gamma)^2} d\gamma - 1 \right]. \tag{248}$$

Using (240), (243) and (244) Eq. $(203)_2$ transforms to

$$\theta\left[-r\left(\cos 2\psi-\sin\varphi\right)\frac{d^2(rv)}{dr^2}+\left(\cos 2\psi+\sin\varphi\right)\frac{d(rv)}{dr}\right]$$
$$+r\left(\cos 2\psi-\sin\varphi\right)\frac{dV}{dr}+2\left[v-\frac{d(rv)}{dr}\right]\sin 2\psi$$
$$-\left(\cos 2\psi+\sin\varphi\right)V+\frac{2r\dot{\psi}}{\dot{R}_1}\sin\varphi=0. \tag{249}$$

Since ψ, $\dot{\psi}$, v and V are independent of θ, this equation may have a solution if and only if

$$r\left(\cos 2\psi-\sin\varphi\right)\frac{d^2(rv)}{dr^2}-\left(\cos 2\psi+\sin\varphi\right)\frac{d(rv)}{dr}=0$$
$$r\left(\cos 2\psi-\sin\varphi\right)\frac{dV}{dr}-\left(\cos 2\psi+\sin\varphi\right)V \tag{250}$$
$$+2\left[v-\frac{d(rv)}{dr}\right]\sin 2\psi+\frac{2r\dot{\psi}}{\dot{R}_1}\sin\varphi=0.$$

Introduce the following notation $Y(\psi)=d(rv)/dr$. Then, Eq. $(244)_1$ transforms to

$$r\left(\cos 2\psi-\sin\varphi\right)\frac{dY}{d\psi}\frac{d\psi}{dr}-\left(\cos 2\psi+\sin\varphi\right)Y=0. \tag{251}$$

Eliminating in this equation the derivative $d\psi/dr$ by means of (237) and integrating yields

$$Y=\frac{d(rv)}{dr}=Y_0\exp\left[\int\limits_{\psi_w}^{\psi}\frac{\left(\cos 2\gamma+\sin\varphi\right)}{\left(c-\sin 2\gamma\right)}d\gamma\right]. \tag{252}$$

Here Y_0 is a constant of integration. Replacing integration with respect to r with integration with respect to ψ in (252) by means of (237) and integrating with the boundary condition (242) result in

$$v(\psi)=\frac{R_1}{r}\left(Y_0\int\limits_{\psi_w}^{\psi}\exp\left[\int\limits_{\psi_w}^{\eta}\frac{\left(\cos 2\gamma+\sin\varphi\right)}{\left(c-\sin 2\gamma\right)}d\gamma\right]\frac{\left(\cos 2\eta-\sin\varphi\right)}{\left(c-\sin 2\eta\right)}\frac{r}{R_1}d\eta+1\right). \tag{253}$$

Here η is a dummy variable of integration. The value of Y_0 is determined from the solution (253) and the boundary condition (241) as

$$Y_0 = -\left\{\int_{\psi_w}^{\psi_f} \exp\left[\int_{\psi_w}^{\eta} \frac{(\cos 2\gamma + \sin\varphi)}{(c - \sin 2\gamma)} d\zeta\right] \frac{(\cos 2\eta - \sin\varphi)}{(c - \sin 2\eta)} \frac{r}{R_1} d\eta\right\}^{-1}. \quad (254)$$

Since the value of c has been found (Fig. 34), eliminating r/R_1 by means of (238) the integrals in (253) and (254) can be evaluated numerically. Since r/R_1 and $d(rv)/dr$ are known functions of ψ due to (238), (252), (253), and (254), it is evident that Eq. $(250)_2$ is a linear ordinary differential equation for $V(\psi)$. Its general solution is

$$V(\psi) = 2Y(\psi)\left[\int_{\psi_w}^{\psi} \frac{\{[Y(\gamma) - v(\gamma)]\sin 2\gamma - \sin\varphi \dot{\psi} r/\dot{R}_1\}}{(c - \sin 2\gamma)\, Y(\gamma)} d\gamma + V_0\right]. \quad (255)$$

Here V_0 is a constant of integration. The integral in (255) can be evaluated numerically. The value of V_0 is found from the condition (69) integrated over the thickness of the layer. Using (238), (245) and (248) the quantity $\dot{\psi} r/\dot{R}_1$ is represented as

$$\frac{\dot{\psi} r}{\dot{R}_1} = \frac{(c - \sin 2\psi)}{(\cos 2\psi - \sin\varphi)} m(\psi),$$

$$m(\psi) = v + \exp\left[\int_{\psi_w}^{\psi} \frac{(\cos 2\gamma - \sin\varphi)}{(c - \sin 2\gamma)} d\gamma\right] \quad (256)$$

$$\times\left\{\left[\int_{\psi_w}^{\psi_f} \frac{(\cos 2\gamma - \sin\varphi)}{(c - \sin 2\gamma)^2} d\gamma\right]^{-1} \int_{\psi_w}^{\psi} \frac{(\cos 2\gamma - \sin\varphi)}{(c - \sin 2\gamma)^2} d\gamma - 1\right\}.$$

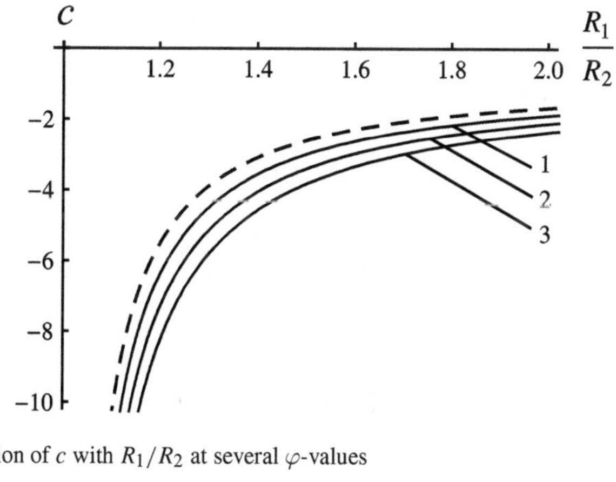

Fig. 34 Variation of c with R_1/R_2 at several φ-values

In order to determine the strain rate intensity factors, it is necessary to find the values of $\dot{\psi} r \dot{R}_1^{-1}$ at the maximum friction surfaces. Moreover, it follows from the boundary conditions (241) and (242) that the right hand side of (256) reduces to the expression 0/0 at $\psi = \psi_w$ and $\psi = \psi_f$. It is evident that the derivative of the denominator with respect to ψ is equal to a finite value at $\psi = \psi_w$ and $\psi = \psi_f$. It follows from (237) and (253) that $\partial r/\partial \psi = 0$ and $\partial v/\partial \psi = 0$ at these points. Therefore, applying l'Hospital's rule,

$$\frac{\dot{\psi} r}{\dot{R}_1} = 0 \quad \text{at} \quad \psi = \psi_w \quad \text{and} \psi = \psi_f. \tag{257}$$

The shear strain rate is determined from (41) where differentiation with respect to r is replaced with differentiation with respect to ψ by means of (237). Then,

$$2\xi_{r\theta} = \frac{\partial u_\theta}{r \partial \psi} \frac{(c - \sin 2\psi)}{(\cos 2\psi - \sin \varphi)} - \frac{u_\theta}{r}. \tag{258}$$

It has been taken into account here that $\partial u_r/\partial \theta = 0$. It is now necessary to find the derivative $\partial u_\theta/\partial \psi$ at $\psi = \psi_w$ and $\psi = \psi_f$. It follows from (244) and (252) that

$$\frac{\partial u_\theta}{U \partial \psi} = -\theta \frac{dY}{d\psi} + \frac{dV}{d\psi}. \tag{259}$$

The derivative $dY/d\psi$ at $\psi = \psi_w$ and $\psi = \psi_f$ is found from (252) as

$$\left.\frac{dY}{d\psi}\right|_{\psi=\psi_w} = \frac{2Y_0 \sin \varphi}{c - \cos \varphi}, \quad \left.\frac{dY}{d\psi}\right|_{\psi=\psi_f} = \frac{2wY_0 \sin \varphi}{c + \cos \varphi}, \tag{260}$$

$$w = \exp\left[\int\limits_{\psi_w}^{\psi_f} \frac{(\cos 2\gamma + \sin \varphi)}{(c - \sin 2\gamma)} d\gamma\right].$$

It is seen from (255) and (260) that the derivative $dV/d\psi$ is finite at $\psi = \psi_w$ and $\psi = \psi_f$. Therefore, it is evident from (258) that $|\xi_{r\theta}| \to \infty$ as $\psi \to \psi_w$ and $\psi \to \psi_f$. The second term in (258) has no effect on this singular behaviour of the shear strain rate since the value of u_θ is finite. Therefore, expanding $\cos 2\psi - \sin \varphi$ in a series in the vicinity of points $\psi = \psi_w$ and $\psi = \psi_f$ Eq. (258) is represented as

$$\begin{aligned}
\xi_{r\theta} &= -\frac{\dot{R}_1 (c - \cos \varphi)}{4R_1 \cos \varphi} \left.\left(\frac{dV}{d\psi} - \theta \frac{dY}{d\psi}\right)\right|_{\psi=\psi_w} (\psi - \psi_w)^{-1} \\
&\quad + o\left[(\psi - \psi_w)^{-1}\right] \quad \text{as} \quad \psi \to \psi_w \\
\xi_{r\theta} &= -\frac{U (c + \cos \varphi)}{4R_2 \cos \varphi} \left.\left(\frac{dV}{d\psi} - \theta \frac{dY}{d\psi}\right)\right|_{\psi=\psi_f} (\psi_f - \psi)^{-1} \\
&\quad + o\left[(\psi_f - \psi)^{-1}\right] \quad \text{as} \quad \psi \to \psi_f.
\end{aligned} \tag{261}$$

Here Eq. (259) has been taken into account. Differentiating (255) with the use of (241), (242), (252), (257), and (260) gives

$$\left.\frac{dV}{d\psi}\right|_{\psi=\psi_w} = \frac{2}{(c-\cos\varphi)}\left[2Y_0V_0\sin\varphi + (Y_0-1)\cos\varphi\right],$$
$$\left.\frac{dV}{d\psi}\right|_{\psi=\psi_f} = \frac{2Y_0w}{(c+\cos\varphi)}\left[2\left(V_0+w_1\right)\sin\varphi - \cos\varphi\right], \tag{262}$$

where

$$w_1 = \int\limits_{\psi_w}^{\psi_f} \frac{\left\{\left[Y(\gamma) - v(\gamma)\right]\sin 2\gamma - \sin\varphi\dot{\psi} r/\dot{R}_1\right\}}{(c-\sin 2\gamma)\,Y(\gamma)}d\gamma. \tag{263}$$

Substituting (260) and (262) into (261) yields

$$\xi_{r\theta} = -\frac{U}{2R_1\cos\varphi}\left[2Y_0V_0\sin\varphi + (Y_0-1)\cos\varphi - \theta Y_0\sin\varphi\right](\psi-\psi_w)^{-1}$$
$$+ o\left[(\psi-\psi_w)^{-1}\right] \quad \text{as} \quad \psi\to\psi_w, \tag{264}$$
$$\xi_{r\theta} = -\frac{UY_0w}{2R_2\cos\varphi}\left[2\left(V_0+w_1\right)\sin\varphi - \cos\varphi - \theta\sin\varphi\right]\left(\psi_f-\psi\right)^{-1}$$
$$+ o\left[\left(\psi_f-\psi\right)^{-1}\right] \quad \text{as} \quad \psi\to\psi_f.$$

Equation (237) in the vicinity of points $\psi=\psi_w$ and $\psi=\psi_f$ is represented as

$$\frac{d\psi}{dr} = -\frac{(c-\cos\varphi)}{2R_1\cos\varphi\,(\psi-\psi_w)} + o\left[(\psi-\psi_w)^{-1}\right] \quad \text{as} \quad \psi\to\psi_w,$$
$$\frac{d\psi}{dr} = -\frac{(c+\cos\varphi)}{2R_2\cos\varphi\left(\psi_f-\psi\right)} + o\left[\left(\psi_f-\psi\right)^{-1}\right] \quad \text{as} \quad \psi\to\psi_f, \tag{265}$$

respectively. Integrating the first of these equations with the boundary condition (232) gives

$$\psi-\psi_w = \sqrt{-\frac{(c-\cos\varphi)}{R_1\cos\varphi}}\sqrt{r-R_1} + o\left(\sqrt{r-R_1}\right) \quad \text{as} \quad r\to R_1. \tag{266}$$

Integrating $(265)_2$ with the boundary condition (233) gives

$$\psi_f-\psi = \sqrt{-\frac{(c+\cos\varphi)}{R_2\cos\varphi}}\sqrt{R_2-r} + o\left(\sqrt{R_2-r}\right) \quad \text{as} \quad r\to R_2. \tag{267}$$

Since the normal strain rates are finite as $r\to R_1$ and $r\to R_2$, Eq. (6) in which ξ_τ should be replaced with $\xi_{r\theta}$ is valid. Substituting (266) into $(264)_1$ and comparing to (6) show that

$$D_1 = \frac{\dot{R}_1}{\sqrt{3R_1 \cos\varphi}} \frac{|2Y_0 V_0 \sin\varphi + (Y_0 - 1)\cos\varphi - \theta Y_0 \sin\varphi|}{\sqrt{\cos\varphi - c}}, \tag{268}$$

where D_1 is the strain rate intensity factor corresponding to the maximum friction surface $r = R_1$. Analogously, it follows from $(265)_2$, (267) and (6) that the strain rate intensity factor corresponding to the maximum friction surface $r = R_2$ is

$$D_2 = \frac{\dot{R}_1 |Y_0 w|}{\sqrt{3R_2 \cos\varphi}} \frac{|2(V_0 + w_1)\sin\varphi - \cos\varphi - \theta \sin\varphi|}{\sqrt{-(c + \cos\varphi)}}. \tag{269}$$

The right hand sides of (268) and (269) can be evaluated using (254) and (263).

It is of interest to introduce the ratio $\Delta = D_1/D_2$. It follows from (268) and (269) that

$$\Delta = \sqrt{\frac{R_2}{R_1}} \frac{|2Y_0 V_0 \sin\varphi + (Y_0 - 1)\cos\varphi - \theta Y_0 \sin\varphi| \sqrt{-(c + \cos\varphi)}}{|Y_0 w| \, |2(V_0 + w_1)\sin\varphi - \cos\varphi - \theta \sin\varphi| \sqrt{\cos\varphi - c}}. \tag{270}$$

The variation of Δ with R_1/R_2 at $\theta = 1$ and $\theta = 1.5$ is shown in Figs. 35 and 36, respectively. The broken line corresponds to the solution for pressure-independent material (see Sect. 3.4) curve 1 corresponds to $\varphi = 0.1$, curve 2 to $\varphi = 0.15$, curve 3 to $\varphi = 0.2$, curve 4 to $\varphi = 0.25$, and curve 5 to $\varphi = 0.3$. The dependence of Δ on θ for several values of R_1/R_2 at $\varphi = 0.1$, $\varphi = 0.2$ and $\varphi = 0.3$ is depicted in Figs. 37, 38 and 39, respectively. Curve 1 corresponds to $R_1/R_2 = 0.5$, curve 2 to $R_1/R_2 = 0.7$, and curve 3 to $R_1/R_2 = 0.9$. In order to illustrate the effect of pressure-dependency of the yield criterion on the strain rate intensity factors, it is convenient to introduce the ratios of the strain rate intensity factors given in (268) and (269) to the respective strain rate intensity factors from the solution for the pressure-independent model. These ratios are denoted by d_1 and d_2 where d_1 is related to the surface $r = R_1$ and d_2 to the surface $r = R_2$. Since the strain rate intensity factors for the model of pressure-independent plasticity are given in (78), the values of d_1 and d_2 can be found with no difficulty. It is evident from (78), (268) and (269) that d_1 and d_2 are linear functions of θ. The dependence of d_1 on θ at $\varphi = 0.1$, $\varphi = 0.2$ and $\varphi = 0.3$ is depicted in Figs. 40, 41 and 42, and the dependence of d_2 on θ at the same values of φ in Figs. 43, 44 and 45 (curve 1 corresponds to $R_1/R_2 = 0.5$, curve 2 to $R_1/R_2 = 0.7$, curve 3 to $R_1/R_2 = 0.9$). The variation of d_1 and d_2 with R_1/R_2 at $\theta = 1$ is depicted in Figs. 46 and 47, respectively. In these figures, curve 1 corresponds to $\varphi = 0.1$, curve 2 to $\varphi = 0.15$, curve 3 to $\varphi = 0.2$, curve 4 to $\varphi = 0.25$, and curve 5 to $\varphi = 0.3$. It is seen from Figs. 40, 41, 42, 43, 44, 45, 46 and 47 that $d_1 < 1$ and $d_2 < 1$. Thus pressure-dependency of the yield criterion leads to a decrease in the magnitude of the strain rate intensity factor in the process under consideration.

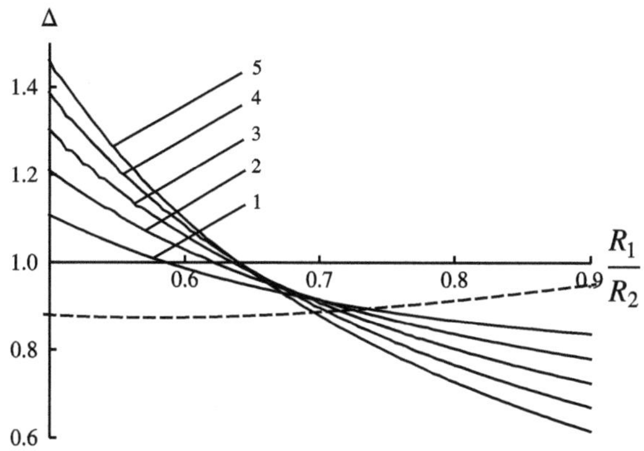

Fig. 35 Variation of Δ with R_1/R_2 at $\theta = 1$

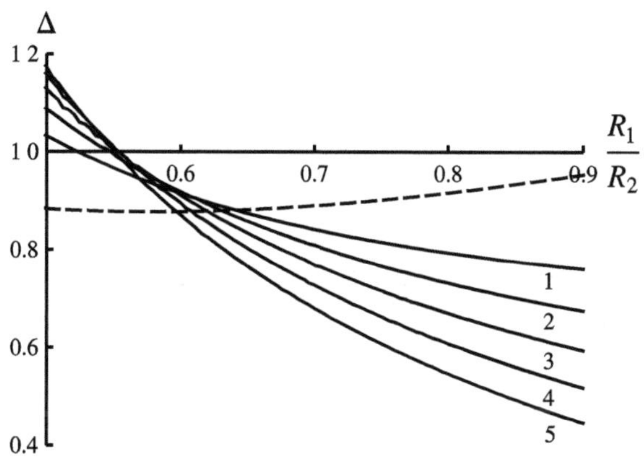

Fig. 36 Variation of Δ with R_1/R_2 at $\theta = 1.5$

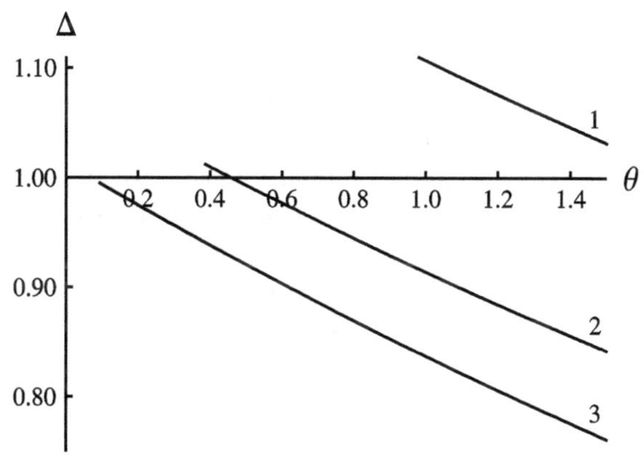

Fig. 37 Dependence of Δ on θ at $\varphi = 0.1$ and several R_1/R_2-values

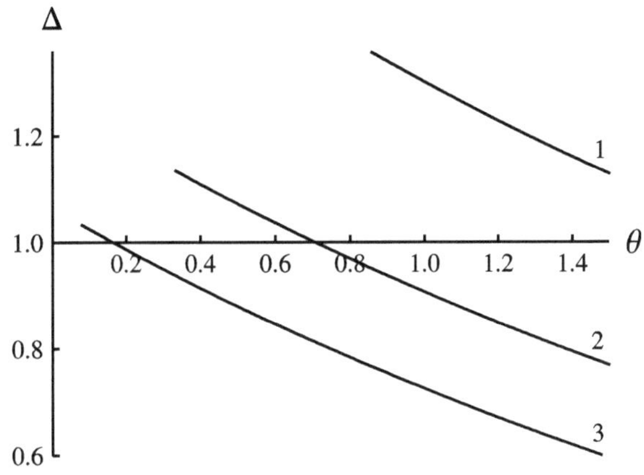

Fig. 38 Dependence of Δ on θ at $\varphi = 0.2$ and several R_1/R_2-values

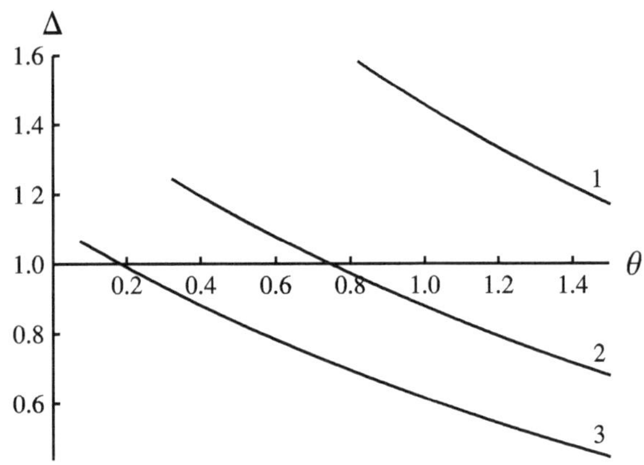

Fig. 39 Dependence of Δ on θ at $\varphi = 0.3$ and several R_1/R_2-values

5.5 Compression of a Plastic Layer Between Rotating Plates I

This boundary value problem has been formulated and solved for pressure-independent material in Sect. 3.5 (see Fig. 11). An extension of this solution to the double shearing model has been proposed in [38].

Let ψ_w be the value of ψ at the maximum friction surface $\theta = \alpha$. The direction of flow (Fig. 11) dictates that $\sigma_{r\theta} < 0$ near the friction surface. Therefore, it follows from (199) that

$$-\frac{\pi}{2} < \psi_w < 0. \tag{271}$$

The maximum friction surface is parallel to the r-axis. Therefore, $\phi_1 = 0$ or $\phi_2 = 0$ in (204). The equation $\phi_1 = 0$ contradicts (271). Therefore, $\phi_2 = 0$ and

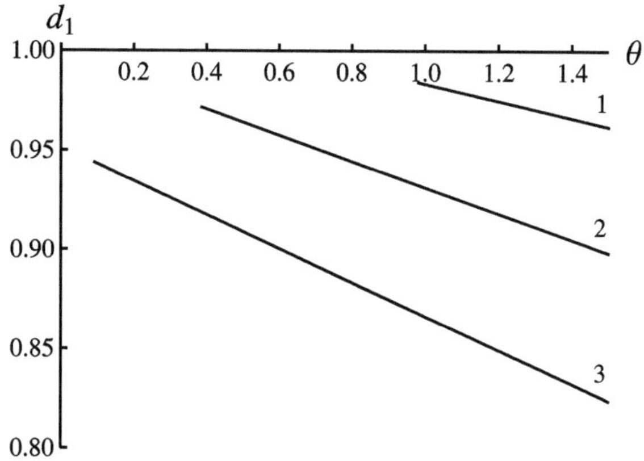

Fig. 40 Dependence of d_1 on θ at $\varphi = 0.1$ and several R_1/R_2-values

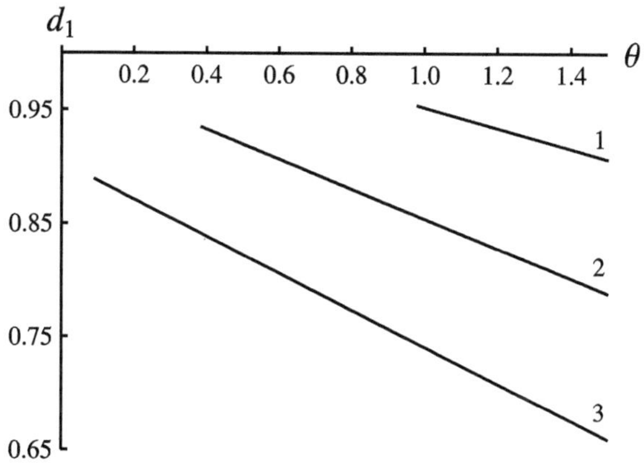

Fig. 41 Dependence of d_1 on θ at $\varphi = 0.2$ and several R_1/R_2-values

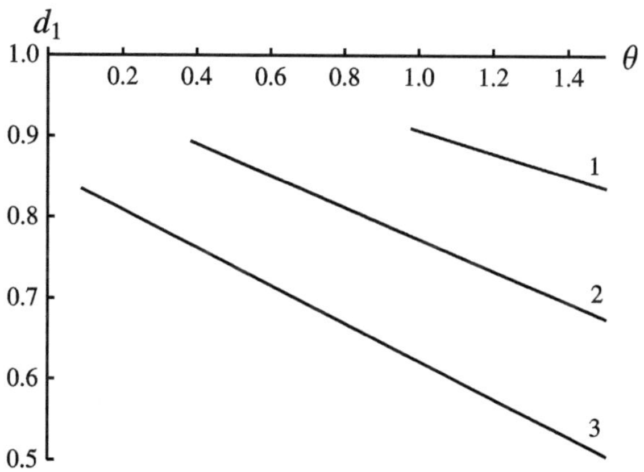

Fig. 42 Dependence of d_1 on θ at $\varphi = 0.3$ and several R_1/R_2-values

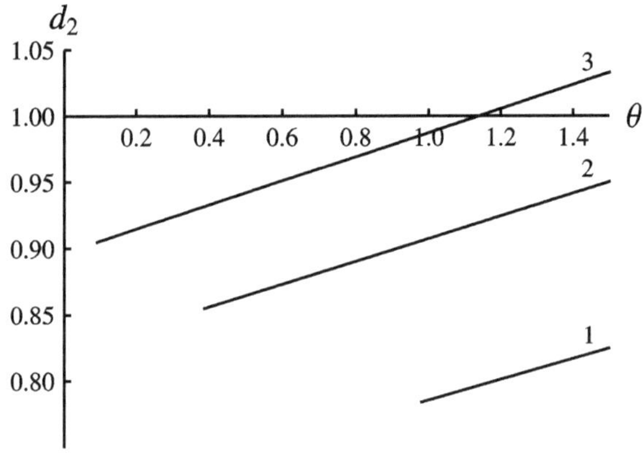

Fig. 43 Dependence of d_2 on θ at $\varphi = 0.1$ and several R_1/R_2-values

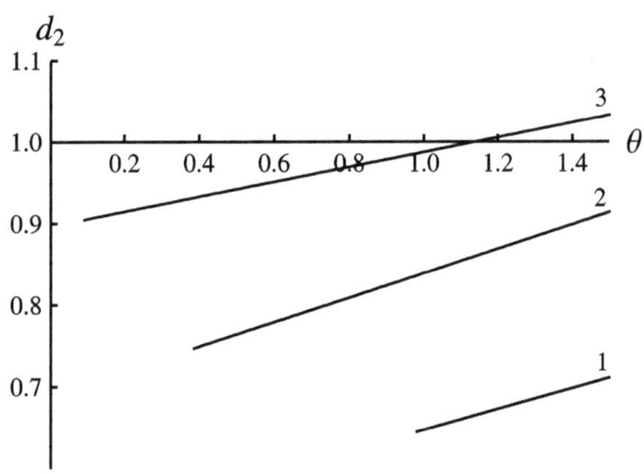

Fig. 44 Dependence of d_2 on θ at $\varphi = 0.2$ and several R_1/R_2-values

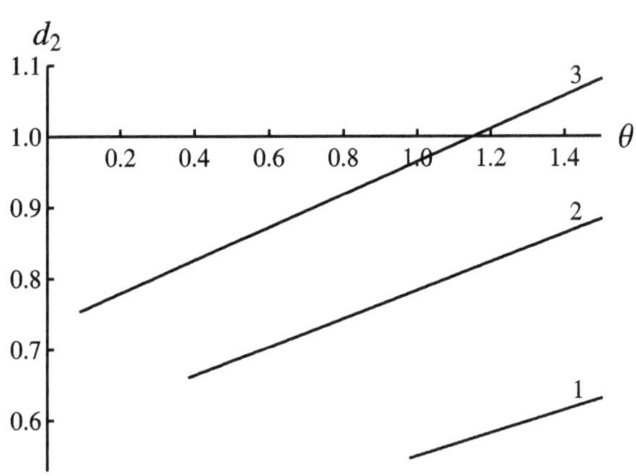

Fig. 45 Dependence of d_2 on θ at $\varphi = 0.3$ and several R_1/R_2-values

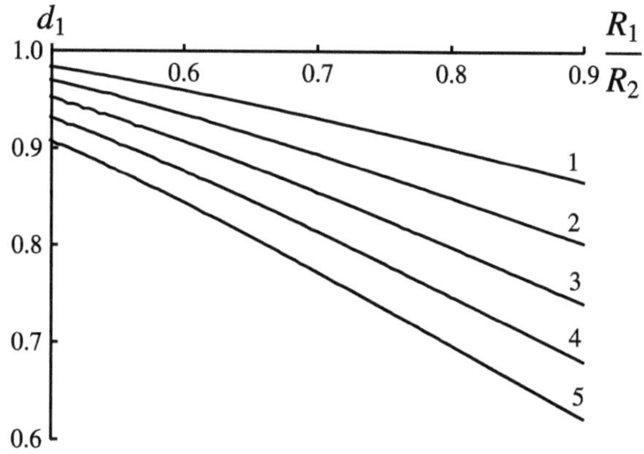

Fig. 46 Variation of d_1 with R_1/R_2 at $\theta = 1$ and at several φ-values

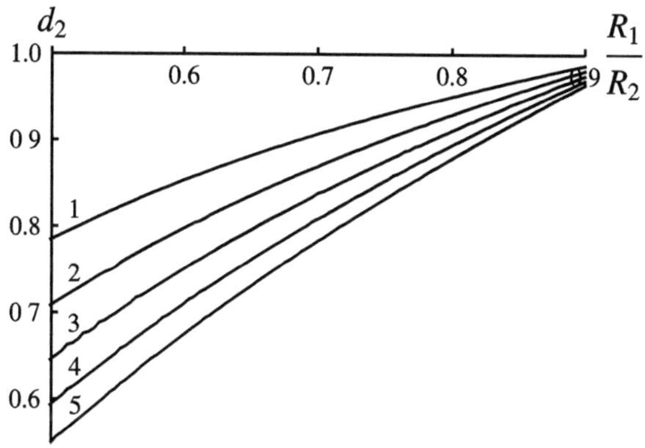

Fig. 47 Variation of d_2 with R_1/R_2 at $\theta = 1$ and at several φ-values

$$\psi = \psi_w = -\frac{\pi}{4} - \frac{\varphi}{2} \quad \text{for} \quad \theta = \alpha. \tag{272}$$

The main assumption accepted in [38] is that ψ is independent of r. Note that this assumption is in agreement with (272). Substituting (191) and (199) into (42) gives

$$\begin{aligned} r\frac{(\cos 2\psi \sin\varphi - 1)}{\sin\varphi}\frac{\partial \ln q}{\partial r} + \sin 2\psi\frac{\partial \ln q}{\partial \theta} + 2\cos 2\psi\left(\frac{d\psi}{d\theta} + 1\right) &= 0, \\ -\frac{(\cos 2\psi \sin\varphi + 1)}{\sin\varphi}\frac{\partial \ln q}{\partial \theta} + r\sin 2\psi\frac{\partial \ln q}{\partial r} + 2\sin 2\psi\left(\frac{d\psi}{d\theta} + 1\right) &= 0. \end{aligned} \tag{273}$$

Eliminating the derivative $\partial \ln q/\partial\theta$ between these equations yields

$$r\frac{\partial \ln q}{\partial r} = \frac{2\sin\varphi\,(\sin\varphi + \cos 2\psi)}{\cos^2\varphi}\left(\frac{d\psi}{d\theta} + 1\right). \tag{274}$$

The right hand side of this equation is independent of r. Therefore, integrating gives

$$\ln q = \frac{2\sin\varphi\left(\sin\varphi + \cos 2\psi\right)}{\cos^2\varphi}\left(\frac{d\psi}{d\theta} + 1\right)\ln r + Q\left(\theta\right), \tag{275}$$

where $Q(\theta)$ is an arbitrary function of θ. Substituting (275) into $(273)_2$ results in

$$-\frac{2\left(\cos 2\psi \sin\varphi + 1\right)}{\cos^2\varphi}\frac{d}{d\theta}\left[\left(\sin\varphi + \cos 2\psi\right)\left(\frac{d\psi}{d\theta} + 1\right)\right]\ln r \tag{276}$$
$$-\frac{dQ}{d\theta}\frac{\left(\cos 2\psi \sin\varphi + 1\right)}{\sin\varphi} + \frac{2\sin 2\psi\left(1 + \sin\varphi\cos 2\psi\right)}{\cos^2\varphi}\left(\frac{d\psi}{d\theta} + 1\right) = 0.$$

This equation can have a solution if and only if the coefficient of $\ln r$ vanishes. Therefore,

$$\left(\sin\varphi + \cos 2\psi\right)\left(\frac{d\psi}{d\theta} + 1\right) = c, \tag{277}$$

where c is constant. Substituting (277) into (276) results in

$$\frac{dQ}{d\theta} = \frac{2c\sin\varphi\sin 2\psi}{\cos^2\varphi\left(\sin\varphi + \cos 2\psi\right)}. \tag{278}$$

Replacing here differentiation with respect to θ with differentiation with respect to ψ by means of (277) and integrating yield

$$Q = \frac{c\sin\varphi}{\cos^2\varphi}\ln\left|c - \sin\varphi - \cos 2\psi\right| + Q_0, \tag{279}$$

where Q_0 is a constant of integration. The solution to Eq. (277) satisfying the boundary condition (272) is

$$\theta = \alpha - \int_{\psi}^{\psi_w}\frac{\left(\sin\varphi + \cos 2\gamma\right)}{\left(c - \sin\varphi - \cos 2\gamma\right)}d\gamma, \tag{280}$$

where γ is a dummy variable of integration. It follows from (37) and (81) that $\psi = 0$ at $\theta = 0$. Using this condition the equation for c is obtained from (280) in the form

$$\alpha = \int_{0}^{\psi_w}\frac{\left(\sin\varphi + \cos 2\gamma\right)}{\left(c - \sin\varphi - \cos 2\gamma\right)}d\gamma. \tag{281}$$

Numerical solution to this equation is illustrated in Fig. 48. The broken line corresponds to pressure-independent material, curve 1 to $\varphi = 0.1$, curve 2 to $\varphi = 0.2$, and curve 3 to $\varphi = 0.3$. In the case of $c = 0$ Eq. (281) is immediately integrated to give $\alpha = \pi/4 + \varphi/2$. It is seen from Fig. 48 that $c < 0$ for $\alpha < \pi/4 + \varphi/2$.

It is necessary to find the derivative $dc/d\alpha$ to determine the strain rate intensity factor. Differentiating (281) gives

$$\frac{dc}{d\alpha} = -\left[\int_0^{\psi_w} \frac{(\sin\varphi + \cos 2\gamma)}{(c - \sin\varphi - \cos 2\gamma)^2} d\gamma\right]^{-1}. \quad (282)$$

At $c = 0$ one gets

$$I(\psi) = \int_0^{\psi} \frac{(\sin\phi + \cos 2\zeta)}{(A - \sin\phi - \cos 2\zeta)^2} d\zeta = \int_0^{\psi} \frac{d\zeta}{(\sin\phi + \cos 2\zeta)}$$
$$= \operatorname{arctanh}\left(\frac{\cos\phi \tan\psi}{1 + \sin\phi}\right)\frac{1}{\cos\phi}.$$

Hence $\lim_{\psi \to \psi_w} I(\psi) = -\infty$ and it follows from (282) that

$$\frac{dc}{d\alpha} = 0 \quad \text{at} \quad c = 0. \quad (283)$$

The velocity field is sought in the form [38]

$$u_r = \frac{\omega r}{2}\frac{\partial g(\psi,\ \alpha)}{\partial \theta}, \qquad u_\theta = -\omega r g(\psi,\ \alpha), \quad (284)$$

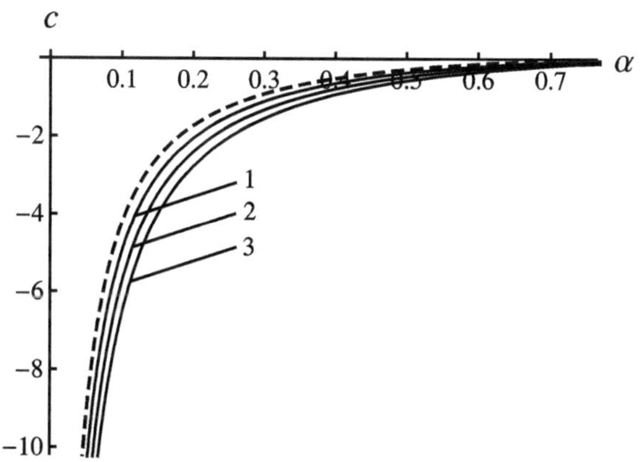

Fig. 48 Variation of c with α at several φ-values

where $g(\psi,\ \alpha)$ is an arbitrary function of ψ and α. The velocity field (284) automatically satisfies the equation of incompressibility $(203)_1$. Using (284) the boundary conditions (79) and (80) transform to

$$g = 0 \quad \text{for} \quad \theta = 0 \quad (\text{or} \quad \psi = 0) \tag{285}$$

and

$$g = 1 \quad \text{for} \quad \theta = \alpha \quad (\text{or} \quad \psi = \psi_w). \tag{286}$$

Substituting (284) into $(203)_2$ gives

$$(\cos 2\psi + \sin\varphi)\frac{d^2 g}{d\theta^2} - 2\sin 2\psi \frac{dg}{d\theta} + 4g\sin\varphi + 4\sin\varphi \frac{\dot{\psi}}{\omega} = 0. \tag{287}$$

The derivative $d\psi/d\theta$ is found from (277). Subsequent differentiation of this derivative with respect to θ gives

$$\frac{d^2\psi}{d\theta^2} = \frac{2c\sin 2\psi\,(c - \sin\varphi - \cos 2\psi)}{(\sin\varphi + \cos 2\psi)^3}. \tag{288}$$

On the other hand,

$$\frac{dg}{d\theta} = \frac{dg}{d\psi}\frac{d\psi}{d\theta}, \quad \frac{d^2 g}{d\theta^2} = \frac{d^2 g}{d\psi^2}\left(\frac{d\psi}{d\theta}\right)^2 + \frac{dg}{d\psi}\frac{d^2\psi}{d\theta^2}. \tag{289}$$

Replacing in (287) differentiation with respect to θ with differentiation with respect to ψ by means of (289) and using (277) and (288) leads to

$$\begin{aligned} &(\sin\varphi + \cos 2\psi)\frac{d^2 g}{d\psi^2} + 2\sin 2\psi \frac{dg}{d\psi} \\ &+ \frac{4\sin\varphi\,(\sin\varphi + \cos 2\psi)^2}{(c - \sin\varphi - \cos 2\psi)^2}\left(g + \frac{\dot{\psi}}{\omega}\right) = 0. \end{aligned} \tag{290}$$

Since $\partial\psi/\partial t = -\omega\partial\psi/\partial\alpha$, it follows from (202) and (277) that

$$\frac{\dot{\psi}}{\omega} = -\frac{\partial\psi}{\partial\theta_0} - \left(\frac{c - \sin\varphi - \cos 2\psi}{\sin\varphi + \cos 2\psi}\right) g. \tag{291}$$

Differentiating (280) gives

$$\frac{(\sin\varphi + \cos 2\psi)}{(c - \sin\varphi - \cos 2\psi)} d\psi$$
$$+ \left\{ 1 + \frac{dc}{d\alpha} \left[\int_{\psi}^{\psi_w} \frac{(\sin\varphi + \cos 2\gamma)}{(c - \sin\varphi - \cos 2\gamma)^2} d\gamma \right] \right\} d\alpha - d\theta = 0. \tag{292}$$

Determining the derivative $\partial\psi/\partial\alpha$ from this equation and substituting it into (291) lead to

$$\frac{\dot{\psi}}{\omega} = \left\{ 1 - g + \frac{dc}{d\alpha} \left[\int_{\psi}^{\psi_w} \frac{(\sin\varphi + \cos 2\gamma)}{(c - \sin\varphi - \cos 2\gamma)^2} d\gamma \right] \right\}$$
$$\times \frac{(c - \sin\varphi - \cos 2\psi)}{(\sin\varphi + \cos 2\psi)}. \tag{293}$$

Eliminating $\dot{\psi}/\omega$ in (290) by means of (293) results in a linear ordinary differential equation for $g(\psi)$ in the form

$$(\sin\varphi + \cos 2\psi) \frac{d^2 g}{d\psi^2} + 2\sin 2\psi \frac{dg}{d\psi} + P_1(\psi)\, g = P_0(\psi),$$
$$P_1(\psi) = \frac{4\sin\varphi (\sin\varphi + \cos 2\psi)(2\sin\varphi + 2\cos 2\psi - c)}{(c - \sin\varphi - \cos 2\psi)^2}, \tag{294}$$
$$P_0(\psi) = -\frac{4\sin\varphi(\sin\varphi + \cos 2\psi)}{(c - \sin\varphi - \cos 2\psi)} \left\{ \frac{dc}{d\alpha} \left[\int_{\psi}^{\psi_w} \frac{(\sin\varphi + \cos 2\gamma)}{(c - \sin\varphi - \cos 2\gamma)^2} d\gamma \right] + 1 \right\}.$$

Expanding the coefficients of this equation in series in the vicinity of $\psi = \psi_w$ gives

$$\sin\varphi + \cos 2\psi = 2\cos\varphi(\psi - \psi_w) + 2\sin\varphi(\psi - \psi_w)^2$$
$$- \frac{4}{3}\cos\varphi(\psi - \psi_w)^3 + o\left[(\psi - \psi_w)^3\right],$$
$$2\sin 2\psi = -2\cos\varphi - 4\sin\varphi(\psi - \psi_w) + 4\cos\varphi(\psi - \psi_w)^2$$
$$+ \frac{8}{3}\sin\varphi(\psi - \psi_w)^3 + o\left[(\psi - \psi_w)^3\right],$$
$$P_1(\psi) = -\frac{4\sin 2\varphi}{c}(\psi - \psi_w) - \frac{8\sin^2\varphi}{c}(\psi - \psi_w)^2 \tag{295}$$
$$+ \frac{8\sin 2\varphi}{c}\left(\frac{1}{3} + \frac{2\cos^2\varphi}{c^2}\right)(\psi - \psi_w)^3 + o\left[(\psi - \psi_w)^3\right],$$

$$P_0(\psi) = -\frac{4\sin 2\varphi}{c}(\psi - \psi_w) - \frac{8\sin\varphi\,(1 + \cos 2\varphi + c\sin\varphi)}{c^2}(\psi - \psi_w)^2$$
$$- \frac{4\sin 2\varphi}{3c^3}\left[2\left(3 - c^2 + 3\cos 2\varphi + 6c\sin\varphi\right) - 3\cos\varphi\frac{dc}{d\alpha}\right](\psi - \psi_w)^3$$
$$+ o\left[(\psi - \psi_w)^3\right]$$

as $\psi \to \psi_w$. It is seen from (295) that $\psi = \psi_w$ is a regular singular point of Eq. (294). Using a standard procedure it is possible to find that the two linearly independent primitive solutions of the corresponding homogeneous equation are represented as

$$g_1 = O\left[(\psi - \psi_w)^2\right] \quad \text{as} \quad \psi \to \psi_w \tag{296}$$

and

$$g_2 = P(\psi) + Cg_1 \ln(\psi - \psi_w), \tag{297}$$

where $P(\psi)$ is a function of ψ represented by a Taylor series in the vicinity of $\psi = \psi_w$ and C is an arbitrary constant. It is possible to show (see [38]) that C must vanish and that $P(\psi)$ does not contain the term $O(\psi - \psi_w)$ as $\psi \to \psi_w$. Then, it follows from (296), (297) and the boundary condition (286) that the function $g(\psi)$ is approximated by

$$g = 1 + c_2(\psi - \psi_w)^2 + c_3(\psi - \psi_w)^3 \tag{298}$$

in the vicinity of point $\psi = \psi_w$. Substituting (295) and (298) into (294) yields

$$c_3 = \frac{2}{3}c_2\tan\varphi - \frac{4\sin 2\varphi}{3c^2}. \tag{299}$$

It has been shown before that $c \le 0$ (Fig. 48). However, it follows from (295) and (299) that the special case $c = 0$ should be treated separately. In order to clarify the general structure of the solution for values of c in the vicinity of $c = 0$, it is necessary to find the radial velocity at $\psi = \psi_w$ (i.e. at the maximum friction surface). It follows from (277), (284), (295), and (298) that

$$u_r = \frac{\omega r}{2}\frac{cc_2}{\cos\varphi} + o(1) \quad \text{as} \quad \psi \to \psi_w. \tag{300}$$

It is seen from this equation that u_r at $\psi = \psi_w$ vanishes if $c_2 = 0$. This means that the regime of sticking occurs. Let α_{cr} be the value of α corresponding to the condition $c_2 = 0$. Assume that

$$0 < \alpha_{cr} < \frac{\pi}{4} + \frac{\varphi}{2}. \tag{301}$$

Then, the general structure of the solution is as follows. The found solution at sliding is valid if $\alpha < \alpha_{cr}$. The condition (301) ensures that $c < 0$ (Fig. 48). Then, since

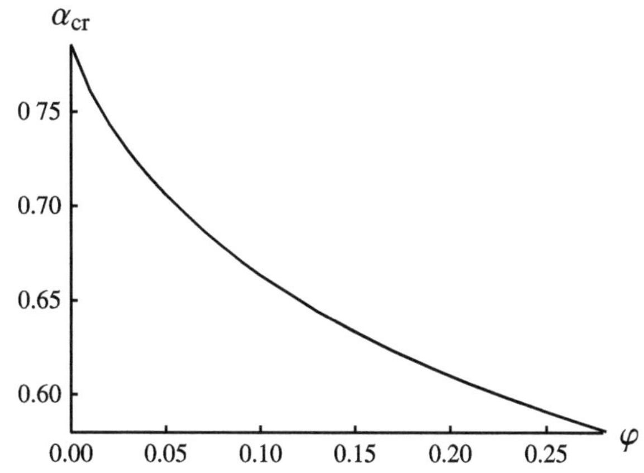

Fig. 49 The variation of α_{cr} with φ

$u_r > 0$, it follows from (300) that

$$c_2 < 0. \tag{302}$$

A limit case occurs at $\alpha = \alpha_{cr}$. In this case the found solution is valid but the regime of sticking takes place at the maximum friction surface. If $\alpha > \alpha_{cr}$, a rigid zone appears near the plate. This case is not important for the present chapter since the solution is not singular. In order to show that the structure of the solution proposed is possible, it is necessary to verify (301). To this end Eq. (294) has been solved numerically at $c_2 = 0$. The representation of the solution given in (298) has been used in the range $\psi_w \geq \psi \geq \psi_w (1 - \delta)$ where $0 < \delta << 1$. The value of c_3 in this representation has been eliminated by means of (299). The solution found must satisfy the boundary condition (285). The value of c corresponding to the limit case is determined from this boundary condition. It is evident that this value is related to α_{cr} by Eq. (281) in which α should be replaced with α_{cr}. The numerical solution is illustrated in Fig. 49. It is seen from this figure that the inequality (301) is satisfied. In what follows, it is assumed that $\alpha \leq \alpha_{cr}$.

Using (277), (284), (288) and (289) the shear strain rate is determined from (41) as

$$\xi_{r\theta} = \frac{\omega (c - \sin\varphi - \cos 2\psi)}{4 (\sin\varphi + \cos 2\psi)^2} \left[(c - \sin\varphi - \cos 2\psi) \frac{d^2 g}{d\psi^2} + \frac{2c \sin 2\psi}{(\sin\varphi + \cos 2\psi)} \frac{dg}{d\psi} \right]. \tag{303}$$

Substituting (298) into (303) and using (295) and (299) yield the representation of the shear strain rate in the vicinity of the maximum friction surface in the form

$$\xi_{r\theta} = -\frac{\omega}{4\cos\varphi} \frac{\left[4\sin\varphi + c\left(c_2 + \tan^2\varphi\right)\right]}{(\psi - \psi_w)} + o\left[(\psi - \psi_w)^{-1}\right] \quad \text{as} \quad \psi \to \psi_w. \tag{304}$$

In the vicinity of the maximum friction surface (277) is written as

$$\frac{d\psi}{d\theta} = \frac{c}{2\cos\varphi\,(\psi - \psi_w)} + o\left[(\psi - \psi_w)^{-1}\right] \quad \text{as} \quad \psi \to \psi_w.$$

Integrating with the use of the boundary condition (272) gives

$$\psi - \psi_w = \frac{\sqrt{-c}}{\sqrt{\cos\varphi}}(\alpha - \theta)^{1/2} + o\left[(\alpha - \theta)^{1/2}\right] \quad \text{as} \quad \theta \to \alpha. \tag{305}$$

Substituting (305) into (304) yields

$$|\xi_{r\theta}| = \frac{\omega}{4\sqrt{\cos\varphi}} \frac{\left[4\sin\varphi + c\left(c_2 + \tan^2\varphi\right)\right]}{\sqrt{-c}\sqrt{\alpha - \theta}} + o\left[(\alpha - \theta)^{-1/2}\right] \quad \text{as} \quad \theta \to \alpha. \tag{306}$$

Equation (6) in which ξ_τ should be replased with $\xi_{r\theta}$ is valid. Therefore, the strain rate intensity factor is given by

$$D = \frac{\omega}{2\sqrt{3}\sqrt{\cos\varphi}} \frac{\left[4\sin\varphi + c\left(c_2 + \tan^2\varphi\right)\right]\sqrt{r}}{\sqrt{-c}}. \tag{307}$$

It is convenient to introduce the dimensionless strain rate intensity factor by

$$d = \frac{D}{\omega\sqrt{r}}. \tag{308}$$

The same definition for the dimensionless strain rate intensity factor has been adopted for the pressure-independent model in Sect. 3.5 (see Fig. 12). The variation of the dimensionless strain rate intensity factor found by means of (307) and (308) as well as the dimensionless strain rate intensity factor for the pressure-independent model is depicted in Fig. 50. The broken line corresponds to pressure-independent material, curve 1 to $\varphi = 0.1$, curve 2 to $\varphi = 0.2$, and curve 3 to $\varphi = 0.3$. The right ends of these curves are determined by the condition $\alpha = \alpha_{\mathrm{cr}}$. It is seen from this figure that the strain rate intensity factor for pressure-independent material is larger than the strain rate intensity factor for pressure-dependent material at smaller values of α and smaller than the strain rate intensity factor for pressure-dependent material at larger values of α.

5.6 Compression of a Plastic Layer Between Rotating Plates II

This boundary value problem has been formulated and solved for pressure-independent material in Sect. 3.6 (see Fig. 14). An extension of this solution to the double shearing model has been proposed in [39].

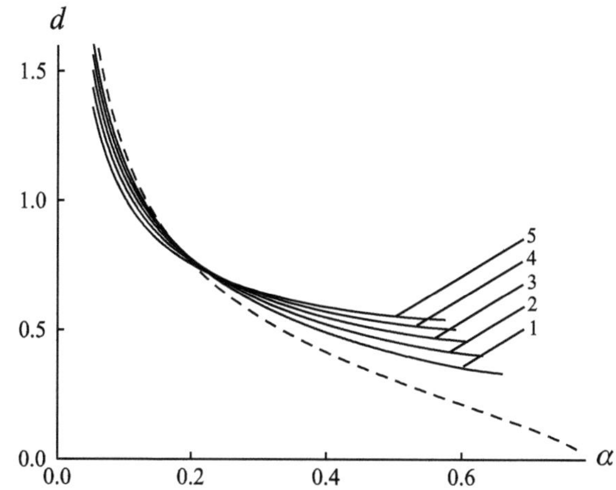

Fig. 50 Variation of the dimensionless strain rate intensity factor with α at several φ-values

Let ψ_w be the value of ψ at the maximum friction surface $\theta = \alpha$. The direction of flow (Fig. 14) dictates that $\sigma_{r\theta} > 0$ near the friction surface. Therefore, it follows from (199) that

$$0 < \psi_w < \frac{\pi}{2}. \tag{309}$$

The maximum friction surface is parallel to the r-axis. Therefore, $\phi_1 = 0$ or $\phi_2 = 0$ in (204). The equation $\phi_2 = 0$ contradicts (309). Therefore, $\phi_1 = 0$ and

$$\psi = \psi_w = \frac{\pi}{4} + \frac{\varphi}{2} \quad \text{for} \quad \theta = \alpha. \tag{310}$$

The main assumption accepted in [39] is that ψ is independent of r. Note that this assumption is in agreement with (310). The general solution for the stress equations given in the previous section is valid. In particular, the dependence of ψ on θ follows from (277) or, after integration, from (280). Then, the value of c involved in (277) is determined from (281) where ψ_w should be eliminated by means of (310). Numerical solution of this equation is illustrated in Fig. 51. The broken line corresponds to the solution for pressure-independent material, curve 1 to $\varphi = 0.1$, curve 2 to $\varphi = 0.2$, and curve 3 to $\varphi = 0.3$. It is seen from this figure that $c > 0$. The derivative $dc/d\alpha$ is given by (282). The velocity field is sought in the form [39]

$$u_r = \frac{\omega r}{2}\frac{\partial g(\psi, \alpha)}{\partial \theta} + \omega \frac{G(\psi, \alpha)}{r}, \quad u_\theta = -\omega r g(\psi, \alpha), \tag{311}$$

where $g(\psi, \alpha)$ and $G(\psi, \alpha)$ are arbitrary functions of ψ and α. It is possible to verify by inspection that $(203)_1$ is automatically satisfied. Using (310) and (311) the boundary conditions (79) and (80) are transformed to

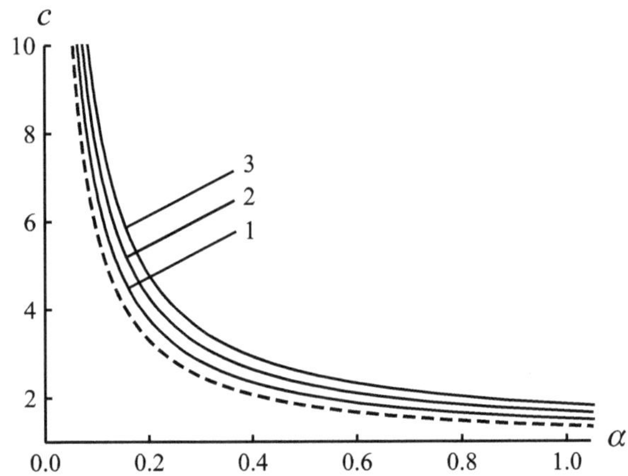

Fig. 51 Variation of c with α

$$g = 0 \quad \text{at} \quad \theta = 0 \quad (\text{or} \quad \psi = 0) \tag{312}$$

and

$$g = 1 \quad \text{at} \quad \theta = \alpha \quad (\text{or} \quad \psi = \psi_w), \tag{313}$$

respectively. Substituting (311) into $(203)_2$ yields

$$\begin{aligned} &(\cos 2\psi + \sin\varphi)\frac{\partial^2 g}{\partial\theta^2} - 2\sin 2\psi\frac{\partial g}{\partial\theta} + 4g\sin\varphi + 4\sin\varphi\frac{\dot{\psi}}{\omega} \\ &+\frac{2}{r^2}\left[(\cos 2\psi + \sin\varphi)\frac{dG}{d\theta} + 2G\sin 2\psi\right] = 0. \end{aligned} \tag{314}$$

Since ψ and $\dot{\psi}$ are independent of r, this equation may have a solution if and only if

$$\begin{aligned} (\cos 2\psi + \sin\varphi)\frac{\partial^2 g}{\partial\theta^2} - 2\sin 2\psi\frac{\partial g}{\partial\theta} + 4g\sin\varphi + 4\sin\varphi\frac{\dot{\psi}}{\omega} = 0, \\ (\cos 2\psi + \sin\varphi)\frac{\partial G}{\partial\theta} + 2G\sin 2\psi = 0. \end{aligned} \tag{315}$$

Replacing here the derivatives $\partial^2 g/\partial\theta^2$ and $\partial g/\partial\theta$ with the derivatives $\partial^2 g/\partial\psi^2$ and $\partial g/\partial\psi$ by means of (289) and eliminating the derivatives $d^2\psi/d\theta^2$ and $d\psi/d\theta$ by means of (277) and (288) result in

$$
\begin{aligned}
&(\sin\varphi + \cos 2\psi)\frac{\partial^2 g}{\partial\psi^2} + 2\sin 2\psi\frac{\partial g}{\partial\psi} \\
&+ \frac{4\sin\varphi\,(\sin\varphi + \cos 2\psi)^2}{(c - \sin\varphi - \cos 2\psi)^2}\left(g + \frac{\dot\psi}{\omega}\right) = 0, \\
&\frac{dG}{d\psi} + \frac{2G\sin 2\psi}{c - \sin\varphi - \cos 2\psi} = 0.
\end{aligned}
\tag{316}
$$

Integrating $(316)_2$ gives

$$
G = \frac{B}{c - \sin\varphi - \cos 2\psi}, \tag{317}
$$

where B is a constant of integration. The derivative $\dot\psi$ is given by (293). Therefore, Eq. $(316)_1$ transforms to (294). However, in contrast to (295), it is now necessary to investigate the coefficients of Eq. (294) in the vicinity of $\psi = \psi_w = \pi/4 + \varphi/2$. Expanding these coefficients in series near this point yields

$$
\begin{aligned}
\sin\varphi + \cos 2\psi =& -2\cos\varphi\,(\psi - \psi_w) + 2\sin\varphi\,(\psi - \psi_w)^2 \\
&+ \frac{4}{3}\cos\varphi\,(\psi - \psi_w)^3 + o\left[(\psi - \psi_w)^3\right], \\
2\sin 2\psi =& \, 2\cos\varphi - 4\sin\varphi\,(\psi - \psi_w) - 4\cos\varphi\,(\psi - \psi_w)^2 \\
&+ \frac{8}{3}\sin\varphi\,(\psi - \psi_w)^3 + o\left[(\psi - \psi_w)^3\right], \\
P_1(\psi) =& \frac{4\sin 2\varphi}{c}(\psi - \psi_w) - \frac{8\sin^2\varphi}{c}(\psi - \psi_w)^2 \\
&- \frac{8\sin 2\varphi}{c}\left(\frac{1}{3} + \frac{2\cos^2\varphi}{A^2}\right)(\psi - \psi_w)^3 + o\left[(\psi - \psi_w)^3\right], \\
P_0(\psi) =& \frac{4\sin 2\varphi}{c}(\psi - \psi_w) - \frac{8\sin\varphi\,(1 + \cos 2\varphi + A\sin\varphi)}{c^2}(\psi - \psi_w)^2 \\
&+ \frac{4\sin 2\varphi}{3c^3}\left[2\left(3 - c^2 + 3\cos 2\varphi + 6c\sin\varphi\right)\right. \\
&\left. + 3\cos\varphi\frac{dc}{d\alpha}\right](\psi - \psi_w)^3 + o\left[(\psi - \psi_w)^3\right]
\end{aligned}
\tag{318}
$$

as $\psi \to \psi_w$. It is seen from (294) and (318) that $\psi = \psi_w$ is a regular singular point of Eq. (294). Using a standard procedure it is possible to find that the solution to this equation in the vicinity of $\psi = \psi_w$ is represented by

$$
g = 1 + c_2(\psi - \psi_w)^2 + c_3(\psi - \psi_w)^3. \tag{319}
$$

It has been taken into account here that the solution must satisfy the boundary condition (313). Substituting (318) and (319) into (294) and collecting the coefficients of $(\psi - \psi_w)^2$ give

$$c_3 = -\frac{2}{3}c_2 \tan\varphi + \frac{4 \sin 2\varphi}{3c^2}. \tag{320}$$

The shear strain rate is determined from (41) and (311) as

$$\frac{\xi_{r\theta}}{\omega} = \frac{1}{4}\frac{\partial^2 g}{\partial \theta^2} + \frac{1}{2r^2}\frac{\partial G}{\partial \theta}. \tag{321}$$

Replacing here differentiation with respect to θ with differentiation with respect to ψ by means of (289), eliminating G by means of (317) and using (277) and (288) result in

$$\begin{aligned} \xi_{r\theta} &= \frac{\omega}{4}\left(\frac{c - \sin\varphi - \cos 2\psi}{\sin\varphi + \cos 2\psi}\right) Q(r, \psi), \\ Q(r, \psi) &= \left(\frac{c - \sin\varphi - \cos 2\psi}{\sin\varphi + \cos 2\psi}\right)\frac{\partial^2 g}{\partial \psi^2} \\ &+ \frac{2c \sin 2\psi}{(\sin\varphi + \cos 2\psi)^2}\frac{\partial g}{\partial \psi} - \frac{4B \sin 2\psi}{r^2 (c - \sin\varphi - \cos 2\psi)^2}. \end{aligned} \tag{322}$$

Substituting (319) into (322), eliminating c_3 by means of (320) and expanding the resulting expression in a series in the vicinity of $\psi = \psi_w$ yield

$$\xi_{r\theta} = -\frac{\omega}{2}\left(\frac{B}{cr^2} + \frac{cc_2}{2\cos\varphi} + \tan\varphi\right)(\psi_w - \psi)^{-1} + o\left[(\psi_w - \psi)^{-1}\right] \quad \text{as} \quad \psi \to \psi_w. \tag{323}$$

In the vicinity of this point Eq. (277) is represented as

$$\frac{d\psi}{d\theta} = \frac{c}{2\cos\varphi\,(\psi_w - \psi)} + o\left[(\psi_w - \psi)^{-1}\right] \quad \text{as} \quad \psi \to \psi_w.$$

Integrating with the boundary condition (310) gives

$$\psi_w - \psi = \sqrt{\frac{c}{\cos\varphi}}\sqrt{\alpha - \theta} + o\left(\sqrt{\alpha - \theta}\right) \quad \text{as} \quad \theta \to \alpha. \tag{324}$$

Substituting (324) into (323) yields

$$\begin{aligned} |\xi_{r\theta}| = &-\frac{\omega}{2}\left(\frac{B}{cr^2} + \frac{cc_2}{2\cos\varphi} + \tan\varphi\right)\sqrt{\frac{\cos\varphi}{c}}(\alpha - \theta)^{-1/2} \\ &+ o\left[(\alpha - \theta)^{-1/2}\right] \quad \text{as} \quad \theta \to \alpha. \end{aligned} \tag{325}$$

Equation (6) in which ξ_τ should be replaced with $\xi_{r\theta}$ is valid. Therefore, the strain rate intensity factor is determined from (325) as

$$D = -\frac{\omega\sqrt{r}}{\sqrt{3}}\left(\frac{B}{cr^2} + \frac{cc_2}{2\cos\varphi} + \tan\phi\right)\sqrt{\frac{\cos\varphi}{c}}. \tag{326}$$

The value of c_2 is found from numerical solution of Eq. (294). The variation of c_2 with α at several values of φ is illustrated in Fig. 52. The broken line corresponds to the model of pressure-independent material. In this case $c_2 = -2$. Curve 1 corresponds to $\varphi = 0.1$, curve 2 to $\varphi = 0.15$, curve 3 to $\varphi = 0.2$, curve 4 to $\varphi = 0.25$, and curve 5 to $\varphi = 0.3$.

The direction of friction stress (Fig. 14) demands $u_r < 0$ at $\theta = \alpha$ (or $\psi = \psi_w$). Taking into account (311), (317) and (319) this inequality is rewritten as

$$u_r|_{\theta=\alpha} = -\frac{rcc_2}{2\cos\varphi} + \frac{B}{rc} < 0.$$

Therefore, the solution found is valid in the range

$$r < r_{cr} = \frac{1}{c}\sqrt{\frac{2B\cos\varphi}{c_2}}. \tag{327}$$

The values of c and c_2 have been already determined (Figs. 51 and 52). In order to find the value of B, it is necessary to formulate an additional condition in integral form (similar to the condition (49) accepted in the classical problem considered in Sect. 3.2). A reasonable condition is

$$\int_0^\alpha u_r|_{r=R}\, d\theta = 0. \tag{328}$$

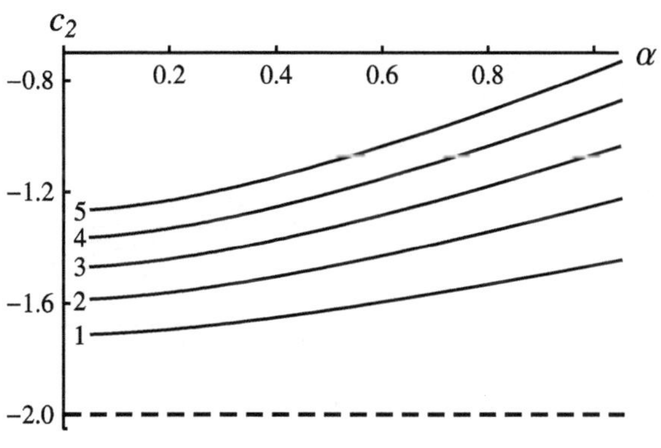

Fig. 52 Variation of c_2 with α at several φ-values

Here R is some prescribed radius. Substituting the radial velocity from (311) into (328), eliminating G by means of (317) and replacing integration with respect to θ with integration with respect to ψ by means of (277) result in

$$B = -\frac{R^2}{2}\left[\int_0^{\psi_w} \frac{(\sin\varphi + \cos 2\psi)}{(c - \sin\varphi - \cos 2\psi)^2} d\psi\right]^{-1}. \tag{329}$$

Eliminating B in (327) by means of (329) determines the dependence of r_{cr}/R on φ and α. The dependence of r_{cr}/R on α at several values of φ is depicted in Fig. 53. The broken line corresponds to the model of pressure-independent material, curve 1 to $\varphi = 0.1$, curve 2 to $\varphi = 0.15$, curve 3 to $\varphi = 0.2$, curve 4 to $\varphi = 0.25$, and curve 5 to $\varphi = 0.3$. Eliminating B in (326) by means of (329), it is possible to conclude that the strain rate intensity factor depends on φ, α and r/R. Consider first the dependence of D on φ and α assuming that $r/R = \mu = \text{constant} < r_{cr}/R$. It is convenient to introduce the dimensionless strain rate intensity factor, d, as the ratio of the strain rate intensity factor given by (326) to the strain rate intensity factor for pressure-independent material given by (105). The variation of d with α at several values of φ is shown in Fig. 54 at $\mu = 0.3$, Fig. 55 at $\mu = 0.2$, and Fig. 56 at $\mu = 0.1$. In these figures, curve 1 corresponds to $\varphi = 0.1$, curve 2 to $\varphi = 0.15$, curve 3 to $\varphi = 0.2$, curve 4 to $\varphi = 0.25$, and curve 5 to $\varphi = 0.3$. In order to illustrate the variation of the strain rate intensity factor with r/R, it is necessary to take into account that r_{cr}/R depends on φ (Fig. 53). Therefore, it is more convenient in this case to introduce the dimensionless strain rate intensity factor by

$$d = D/\left(\omega\sqrt{r}\right).$$

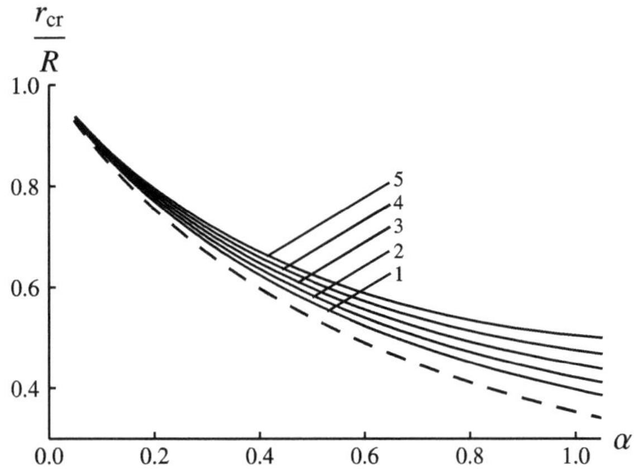

Fig. 53 Variation of r_{cr}/R with α at several φ-values

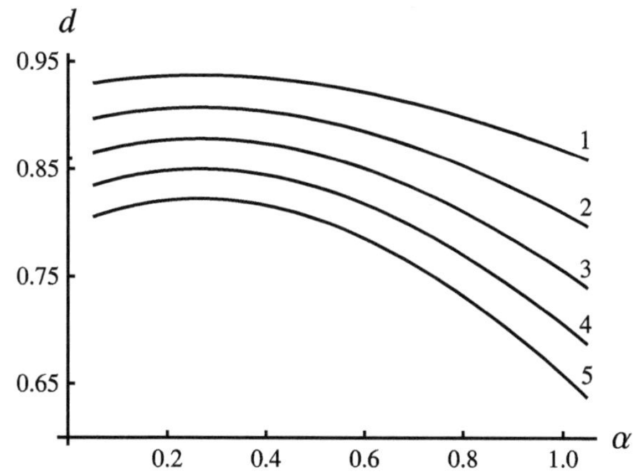

Fig. 54 Variation of the dimensionless strain rate intensity factor with α at several φ-values and $\mu = 0.3$

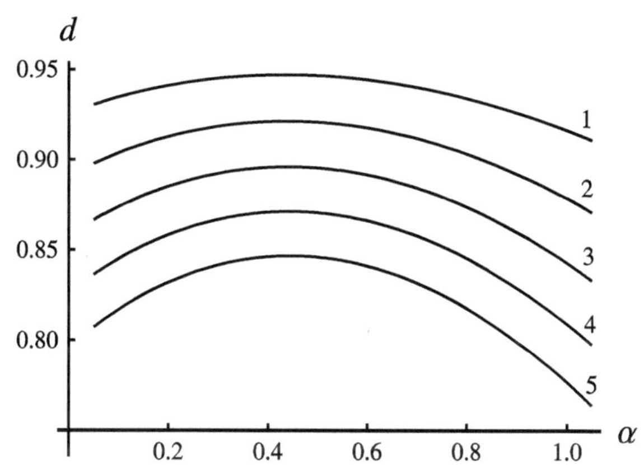

Fig. 55 Variation of the dimensionless strain rate intensity factor with α at several φ-values and $\mu = 0.2$

Note that the same definition has been adopted for the dimensionless strain rate intensity factor depicted in Fig. 15. The variation of d with r/R in the range $0.2 \le r/R \le r_{cr}/R$ is illustrated in Figs. 57, 58, 59 and 60. In Figs. 57 and 58 the value of α is fixed ($\alpha = \pi/36$ in Fig. 57 and $\alpha = \pi/6$ in Fig. 58) and the curves correspond to different values of φ (the broken line corresponds to the model of pressure-independent material, curve 1 to $\varphi = 0.1$, curve 2 to $\varphi = 0.2$ and curve 3 to $\varphi = 0.3$). In Figs. 59 and 60 the value of φ is fixed ($\varphi = 0.1$ in Fig. 59 and $\varphi = 0.3$ in Fig. 60) and the curves correspond to different values of α (curve 1 corresponds to $\alpha = \pi/36$, curve 2 to $\alpha = \pi/18$, curve 3 to $\alpha = \pi/12$, curve 4 to $\alpha = \pi/9$, and curve 5 to $\alpha = \pi/6$).

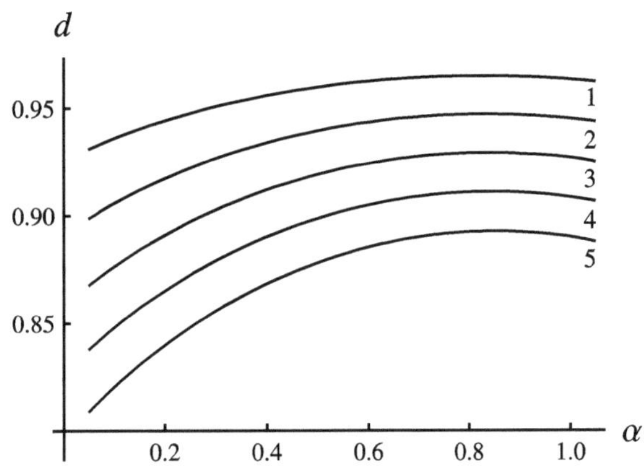

Fig. 56 Variation of the dimensionless strain rate intensity factor with α at several φ-values and $\mu = 0.1$

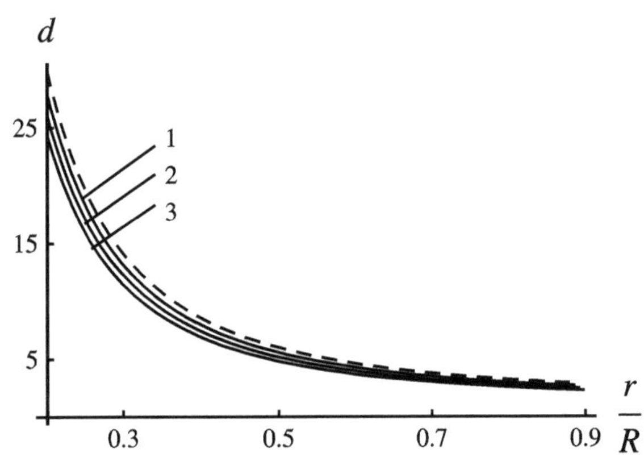

Fig. 57 Variation of the dimensionless strain rate intensity factor with r/R at several φ-values and $\alpha = \pi/36$

It is seen from Figs. 54, 55, 56, 57 and 58 that the strain rate intensity factor for pressure-dependent material is smaller than the strain rate intensity factor for pressure-independent material.

5.7 Simultaneous Shearing and Expansion of a Hollow Cylinder

This boundary value problem has been formulated and solved for pressure-independent material in Sect. 3.7 (see Fig. 17). An extension of this solution to the double shearing model has been proposed in [5].

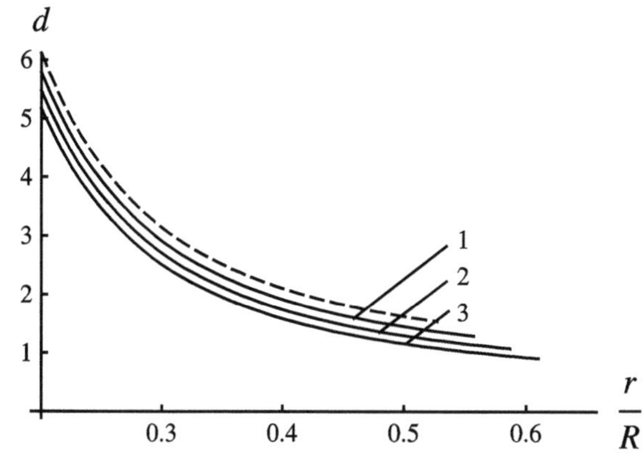

Fig. 58 Variation of the dimensionless strain rate intensity factor with r/R at several φ-values and $\alpha = \pi/6$

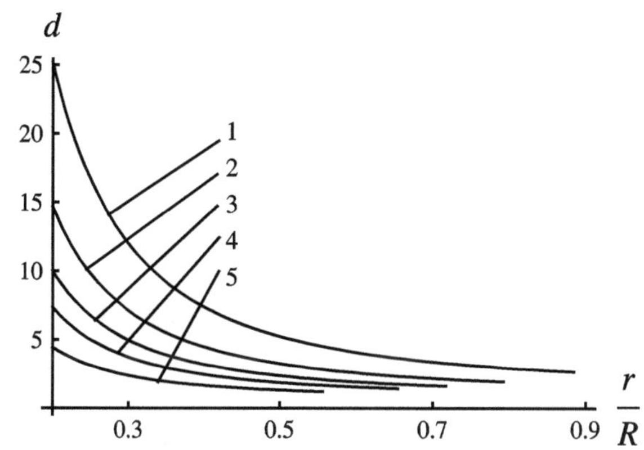

Fig. 59 Variation of the dimensionless strain rate intensity factor with r/R at several α-values and $\varphi = 0.1$

Let ψ_w be the value of ψ at the maximum friction surface $r = a$. The direction of rotation of the rigid rod (Fig. 17) dictates that $\sigma_{r\theta} > 0$ near the friction surface. Therefore, it follows from (199) that

$$0 < \psi_w < \frac{\pi}{2}. \tag{330}$$

The maximum friction surface is perpendicular to the r-axis. Therefore, $\phi_1 = \pi/2$ or $\phi_2 = \pi/2$ in (204). The equation $\phi_1 = \pi/2$ contradicts (330). Therefore, $\phi_2 = \pi/2$ and

$$\psi = \psi_w = \frac{\pi}{4} - \frac{\varphi}{2} \quad \text{for} \quad r = a. \tag{331}$$

The solution is independent of θ. Therefore, the equilibrium Eq. (42) become

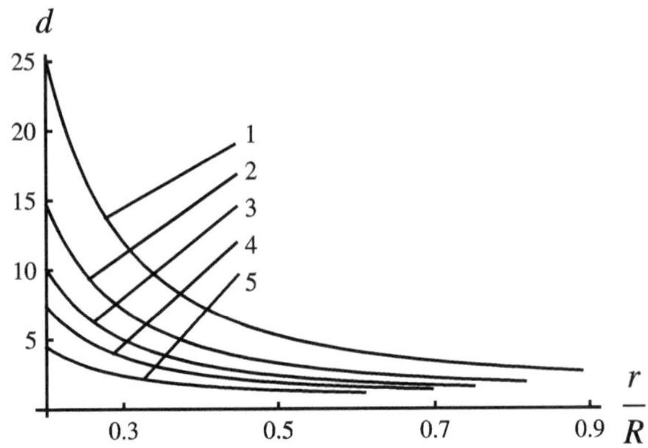

Fig. 60 Variation of the dimensionless strain rate intensity factor with r/R at several α-values and $\varphi = 0.3$

$$\frac{\partial \sigma_{rr}}{\partial r} + \frac{\sigma_{rr} - \sigma_{\theta\theta}}{r} = 0, \quad \frac{\partial \sigma_{r\theta}}{\partial r} + \frac{2\sigma_{r\theta}}{r} = 0. \tag{332}$$

The general solution to Eq. $(332)_2$ is

$$\frac{\sigma_{r\theta}}{k} = \frac{C^2}{r^2}, \tag{333}$$

where C is a constant of integration. Using (191), (200) and (333) the normal stresses are expressed as

$$\begin{aligned} \frac{\sigma_{rr}}{k} &= \cot\varphi - \frac{C^2\left(1 - \cos 2\psi \sin\varphi\right)}{r^2 \sin 2\psi \sin\varphi}, \\ \frac{\sigma_{\theta\theta}}{k} &= \cot\varphi - \frac{C^2\left(1 + \cos 2\psi \sin\varphi\right)}{r^2 \sin 2\psi \sin\varphi}. \end{aligned} \tag{334}$$

Substituting (334) into $(332)_1$ gives the following equation for ψ

$$\frac{\partial \psi}{\partial r} = -\frac{\sin 2\psi}{r\left(\cos 2\psi - \sin\varphi\right)}. \tag{335}$$

Integrating this equation with the boundary condition (331) results in

$$\frac{r}{a} = \frac{\cos\psi_w \tan^m \psi_w}{\cos\psi \tan^m \psi}, \quad m = \frac{1 - \sin\varphi}{2}. \tag{336}$$

This equation determines ψ as a function of r in implicit form. Let ψ_b be the value of ψ at $r = b$. Then, it follows from (336) that

$$\frac{b}{a} = \frac{\cos\psi_w \tan^m \psi_w}{\cos\psi_b \tan^m \psi_b}. \tag{337}$$

This equation should be solved numerically to find the value of ψ_b.

Since the solution is independent of θ, the solution to the incompressibility Eq. $(302)_1$ satisfying the boundary condition (106) is

$$u_r = \frac{\dot{a}a}{r}. \tag{338}$$

Since ψ is independent of θ, $\partial\psi/\partial r$ is given by (335) and $\partial\psi/\partial t = \dot{a}\partial\psi/\partial a$, Eq. $(301)_2$ becomes

$$\dot{\psi} = \dot{a}\frac{\partial\psi}{\partial a} - \frac{\dot{a}a\sin 2\psi}{r^2(\cos 2\psi - \sin\varphi)}. \tag{339}$$

Here the radial velocity has been eliminated by means of (338). Substituting (338) and (339) into $(302)_2$ gives

$$\sin 2\psi\frac{\partial u_\theta}{\partial\psi} + (\cos 2\psi + \sin\varphi)\,u_\theta - \frac{2a\dot{a}\sin 2\psi}{r} - 2\dot{a}r\sin\varphi\frac{\partial\psi}{\partial a} + \frac{2a\dot{a}\sin\varphi}{r}\frac{\sin 2\psi}{(\cos 2\psi - \sin\varphi)} = 0. \tag{340}$$

Since ψ_w is constant, it follows from (336) that

$$\frac{\partial\psi}{\partial a} = \frac{\sin 2\psi}{a(\cos 2\psi - \sin\varphi)}. \tag{341}$$

Substituting (336) and (341) into (340) leads to

$$\frac{\partial u_\theta}{\partial\psi} + \frac{(\cos 2\psi + \sin\varphi)}{\sin 2\psi}u_\theta = P(\psi),$$
$$P(\psi) = \frac{2\dot{a}\tan^m\psi\cos\psi}{\cos\psi_w\tan^m\psi_w(\cos 2\psi - \sin\varphi)} \times \left(\frac{\sin\varphi\cos\varphi\tan^{\sin\varphi}\psi}{\tan^{\sin\varphi}\psi_w\sin 2\psi} + \cos 2\psi - 2\sin\varphi\right). \tag{342}$$

Applying l'Hospital's rule it is possible to find that $\lim\limits_{\psi\to\psi_w} P(\psi) = 2\dot{a} > 0$. Then, it follows from (342) that

$$\left.\frac{\partial u_\theta}{\partial\psi}\right|_{\psi=\psi_w} = 2(\dot{a} - u_a\tan\varphi). \tag{343}$$

Here u_a is the value of the circumferential velocity at $r = a$. Taking into account that the solution is independent of θ and replacing differentiation with respect to r with differentiation with respect to ψ by means of (335) the shear strain rate is determined from (41) as

$$\xi_{r\theta} = \frac{(\dot{a} - u_a \tan\varphi)}{2a(\psi - \psi_w)} + o\left[(\psi - \psi_w)^{-1}\right] \quad \text{as} \quad \psi \to \psi_w. \tag{344}$$

Here Eq. (343) has been used to eliminate the derivative $\partial u_\theta/\partial\psi$. Equation (335) in the vicinity of $r = a$ (or $\psi = \psi_w$) is represented as

$$\frac{\partial\psi}{\partial r} = \frac{1}{2a(\psi - \psi_w)} + o\left[(\psi - \psi_w)^{-1}\right] \quad \text{as} \quad \psi \to \psi_w. \tag{345}$$

Integrating this equation with the boundary condition (331) gives

$$\psi - \psi_w = \frac{\sqrt{r-a}}{\sqrt{a}} + o\left(\sqrt{r-a}\right) \quad \text{as} \quad r \to a. \tag{346}$$

Substituting (346) into (344) leads to

$$\xi_{r\theta} = \frac{(\dot{a} - u_a \tan\varphi)}{2\sqrt{a}\sqrt{r-a}} + o\left[(r-a)^{-1/2}\right] \quad \text{as} \quad r \to a. \tag{347}$$

Equation (6) in which ξ_τ should be replaced with $\xi_{r\theta}$ is valid. Therefore, the strain rate intensity factor is determined using (347) as

$$D = \frac{(\dot{a} - u_a \tan\varphi)}{\sqrt{3a}}. \tag{348}$$

It is convenient to introduce the dimensionless strain rate intensity factor, d, as the ratio of the strain rate intensity factor given by (348) to the strain rate intensity factor given by (112). As a result,

$$d = 1 - \frac{u_a}{\dot{a}} \tan\varphi. \tag{349}$$

The solution of Eq. (342) satisfying the boundary condition (107) is

$$\frac{u_\theta}{\dot{a}} = \frac{\tan^m \psi}{\cos\psi_w \tan^{m+\sin\varphi}\psi_w \sin\psi} \tag{350}$$

$$\times \int\limits_{\psi_b}^{\psi} \frac{\left[\sin\varphi\cos\varphi\tan^{\sin\varphi}\gamma + \tan^{\sin\varphi}\psi_w \sin 2\gamma\,(\cos 2\gamma - 2\sin\varphi)\right]}{(\cos 2\gamma - \sin\varphi)} d\gamma.$$

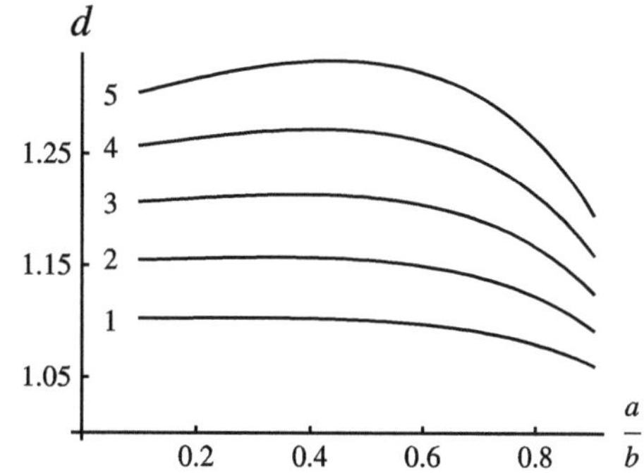

Fig. 61 Variation of the dimensionless strain rate intensity factor with α at several φ-values

Here γ is a dummy variable of integration and ψ_b is determined from the solution of Eq. (337). It follows from the definition for u_a and (350) that

$$-\frac{u_a}{\dot{a}} = \frac{2}{\cos\varphi}\int_{\psi_w}^{\psi_b} \frac{\left[\sin\varphi\cos\varphi\,(\tan\gamma/\tan\psi_w)^{\sin\varphi} + \sin 2\gamma\,(\cos 2\gamma - 2\sin\varphi)\right]}{(\cos 2\gamma - \sin\varphi)}d\gamma. \tag{351}$$

The integrand reduces to the expression $0/0$ at $\gamma = \psi_w$. Applying l'Hospital's rule yields

$$\lim_{\gamma\to\psi_w}\left\{\frac{\sin\varphi\cos\varphi\,(\tan\gamma\cot\psi_w)^{\sin\varphi} + \sin 2\gamma\,(\cos 2\gamma - 2\sin\varphi)}{(\cos 2\gamma - \sin\varphi)}\right\} = \cos\varphi. \tag{352}$$

Using (352) the integral in (351) can be evaluated numerically with no difficulty. Substituting the value of u_a found into (349) gives the dimensionless strain rate intensity factor. Its variation with the ratio a/b at several values of φ is depicted in Fig. 61 (curve 1 corresponds to $\varphi = 0.1$, curve 2 to $\varphi = 0.15$, curve 3 to $\varphi = 0.2$, curve 4 to $\varphi = 0.25$, and curve 5 to $\varphi = 0.3$). It is seen from this figure that the strain rate intensity factor for pressure-independent material is smaller than the strain rate intensity factor for pressure-dependent material.

6 Axisymmetric Solutions for the Double-Shearing Model

6.1 Basic Equations

Section 6 is concerned with axisymmetric solutions for the double-shearing model. In this section, a spherical coordinate system (r, θ, ϑ) will be employed. The solutions are independent of ϑ, the stress $\sigma_{\vartheta\vartheta}$ is one of the principal stresses and the velocity

u_ϑ vanishes. A cross-section of the Mohr-Coulomb yield surface by a plane $\sigma_{\vartheta\vartheta} =$ constant is shown in Fig. 62. In general, various regimes of flow arise depending on the relative magnitudes of $\sigma_{\vartheta\vartheta}$, σ_1 and σ_2. However, in what follows, regime A only is of interest. In this regime

$$\sigma_1 (1 + \sin\varphi) = 2k\cos\varphi + \sigma_{\vartheta\vartheta}(1 - \sin\varphi), \quad \sigma_2 = \sigma_{\vartheta\vartheta}. \tag{353}$$

Using the transformation equations for stress components in $r\theta$-planes the stress components in the spherical coordinate system are expressed as (Fig. 21)

$$\sigma_{rr} = -p + q\cos 2\psi, \quad \sigma_{\theta\theta} = -p - q\cos 2\psi, \quad \sigma_{r\theta} = q\sin 2\psi. \tag{354}$$

Here p and q are given by (192). Using (354) the yield criterion (353) transforms to

$$q - p\sin\varphi = k\cos\varphi, \quad \sigma_{\vartheta\vartheta} = -p - q. \tag{355}$$

The velocity equations have been given in [10]. Those are

$$\begin{gathered}\xi_{rr} + \xi_{\theta\theta} + \xi_{\vartheta\vartheta} = 0, \\ 2\xi_{r\theta}\cos 2\psi - (\xi_{rr} - \xi_{\theta\theta})\sin 2\psi + \sin\varphi\left(\omega_{r\theta} + \dot{\psi}\right) = 0.\end{gathered} \tag{356}$$

The components of the strain rate tensor are given by (124). Equation $(356)_1$ is equivalent to (128). The only non-zero spin component is

$$\omega_{r\theta} = \frac{1}{2r}\left(\frac{\partial u_r}{\partial \theta} - r\frac{\partial u_\theta}{\partial r} - u_\theta\right). \tag{357}$$

Since $u_\vartheta = 0$, the derivative $\dot{\psi}$ is given by

$$\dot{\psi} = \frac{\partial \psi}{\partial t} + u_r\frac{\partial \psi}{\partial r} + \frac{u_\theta}{r}\frac{\partial \psi}{\partial \theta}. \tag{358}$$

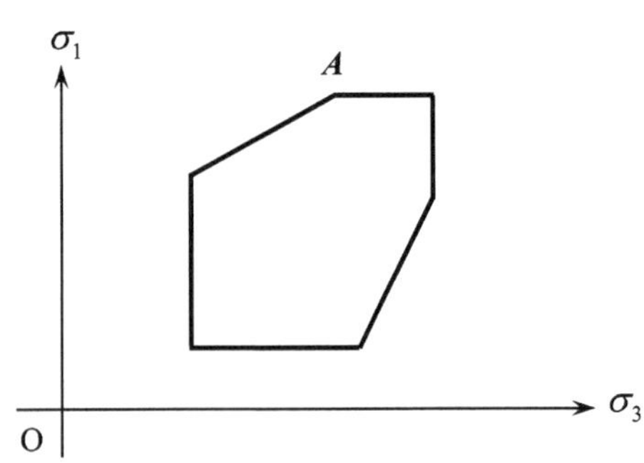

Fig. 62 Cross-section of the Mohr-Coulomb yield surface by plane $\sigma_{\vartheta\vartheta} =$ constant

Equation (131) is replaced with

$$\phi_1 = \psi - \frac{\pi}{4} - \frac{\varphi}{2}, \quad \phi_2 = \psi + \frac{\pi}{4} + \frac{\varphi}{2}. \tag{359}$$

The equilibrium equations are given by (125).

6.2 Flow of Plastic Material Through an Converging Conical Channel

This boundary value problem has been formulated and solved for pressure-independent material in Sect. 4.3 (see Fig. 23). An extension of this solution to the double shearing model has been outlined in [10].

Let ψ_w be the value of ψ at the maximum friction surface $\theta = \alpha$. The direction of flow (Fig. 23) dictates that $\sigma_{r\theta} > 0$ near the friction surface. Therefore, it follows from (354) that

$$0 < \psi_w < \frac{\pi}{2}. \tag{360}$$

The orientation of the maximum friction surface shows that $\phi_1 = 0$ or $\phi_2 = 0$ in (359). The equation $\phi_2 = 0$ contradicts (360). Therefore, $\phi_1 = 0$ and

$$\psi = \psi_w = \frac{\pi}{4} + \frac{\varphi}{2} \quad \text{for} \quad \theta = \alpha. \tag{361}$$

The main assumptions accepted in [10] are that ψ is independent of r and $u_\theta = 0$. Note that the former is in agreement with (361) and the latter automatically satisfies the boundary conditions (147). Equation (128) reduces to

$$\frac{\partial u_r}{\partial r} + 2\frac{u_r}{r} = 0.$$

The general solution of this equation is

$$u_r = -\frac{U(\theta)}{r^2}, \tag{362}$$

where $U(\theta) > 0$ is an arbitrary function of θ. The solution for stresses is sought in the form [40]

$$q = \exp[f(\theta)]\, r^n. \tag{363}$$

Here $f(\theta)$ is an arbitrary function of θ and $n =$ constant. Using (355) and (363) it is possible to find that

$$p = \frac{\exp\left[f\left(\theta\right)\right] r^n}{\sin\varphi} - k\cot\varphi, \tag{364}$$

$$\sigma_{\vartheta\vartheta} = -\exp\left[f\left(\theta\right)\right] r^n \left(\frac{1}{\sin\varphi} + 1\right) + k\cot\varphi.$$

Substituting (363) and (364) into (354) gives the stresses in terms of $f\left(\theta\right)$ and r. Substituting these stresses and $\sigma_{\vartheta\vartheta}$ from (364) into the equilibrium Eq. (125) results in the following equations for $\psi\left(\theta\right)$ and $f\left(\theta\right)$

$$\frac{d\psi}{d\theta} = \frac{n\cos^2\varphi - \sin\varphi\omega_0(\psi,\theta)}{2\sin\varphi\left(\sin\varphi + \cos 2\psi\right)}, \tag{365}$$

$$\omega_0\left(\psi,\theta\right) = 3\sin\varphi + 1 + \cos 2\psi\left(3 + \sin\varphi\right) + \cot\theta\sin 2\psi\left(1 + \sin\varphi\right),$$

and

$$\frac{df}{d\theta} = \frac{n\sin 2\psi - \sin\varphi\left[\cot\theta\left(1 - \cos 2\psi\right) + \sin 2\psi\right]}{\sin\varphi + \cos 2\psi}, \tag{366}$$

respectively. Using (365) Eq. (366) can be rewritten as

$$\frac{df}{d\psi} = \frac{2\sin\varphi\left\{n\sin 2\psi - \sin\varphi\left[\cot\theta\left(1 - \cos 2\psi\right) + \sin 2\psi\right]\right\}}{n\cos^2\varphi - \sin\varphi\omega_0(\psi,\theta)}. \tag{367}$$

An advantage of Eq. (367) over Eq. (366) is that the denominator in (366) vanishes at $\psi = \psi_w$. For the same reason, it is advantageous to rewrite Eq. (365) as

$$\frac{d\theta}{d\psi} = \frac{2\sin\varphi\left(\sin\varphi + \cos 2\psi\right)}{n\cos^2\varphi - \sin\varphi\omega_0(\psi,\theta)}. \tag{368}$$

The process is stationary. Therefore, $\partial\psi/\partial t = 0$. Moreover, by assumption, $u_\theta = 0$ and $\partial\psi/\partial r = 0$. Therefore, it follows from (358) that $\dot{\psi} = 0$. Then, using (124) and (357) it is possible to simplify Eq. $(356)_2$ to

$$\left(\cos 2\psi + \sin\varphi\right)\frac{\partial u_r}{\partial\theta} - \sin 2\psi\left(r\frac{\partial u_r}{\partial r} - u_r\right) = 0.$$

Substituting (362) into this equation gives

$$\left(\cos 2\psi + \sin\varphi\right)\frac{dU}{d\theta} + 3U\sin 2\psi = 0. \tag{369}$$

The coefficient of the derivative vanishes at $\psi = \psi_w$. Therefore, using (368) it is advantageous to rewrite (369) as

$$\frac{dU}{d\psi} = \frac{-6U\sin\varphi\sin 2\psi}{n\cos^2\varphi - \sin\varphi\omega_0(\psi,\theta)}. \tag{370}$$

In the case under consideration, the shear strain rate is determined from (124), (362) and (369) as

$$\xi_{r\theta} = -\frac{1}{2r^3}\frac{dU}{d\theta} = \frac{3U\sin 2\psi}{2r^3\left(\cos 2\psi + \sin\varphi\right)}. \tag{371}$$

Let U_w be the value of U at $\theta = \alpha$ (or $\psi = \psi_w$). Then, expanding the right hand side of (371) in a series in the vicinity of $\psi = \psi_w$ yields

$$\xi_{r\theta} = \frac{3U_w}{4r^3\left(\psi_w - \psi\right)} + o\left(\frac{1}{\psi_w - \psi}\right) \quad \text{as} \quad \psi \to \psi_w. \tag{372}$$

Equation (368) in the vicinity of $\psi = \psi_w$ (or $\theta = \alpha$) becomes

$$\frac{d\theta}{d\psi} = \frac{4\sin\varphi\left(\psi_w - \psi\right)}{n\cos\varphi - \sin\varphi\left[\cos\varphi + \cot\alpha\left(1 + \sin\varphi\right)\right]} + o\left(\psi_w - \psi\right) \quad \text{as} \quad \psi \to \psi_w.$$

Integrating this equation gives

$$\left(\psi_w - \psi\right)^2 = \frac{\left\{n\cos\varphi - \sin\varphi\left[\cos\varphi + \cot\alpha\left(1 + \sin\varphi\right)\right]\right\}}{2\sin\varphi}\left(\alpha - \theta\right) \tag{373}$$

to leading order. Substituting (373) into (372) gives

$$\xi_{r\theta} = \frac{3\sqrt{2\sin\varphi}\,U_w}{4r^3\sqrt{n\cos\varphi - \sin\varphi\left[\cos\varphi + \cot\alpha\left(1 + \sin\varphi\right)\right]}\sqrt{\alpha - \theta}} + o\left(\frac{1}{\sqrt{\alpha - \theta}}\right) \quad \text{as} \quad \theta \to \alpha. \tag{374}$$

The normal strain rates are bounded in the vicinity of the maximum friction surface and $|\xi_{r\theta}| \to \infty$ as $\theta \to \alpha$. Therefore, Eq. (6) in which ξ_τ should be replaced with $\xi_{r\theta}$ is valid. Then, it follows from (6) and (374) that the strain rate intensity factor is

$$D = \frac{\sqrt{6\sin\varphi}\,U_w}{2r^{5/2}\sqrt{n\cos\varphi - \sin\varphi\left[\cos\varphi + \cot\alpha\left(1 + \sin\varphi\right)\right]}}. \tag{375}$$

A difficulty with solving (367), (368) and (370) numerically is that the right hand sides of these equations contain the expression $0\cdot\infty$ as $\theta \to 0$ and $\psi \to 0$. Represent the function $\psi(\theta)$ in the form

$$\psi(\theta) = A\theta + B\theta^2 + O\left(\theta^3\right) \quad \text{as} \quad \theta \to 0. \tag{376}$$

Substituting this representation into (365) gives

$$A + 2B\theta = \frac{1}{2}\left(\frac{n}{\sin\varphi} - n - 4 - 2A\right) - B\theta \quad \text{as} \quad \theta \to 0. \tag{377}$$

Therefore,

$$A = -1 + \frac{n}{4}\left(\frac{1}{\sin\varphi} - 1\right), \quad B = 0$$

and Eq. (376) becomes

$$\psi(\theta) = \left[\frac{n}{4}\left(\frac{1}{\sin\varphi} - 1\right) - 1\right]\theta + o\left(\theta^3\right) \quad \text{as} \quad \theta \to 0. \tag{378}$$

Substituting this representation into (367) and (370) yields

$$\frac{df}{d\psi} = n\left(\frac{1}{2} + \frac{1}{1 + \sin\varphi}\right)\theta + o\left(\theta^3\right) \quad \text{as} \quad \theta \to 0 \tag{379}$$

and

$$\frac{dU}{d\psi} = -\frac{6U}{(1 + \sin\varphi)}\theta + o\left(\theta^3\right) \quad \text{as} \quad \theta \to 0, \tag{380}$$

respectively. It follows from (378)-(380) that

$$f = f_0 + \frac{n\sin\varphi\,(3 + \sin\varphi)}{(1 + \sin\varphi)\,[n\,(1 - \sin\varphi) - 4\sin\varphi]}\psi^2 + o\left(\psi^2\right) \quad \text{as} \quad \psi \to 0,$$

$$\ln\frac{U}{U_0} = -\frac{12\sin\varphi}{\left[n\cos^2\varphi - 4\sin\varphi\,(1 + \sin\varphi)\right]}\psi^2 + o\left(\psi^2\right) \quad \text{as} \quad \psi \to 0, \tag{381}$$

where f_0 and U_0 are constants of integration. It is seen from (375) that the function $f(\theta)$ is not involved in the expression for the strain rate intensity factor. Therefore, Eq. (367) is not solved here. The solution of Eq. (370) is

$$U = U_w w(\psi), \tag{382}$$

$$w(\psi) = \exp\left\{-6\sin\varphi\int\limits_{\psi_w}^{\psi}\frac{\sin 2\gamma}{n\cos^2\varphi - \sin\varphi\,\omega_0\,[\gamma, \theta(\gamma)]}d\gamma\right\}.$$

Equation (168) should be used to determine U_w. In particular, substituting (362) and (382) into this equation gives

$$Q = 2\pi U_w\int\limits_0^{\alpha} w(\psi)\sin\theta d\theta.$$

Replacing here integration with respect to θ with integration with respect to ψ by means of (368) yields

$$U_w = \frac{Q}{4\pi \sin\varphi}\left[\int_0^{\psi_w} \frac{\sin\theta\,(\sin\varphi + \cos 2\psi)\,w(\psi)}{n\cos^2\varphi - \sin\varphi\omega_0(\psi,\theta)}d\psi\right]^{-1}. \qquad (383)$$

In order to make comparison with the solution for pressure-independent material (Sect. 4.3), it is convenient to introduce the dimensionless strain rate intensity factor, d, by (170).

Using (378) Eq. (368) can be solved numerically if a value of n is specified. An iterative procedure should be adopted to determine the value of n using the boundary condition (361). The variation of n with α at several values of φ is depicted in Fig. 63 (curve 1 corresponds to $\varphi = 0.1$, curve 2 to $\varphi = 0.15$, curve 3 to $\varphi = 0.2$, curve 4 to $\varphi = 0.25$, and curve 5 to $\varphi = 0.3$). Having the solution to Eq. (368) the integrals in (382) and (383) can be evaluated. Finally, the dimensionless strain rate intensity factor is determined from (170) and (375). The variation of d with α at several values of φ is depicted in Fig. 64. The broken line corresponds to the model of pressure-independent plasticity. It is seen that the effect of φ on d is not significant (the solid curves cover the range $0.1 \le \varphi \le 0.3$). It is more pronounced at smaller α-angles (Fig. 65). In this figure, curve 1 corresponds to $\varphi = 0.1$, curve 2 to $\varphi = 0.15$, curve 3 to $\varphi = 0.2$, curve 4 to $\varphi = 0.25$, and curve 5 to $\varphi = 0.3$. It is seen that the dimensionless strain rate intensity factor increases with φ. It is interesting to mention that the opposite tendency occurs at larger α-angles (Fig. 66).

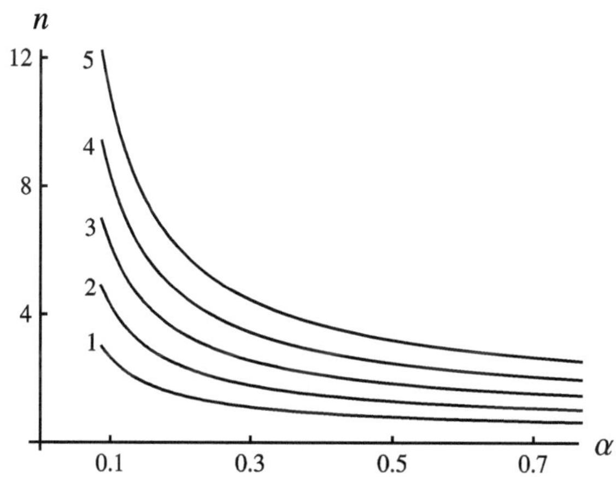

Fig. 63 Variation of n with α at several φ-values

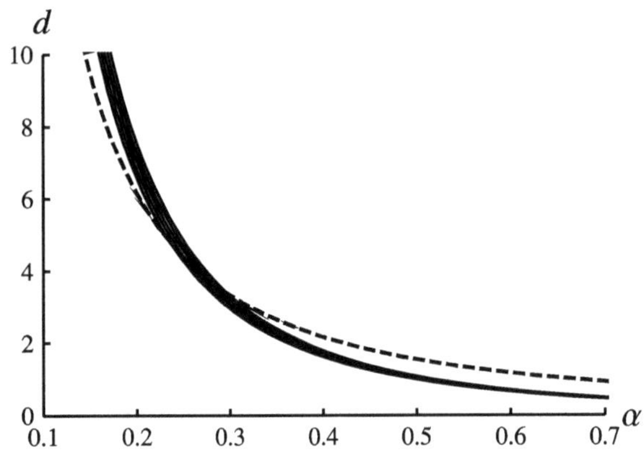

Fig. 64 Variation of the dimensionless strain rate intensity factor with α at several φ-values

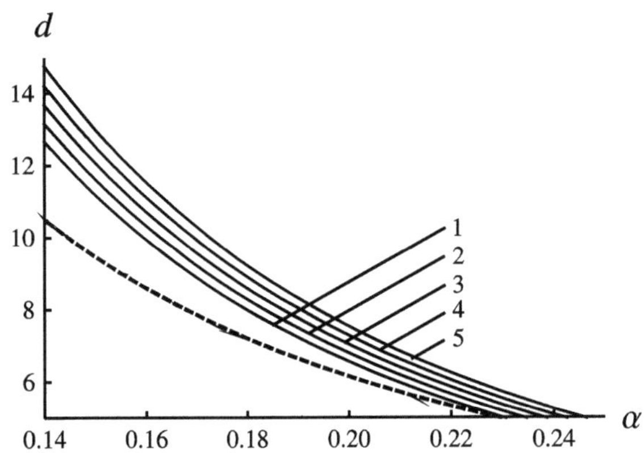

Fig. 65 Variation of the dimensionless strain rate intensity factor with α at several φ-values (small α-values)

6.3 Radial Flow Between Two Conical Surfaces

This boundary value problem has been formulated and solved for pressure-independent material in Sect. 4.4 (see Fig. 26). An extension of this solution to the double shearing model has been given in [40].

Let ψ_f be the value of ψ at the maximum friction surface $\theta = \theta_0$ and ψ_w be the value of ψ at the maximum friction surface $\theta = \theta_1$. The direction of flow (Fig. 26) dictates that $\sigma_{r\theta} < 0$ near the friction surface $\theta = \theta_0$ and $\sigma_{r\theta} > 0$ near the friction surface $\theta = \theta_1$. Therefore, it follows from (354) that

$$-\frac{\pi}{2} < \psi_f < 0 \quad \text{and} \quad 0 < \psi_w < \frac{\pi}{2}. \tag{384}$$

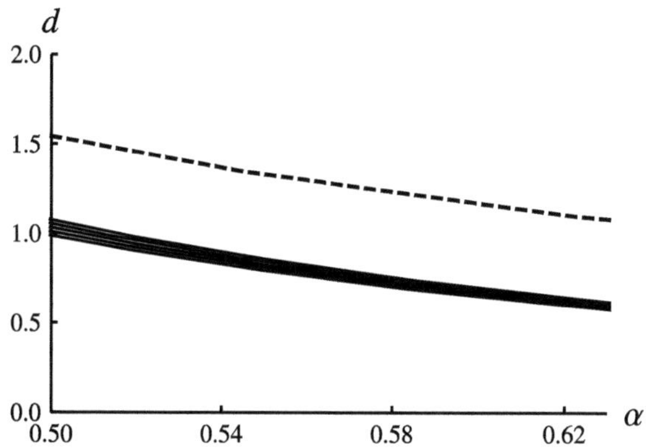

Fig. 66 Variation of the dimensionless strain rate intensity factor with α at several φ-values (large α-values)

The orientation of the normal to the maximum friction surfaces shows that $\phi_1 = 0$ or $\phi_2 = 0$ in (359). Equation $\phi_1 = 0$ contradicts $(384)_1$ and equation $\phi_2 = 0$ to $(384)_2$. Therefore,

$$\psi = \psi_f = -\frac{\pi}{4} - \frac{\varphi}{2} \quad \text{for} \quad \theta = \theta_0 \tag{385}$$

and

$$\psi = \psi_w = \frac{\pi}{4} + \frac{\varphi}{2} \quad \text{for} \quad \theta = \theta_1. \tag{386}$$

The velocity boundary conditions (172) are automatically satisfied assuming that $u_\theta = 0$. Then, the radial velocity is given by (362). The general solution for stresses given in the previous section is also valid. In particular, the dependence of ψ on θ and the value of n are determined from (368) and the boundary conditions (385) and (386). Equations (370) and (371) are also valid. Since the boundary condition (386) coincides with (361), the strain rate intensity factor corresponding to the maximum friction surface $\theta = \theta_1$ is given by (375). Using the nomenclature of the present section the expression for this strain rate intensity factor becomes

$$D_{\text{ex}} = \frac{\sqrt{6 \sin \varphi} U_w}{2r^{5/2}\sqrt{n \cos \varphi - \sin \varphi \left[\cos \varphi + \cot \theta_1 \left(1 + \sin \varphi\right)\right]}}. \tag{387}$$

Here U_w is the value of U at $\theta = \theta_1$. Expanding the right hand side of (371) in a series in the vicinity of $\psi = \psi_f$ (or $\theta = \theta_0$) yields

$$\xi_{r\theta} = -\frac{3U_f}{4r^3\left(\psi - \psi_f\right)} + o\left(\frac{1}{\psi - \psi_f}\right) \quad \text{as} \quad \psi \to \psi_f. \tag{388}$$

Here U_f is the value of U at $\theta = \theta_0$. Equation (368) in the vicinity of $\psi = \psi_f$ becomes

$$\frac{d\theta}{d\psi} = \frac{4\sin\varphi\left(\psi - \psi_f\right)}{n\cos\varphi - \sin\varphi\left[\cos\varphi - \cot\theta_0\left(1 + \sin\varphi\right)\right]} + o\left(\psi - \psi_f\right) \quad \text{as} \quad \psi \to \psi_f. \tag{389}$$

Integrating this equation gives

$$\left(\psi - \psi_f\right)^2 = \frac{\{n\cos\varphi - \sin\varphi\left[\cos\varphi - \cot\theta_0\left(1 + \sin\varphi\right)\right]\}}{2\sin\varphi}\left(\theta - \theta_0\right) \tag{390}$$

to leading order. Substituting (390) into (388) yields

$$\xi_{r\theta} = -\frac{3\sqrt{2\sin\varphi}U_f}{4r^3\sqrt{n\cos\varphi - \sin\varphi\left[\cos\varphi - \cot\theta_0\left(1 + \sin\varphi\right)\right]}\sqrt{\theta - \theta_0}} + o\left(\frac{1}{\sqrt{\theta - \theta_0}}\right) \quad \text{as} \quad \theta \to \theta_0. \tag{391}$$

The normal strain rates are bounded in the vicinity of the maximum friction surface $\theta = \theta_0$ and $\xi_{r\theta} \to \infty$ as $\theta \to \theta_0$. Therefore, Eq. (6) in which ξ_τ should be replaced with $\xi_{r\theta}$ is valid. Then, it follows from (6) and (391) that the strain rate intensity factor corresponding to the friction surface $\theta = \theta_0$ is

$$D_{in} = \frac{\sqrt{6\sin\varphi}U_f}{2r^{5/2}\sqrt{n\cos\varphi - \sin\varphi\left[\cos\varphi - \cot\theta_0\left(1 + \sin\varphi\right)\right]}}. \tag{392}$$

The solution of Eq. (370) is given by (382). The condition (187) should be used to determine U_w. In particular, substituting (362) and (382) into this condition results in

$$Q = 2\pi U_w \int_{\theta_0}^{\theta_1} w\left(\psi\right)\sin\theta d\theta.$$

Replacing here integration with respect to θ with integration with respect to ψ by means of (368) yields

$$\frac{U_w}{Q} = \frac{1}{4\pi\sin\varphi}\left[\int_{\psi_f}^{\psi_w} \frac{\sin\theta\left(\sin\varphi + \cos 2\psi\right) w\left(\psi\right)}{n\cos^2\varphi - \sin\varphi\omega_0\left(\psi, \theta\right)} d\psi\right]^{-1}, \tag{393}$$

where $\omega_0(\psi, \theta)$ is defined in (365). The value of U_f is found from (382) as

$$U_f = U_w w\left(\psi_f\right). \tag{394}$$

In order to make comparison with the solution for pressure-independent material (Sect. 4.4), it is convenient to introduce the dimensionless strain rate intensity factors, $d_{\rm ex}$ and d_{in}, by Eq. (170). Then, using Eqs. (387), (392) and (394) yields

$$d_{\rm ex} = \frac{\sqrt{6}\sin\varphi}{2\sqrt{n\cos\varphi - \sin\varphi\left[\cos\varphi + \cot\theta_1\left(1+\sin\varphi\right)\right]}}\left(\frac{U_w}{Q}\right) \tag{395}$$

and

$$d_{in} = \frac{\sqrt{6}\sin\varphi w\left(\psi_f\right)}{2\sqrt{n\cos\varphi - \sin\varphi\left[\cos\varphi - \cot\theta_0\left(1+\sin\varphi\right)\right]}}\left(\frac{U_w}{Q}\right). \tag{396}$$

The ratio U_w/Q in Eqs. (395) and (396) should be eliminated by means of (393).

7 Concluding Remarks

The present chapter provides a comprehensive review of solutions for the strain rate intensity factor for the model of classical pressure-independent plasticity and the double-shearing model of pressure-dependent plasticity. Comparison made allows one to estimate the effect of pressure-dependency of the yield criterion on the magnitude of the strain rate intensity factor. The importance of this quantity for applications is that it controls the intensity of plastic deformation and, as a consequence, the intensity of physical processes in a narrow layer near frictional interfaces. Some theories have been already proposed to use the strain rate intensity factor in constitutive equations [41, 42]. In certain sense, these theories are similar to some theories in the mechanics of cracks based on the stress intensity factor [43]. Since the latter are very successful in engineering applications, it is expected that the theories based on the strain rate intensity factor can be successful as well.

Acknowledgments The research described in this chapter has been supported by the grant RFBR-11-01-00987. A part of this work was done while the first author was with National Chung Cheng University (Taiwan) as a research scholar under the recruitment program supported by the National Science Council of Taiwan (contact 99-2811-E-194-009).

References

1. Alexandrov, S., Richmond, O.: Int. J. Non-Linear Mech. **36**(1), 1 (2001)
2. Alexandrov, S., Lyamina, E.: Dokl. Phys. **47**(4), 308 (2002)
3. Alexandrov, S., Mishuris, G.: J. Eng. Math. **65**(2), 143 (2009)
4. Alexandrov, S.: In: Proceedings of 10th Biennial ASME Conference on Engineering Systems Design and Analysis (ESDA 2010) ESDA2010-24021 (2010)
5. Alexandrov, S., Harris, D.: Int. J. Mech. Sci. **48**(7), 750 (2006)
6. Moylan, S., Kompella, S., Chandrasekar, S., Farris, T.: Trans. ASME J. Manuf. Sci. Eng. **125**(2), 310 (2003)

7. Trunina, T., Kokovkhin, E.: J. Mach. Manuf. Reliab. **37**(2), 160 (2008)
8. Alexandrov, S., Grabko, D., Shikimaka, O.: J. Mach. Manuf. Reliab. **38**(3), 277 (2009)
9. Fries, T.P., Belytschko, T.: Int. J. Numer. Meth. Eng. **84**, 253 (2010)
10. Spencer, A.: Mechanics of Solids, pp. 607–652. Pergamon Press, Oxford (1982)
11. Hill, R.: The Mathematical Theory of Plasticity. Clarendon Press, Oxford (1950)
12. Rees, D.: Basic Engineering Plasticity. Butterworth-Heinemann, Oxford (2006)
13. Shield, R.: Proc. Roy. Soc. **233A**(1193), 267 (1955)
14. Druyanov, B., Nepershin, R.: Problems of Technological Plasticity. Elsevier, Amsterdam (1994)
15. Menrabadi, M., Cowin, S.: J. Mech. Phys. Solids **26**, 269 (2001)
16. Alexandrov, S., Druyanov, B.: Mech. Solids **27**(4), 110 (1992)
17. Druyanov, B., Alexandrov, S.: Int. J. Plast. **8**(7), 819 (1992)
18. Alexandrov, S.: Sov. Phys. Dokl. **37**(6), 283 (1992)
19. Alexandrov, S.: Mech. Solids **30**(5), 111 (1995)
20. Alexandrov, S., Richmond, O.: Dokl. Phys. **43**(6), 362 (1998)
21. Alexandrov, S.: J. Appl. Mech. Tech. Phys. **46**(5), 766 (2005)
22. Alexandrov, S., Mishuris, G.: Arch. Appl. Mech. **77**(1), 35 (2007)
23. Alexandrov, S.: J. Appl. Mech. Tech. Phys. **50**(5), 886 (2009)
24. Alexandrov, S., Harris, D.: Eur. J. Mech. A. Solids **29**(6), 966 (2010)
25. Alexandrov, S.: J. Appl. Mech. Tech. Phys **52**(3), 483 (2011)
26. Alexandrov, S.: Mater. Sci. Forum **623**, 1 (2009)
27. Alexandrov, S., Jeng, Y.R.: J. Eng. Math. **71**(4), 339 (2011)
28. Alexandrov, S., Lyamina, E.: J. Appl. Mech. Tech. Phys. **50**(3), 504 (2009)
29. Alexandrov, S., Kocak, M.: Fatigue Fract. Eng. Mater. Struct. **30**(4), 351 (2007)
30. Alexandrov, S., Richmond, O.: Fatigue Fract. Eng. Mater. Struct. 723–728 (2000)
31. Spencer, A.: Int. J. Mech. Sci. **7**, 197 (1965)
32. Shield, R.: J. Mech. Phys. Solids **3**, 246 (1955)
33. Alexandrov, S., Lyamina, E.: Mech. Solids **43**(5), 751 (2008)
34. Spencer, A.: J. Mech. Phys. Solids **12**, 337 (1964)
35. Marshall, E.: Acta Mech. **3**, 82 (1967)
36. Pemberton, C.: J. Mech. Phys. Solids **13**(6), 351 (1965)
37. Alexandrov S., Lyamina, E.: Mech. Solids (in press)
38. Alexandrov, S., Lyamina, E.: Mech. Solids **38**(6), 40 (2003)
39. Alexandrov, S., Lyamina, E.: Int. J. Mech. Sci. **45**(9), 1505 (2003)
40. Alexandrov, S., Lyamina, E.: Acta Mech. **187**, 37 (2006)
41. Alexandrov, S., Lyamina, E.: J. Appl. Mech. Tech. Phys. **47**(5), 757 (2006)
42. Alexandrov, S., Lyamina, E.: J. Appl. Mech. Tech. Phys. **52**(4), 657 (2011)
43. Kanninen, M., Popelar, C.: Advanced Fracture Mechanics. Oxford University Press, New York (1985)

Mechanical Response of Porous Materials: The Gurson Model

Luiz. A. B. da Cunda and Guillermo J. Creus

Abstract In this chapter the formulation for damage known as Gurson model is presented. The original formulation, set in a micro-mechanical context, and different adjustments of phenomenological nature are described. The range of the parameters of the model and their influence on the representation are described. The main computational details for the implementation of the model by means of the finite element method are presented and examples of application are given.

1 Introduction

Fracture Mechanics that uses global fracture parameters, such as J integral or Crack Tip Opening Displacement (CTOD) [34], only in special situations represents the behavior of ductile solids. In polycrystalline metals, ductile fracture is controlled by nucleation, growth and coalescence of microvoids and a local approach provides a clearer picture.

Voids can nucleate from large inclusions and second phase particles, by particle fracture or interfacial decohesion [47]. Once a void has been nucleated, it will grow under plastic deformation and hydrostatic stress. Eventually the voids will connect and ductile fracture by void coalescence will take place.

Thus, four stages, as indicated in Fig. 1 are observed: homogenous deformation with void nucleation and growth, localized deformation and void coalescence.

L. A. B. da Cunda (✉)
Escola de Engenharia, Universidade Federal do Rio Grande-FURG, Av. Itália, km 8, Campus Carreiros, Rio Grande, RS 96203-900, Brazil
e-mail: luizcunda@furg.br

G. J. Creus
Instituto Mercosul de Estudos Avançados, Universidade Federal da Integração Latino-Americana, Foz do Iguaçu, PR, Brazil
e-mail: creus@ufrgs.br

H. Altenbach and A. Öchsner (eds.), *Plasticity of Pressure-Sensitive Materials*, Engineering Materials, DOI: 10.1007/978-3-642-40945-5_6,

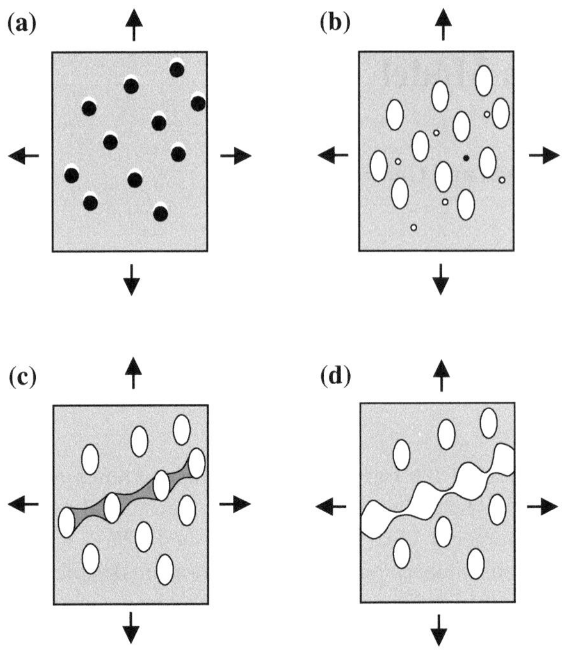

Fig. 1 Stages in the development of damage: **a** nucleation, **b** growth, **c** coalescence and **d** fracture

The best known micro-mechanical model for void related damage and fracture is due to Gurson [25, 26] and includes a plastic damage yield condition that depends on a damage parameter (porosity) and a growth law for this damage variable. The original Gurson model has been subjected to many analysis, criticisms and improvements, some of which are reviewed in the present work. The original model assumes a homogenous deformation field and thus is not able to describe interaction effects, void shape changes and the non-homogeneous transformations that lead to coalescence and rupture. Some important modifications are due to Tvergaard [76, 79] who introduced adjustment parameters and to Chu and Needleman [10] who proposed improved nucleation laws for porosity. For this reason the model is sometimes referred to as GTN (for Gurson, Tvergaard, Needleman). Other modifications will be described in Sect. 4. Many of them maintain the same basic variables and general form of the equations and have only a phenomenological (i.e. no micromechanical) base. Application of the Gurson model to practical problems is only possible in a computational context. So, diverse studies have been devoted to the numerical implementation (usually through finite element techniques) of the model. Details of some procedures and application examples are given in Sects. 5 and 6.

2 Gurson Damage Model

2.1 Yield Criterion

This theory—originally presented by Gurson [25, 26] in a PhD dissertation supervised by J. R. Rice—proposed yield criteria and flow rules for porous materials, focusing the effect of void nucleation and growth, as observed in ductile fracture. Figure 2 (from Gurson [26]), defines macroscopic and microscopic stresses and strains and the spherical model of a unit cell. The isotropic damage variable (in the framework of continuum damage mechanics [32, 33]) is the volumetric void fraction or porosity $f = V_v/V$, with V_v being the volume of voids in a representative small volume V. The volumetric void fraction f is assumed as defined at each point of the continuum.

The macroscopic yield criterion was approximated with an upper bound approach [28]. Aggregates of cells representing voids in a ductile matrix were employed, with the matrix material idealized as rigid-perfectly plastic obeying von Mises yield criterion. Using a distribution of macroscopic flow fields and working with a dissipation integral, upper bounds for the macroscopic stress fields required for yield were determined. Their locus in the stress space determines the yield surface. It was shown that normality of plastic flow holds for this yield surface. The expression proposed by Gurson is

$$\begin{aligned} \Phi &= \frac{\Sigma_{\mathrm{eq}}^2}{\sigma_0^2} + 2f\cosh\frac{3\Sigma_{\mathrm{m}}}{2\sigma_0} - 1 - f^2 = 0 \\ \Sigma_{\mathrm{eq}} &= \sqrt{\frac{3}{2}\Sigma' : \Sigma'}, \quad \Sigma_{\mathrm{m}} = p = \frac{1}{3}\mathrm{tr}\,\Sigma, \quad \Sigma' = \Sigma - \Sigma_{\mathrm{m}}, \end{aligned} \tag{1}$$

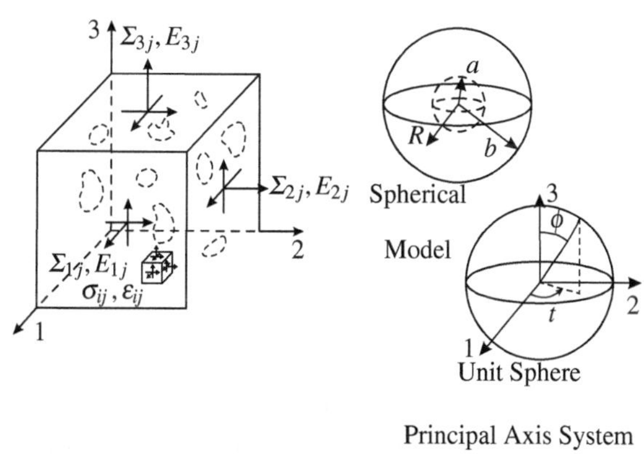

Fig. 2 Void-matrix aggregate, with random void shapes and orientations, evidencing macroscopic and microscopic tensor quantities, and also the unit cell model studied by Gurson [26]

where Σ_{eq} is the macroscopic von Mises equivalent stress, Σ' is the deviator of the macroscopic stress, Σ_m is the mean macroscopic stress or pressure, Σ_0 is the yield stress of the matrix undamaged material and f is the porosity. For $f = 0$ Gurson's model reduces to the von Mises criterion.

A prior model, the Drucker-Prager theory [18], had already proposed a yield criterion dependent on hydrostatic stress in the general form (the macroscopic stress being now indicated with σ)

$$\Phi = \sigma_{eq} - a\sigma_m - b = 0. \tag{2}$$

Gurson seminal contribution resides in the establishment of a microstructural relation for the effect of the hydrostatic stress p on the yield function through the consideration of a porosity variable f.

In the original Gurson model (Eq. 1), the softening of the material with the increase of void volume fraction is a continuous process, and a complete loss of load carrying capacity would occur only when the void has grown to the ultimate value $f = 1$. Tvergaard [76, 79] compared the bifurcation predictions based on the Gurson model and his own numerical studies for material containing periodic distribution of voids and suggested a modification of the Gurson model. In the most usual notation

$$\Phi = \frac{\sigma_{eq}}{\sigma_y} + \sqrt{2\alpha_1 f \cosh \alpha_2 \frac{3\Sigma_m}{2\sigma_y} - 1 - \alpha_1^2 f^2} = 0 \tag{3}$$

or

$$\Phi = \sigma_{eq} - \varpi\sigma_y = 0, \tag{4}$$

with

$$\varpi = \sqrt{1 - 2\alpha_1 f \cosh\left(\alpha_2 \frac{3\Sigma_m}{2\sigma_y}\right) + \alpha_1^2 f^2}. \tag{5}$$

The substitution of Σ by σ is coherent with the fact that the original micromechanical model has been altered to include phenomenological parameters.

In Fig. 3, yield surfaces for different levels of void content are shown, in a plot of normalized macroscopic deviatory stress versus normalized pressure. It can be seen that the elastic domain depends on the hydrostatic pressure. When the volumetric void fraction f decreases, the influence of pressure also decreases, leading to a larger elastic domain. For $f = 0$, the model reduces to the von Mises criterion, which is independent of the hydrostatic pressure.

The parameter α_1 in Eq. (3) is a coefficient multiplying the porosity f, to be adjusted by comparing numerical simulations of RVE aggregates and the predictions of the model. Since the element studied by Gurson was a single hollow sphere and thus disregarded interaction among voids, this coefficient introduces somehow the interaction effect. The parameter α_2 in Eq. (3) can be understood as a calibrating coefficient acting on the pressure.

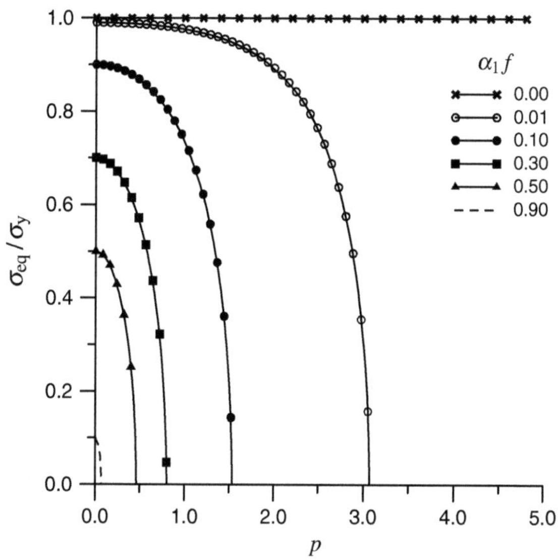

Fig. 3 Gurson yield surface for a porous material, showing the influence of pressure and volumetric void fraction

The parameter α_1 is the inverse of the value f_U, being f_U the volumetric void fraction that corresponds to rupture in the absence of hydrostatic pressure. From $\varpi(p = 0, \sigma_y, f) = 0$, results $f_U = 1/\alpha_1$.

Rupture occurs at a porosity level f_U in the absence of pressure. If pressure is present, rupture takes place at a porosity value lower than f_U. The combinations of pressure and porosity that lead to rupture are given by

$$\varpi = \frac{2}{3\alpha_2} \operatorname{arccosh} \frac{1 + \alpha_1^2 f^2}{2\alpha_1 f} - \frac{p}{\sigma_y} = 0. \tag{6}$$

Figure 4 shows plots corresponding to Eq. (6) for $\alpha_2 = 1.0$ and $\alpha_2 = 0.7$. Even with this modification, the void volume fraction at which the Gurson model will lose load carrying capacity is still unrealistically large. Both experimental observations [9] and results of cell model analysis by Koplik and Needleman [37] show that the volume fraction of voids at which void coalescence starts is usually less than 15 %.

Flow rule: As shown by Gurson [26] the plastic strain rate tensor D_{ij}^{p} obeys the normality rule

$$D_{ij}^{p} = \dot{\lambda} \frac{\partial \Phi}{\partial \sigma_{ij}}. \tag{7}$$

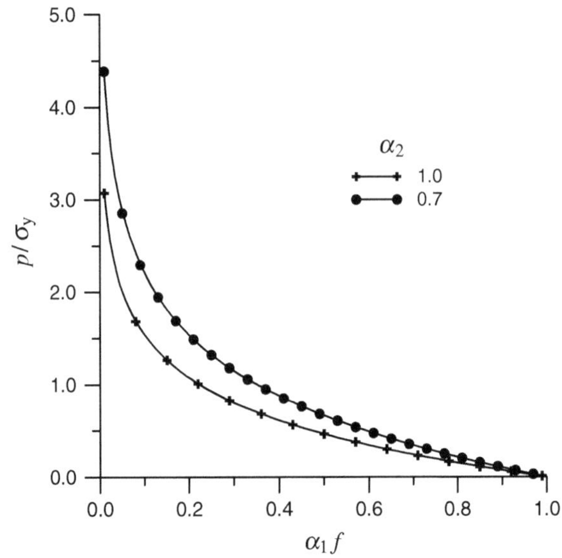

Fig. 4 Combinations of pressure and porosity values that correspond to loss of strength ($\varpi = 0$)

The equivalent plastic strain rate is defined as

$$\dot{\varepsilon}^{\mathrm{p}} = \sqrt{\frac{2}{3} D^{\mathrm{p}}_{ij} D^{\mathrm{p}}_{ij}}. \tag{8}$$

As in classical plasticity, the relation giving the plastic strain rates as a function of the stress rates is obtained using the consistency condition $\dot{\Phi} = 0$.

2.2 Evolution Law for the Porosity

In a plastic damage theory it is necessary to have, in addition to the yield criterion and the flow rule, an evolution law for damage (porosity, in this case). The mechanism of damage evolution considered in the original Gurson model was growth. Growth occurs when the voids (pre-existent or nucleated) change their size according to the volume change in the continuum and is controlled by mass conservation through the expression

$$\dot{f}_{\mathrm{g}} = (1 - f) D^{\mathrm{p}}_{ii}, \tag{9}$$

which determines that voids increase or decrease their volume according to the volume variation in the continuum.

In the GTN version, two additional mechanisms are included: nucleation and coalescence. Nucleation occurs mainly due to material defects, in the presence of tension. Coalescence is related to the fast rupture process that occurs after the volumetric void fraction reaches a limit, usually indicated by f_C. Coalescence is the union of neighbor voids due to the rupture of a ligament among them (see Fig. 1).

The equations that govern damage evolution are modeled in a simplified form as follows. First, it is assumed that the total void growth rate is given by

$$\dot{f} = \begin{cases} \dot{f}_n + \dot{f}_g & f \leq f_C, \\ \dot{f}_c & f > f_C, \end{cases} \tag{10}$$

where $\dot{f}_n$ is the void nucleation rate, $\dot{f}_g$ is the void growth rate and $\dot{f}_c$ is the void coalescence rate. Thus, as long as f is smaller than a characteristic value f_C, only nucleation and growth develop. Above f_C, only coalescence takes place. The nucleation rate is proportional to the rate of the equivalent plastic strain

$$\dot{f}_n = A(\varepsilon^p)\dot{\varepsilon}^p. \tag{11}$$

For $A(\varepsilon^p)$ Chu and Needleman [10] proposed the statistical distribution

$$A(\varepsilon^p) = \frac{f_N}{s_N\sqrt{2\pi}} \exp\left[-\frac{1}{2}\left(\frac{\varepsilon^p - \varepsilon_N}{s_N}\right)^2\right], \tag{12}$$

where f_N is the proposed final nucleation void volumetric fraction, ε_N is the mean plastic strain value for nucleation and s_N is the standard deviation for the distribution (see Fig. 5).

Sometimes it is assumed that nucleation does not take place when the material is in compression. The compression state is indicated by a negative pressure p, and thus $A(\varepsilon^p) = 0$ if $p < 0$.

Coalescence is governed [78] by the relation

$$\dot{f}_c = \frac{f_U - f_C}{\Delta\varepsilon}\dot{\varepsilon}^p, \tag{13}$$

where $\Delta\varepsilon$ is a material parameter that controls how fast the coalescence happens. An alternative way of taking coalescence into account [80] is to replace the volumetric void fraction f in the Gurson yield surface (3) by a corrected volumetric void fraction f^* given by

$$f^* = \begin{cases} f & f < f_C, \\ f_C + \dfrac{1.0 - f_C}{f_F - f_C}(f - f_C) & f > f_C, \end{cases} \tag{14}$$

with f_F being the rupture volumetric void fraction. In this case, only nucleation and growth are considered in Eq. (10). Nucleation and coalescence are irreversible. Thus, it seems natural to model them (Eqs. (11) and (13)) as governed by equivalent plastic

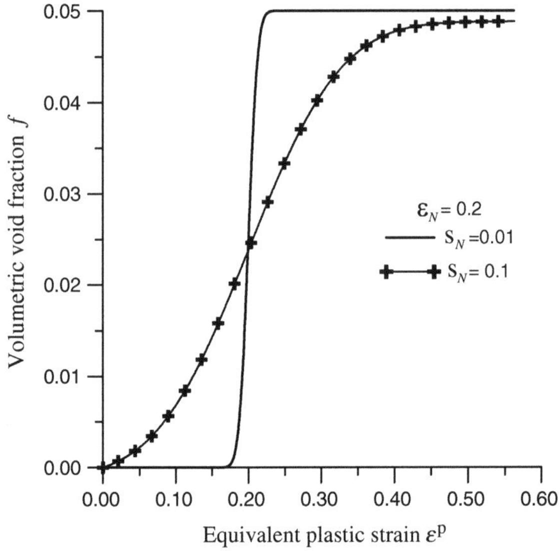

Fig. 5 Evolution of porosity in the nucleation stage considering $\varepsilon_N = 0.2$ and two different s_N values. As a consequence of the combinations of ε_N and s_N adopted, nucleation is almost complete in both cases

strain. The growth mechanism is reversible, so it is modeled (Eq. (9)) as volume change dependent.

In Fig. 6, the evolution of porosity is shown. The occurrence of the three stages of porosity evolution can be observed: initially, a relative fast nucleation, according to the standard deviation s_N imposed, placed around equivalent plastic strain 0.1; afterwards, a growth stage until an equivalent plastic strain around 0.5, 0.7 or 0.9 is reached. And finally, the coalescence stage, in which the porosity level changes abruptly, according to the parameter employed. The start of the coalescence stage is defined based on porosity, according to Eq. (13).

2.3 Elastic Constants for the Damaged Material

The presence of embedded voids in a metallic matrix alters also the elastic behavior. Mori and Tanaka [49] relations are usually adopted

$$K = \frac{4K_0G_0(1-f)}{4G_0 + 3K_0 f}, \qquad G = \frac{G_0(1-f)}{1 + \dfrac{6K_0 + 12G_0}{9K_0 + 8G_0} f}, \tag{15}$$

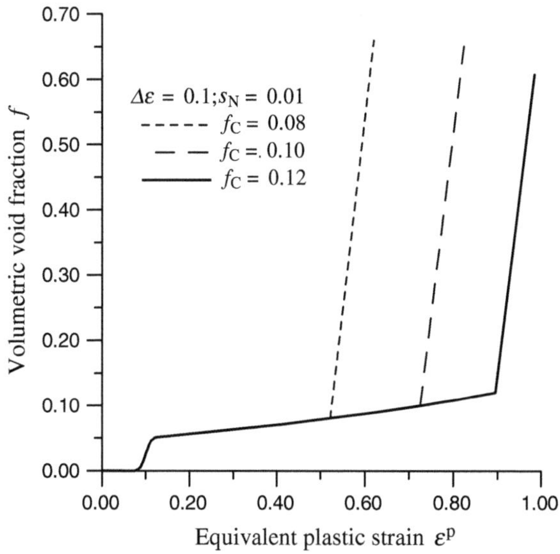

Fig. 6 Evolution of porosity for a tensioned material obeying Gurson model. The three mechanisms of voids evolution: nucleation, growth and coalescence can be identified. It is observed that the parameter $\Delta\varepsilon$ is the equivalent plastic strain increment from the beginning of coalescence until the final rupture of the material

with K_0 and G_0 being the undamaged values of compressibility and shear modulus, respectively. There are other proposals to include the effect of porosity on elastic constants [31, 45, 59], leading to similar results for low porosities. Figure 7 shows the dependence of damaged Young's modulus E on the porosity, evaluated from Eqs. (15) and from [45]. E_0 is the undamaged Young's modulus.

2.4 Assessment of Gurson Model

The Gurson model has been assessed using numerical micromechanical techniques. Trillat and Pastor [75] use the static and kinematic methods of limit analysis to check both for the 2D and 3D Gurson expressions. For spherical cavities Gurson criterion seems to be a good analytical expression; in the case of cylindrical cavities Gurson expression seems too restrictive. The 2D formulation is also analyzed in [20], where a modification of the Gurson yield function is proposed.

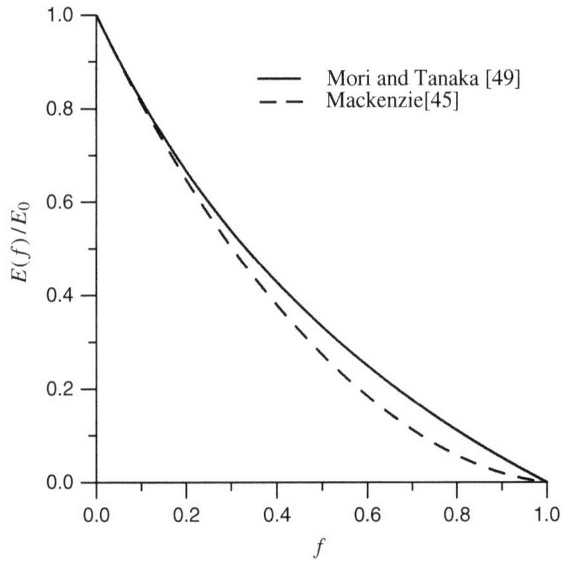

Fig. 7 Dependence of Young's modulus on the void volumetric fraction as introduced by Mackenzie [45] and Mori and Tanaka [49] equations

3 Influence of the Parameter Values on the Behavior of the Damage Model

The material parameters in the Gurson model can be classified as:

(a) constitutive parameters, related to the Gurson yield surface (α_1, α_2);
(b) initial parameters, associated to the origin of the porosity, whether present in the virgin material (f_0) or nucleated by plastic straining (f_N, s_N and ε_N);
(c) critical parameters, related to the interaction between neighbor voids, describing the coalescence stage and the final rupture of the material.

The constitutive parameters α_1 and α_2 act as multipliers on the volumetric void fraction and on the pressure respectively, introducing the possibility of adjusting the the Gurson yield surface with available experimental or numerical data. To larger values of α_1 and α_2 correspond a smaller elastic domain. Figure 8 shows the dependence of the Gurson yield surface on the α_2 parameter.

The second group of parameters is related to the origin of the voids. An initial porosity f_0 can be employed in two situations: when the material has actually an initial porosity or when voids are developed from inclusions that break or debond from the matrix at a very low strain level. Otherwise, the strain governed nucleation relation proposed by Chu and Needleman [10], described by Eqs. (11) and (12), is used.

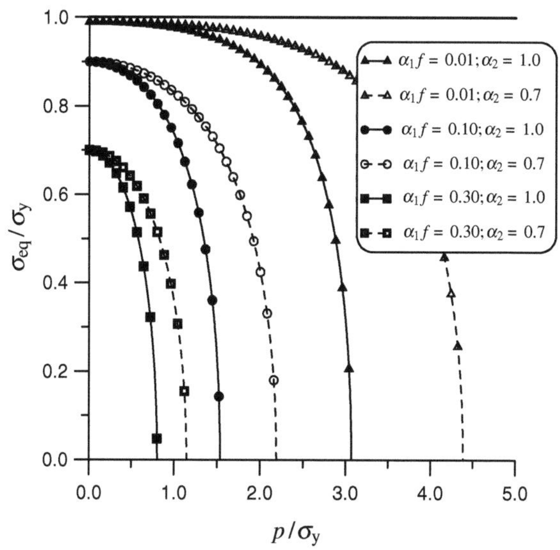

Fig. 8 Dependence of Gurson yield surface on α_2 parameter presented for $\alpha_2 = 0.7$ and $\alpha_2 = 1.0$

The value adopted for f_N determines the proposed level of nucleated voids. The parameter ε_N corresponds to the mean equivalent plastic strain for which nucleation is developed. The nucleation standard deviation s_N controls the localization of nucleation around ε_N. Figures 5 and 9 shows two nucleation processes with different ε_N: to the smaller value (Fig. 9) corresponds an earlier nucleation. In both the figures, plots with different s_N are presented: to smaller values of s_N corresponds a faster nucleation, with the nucleation localized around the mean equivalent plastic strain nucleation ε_N.

Figure 9 shows that the proposed level f_N is not reached, if $s_N = 0.1$ is employed. This is because the porosity evolution law in the nucleation stage $\dot{f}_n$ is given in a rate form that must be integrated while the material is plastically deformed. If an inadequate relation between s_N and ε_N is employed, a significant part of the porosity evolution rate $\dot{f}_n$ will take place in the "fictitious negative" part of the equivalent plastic strain domain (Fig. 10), and not be integrated, since the equivalent plastic strain is always positive, determining an incomplete nucleation.

To avoid this problem, a relation between s_N and ε_N in the form $\varepsilon_N > z s_N$ must be respected. So, to each value of z corresponds a different level of nucleation. To ensure a nucleation level of at least 95 % of f_N it is necessary to employ a z value of 1.645. To obtain a nucleation level of 97 % of f_N, $z = 1.882$ must be used and to attain a nucleation level of 99 % of f_N, $z = 2.337$ must be employed.

Another important choice concerns the influence of the pressure sign on the nucleation of voids. One approach is to consider nucleation completely independent of the pressure sign [71, 77]. For a material without initial porosity, this choice leads to a Gurson yield surface which is symmetric with respect to pressure, as indicated

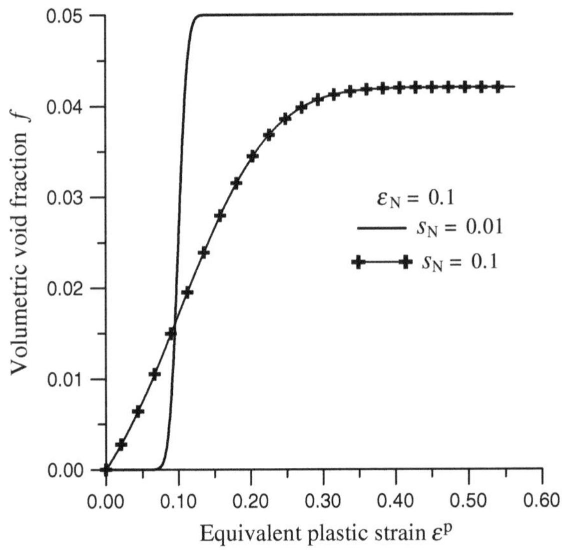

Fig. 9 Evolution of porosity in the nucleation stage considering $\varepsilon_N = 0.1$ and two different s_N values, showing localized nucleation around ε_N if a smaller value of s_N is employed. Incomplete nucleation can be observed if an inadequate combination of ε_N and s_N is adopted, i.e., a large s_N value with relation to the ε_N value

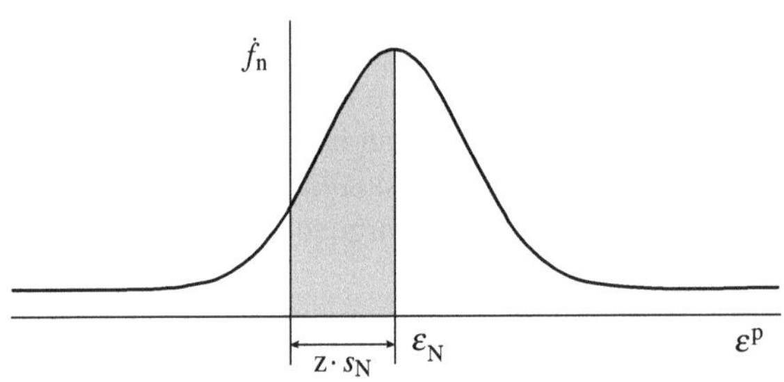

Fig. 10 Dependence of nucleation evolution rate on equivalent plastic strain. Part of the area under the curve is placed on the "fictitious negative" domain of the equivalent plastic strain

in Fig. 11 . On the other hand, if nucleation is associated to debonding between inclusions and metallic matrix, this debonding will decrease whenever the region is submitted to compression. To avoid this contradiction, some proposals consider nucleation only for $p > 0$ (tension), imposing $A(\varepsilon^p) = 0$ if $p < 0$ [69]. For a material initially free of voids, we will have Gurson behavior for $p > 0$ and von Mises behavior for $p < 0$ as indicated in Fig. 11. This approach also has drawbacks. Considering a material initially compressed and plastically deformed to an equivalent plastic strain level higher than ε_N, if the load reversal happens and the hydrostatic stress state on the material changes to tension, nucleation will not take place. In the absence of nucleated voids, there will be no void evolution and the material will

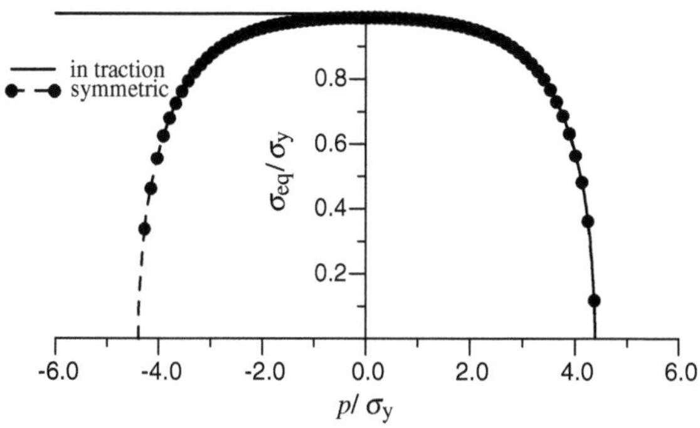

Fig. 11 Gurson yield surface associated to a material initially voids free, under different nucleation approaches: nucleation only in tension or pressure independent nucleation, symmetric with relation to the pressure sign. If the material is tensioned, both approaches lead to similar results. If the material is compressed, with nucleation only in tension, von Mises behavior is reproduced

continue obeying von Mises yield criterion even for a high level of plastic straining and positive hydrostatic tension [13]. Both nucleation approaches give the same response to monotonic positive hydrostatic pressure.

The third class of parameters includes those related to the coalescence of voids. Considering the coalescence rate of voids described by the Eq. (13), two material parameters are present: f_C, that indicates the initial level of voids at which coalescence takes place, and $\Delta\varepsilon$, that indicates how fast coalescence occurs. Considering a material submitted to uniform tension, it can be seen that the coalescence parameters control the final branch of the load-displacement relation presented in Fig. 12: f_C controls the start of the branch and $\Delta\varepsilon$ controls its slope. To a small f_C corresponds a final branch that starts earlier and to a small $\Delta\varepsilon$ corresponds a steeper one.

4 Further Developments and New Trends for Gurson Model

The Gurson model has been modified by several authors, particularly in reference to its parameters. There are proposals to make these parameters function of porosity [23, 81], triaxiality [40], shape of voids [36], etc. There are proposals to use kinematical hardening [8, 42]. A thermo mechanical coupling [84] includes the possibility of using parameters dependent on temperature.

Thomason [74] proposes a model that incorporates Rice and Tracey [66] equations, and is able to represent both the growth and the change of void shapes. For this model, the α_i parameters proposed by Tvergaard [76, 79] would be unnecessary. Klöecker and Tvergaard [36] also propose a modification of the Gurson model to take into account changes in void shapes and the coalescence process. Zhang and Niemi [88] present a mechanism of coalescence that avoids the need for determining experimentally the critical value of the coalescence beginning f_C. In this model, the

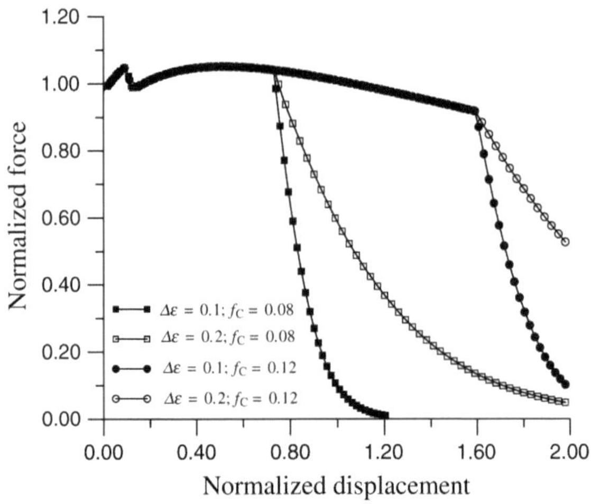

Fig. 12 Force versus displacement relations for a tensioned material obeying Gurson behavior. It can be seen that after yielding there is a decay in tension corresponding to nucleation, and that the final branch corresponding to rupture starts earlier or later depending on f_C and with its slope depending on $\Delta\varepsilon$

void remains spherical during growth, and the initiation of coalescence is controlled by the triaxiality level.

Voyiadjis and Kattan [81] propose a formulation that introduces damage through a damage tensor. Applying this formulation to simulate Gurson model, they obtain a yield surface with porosity dependent parameters. Wen et al. [82] propose a modification of the Gurson model to take into account void size. They show that the yield surface is larger for materials with very small voids. The effect becomes important for high porosities.

The Gurson model has been used in combination with Fracture Mechanics by Kikuchi et al. [35], Needleman and Tvergaard [52], Koppenhoefer and Dodds Jr. [38] and Skallerud and Zhang [70] employing J integral and by Aravas and McMeeking [3] employing J integral and COD.

Subjects that deserve particular attention are situations with low triaxility and shear stresses, the effect of hardening and the consideration of cyclic loading.

4.1 Non-Spherical Voids

The evolution of the void shape and its effect on the mechanical behavior has been considered by Gologanu et al. [22] and Pardoen and Hutchinson [58].

Gologanu et al. [22] extended the Gurson model to prolate and oblate voids in a plastic material. The stress potential proposed corresponds to an ellipsoidal volume of perfectly plastic material containing a confocal ellipsoidal void,

$$\Phi = C\left(\frac{\Sigma_{\mathrm{eq}} + \eta\Sigma_{\mathrm{h}}}{\sigma_0}\right)^2 - 2q(g+1)(g+f)\cosh\left(\kappa\frac{\Sigma_{\mathrm{h}}}{\Sigma_0}\right) - q^2(g+f)^2 = 0, \quad (16)$$

with

$$\Sigma_{\mathrm{h}} = 2\alpha_2\left(\frac{\Sigma_{11} + \Sigma_{22}}{2}\right) + (1 - 2\alpha_2)\Sigma_{33}.$$

The parameters C, η, q, α_2 depend on the current void and cell shape. Pardoen and Hutchinson [58] use a potential function similar to Eq. (16) and add a coalescence function that determines the initiation of the coalescence, based on the work of Thomason [74]. Two new variables determine void shape and the relative void spacing. The analysis seems to give satisfactory results for the overall cell behavior (i.e. equivalent stress-strain curves), but a precise prediction of the shape evolution needs the introduction of correction functions.

4.2 Shear Effects

The difficulty of the Gurson formulation to model damage under pure shear has been tackled in a phenomenological form by Nahshon and Hutchinson [50]. An extension of the Gurson model that incorporates damage growth under low triaxial straining in shear-dominated states is proposed. This extension retains the isotropy of the original Gurson model by making use of the third invariant of stress to introduce shear dependence. This extension opens the possibility for computational approaches based on the Gurson model to be extended to shear-dominated failures. This modification assumes that the volume of voids undergoing shear may not increase, but void deformation and reorientation contribute to damage and softening increase. Thus, f is no longer directly tied to the plastic volume change. Instead, it is regarded as an effective damage parameter. The modification, while phenomenological, is nevertheless formulated to be consistent with the mechanism of softening in shear. Specifically, it is proposed that the growth rate expression be written as

$$\dot{f} = (1-f)D^{\mathrm{p}}_{ii} + A\dot{\varepsilon}^{\mathrm{p}} + k_{\omega} f \omega_0 \frac{s_{ij}D^{\mathrm{p}}_{ij}}{\sigma_{\mathrm{y}}}, \quad (17)$$

where s_{ij} is the deviatoric part of the stress tensor and the invariant measure $\omega_0 = \omega(\sigma)$ is defined as

$$\omega_0 = \omega(\sigma) = 1 - \left(\frac{27III_s}{2\sigma_{\mathrm{y}}^3}\right)^2, \quad (18)$$

in which III_s is the third invariant of the deviator. In Eq. (17), the first term representing growth of existing voids follows from plastic incompressibility, the second term describes nucleation of new voids, while the last term, introduced by Nahshon and

Hutchinson [50], is formulated to be consistent with the mechanism of void softening in shear. Void nucleation is here taken to be plastic strain controlled as suggested by Chu and Needleman [10], so that the coefficient A in Eq. (17) takes the form of Eq. (12).

The model approximates experimental results obtained for various structural alloys that show a marked difference between fracture strains under axisymmetric stress and those under a pure shear stress plus a hydrostatic component or under plane stress states. The shear contribution added to the damage growth rate in Eqs. (17) and (18) does not affect the normality of the plastic flow rule [50].

The proposal above has been critically analyzed by Nielsen and Tvergaard [53], which claims the modification represents damage development in shear, but also gives a contribution to the damage development at plane strain uniaxial tension, even though the stress triaxiality is far from zero.

4.3 Hardening

Gurson [25, 26] had already considered the case of a matrix with isotropic hardening, writing Eq. (1) with $\sigma_0 = \sigma(\bar{E})$, where $\bar{E}$ is given by the evolution law $(1-f)\sigma_0\dot{\bar{E}} = \Sigma_{ij}D^{\mathrm{p}}_{ij}$. Mear and Hutchinson [48] and Becker and Needleman [6], were the first to introduce linear kinematic hardening into the Gurson yield function. In the case of purely kinematic hardening [48], the proposed criterion is

$$\Phi = \frac{(\Sigma - A)^2_{\mathrm{eq}}}{\sigma_{\mathrm{y}}^2} + 2f\cosh\left(\frac{3(\Sigma - A)_{\mathrm{m}}}{2\sigma_{\mathrm{y}}}\right) - 1 - f^2 = 0, \tag{19}$$

where A denotes now the back stress (center of the macroscopic elastic domain).

The extensions mentioned above are purely phenomenological. Leblond et al. [39] derived another yield function (the extended Leblond-Perrin-Devaux—LPD model) based on the analysis of a spherical void in a spherical volume element assuming an incompressible isotropic and kinematic hardening matrix material:

$$\Phi = \frac{(\Sigma - A)^2_{\mathrm{eq}}}{\Sigma_1^2} + 2\alpha_1 f\cosh\left(\frac{3\alpha_2(\Sigma - A)_{\mathrm{m}}}{2\Sigma_2}\right) - 1 - (\alpha_1 f)^2 = 0. \tag{20}$$

The two variables Σ_1 and Σ_2 replace the isotropic flow stress in the Gurson equation.

Steglich et al. [72] carried out unit cell calculations assuming a non-linear kinematic hardening matrix material surrounding spherical voids. They compared the unit cell results to predictions of the LPD model with non-linear kinematic hardening and found that, in principle, the behavior under constant triaxiality, as observed in the cell calculations, can be described with the model.

5 Computational Details

5.1 Numerical Implementation

The Gurson damage model computational formulation is usually presented in rate form (hypoelasticity). Thus, a numerical scheme must be adopted to integrate the rate equations following the evolution of the internal variables (stresses, damage and plastic strain). A time discretization procedure is adopted, associating to each time t a specific load level. Then, the evolution of internal variables is obtained for subsequent times $t + \Delta t, t + 2\Delta t$, etc. The correct integration of rate constitutive equations has a direct implication on the precision of the solution.

Physical and geometrical nonlinearities must be taken into account during the integration process. A convenient way is to consider the physical and geometrical nonlinearities in separate levels. Therefore, the integration of constitutive equations is organized in two stages: first, the evaluation of corotational Cauchy stresses (material nonlinearity); and secondly, the evaluation of stresses, from the previously obtained corotational stresses (geometrical nonlinearity). A good integration scheme must provide incremental stability, precision and incremental objectivity.

Incremental objectivity is assured by the corotational formulation. An integration procedure that has been widely employed is the so called split-operator scheme, with which all strain increase is initially considered as elastic. If the yield surface is exceeded, a plastic corrector is applied.

Ortiz and Popov [57], Runesson et al. [67], Gratacos et al. [24], Lee [41] and Zhang [86] present studies on the stability and precision of different integration schemes. Zhang [85] and Zhang and Niemi [87] present a generalized mid-point algorithm that is an evolution of the algorithm presented by Aravas [2] to integrate constitutive equations with internal variables and isotropic hardening. The algorithm determines the change in corotational stresses and internal variables such as porosity and plastic strain. Beginning at a time t (characterized by a subindex n), at which all the stresses and internal variables are known, the algorithm provides the updated values at time $t + \Delta t$ (subindex $n+1$). It uses a predictor-corrector strategy, partitioning volumetric and deviatory plastic strains. The parameter α controls whether the integration scheme is explicit ($\alpha = 0$) or implicit ($\alpha > 0$). The process is as follows:

(a) the logarithmic strain E_{ij}^{N} is decomposed into the sum of an elastic part $E_{ij}^{N,\mathrm{e}}$ and a plastic part $E_{ij}^{N,\mathrm{p}}$; such additive decomposition is adequate when an Updated Lagrangian description (with reference and updated configurations close to updated each other) is employed;
(b) stresses are determined from the elastic strains;
(c) the yield surface is determined using the conventional von Mises equivalent stress q, the pressure p and the internal variables H_t;
(d) the plastic strain rate is determined using a potential function g. In the present case the potential function is the Gurson yield surface, that is associated to Eq. (7);

(e) the rate h_t of the internal variables H_t is a function of the stress increment and on the value of the internal variables.

Initially, all the strain increment is supposed as being elastic. In this case, plastic strain and porosity remains unchanged in the time step. If the yield surface is violated, a plastic corrector is applied. The integration algorithm can be viewed as a system of non-linear equations to be solved.

$$\Delta E_{\mathrm{p}} P + \Delta E_{\mathrm{q}} Q = 0, \tag{21}$$

$$\Phi_{n+1} = \phi\left\{p_{n+1}, (s_{ij})_{n+1}, (H_t)_{n+1}\right\} = 0, \tag{22}$$

$$p_{n+1} = p^T_{n+1} + K\{f\}\Delta E_{\mathrm{p}} \tag{23}$$

$$(s_{ij})_{n+1} = (s_{ij})^T_{n+1} - 3G\{f\}\Delta E_{\mathrm{q}} \frac{(s_{ij})_{n+\alpha}}{q_n}, \tag{24}$$

$$(\Delta H_t)_{n+1} = (h_t)_{n+\alpha}\left\{\Delta E_{\mathrm{p}}, \Delta E_{\mathrm{q}}, p_{n+\alpha}, (s_{ij})_{n+\alpha}, (H_t)_{n+\alpha}\right\} \tag{25}$$

with $K\{f\}$ and $G\{f\}$ being the bulk modulus and shear modulus respectively, both corrected by the porosity. ΔE_{p} and ΔE_{q} are related, respectively, to the volumetric and deviatory part of the logarithmic plastic strain increment [87].

The non-linear equations system can be solved by the Newton-Raphson method, taking as variables ΔE_{p} and ΔE_{q}. Once obtained their values, the next step is to update stresses and internal variables (Eqs. (23), (24) and (25)). The process is repeated until $|(\Phi)_{n+1}| \leq 10{-}7$.

Some additional strategies can be used to enhance the robustness of the integration scheme. Worswick and Pick [83] recommends to employ sub-incrementation, breaking a time step into sub-steps. The number of sub-steps *NSI* may be chosen as a function of elastic predictor values, determining

$$NSI_1 = \mathrm{INT}(\Phi^T_{n+1}/\sigma_{\mathrm{y}}), \tag{26}$$

$$NSI_2 = \mathrm{INT}(\mathrm{ABS}(p^T_{n+1}/\sigma_{\mathrm{y}})), \tag{27}$$

and choosing

$$NSI = \mathrm{MAX}(NSI_1, NSI_2). \tag{28}$$

5.2 Mesh-Size Dependence

A subject that deserves particular attention is the influence of mesh size in finite element analysis with the Gurson model as in all situations that involve softening. Then the results become strongly mesh dependent, unless special procedures are employed. Procedures proposed include non-local models, viscoplasticity, gradient plasticity, etc., in order to introduce a characteristic length, related not to the mesh size but to the material structure [1, 5, 15, 16, 46, 51, 60].

Non-local strategies for the Gurson model have been proposed by Leblond et al. [39], Needleman and Tvergaard [52] and Reusch et al. [64, 65].

The use of gradient plasticity formulations considers that the yield surface depends not only on internal variables but also on its gradients. This dependence leads to behavior similar to the nonlocal approach. Gologanu et al. [21] and Ramaswamy and Aravas [62, 63] have studied the application of gradient plasticity together with the Gurson model. A viscoplastic formulation also introduces a characteristic length. In the context of the Gurson damage, viscoplascity is used by Needleman and Tvergaard [52] and Stainier [71].

5.3 Arbitrary Lagrangian-Eulerian Alternative

The Gurson damage model involves two major components: a yield surface that depends on the stress state, virgin yield stress and porosity level, and a law for the evolution of porosity, also dependent on stresses and strains. Thus, the results obtained applying the damage model are only as good as the displacements, stresses and strains used as input data. Modeling the problem with finite elements, the results obtained are in strong dependence on the quality of the mesh employed. In the presence of the finite strains allowed by ductile behavior, errors due to high mesh distortion can be expected. In order to improve the quality of the results some action must be taken. One possibility is to employ remeshing [11, 73]. This is a good option, but usually expensive, because it needs continuous error monitoring to define the exact moment to remesh, a good mesh generator and an experienced user to control the process. Another way of minimizing the mesh distortion is to employ an ALE formulation, in which the mesh is redefined at arbitrary steps, in an automatic way. Both methods may be combined.

Arbitrary Lagrangian-Eulerian formulation (ALE) is a strategy initially developed for hydro-codes [17, 30], and after extended to solid mechanics problems [4, 27, 29, 44, 68], enhancing the quality of the meshes in processes that occur with large deformations. The main characteristic of ALE formulation is the relative movement between finite element mesh and material points. Considering this characteristic, Eulerian and Lagrangian formulations can be understood as particular cases of ALE formulation.

As mesh and material displace independently, a value relating material velocity v_i and mesh velocity $\hat{v}_i$, the convective velocity c_i, can be established as $c_i = v_i - \hat{v}_i$. A material rate of a function g_i is defined as $g_i^{\bullet} = g_i^o + c_j g_{i,j}$, where g_i^o represents the local variation of f_i and $c_j g_{i,j}$ represent the convective effects. Considering the balance of momentum equation in the Lagrangian form and applying equilibrium conditions in the Lagrangian-Eulerian form results

$$\sigma_{ij,j} + \rho b_i = \rho \left(v_i^o + c_j v_{i,j} \right), \tag{29}$$

where ρ is the specific mass and b_i is the body force per unit mass. Applying the virtual work principle to Eq. (29) results in

$$\int_V \delta u_i \left[\sigma_{ij,j} + \rho b_i - \rho \left(v_i^o + c_j v_{i,j} \right) \right] \mathrm{d}V = 0. \tag{30}$$

A weak form of the equilibrium conditions can be obtained,

$$\int_V \rho v_i^o \delta u_i \mathrm{d}V + \int_V \rho c_j v_{i,j} \delta u_i \mathrm{d}V + \int_V \delta u_{i,j} \sigma_{ji} \mathrm{d}V = \int_V \rho b_i \delta u_i \mathrm{d}V + \delta \int_S t_i \delta u_i \mathrm{d}S. \tag{31}$$

It is easy to see that in Eq. (31) both the mesh and material velocities are involved. In a finite element implementation, the system of equations resulting of Eq. (31) can be solved by means of two alternative strategies: (a) to define a system considering as the degrees of freedom, those corresponding to the displacements of both the mesh and the material; (b) to solve the problem in a staggered manner [7, 61]. The second alternative considers two stages at each load increment. First, the Updated Lagrange (UL) stage, with the mesh attached to the material, that ends after equilibrium is obtained. Afterwards, in the Eulerian stage, the new mesh position is defined, trying to reduce distortion, and the relevant information is transferred from the old to the new mesh.

6 Numerical Examples

The numerical examples presented in this section were obtained employing two different finite element codes. The first one [69] is a well-known commercial software. The second one, MetaFor, was developed at the University of Liège by Ponthot and Hogge [61], to treat problems of metal forming, and was used under a courtesy license. Both codes have an adequate treatment of geometrical non-linearity and contact. The Gurson model using the ALE formulation and implicit time-integration was implemented on MetaFor [12].

6.1 Indentation of a Block by a Sphere

This example analizes the punching of a block of square section by a sphere [12]. The height of the block is 100 mm and the transversal section is 140 × 140 mm. The sphere has a radius of 50 mm, and travels 50 mm in the vertical downward direction. Because of symmetry, a quarter of the problem is modeled. The contact between

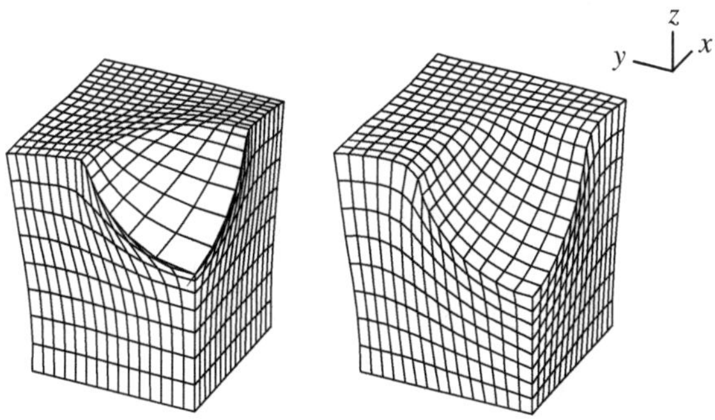

Fig. 13 Last configuration achieved: **a** UL formulation and **b** ALE formulation

block and sphere is considered as sliding, with normal penalty factor of 618 N/m. The material of the block has elastic modulus $E = 206$ GPa, Poisson's ratio $\nu = 0.3$, density $\rho = 7500$ kg/m^3. Linear hardening is considered with $\sigma_y = \sigma_y^0 + h\varepsilon^p$ being $\sigma_y^0 = 346.4$ MPa, $h = 138$ MPa. For the Gurson model $\alpha_1 = 1.5$, $\alpha_2 = 1.0$, $f_N = 0.04$, $\varepsilon_N = 0.5$, $s_N = 0.1$, $f_C = 0.15$ and $\Delta\varepsilon = 0.3$. Results obtained employing MetaFor [61] are presented in Fig. 13, considering both UL and ALE formulations.

Figure 13a shows that in the case of the UL formulation the elements in the contact zone are highly distorted. The nodes on the contact surface are distant one from another and the contact surface gets far from the spherical surface and closer to a polyhedron. The simulation came to a stop at 82 % of the proposed punch displacement with a message of negative Jacobian. With the ALE formulation, the total proposed punch displacement was attained with a good quality mesh. Figure 14 shows the final porosity distribution obtained with pressure independent void nucleation model [13].

6.2 Analysis of Metallic Foams

Analysis of a single sphere: the finite element analysis in this section follows that in [55]. The sphere is modeled as an axisymmetric body with 375 linear quadrilateral elements. Because of symmetry considerations only one half of the sphere is modeled. The platen of the test machine is modeled as a rigid plane with prescribed displacements and the contact procedure is activated.

The sphere analyzed has an external radius of 1.0 mm and wall thickness of 0.1 mm. The material constants used are: elastic modulus $E = 200$ GPa, initial yield stress 200 MPa and Poisson's ratio $\nu = 0.3$. In the cases without damage (von Mises yield criterion), kinematic hardening is employed, with a hardened yield stress of

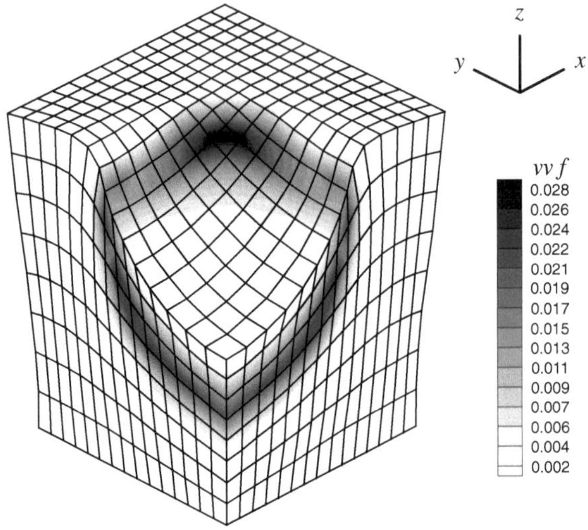

Fig. 14 Final porosity (void volumetric fraction *vvf*) distribution obtained with ALE formulation and pressure independent void nucleation model

250 MPa to a unitary plastic strain. When Gurson model is employed, hardening is considered as isotropic with the same magnitude.

Plots of macroscopic stress (defined as the ratio between the sum of the reactions in the compression direction and the surface of the circle corresponding to the projection of the undeformed sphere onto a plane) versus normalized displacement (defined as the relation between the imposed displacement to the top plane and the original radius) are given in Fig. 15.

The plot corresponding to elasto-plastic behavior coincides with Lim et al. [43] results. Two other plots, obtained with the consideration of damage are shown in Fig. 15. One simulation considers 5 % of initial porosity with no void nucleation and the other considers only 5 % void nucleation without initial porosity. In both alternatives damage accumulates in the same region of the sphere.

Figures 16 and 17 show the distribution of porosity at the end of the compression process, corresponding to a normalized displacement of 0.9. The damage parameter reaches 13.4 % when initial porosity is used (Fig. 16) and 9.8 % if only nucleation is considered (Fig. 17). These are fairly high values, but as the damage region is localized, only in small changes in the load-displacement relation (Fig. 15) are observed.

Analysis of a Representative Volume Element (RVE) representing a Metallic Hollow Sphere Structure (MHSS): In this analysis [56] two geometries are studied. The first one, in which the space among the spheres is fully occupied by resin is called *syntactic*. The second one, in which the space between adjacent spheres is only partially occupied by the resin is called *partial*. The models used in the analysis are made to fit global densities for the set resin-metal of 1.2 g/cm^3 (syntactic) and 0.6 g/cm^3 (partial). The metal spheres have an external radius of 1.5 mm and the

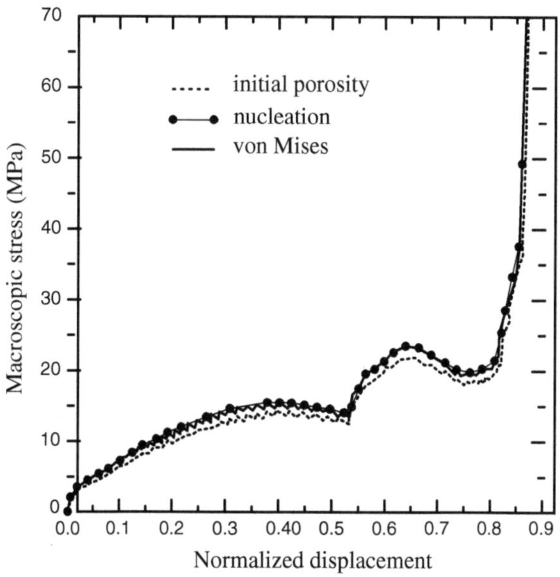

Fig. 15 Macroscopic stress versus normalized displacement with and without damage

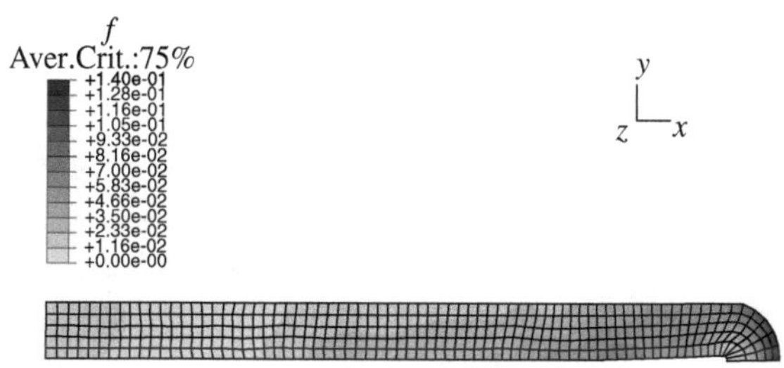

Fig. 16 Final distribution of porosity starting with an initial porosity of 5 %

resin thickness between spheres is 0.36 mm. The boundary conditions employed in both RVEs are shown in Fig. 18.

Materials constants used are $E = 110$ GPa, $\nu = 0.30$, virgin yield stress $\sigma_y^0 = 300$ MPa and $\rho = 6.95$ g/cm^3 for the metal of the sphere and $E = 24.6$ GPa, $\nu = 0.34$, compression yield stress $\sigma_y^0 = 113$ MPa, traction yield stress $\sigma_y^0 = 61.5$ MPa and $\rho = 1.13$ g/cm^3 for the resin. Both metal sphere and matrix were modeled as elasto-plastic. Damage is considered only for the metallic spheres. The meshes presented in Fig. 19-left (partial) and in Fig. 19-right (syntactic) are employed.

Figure 20 shows macroscopic stresses versus normalized displacement obtained simulating the partial geometry behavior. Experimental results [19] are also given. The abrupt changes in stiffness (points 1, 2 and 3 of Fig. 20) observed in the numerical results for a single sphere, free or confined, are due to the new contact zones that

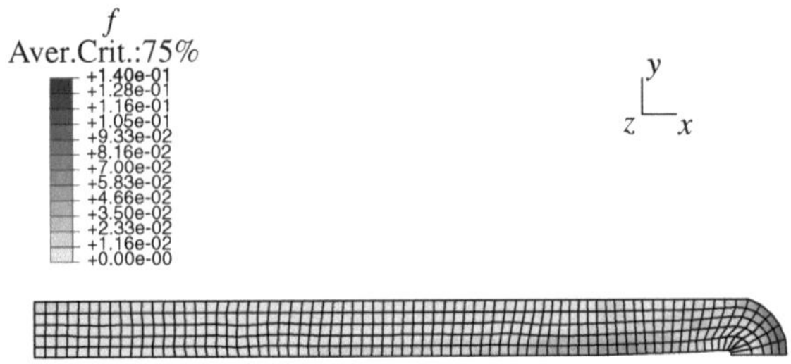

Fig. 17 Final distribution of porosity starting with nucleation

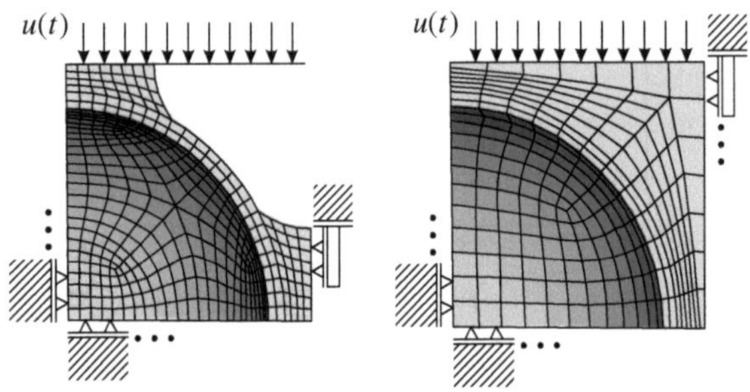

Fig. 18 Boundary conditions adopted in partial and syntactic morphologies

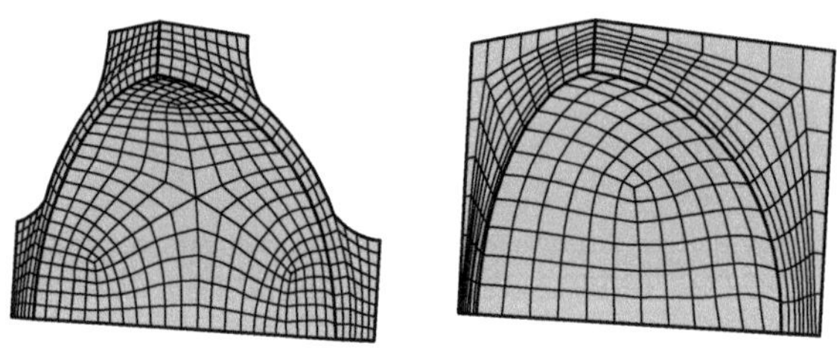

Fig. 19 Meshes employed to analyze partial and syntactic morphologies

appear. This effect is not so apparent in the experimental results which correspond to a conglomerate of spheres with random geometries and properties that smoothed up such details.

Additional results may be found in [14, 54].

Acknowledgments Financial support of Brazilian agencies CNPq and CAPES is gratefully acknowledged.

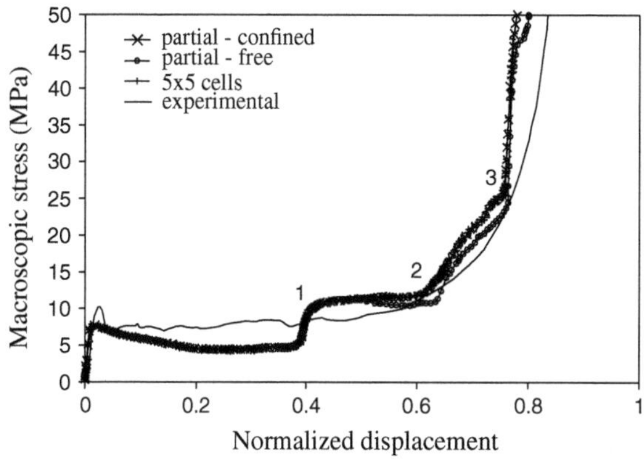

Fig. 20 Load-displacement plots for the case of partial geometry (0.6 g/cm^3) considering a RVE confined that keeps on a vertical plane the lateral sides of the cell and a free RVE without restrictions on the lateral sides

References

1. Abu Al-Rub, R., Voyiadjis, G.Z.: A direct finite element implementation of the gradient-dependent theory. Int. J. Numer. Meth. Eng. **63**(4), 603–629 (2005)
2. Aravas, N.: On the numerical integration of a class of pressure-dependent plasticity models. Int. J. Numer. Meth. Eng. **24**, 1395–1416 (1987)
3. Aravas, N., McMeeking, R.M.: Microvoid growth and failure in the ligament between a hole and a blunt crack tip. Int. J. Fracture **29**, 21–38 (1985)
4. Aymone, J., Bittencourt, E., Creus, G.: Simulation of 3d metal-forming using an arbitrary Lagrangian-Eulerian finite element method. ASME J. Mater. Process Technol. **110**, 218–232 (2001)
5. Bazant, Z., Belytschko, T., Chang, T.: Continuum model for strain softening. J. Eng. Mech. ASCE **110**, 1666–1692 (1984)
6. Becker, B., Needleman, A.: Effect of yield surface curvature on necking and failure in porous plastic solids. ASME J. Appl. Mech. **53**, 491–499 (1986)
7. Benson, D.: An efficient, accurate, simple ale method for nonlinear fe programs. Comput. Method Appl. M **72**, 305–350 (1989)
8. Besson, J., Guillemer-Neel, C.: An extension of the Green and Gurson models to kinematic hardening. Mech. Mater. **35**, 1–18 (2003)
9. Brown, L.M., Embury, J.D.: The initiation and growth of voids at second phase particles. In: Proceedings of the 3rd International Conference on Strength of Metals and Alloys, London, Institute of Metals, pp. 164–169 (1973)
10. Chu, C., Needleman, A.: Void nucleation effects in biaxially stretched sheets. J. Eng. Mater-T ASME **102**, 249–256 (1980)
11. Coupez, T., Soyriz, N., Chenot, J.L.: 3-d finite element modeling of the forging process with automatic remeshing. J. Mater. Process Technol. **27**, 119–133 (1991)
12. Cunda, L.: Gurson model for ductile damage: computational approach and applications (in Portuguese). PhD thesis, Universidade Federal do Rio Grande do Sul, Porto Alegre, Brazil (2006)
13. Cunda, L.A.B., Creus, G.J.: A note on damage analyses in processes with nonmonotonic loading. Comput. Model. Simul. Eng. **4**(4), 300–303 (1999)
14. Cunda, L.A.B., Oliveira, B.F., Creus, G.J.: Plasticity and damage analysis of metal foams under dynamic loading. Materialwiss Werkst **42**, 356–364 (2011)

15. de Borst, R., Muhlhaus, H.: Gradient-dependent plasticity: formulation and algorithmic aspects. Int. J. Numer. Meth. Eng. **35**, 521–539 (1992)
16. de Borst, R., Sluys, L.: Localization in a Cosserat continuum under static and loading conditions. Comput. Meth. Appl. M **90**, 805–827 (1991)
17. Donea, J., Giuliani, S., Halleux, J.: An arbitrary Lagrangian-Eulerian finite element method for transient dynamic fluid-structure interactions. Comput. Meth. Appl. M **33**, 689–723 (1982)
18. Drucker, D.C., Prager, W.: Soil mechanics and plastic analysis for limit design. Q. Appl. Math. **10**(2), 157–165 (1952)
19. Fiedler, T.: Numerical and experimental investigation of hollow sphere structures in sandwich panels. PhD thesis, University of Aveiro, Aveiro, Portugal (2007)
20. Giusti, S., Blanco, P., Souza Neto, E., Feijó, R.: An assessment of the Gurson yield criterion by a computational multi-scale approach. Eng. Comput. **26**(3–4), 281–301 (2009)
21. Gologanu, M., Leblond, J., Perrin, G.: A micromechanically based Gurson-type model for ductile porous metals including strain gradient effects. ASME Net Shape Process. Powder Mater. Appl. Mech. Div. AMD **216**, 47–56 (1995)
22. Gologanu, M., Leblond, J., Perrin, G., Devaux, J.: Recent extensions of Gurson model for porous ductile metals. In: Suquet, P. (ed.) Continuum Micromechanics, CISM Courses and Lectures, vol. 377, pp. 61–130. Springer, Wien (1997)
23. Goya, M., Hagaki, S., Sowerby, R.: Yield criteria for ductile porous solids. JSME Int. J. **35**(3), 310–318 (1992)
24. Gratacos, P., Montmitonet, P., Chenot, J.: An integration scheme for Prandtl-Reuss elastoplastic constitutive relations. Int. J. Numer. Meth. Eng. **33**, 943–961 (1992)
25. Gurson, A.: Plastic flow and fracture behavior of ductile materials incorporating void nucleation, growth and interaction. PhD thesis, Brown University, Providence, RI (1975)
26. Gurson, A.: Continuum theory of ductile rupture by void nucleation and growth. Part I. Yield criteria and flow rules for porous ductile media. J. Eng. Mater-T ASME **99**, 2–15 (1977)
27. Haber, R.: A mixed Eulerian-Lagrangian displacement model for large-deformation analysis in solid mechanics. Comput. Meth. Appl. M **43**, 277–292 (1984)
28. Hill, R.: The Mathematical Theory of Plasticity. Oxford Press, Oxford (1950)
29. Huetnik, J.: On the simulation of thermo-mechanical forming processes: A mixed Eulerian-Lagrangian finite element method. PhD thesis, Twente University of Technology, Holland (1986)
30. Hughes, T., Liu, W., Zimmermann, T.: Lagrangian-Eulerian finite element formulation for incompressible viscous flows. Comput. Meth. Appl. M **29**, 329–349 (1981)
31. Johnson, J.: Dynamic fracture and spallation in ductile solids. J. Appl. Phys. **52**, 2812–2825 (1981)
32. Kachanov, L.M.: On the time to rupture under creep conditions (in Russ.). Izv AN SSSR Otdelenie tekhnicheskich nauk **8**, 26–31 (1958)
33. Kachanov, L.M.: Rupture time under creep conditions. Int. J. Fracture **97**(1–4), 11–18 (1999)
34. Kanninen, M., Popelar, C.: Advanced Fracture Mechanics. Oxford University Press, Oxford (1985)
35. Kikuchi, M., Miyamoto, H., Otoyo, H., Kuroda, M.: Ductile fracture of aluminum alloys. JSME Int. J. **34**(1), 90–97 (1991)
36. Klöcker, H., Tvergaard, V.: Growth and coalescence of non-spherical voids in metals deformed at elevated temperature. Int. J. Mech. Sci. **45**, 1283–1308 (2003)
37. Koplik, J., Needleman, A.: Void growth and coalescence in porous plastic solids. Int. J. Solids Struct. **24**, 835–853 (1988)
38. Koppenhoefer, K., Dodds Jr, R.: Ductile crack growth in pre-cracked cvn specimens: numerical studies. Nucl. Eng. Des. **180**, 221–241 (1998)
39. Leblond, J., Perrin, G., Devaux, J.: Bifurcation effects in ductile metals with nonlocal damage. J. Appl. Mech. **61**(2), 236–242 (1994)
40. Lee, B., Mear, M.: An evalution of gurson's theory of dilatational plasticity. J. Eng. Mater-T ASME **115**, 339–344 (1993)

41. Lee, J.: Accuracies of numerical solution methods for the pressure-modified von mises model. Int. J. Numer. Meth. Eng. **26**, 453–465 (1988)
42. Lee, J., Zhang, Y.: A finite-element work-hardening plasticity model of the uniaxial compression and subsequent failure of porous cylinders including effects of void nucleation and growth - part 1: plastic flow and damage. J. Eng. Mater-T ASME **116**, 69–79 (1994)
43. Lim, T., Smith, B., McDowell, D.: Behaviour of a random hollow sphere metal foam. Acta Mater. **50**, 2867–2879 (2002)
44. Liu, W., Belytschko, T., Chang, H.: An arbitrary Lagrangian-Eulerian finite element method for path-dependent materials. Comput. Meth. Appl. M **58**, 227–245 (1986)
45. Mackenzie, J.: The elastic constants of a solid containing spherical holes. Proced. Phys. Soc. **63B**, 2–11 (1959)
46. Mazars, J., Bazant, Z.: Cracking and Damage: Strain Localization and Size Effects. Elsevier, Amsterdam (1989)
47. McClintock, F.: A criterion for ductile fracture by the enlargement of holes. J. Appl. Mech. **35**, 363–371 (1968)
48. Mear, M.E., Hutchinson, J.W.: Influence of yield surface curvature on flow localization in dilatant plasticity. Mech. Mat. **4**, 395–407 (1985)
49. Mori, T., Tanaka, K.: Average stress in matrix and average elastic energy of materials with misfitting inclusion. Acta Metall. Mater. **21**, 571–579 (1973)
50. Nahshon, K., Hutchinson, J.: Modification of the gurson model for shear failure. Eur. J. Mech. A-Solid **27**, 1–17 (2008)
51. Needleman, A.: A material rate dependence and mesh sensitivity in localization problems. Comput. Meth. Appl. Mech. Eng. **67**, 69–86 (1988)
52. Needleman, A., Tvergaard, V.: Mesh effects in the analysis of dynamic ductile crack growth. Eng. Fract. Mech. **47**(1), 75–91 (1994)
53. Nielsen, K.L., Tvergaard, V.: Ductile shear failure or plug failure of spot welds modelled by modified gurson model. Eng. Fract. Mech. **77**(7), 1031–1047 (2010)
54. Oliveira, B.F., Cunda, L.A.B., Öchsner, A., Creus, G.J.: Comparison between rve and full mesh approaches for the simulation of compression tests on cellular metals. Materialwiss Werkst **39**, 1–6 (2008)
55. Oliveira, B.F., Cunda, L.A.B., Öchsner, A., Creus, G.J.: Hollow sphere structures: a study of mechanical behaviour using numerical simulation. Materialwiss Werkst **40**, 144–153 (2009)
56. Oliveira, B.F., Cunda, L.A.B., Creus, G.J.: Modeling of the mechanical behavior of metallic foams: Damage effects at finite strains. Mech. Adv. Mater. Struct. 16, 110–119 (2009)
57. Ortiz, M., Popov, E.: Accuracy and stability of integration algorithms for elastoplastic constitutive relations. Int. J. Numer. Meth. Eng. **21**, 1561–1576 (1985)
58. Pardoen, T., Hutchinson, J.: An extended model for void growth and coalescence. J. Mech. Phys. Solids **48**, 2467–2512 (2000)
59. Perzyna, P.: Internal state variable description of dynamic fracture of ductile solids. Int. J. Solids Struct. **22**(7), 797–818 (1986)
60. Pijaudier-Cabot, G., Bazant, Z.: Nonlocal damage theory. J. Eng. Mech.-ASCE **113**, 1512–1533 (1987)
61. Ponthot, J.P., Hogge, M.: The use of the Eulerian-Lagrangian fem in metal forming applications including contact and adaptive mesh. In: Chandra, N., Reddy, J.N. (eds.) Advances in Finite Deformation Problems in Materials Processing and Structures, ASME, Atlanta, USA, vol ASME Winter Annual Meeting, pp 44–64 (1991)
62. Ramaswamy, S., Aravas, N.: Finite element implementation of gradient plasticity models. Part I: Gradient-dependent yield functions. Comput. Meth. Appl. M **163**, 11–32 (1998)
63. Ramaswamy, S., Aravas, N.: Finite element implementation of gradient plasticity models Part II: Gradient-dependent evolution equations. Comput. Meth. Appl. M **163**, 33–53 (1998)
64. Reusch, F., Svendsen, B., Klingbeil, D.: Local and non-local Gurson-based ductile damage and failure modelling at large deformation. Eur. J. Mech. A-Solid **22**, 779–792 (2003)
65. Reusch, F., Svendsen, B., Klingbeil, D.: A non-local extension of Gurson-based ductile damage modeling. Comput. Mater. Sci. **26**, 219–229 (2003)

66. Rice, J., Tracey, D.: On the ductile enlargement of voids in triaxial stress fields. J. Mech. Phys. Solids **17**, 201–217 (1969)
67. Runesson, K., Sture, S., Willam, K.: Integration in computational plasticity. Comput. Struct. **30**(1–2), 119–130 (1988)
68. Schreurs, P., Veldpaus, F., Brekelmans, W.: Simulation of forming processes, using the arbitrary Eulerian-Lagrangian formulation. Comput. Meth. Appl. M **58**, 19–36 (1986)
69. Simulia (2009) Abaqus, 6.9. Abaqus Theory Manual. Dassault Systèmes Simulia Corp., Providence, RI, USA
70. Skallerud, B., Zhang, Z.: On numerical analysis of damage evolution in cyclic elastic-plastic crack growth problems. Fatigue Fracture Eng. Mater. Struct. **23**, 81–86 (2001)
71. Stainier, L.: Modélisation numérique du comportement irréversible des métaux ductiles soumis à grandes déformations avec endommagement. PhD thesis, Universitè de Liège, Liège, Belgium (1996)
72. Steglich, D., Pirondi, A., Bonora, N.: Micromechanical modelling of cyclic plasticity incorporating damage. Int. J. Solids Struct. **42**(2), 337–351 (2005)
73. Szentmihali, V., Lange, K., Tronel, Y., Chenot, J.L., Ducloux, R.: 3-d finite element simulation of the cold forging of helical gears. J. Mater. Process Technol. **43**, 279–291 (1994)
74. Thomason, P.: Three-dimensional models for the internal neckings at incipient failure of intervoid matrix in ductile porous solids. Acta Metall. Mater. **33**(6), 1079–1085 (1985)
75. Trillat, M., Pastor, J.: Limit analysis and gurson model. Eur. J. Mech. A-Solid **24**, 800–819 (2005)
76. Tvergaard, V.: Influence of voids on shear band instabilities under plane strain conditions. Int. J. Fracture **17**, 389–407 (1981)
77. Tvergaard, V.: Ductile fracture by cavity nucleation between larger voids. J. Mech. Phys. Solids **30**(4), 265–286 (1982)
78. Tvergaard, V.: Material failure by void coalescence in localized shear bands. Int. J. Solids Struct. **18**(8), 659–672 (1982)
79. Tvergaard, V.: On localization in ductile materials containing spherical voids. Int. J. Fract. **18**, 237–252 (1982)
80. Tvergaard, V., Needleman, A.: Analysis of the cup-cone fracture in a round tensile bar. Acta Metall. Mater. **32**(1), 157–169 (1984)
81. Voyiadjis, G., Kattan, P.: A plasticity-damage theory for large deformation of solids. I. Theoretical formulation. Int. J. Eng. Sci. **30**(9), 1089–1108 (1992)
82. Wen, J., Huang, Y., Hwang, K., Liu, C., Li, M.: The modified gurson model accounting for the void size effect. Int. J. Plast. **21**, 381–395 (2005)
83. Worswick, M., Pick, R.: Void growth in plastically deformed free-cutting brass. J. Appl. Mech. **58**, 631–638 (1991)
84. Zavaliangos, A., Anand, A.L.: Thermal aspects of shear localization in microporous viscoplastic solids. Int. J. Numer. Meth. Eng. **33**, 595–634 (1992)
85. Zhang, Z.: Explicit consistent tangent moduli with a return mapping algorithm for pressure dependent elastoplasticity models. Comput. Meth. Appl. M **121**, 29–44 (1995)
86. Zhang, Z.: On the accuracies of numerical integration algorithms for Gurson-based pressure-dependent elastoplastic constitutive models. Comput. Meth. Appl. M **121**, 15–28 (1995)
87. Zhang, Z., Niemi, E.: A class of generalized mid-point algorithms for the Gurson-Tvergaard material model. Int. J. Numer. Meth. Eng. **38**, 2033–2053 (1995)
88. Zhang, Z., Niemi, E.: A new failure criterion for the Gurson-Tvergaard dilatational constitutive model. Int. J. Fract. **70**, 321–334 (1995)

MIX
Papier aus verantwortungsvollen Quellen
Paper from responsible sources
FSC® C105338

If you have any concerns about our products,
you can contact us on
ProductSafety@springernature.com

In case Publisher is established outside the EU,
the EU authorized representative is:
Springer Nature Customer Service Center GmbH
Europaplatz 3, 69115 Heidelberg, Germany

Printed by Libri Plureos GmbH
in Hamburg, Germany